ENVIRONMENTAL MONITORING AND CHARACTERIZATION

ENVIRONMENTAL MONITORING AND CHARACTERIZATION

JANICK F. ARTIOLA
Department of Soil, Water and Environmental Science
University of Arizona
Tucson, AZ

IAN L. PEPPER
Department of Soil, Water and Environmental Science
University of Arizona
Tucson, AZ

MARK L. BRUSSEAU
Department of Soil, Water and Environmental Science
University of Arizona
Tucson, AZ

ELSEVIER
ACADEMIC
PRESS

Amsterdam • Boston • Heidelberg • London
New York • Oxford • Paris • San Diego
San Francisco • Singapore • Sydney • Tokyo

Acquisition Editor: Charles Crumly
Editorial Assistant: Kelly Sonnack
Project Manager: Julio Esperas/Troy Lilly
Marketing Manager: Linda Beattie
Marketing Manager: Clare Fleming
Cover Design: Eric DeCicco
Full-Service Provider: Kolam USA
Composition: Kolam India
Printer: Paramount

Elsevier Academic Press
200 Wheeler Road, Burlington, MA 01803, USA
525 B Street, Suite 1900, San Diego, California 92101-4495, USA
84 Theobald's Road, London WC1X 8RR, UK

This book is printed on acid-free paper. ∞

Library of Congress Cataloging-in-Publication Data

Environmental monitoring and characterization / [edited by] Janick F. Artiola, Ian L. Pepper, Mark L. Brusseau.
    p. cm.
Includes bibliographical references and index.
  ISBN 0-12-064477-0 (acid-free paper)
  1. Environmental monitoring. I. Artiola, Janick F. II. Pepper, Ian L. III. Brusseau, Mark L.
  QH541.15.M64E5864 2003
  628–dc22   2003020654

British Library Cataloguing in Publication Data
A catalogue record for this book is available from the British Library

For all information on all Academic Press publications
visit our website at www.academicpressbooks.com

PRINTED IN CHINA
04 05 06 07 08 09 9 8 7 6 5 4 3 2 1

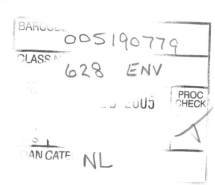

Contents

PREFACE

In the 21st century, the fate of the environment has become a critical issue in both developed and developing countries throughout the world. Population increases and technological advances are creating a burden on society by requiring continued expansion and concomitant resource use. Substantial evidence exists showing that such development has led to detrimental impacts on the environment. We also know that increased societal activities and demands are changing soil, water, air, climate, and resources in unexpected ways. This in turn has led to a renewed interest in protecting the environment and has focused attention on the concept of environmental monitoring and site characterization, including an evaluation of the physical, chemical, and biological factors that impact the environment. This information is necessary for researchers, decision-makers, and the community as a whole, to implement social changes needed to preserve and sustain a healthy environment for future generations.

The purpose of this textbook is to document the latest methodologies of environmental monitoring and site characterization important to society and human health and welfare. We know that the environment exists as a continuum of biosystems and physio-chemical processes that help sustain life on earth. Therefore environmental monitoring should ideally consist of examining the integrative nature of these processes. To this end, basic principles of monitoring and characterization are described for different environments, considering their most relevant processes. Initially, sampling protocols are described, followed by documentation of quality control issues and statistical methods for data analysis. Methods for making field measurements in soil, vadose zone, water, and atmospheric environments are described. This includes real-time monitoring, temporal and spatial issues, and the issues of scale of measurement. The book advances the state-of-the-art by not only documenting how to monitor the environment, but also by developing active strategies that allow for efficient characterization of specific environments. In addition we provide approaches to evaluate and interpret data efficiently, with significant processes being documented via statistical analyses and, where appropriate, model development. A particularly unique feature of the text is the discussion of physical, chemical, and microbial processes that effect beneficial as well as detrimental influences on the environment. The text also puts into perspective site-specific remediation techniques that are appropriate for localized environments as well as full-scale ecosystem restoration. Finally, the role of risk assessment and environmental regulations in environmental monitoring is assessed.

In summary, this book attempts to answer these questions: "How should samples be taken, including why, when, and where? How should the samples be analyzed? How should the data be interpreted?" This book should be useful at the senior undergraduate level, as well as to students initiating graduate studies in the environmental science arena. The fact that contributions come from national experts all located at the University of Arizona ensures that the book is well integrated and uniform in its level of content.

Key features of the book include:

- The concept of integrating environmental monitoring into site characterization
- Numerous real-life case studies
- The use of numerous computer graphics and photographs
- The integration of physical, chemical, and biological processes
- Key references relevant to each topic
- Examples of problems, calculations, and thought-provoking questions

CONTRIBUTORS

Janick F. Artiola
Department of Soil, Water and Environmental Science
University of Arizona
Tucson, AZ

Martin Barackman
CH₂M Hill
Redding, CA

Paul W. Brown
Department of Soil, Water and Environmental Science
University of Arizona
Tucson, AZ

Mark L. Brusseau
Department of Soil, Water and Environmental Science
Department of Hydrology and Water Resources
University of Arizona
Tucson, AZ

Gale B. Famison
Department of Soil, Water and Environmental Science
University of Arizona
Tucson, AZ

Charles P. Gerba
Department of Soil, Water and Environmental Science
University of Arizona
Tucson, AZ

Vanda J. Gerhart
Environmental Research Laboratory
University of Arizona
Tucson, AZ

Edward P. Glenn
Environmental Research Laboratory
University of Arizona
Tucson, AZ

David M. Hendricks
Department of Soil, Water and Environmental Science
University of Arizona
Tucson, AZ

Alfredo R. Huete
Department of Soil, Water and Environmental Science
University of Arizona
Tucson, AZ

Raina M. Maier
Department of Soil, Water and Environmental Science
University of Arizona
Tucson, AZ

Robert MacArthur
Educational Communications & Technologies
University of Arizona
Tucson, AZ

Allan D. Matthias
Department of Soil, Water and Environmental Science
University of Arizona
Tucson, AZ

Sheri A. Musil
Department of Soil, Water and Environmental Science
University of Arizona
Tucson, AZ

Donald E. Myers
Professor Emeritus
University of Arizona
Tucson, AZ

Ian L. Pepper
Environmental Research Laboratory
University of Arizona
Tucson, AZ

Chris Rensing
Department of Soil, Water and Environmental Science
University of Arizona
Tucson, AZ

José A. Vargas-Guzmán
Physical Sciences and Architecture
University of Queensland
Brisbane, Australia

Arthur W. Warrick
Department of Soil, Water and Environmental Science
University of Arizona
Tucson, AZ

W.J. Waugh
U.S. Department of Energy
Grand Junction, CO

Peter J. Wierenga
Department of Soil, Water and Environmental Science
University of Arizona
Tucson, AZ

Lorne Graham Wilson
Department of Hydrology and Water Resources
University of Arizona
Tucson, AZ

Irfan Yolcubal
Department of Hydrology and Water Resources
University of Arizona
Tucson, AZ

REVIEWERS

Dr. Michael J. Barcelona
Research Professor
Department of Civil and Environmental Engineering
Environmental and Water Resources Engineering
University of Michigan
Ann Arbor, MI

Dr. Kirk W. Brown
Professor Emeritus
Texas A & M University
College Station, TX

Dr. Karl Enfield
R.S. Kerr, U.S. EPA Laboratory
Ada, OK

Dr. William T. Frankenberger Jr.
University of California, Riverside
Department of Environmental Sciences
Riverside, CA

Dr. Charles Haas
Drexel University
Department of Civil, Architectural and Environmental
Engineering
Philadelphia, PA

Dr. Arthur G. Hornsby
University of Florida
Soil and Water Science Department
Gainesville, FL

Dr. Lawrence H Keith
Instant Reference Sources, Inc.
Monroe, GA

Dr. Ronald Turco
Purdue University
Department of Agronomy
West Lafayette, IN

1

MONITORING AND CHARACTERIZATION OF THE ENVIRONMENT

J.F. ARTIOLA, I.L. PEPPER, AND M.L BRUSSEAU

THE ENVIRONMENT

Environmental changes occur naturally and are a part of or the result of multiple cycles and interactions. Numerous natural cycles of the earth's environment have been studied within the framework of three major scientific disciplines: chemistry, physics, and biology. Environmental scientists study the dynamics of cycles, such as the nitrogen and water cycles, and their relationships to soil-geologic materials, surface waters, the atmosphere, and living organisms. The untrained observer may see the atmosphere as being separated from the earth's surface. However, to the trained observer the environment is composed of integrated and interconnected cycles and domains. We now know that the environment is a continuum of physical, chemical, and biological processes that cannot be easily separated from one another. Water, for example, exists in three states and is found inside and on the surface of earth's crust, in the atmosphere, and within living organisms. It is difficult to separate the physical, chemical, and biological processes of water within any particular environment, because water is transferred across boundaries.

Humans now have a more holistic view of the environment and recognize that many factors determine its health and preservation. This in turn has led to the new term *biocomplexity,* which is defined as "the interdependence of elements within specific environmental systems, and the interactions between different types of systems." Thus, research on the individual components of environmental systems provides limited information on the system itself. We are now also concerned with sustainable and renewable versus non-renewable natural resources as well as with biodiversity in relation to our own survival.

1

ENVIRONMENTAL MONITORING

Environmental monitoring is the observation and study of the environment. In scientific terms, we wish to collect data from which we can derive knowledge (Figure 1.1). Thus, environmental monitoring has its role defined in the first three steps of the staircase and is rooted in the *scientific method*. Objective observations produce sound data, which in turn produce valuable information. Information-derived knowledge usually leads to an enhanced understanding of the problem/situation, which improves the chances of making informed decisions. However, it is important to understand that other factors, including political, economic, and social factors, influence decision making.

The information generated from monitoring activities can be used in a myriad of ways, ranging from understanding the short-term fate of an endangered fish species in a small stream, to defining the long-term management and preservation strategies of natural resources over vast tracts of land. Box 1.1 lists some recognizable knowledge-based regulations and benefits of environmental monitoring.

Although Box 1.1 is not exhaustive, it does give an idea of the major role that environmental monitoring plays in our lives. Many of us are rarely aware that such regulations exist and that these are the result of ongoing monitoring activities. Nonetheless, we all receive the benefits associated with these activities.

Recently, environmental monitoring has become even more critical as human populations increase, adding ever-increasing strains on the environment. There are numerous examples of deleterious environmental changes that result from population increases and concentrated human activities. For example, in the United States, the industrial and agricultural revolutions of the last 100 years have produced large amounts of waste by-products that, until the late 1960s, were released into the environment without regard to consequences. In many parts of the developing world, wastes are still disposed of without treatment. Through environmental monitoring we know that most surface soils, bodies of waters, and even ice caps contain trace and ultratrace levels of synthetic chemicals (e.g., dioxins) and nuclear-fallout components (e.g., radioactive cesium). Also, many surface waters, including rivers and lakes, contain trace concentrations of pesticides because of the results of agricultural runoff and rainfall tainted with atmospheric pollutants. The indirect effects of released chemicals into the environment are also a recent cause of concern. Carbon dioxide gas from automobiles and power plants and Freon (refrigerant gas) released into the atmosphere may be involved in deleterious climatic changes.

Environmental monitoring is very broad and requires a multi-disciplinary scientific approach. Environmental scientists require skills in basic sciences such as chemistry, physics, biology, mathematics, statistics, and computer science. Therefore, all science-based disciplines are involved in this endeavor.

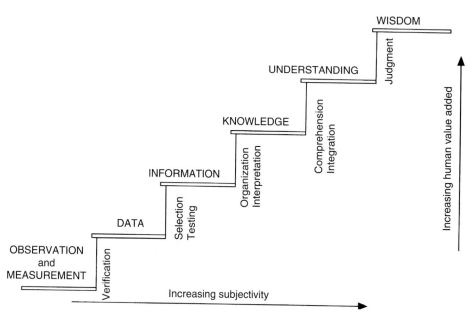

FIGURE 1.1 The staircase of knowing. Science-based observations and measurements improve our understanding of the environment and lead to wise decision-making. (From Roots, E.F. (1997) Inclusion of different knowledge systems in research. In: Terra Borealis. Traditional and Western Scientific Environmental Knowledge. Workshop Proceedings, Northwest River, Labrador 10 & 11 Sept. 1997. No. 1. Manseau M. (ed), Institute for Environmental Monitoring and Research, P.O. Box 1859, Station B Happy Valley–Goose Bay Labrador, Newfoundland, AOP E10. Terra Borealis 1:42–49, 1998.)

BOX 1.1 *Knowledge-Based Regulation and Benefits of Environmental Monitoring*

Protection of public water supplies: Including surface and groundwater monitoring; sources of water pollution; waste and wastewater treatment and their disposal and discharge into the environment

Hazardous, nonhazardous and radioactive waste management: Including disposal, reuse, and possible impacts to human health and the environment

Urban air quality: Sources of pollution, transportation, and industrial effects on human health

Natural resources protection and management: Land and soil degradation; forests and wood harvesting; water supplies, including lakes, rivers, and oceans; recreation; food supply

Weather forecasting: Anticipating weather, long- and short-term climatic changes, and weather-related catastrophes, including floods, droughts, hurricanes, and tornadoes

Economic development and land planning: Resources allocation; resource exploitation

Population growth: Density patterns, related to economic development and natural resources

Delineation: Mapping of natural resources; soil classification; wetland delineation; critical habitats; water resources; boundary changes

Endangered species and biodiversity: Enumeration of species; extinction, discovery, protection

Global climate changes: Strategies to control pollution emissions and weather- and health-related gaseous emissions

ENVIRONMENTAL REMEDIATION AND RESTORATION

Environmental remediation and restoration focus on the development and implementation of strategies geared to reverse negative environmental impacts. Anthropogenic activities often perturb environments and severely limit their capacity for regeneration. For example, metal-contaminated soils often have restrictive physical, chemical, and biological characteristics that hinder self-regenerating mechanisms. High metal concentrations are toxic to plants and microbes such as beneficial soil bacteria. Low-soil microbial populations in turn slow down the rates of microbially-mediated decomposition of organic matter and nutrient cycling. Limited plant nutrient availability leads to poor or non-existent vegetative plant cover. This is turn increases the chances for wind and water soil erosion that further degrades the ecosystem, which also can generate off-site metal contamination. Remediation activities are focused on removing or treating the contamination, whereas restoration activities are focused on rehabilitating the ecosystem.

An interdisciplinary approach is critical for the success of any remediation or restoration activity. Environmental remediation and restoration activities involve contributions from environmental scientists and engineers, soil and water scientists, hydrologists, microbiologists, computer scientists, and statisticians. To develop and implement effective environmental monitoring and restoration programs, it is necessary to understand the major physical, chemical, and biological processes operative at the site and to characterize the nature and extent of the problem. This information is gathered with environmental monitoring activities.

SCALES OF OBSERVATION

At the heart of environmental monitoring are the definitions of observation, sample, and measurement, and their relationships to scale. Modern science and engineering allow us to make observations at the micro and global scales. For example, scientists can use subatomic particles as probes to determine atomic and molecular properties of solids, liquids, and gases. Using this technology, scientists can now measure minute quantities of chemicals in the environment. At the other end of the scale, space-based satellite sensors now routinely scan and map the entire surface of the earth several times a day. However, all observations have a finite resolution in either two or three dimensions, which further complicates the definition of scale. For example, consider a satellite picture of a $100\,km^2$ watershed, taken with a single exposure, that has a resolution of $100\,m^2$. What is the scale of the observation: $100\,km^2$ or $100\,m^2$? Time is another variable that often defines the scale of an observation. Often, temporal environmental data are reported within a defined time frame because most data (values) are not collected instantaneously. Small-scale or short-interval measurements can be combined to obtain measurements of a larger temporal scale. Therefore, the scale of a "single" observation is not always self-evident. Quite often the scale of a measurement has a hidden area space and a time component. Figure 1.2 shows scale definitions for spatial and temporal domains, respectively. The actual scales may seem arbitrary, but they illustrate the range of scales that environmental data can comprise. Example 1.1 illustrates the scales of environmental measurements.

EXAMPLE 1.1 A city air quality–monitoring station near a busy intersection collects air samples from an inlet 3 m above ground at a flow rate of 1 L min^{-1}. The stream of air is passed through an infrared (IR) analyzer, and carbon monoxide (CO) concentrations are measured every second. One-second–interval data are stored in the instrument memory and every hour the mean value of the 1200 data points is sent by the instrument to a data logger (see Chapter 9). Subsequently, the data logger stores the 24 data points and computes a mean to obtain daily CO averages. The data logger sends the stored hourly and daily data to a central repository location for permanent storage and further statistical analysis. Figure 2.3a shows an example of mean 24-hour hourly data CO concentrations during a winter month at Station #3. Daily values are then averaged monthly (Figure 2.3B) and finally mean annual values collected from three other city CO monitoring stations are compared (Table 1.1). Table 1.1 also shows maximum 1-hour and 8-hour CO concentrations that can be used to determine compliance with air-quality standards (see Chapter 6) at four different city locations. The true scale and effort spent to collect these data often escapes the end user. The annual values are not the result of one large-scale (1 year long) measurement. They are the means of thousands of small-scale (1-second interval) measurements.

GLOBAL–Earth(>10,000km)
MESO–Continent, country, state (>100km)
INTERMEDIATE–Watershed, river, lake(>1km)
FIELD–Agric. field, waste site (>1m)
MACRO–Animal, plant, soil clod (>1mm)
MICRO—Soil particle, fungi, bacteria (>1μm)
ULTRA-MICRO—Virus, molecules (>1nm)

A
ATOMIC–Atoms, subatomic particles (<1nm)

GEOLOGIC (> 10,000 years)
GENERATION-LIFETIME (20-100 years)
ANNUAL (>1 year)
SEASONAL (>4 months)
DAILY (>24 hours)
HOURLY (>60 minutes)

B
INSTANTANEOUS (<1second)

FIGURE 1.2 **(A)** Scales of space. Observations and measurements can be made at multiple scales. Satellites take pictures of entire earth, whereas atoms are probed with light and subatomic particles. Intermediate, field, and macro scales of observations dominate environmental monitoring and the remediation strategies. **(B)** Scales of time. Time-based observations and measurements can be made at many intervals. Geologic time changes are usually inferred with present time observations from known time-based changes. For example, the chronological sequencing of soils (usually measured in thousands of years) may be inferred from the appearance or disappearance of key minerals. Observations done over 1 year and each season are very common and useful in monitoring critical water- and air-quality changes.

TABLE 1.1

Carbon Monoxide (CO) Concentrations Data Summary for 1998 from Stations #1 to #4, in Tucson, Arizona

Station #	Annual Average ($\mu g\, g^{-1}$)	Maximum 1-hour CO ($\mu g\, g^{-1}$)	Maximum 8-hour CO ($\mu g\, g^{-1}$)
1	0.80	7.6 (11–16)	4.3 (11–16)
2	0.50	4.8 (12–03)	2.6 (01–08)
3	1.20	7.8 (11–14)	4.0 (11–27)
4	0.50	5.9 (01–08)	4.3 (01–09)

Adapted from Pima County Department of Environmental Quality 1998 Annual Data Summary of Air Quality in Tucson, AZ. Report AQ–309.

Note: Numbers in parentheses indicate date of recorded value.

How are measurements and scales related? The answer is through the use of statistics. Scientists have recognized the limits of their powers of observation. Essentially, it is impossible to be everywhere all the time, and it is impossible to see and observe everything. Statistics help environmental scientists interpolate and extrapolate information from a few sample observations (see Chapter 3) to an entire environment or population. These concepts will be discussed in subsequent chapters.

AGENCIES

Many government, commercial, and private institutions are involved in the collection, storage, and evaluation of environmental data. Local and state institutions are becoming increasingly involved in environmental monitoring and remediation activities. Often agencies represent and/or enforce laws and regulations that have their roots in much larger governmental institutions. For example, the Arizona Department of Environmental Quality is in charge of enforcing air quality and groundwater protection laws by routinely collecting data on Arizona's air and groundwater quality. However, most of the pollutant limits this agency regulates in Arizona come from federal regulations. These government institutions are agencies and commissions of the federal executive government of the United States. They originate at various U.S. departments that include the agencies and some of their bureaus, offices, and services shown in Box 1.2.

The roles of these agencies are well defined, but they may not be mutually exclusive. When there is overlap, agencies often establish cooperative programs to reduce duplication of efforts. For example, NOAA is in charge of weather forecasting and severe storm predictions; EPA monitors pollution derived from fossil fuel consumption, waste management/disposal, natural resources, and remediation activities at abandoned landfills and industrial sites. The Department of Energy (DOE) has a nuclear dump and radiation release monitoring program;

BOX 1.2 *U.S. Departments with Ties to Environmental Monitoring*

U.S. Department of Commerce

National Oceanic and Atmospheric Administration (NOAA)
    National Weather Service (NWS)
    National Environmental Satellite, Data & Information Service (NESDIS)
    National Oceanic Data Center (NODC)
U.S. Census Bureau (USCB)
Economics and Statistics Administration (ESA)

Other federal independent organizations involved in environmental monitoring include:

Department of Health and Human Services

Food and Drug Administration (FDA)
Centers for Disease Control and Prevention (CDC)

Department of Defense (DOD)

U.S. Army Corp of Engineers

National Aeronautics and Space Administration (NASA)

Environmental Protection Agency (EPA)

Office of Solid Waste and Emergency Response
Office of Air and Radiation
Office of Water

Department of Interior (DOI)

U.S. Geological Survey (USGS)
National Biological Information Service (NBIS)
Bureau of Land Management (BLM)
U.S. Fish and Wildlife
National Park Service (NPS)
Bureau of Reclamation (BR)
Office of Surface Mining (OSM)

Department of Energy (DOE)

Federal Energy Regulatory Commission (FERC)

U.S. Department of Agriculture (USDA)

National Resources & Environment (NRE)

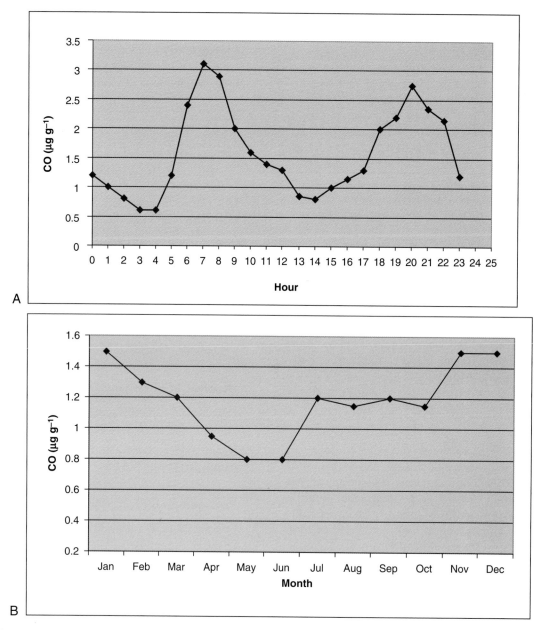

FIGURE 1.3 **(A)** Hourly carbon monoxide (CO) concentrations obtained from averaged measurements taken every second for 24 hours. This figure shows the December 1998 average in Station #3. **(B)** Monthly CO concentrations obtained by averaging daily values (see [A]) during 12 months in 1998 in Station #3. Adapted from Pima County Department of Environmental Quality 1998 Annual Data Summary of Air Quality in Tucson, AZ. Report AQ-309.

DOI characterizes the earth and water resources and public land management of the United States; NASA has a space-based atmosphere/ocean and land science research program; DOD has a global weather and ocean prediction system in support of national security operations; and the U.S. Department of Agriculture (USDA) monitors agricultural food production and quality and soil resources.

There are also several world organizations that collect and distribute environmental data globally. The United Nations (UN) oversees several organizations that monitor weather, food supplies, population, and health. The Food and Agricultural Organization (FAO) monitors world food production, inputs/outputs, pesticide consumption, agricultural indices, commodities, land use, soil degradation, livestock, forests, and fisheries with significant help from the USDA's Foreign Agricultural Service (FAS).

Other UN agencies such as the World Meteorological Organization (WMO), the World Weather Watch

(WWW), and the World Health Organization (WHO) manage the Global Environment Monitoring System (GEMS), which monitors and reports on the global state of water, air, climate, atmosphere, and food contamination. Other organizations affiliated with GEMS are the International Atomic Energy Agency (IAEA), which monitors isotope fallout, and the Global Atmosphere Watch, which monitors atmospheric pollutants, chlorofluorocarbons (CFCs), and ozone.

Much of the environmental data collected by UN organizations is disseminated in the form of statistical reports generated by the United Nations Statistical Office (UNSO) and the International Energy Administration (IEA). In addition, private institutions such as the World Conservation Monitoring Center (WCMC) store and manage an extensive worldwide database on biodiversity, endangered species, and protected habitats. Carter and Diamondstone (1990) show a more complete list of international agencies involved in global monitoring and data storage banks.

CURRENT AND FUTURE STATUS OF ENVIRONMENTAL MONITORING

CURRENT STATUS

Many national agencies collect data in the United States and in other industrialized nations. Nonetheless, the United States has the largest environmental monitoring network and data repository in the world. Additionally, many environmental monitoring programs of the UN depend on the collaboration and funds from U.S. agencies. Consequently, the level of regional knowledge on the environmental varies widely across countries and continents. Non-industrial countries have limited environmental research and few if any environmental monitoring programs relative to industrialized countries. Therefore, critical intermediate scale information is scarce and often outdated. On a regional scale, the development of space-based monitoring systems by the United States and other industrial nations is providing much needed surface information about remote and underdeveloped regions of the earth. This information can be made available universally via the Internet.

Environmental data are often not easily transferred nor integrated. That is, data cannot be used by other agencies nor can they be easily incorporated into other data sets. New ways to integrate data sets must be developed. In addition, there is a need to develop and apply common equivalency standards for all environmental monitoring data. That is, we should all use the same communications software and protocols to exchange data and the same units to define the data. Universal adoption of the Systeme Internationale d'Unites (SI) system of units would ease the exchange of information on a global level and reduce costly unit conversion time and errors. Also, major advances in computer processing, telecommunication, and networking allow for rapid data processing and transfer, to agencies and research institutions throughout the world. In the United States, the National Information Infrastructure (NII) plays an important role in transferring environmental monitoring data among generators such as NOAA, EPA, DOI-USGS USDA, and other world organizations. At the global scale, harmonization of environmental data is being sought by the United Nations Environment Program (UNEP), which provides the world community with environmental data, including trend forecasting in areas that include environmental assessment, atmosphere, fresh water, biodiversity, energy, and chemicals. The UNEP provides a repository of environmental data, ensuring data quality and its worldwide compatibility.

It can also be argued that until recently, most environmental monitoring programs have targeted short-term issues related only to human welfare while ignoring long-term changes to the environment. Case Study 1.1 illustrates this issue. In many industrialized countries, intensive monitoring programs of air and water quality exist to protect human health from immediate danger. Yet, we have been far less diligent in monitoring and protecting the long-term health of other species and the environment overall.

The White House National Science & Technology Council proposed a National Environmental Monitoring Initiative to integrate the nation's environmental monitoring and research networks and programs (NSTC, 1997). Box 1.3 lists some of these recommendations; they provide a good summary of the needs and challenges of environmental monitoring in the 21st century. This has led to the start of an Environmental Monitoring and Assessment Program (EMAP), which seeks to monitor the conditions of the nation's ecological resources to evaluate the cumulative success of current policies and programs, and to identify emerging problems before they become widespread or irreversible. One goal of EMAP is to bring together all government agencies involved in environmental monitoring of natural resources. The EMAP supports the National Monitoring Initiative that brings together the 13 major federal agencies of the United States for the following purposes: (1) to create partnerships among these agencies and combine resources, (2) to develop a national repository of information, and (3) to coordinate research efforts from as many as 34 national research and monitoring programs. For further information, see www.epa.gov/

CASE STUDY 1.1 Marine environments have helped nurture and develop the evolution of humankind since time immemorial. We have in turn repaid this kindness by loading our wastes near shores and overfishing the oceans. Now many of these coastal marine ecosystems are under serious stress. Nevertheless, we lack the information and often the resources to stop this deterioration. The following is a quote from the NRC report "Managing Troubled Waters" (1990):

Marine environmental monitoring has been successfully employed to protect public health through systematic measurement of microbial indicators of fecal pathogens in swimming and shellfish growing areas, to validate water quality models, and to assess the effectiveness of pollution abatement. Nevertheless, despite these considerable efforts and expenditures, most environmental monitoring programs fail to provide the information needed to understand the condition of the marine environment or to assess the effects of human activity. . . . "

Recent regulations limiting or eliminating ocean dumping of wastes have helped reduce the deterioration of some coastal environments. However, these policies have often been driven by concerns for human health rather than environmental deterioration. In the United States, about nine federal agencies and numerous state and local agencies monitor several aspects of the marine environment (NRC, 1990).

This case also illustrates the need for integration among monitoring programs and agencies to reduce redundancy and costs.

BOX 1.3 *Selected Major Recommendations for a National Environmental Monitoring Framework*[*]

1. Make integration of environmental monitoring and research networks and programs across temporal and spatial scales, and among resources the highest priority of the Framework.
2. Increase the use of remotely sensed information obtained for detecting and evaluating environmental status, and change by coordinating these analyses with ongoing *in situ* monitoring and research efforts.
3. Select variables that are responsive to policy needs.
4. Ensure that the variables being measured and the locations where they are measured are sensitive to environmental change.
5. Establish standards and protocols for data comparability and quality as integral components of the Framework.
6. Adopt performance-based protocols for quality control and data and information management that apply to all components of the Framework, and establish a national quality control program.

[*]List is incomplete. See www.epa.gov/cludygxb/html/pubs. htm.

emap and www.epa.gov/cludygxb/Pubs/factsheet.html Websites.

THE FUTURE OF ENVIRONMENTAL MONITORING

There are gaps in our knowledge of the environment. This lack of information extends to past, present, and even future events. Most social, political, and economic decisions are made at the local, national, and global levels, and are tied to the expectation that growth will continue unhindered. Yet humans often abuse or destroy entire habitats, harvest some animal and plant species to near extinction, and increasingly stress land and water resources through pollution and overuse. The rates of environmental impacts have increased dramatically during the 20th century. On a geopolitical scale, we now know that the impact of human activities in one country or region is often felt in other countries. There are many examples of this, including acid rain in Canada that results from coal-burning plants in the United States, and particulate pollution in the United States that originates from wind erosion in Asia and Africa.

Despite social and political considerations, we cannot quantify or predict the short- and long-term implications of many human activities without adequate information. More data generated from environmental monitoring are needed to anticipate future changes. If the earth's environment and human institutions are to be preserved for future generations, more evenly distributed environmental data must be generated in the future. These data should be generated at all scales and should be of high quality, as well as useable and exchangeable.

High expectations are associated with space (satellite) based global monitoring systems such as the Landsat series (see Chapter 11). Satellite monitoring systems have been used and are increasingly being used to collect voluminous amounts of data on the earth's surface and atmosphere in real time. These systems can provide more uniform access to remote areas of the world. Nevertheless, these systems do not always allow us to measure at the intermediate or smaller scales, and many physical, chemical, and biological properties still cannot be monitored adequately. Therefore, for the time being, we have to continue to rely on "hands on" sampling and measurement techniques to assess the state of the environment. As the state of scientific knowledge advances, so does our ability to monitor the environment with more efficient, accurate, and precise techniques and instrumentation.

PURPOSE OF THIS TEXTBOOK

This textbook is composed of chapters that cover environmental monitoring from all aspects, including sampling methods, environmental characterization, and associated applications. Chapters 1 through 4 cover basic information central to environmental monitoring, including objectives and definitions, statistics and geostatistics, field surveys and mapping, and automated data acquisition. Chapters 5 through 11 cover techniques of sample collection with emphasis on field methodology used in soil, vadose zone, water, and air sampling, including remote sensing. With Chapters 12 through 17, a general approach to monitoring and characterization of physical, chemical, and biological properties and processes is presented. Finally, in Chapters 18 through 20, general applications of environmental monitoring are presented and discussed, including risk assessment and environmental regulations.

The approach used in the development of each chapter is scientific and objective. It presents the facts based on well-established and accepted scientific principles, and it gives the reader basic underlying theory on each method or process and refers the reader to other more detailed comprehensive textbooks when needed. The intended target audience is for junior and senior undergraduates majoring in Environmental Sciences and for graduate students who wish to have a comprehensive introduction into monitoring and characterizing the environment. The focus of this textbook is on methods and strategies for environmental monitoring with emphasis on field methods. Laboratory methods are also presented in each chapter as needed to complement field methodology or to illustrate a principle or an application.

This textbook covers the following subjects:

- Types of data required to meet objectives
- Equipment needed and necessary measurements
- Field sample collection and real-time sampling
- Direct (destructive) and indirect (non-destructive) methods
- Statistics to decide numbers and locations and to evaluate field data
- Data interpretation for characterization of environmental processes and ecosystems
- Environmental monitoring information needed to develop remediation and restoration strategies

REFERENCES AND ADDITIONAL READING

1996 Annual Data Summary of Air Quality in Tucson, AZ. PCDEQ AQ–299. Nov. 1997.

Carter, G.C. and Diamondstone, B.I. (1990) *Directions for Internationally Compatible Environmental Data.* Hemisphere Publishing Corporation. New York.

National Research Council (NRC). (1990) *Managing Troubled Waters.* The role of marine environmental monitoring. National Academic Press. Washington, D.C.

National Science and Technology Council (NSTC). (1997) Integrating the Nation's Environmental Monitoring and Related Research Networks and Programs. Available at: www.epa.gov/cludygxb/Pubs/factsheet. html.

Roots, E.F. (1997) Inclusion of different knowledge systems in research. In: Terra Borealis. Traditional and Western Scientific Environmental Knowledge. Workshop Proceedings, Northwest River, Labrador 10 & 11 Sept. 1997. No. 1. Manseau M. (ed), Institute for Environmental Monitoring and Research, P.O. Box 1859, Station B Happy Valley–Goose Bay Labrador, Newfoundland, AOP E10. Terra Borealis 1:42–49, 1998.

Schröder, W., Franzle, O., Keune, H., and Mandy, P. (1996) *Global Monitoring of Terrestrial Ecosystems.* Ernst & Sohn. Berlin.

US EPA ORD. (1997) Environmental Monitoring and Assessment Program (EMAP). See www.epa.gov/emap.

2

SAMPLING AND DATA QUALITY OBJECTIVES FOR ENVIRONMENTAL MONITORING

J.F. ARTIOLA AND A.W. WARRICK

Environmental monitoring involves the collection of one or more measurements that are used to assess the status of an environment. However, the goals, sample collection strategies, and methods of analysis used in monitoring must be well defined in advance to obtain robust results. In the preparation of a sampling plan, goals, strategies, and methods must be considered in conjunction with an understanding of the target environment, including the physical, chemical, and biological variables and processes involved. Existing knowledge of the environment is used to help develop the monitoring plan. Box 2.1 lists general definitions of the three components associated with environmental monitoring. The reader may find these definitions self-evident, but each component must be carefully considered in relation to the others, if environmental monitoring efforts are to succeed.

This chapter presents general concepts about environmental sampling and data quality objectives, including definitions of sampling units, environmental patterns, and basic statistical concepts used in monitoring. Instrument measurements and basic analytical data quality requirements are also introduced in this chapter. Statistical principles of sampling, data processing, and specific

---

monitoring methods are discussed in more detail in subsequent chapters. A list of terms is presented in Box 2.2 to assist in the discussion of the topics covered in this and other chapters.

ENVIRONMENTAL CHARACTERISTICS

Most environments have unique features or special characteristics that help environmental scientists choose and ultimately select one sampling approach over another. On a global scale, one can distinguish between land- and water-covered areas and separate them with ease. On a watershed scale, aerial photographic and topographic maps may be used to identify the location of streams, agricultural fields, or industrial activities (Figure 2.1) that further subdivide the land environment. At the field scale, information on soil series and soil horizons can be used by scientists to design soil sampling plans for waste-contaminated sites (see Chapter 7). These examples illustrate that a priori knowledge of the general physical, chemical, and biological characteristics of an environment is indispensable in environmental monitoring.

Environmental monitoring often has a temporal component. Therefore knowledge of the dominant cycles that affect an environment or a parameter of interest is also indispensable. For example, information about the degradation rate of a pesticide in soil may help scientists design a cost-effective series of soil sampling events. If the estimated half-life (that is, the time it takes for the chemical concentration to decrease by 50%) of the pesticide in the soil environment is known to be approximately 6 months, it may be sufficient to collect a soil sample every 3 months over 2 to 3 years to monitor and quantify degradation rates. However, if the pesticide's half-life is closer to 30 days in the soil environment, then weekly sampling for up to 6 months may be needed to obtain useful results.

FIGURE 2.1 Environmental features. *Top*, agricultural field; *left*, stock pile; *center*, landfill. (Corel CD photo collection, public domain.)

SPATIAL PROPERTIES

The earth environment is defined by two or three spatial dimensions. Measurements at the interface between two environments have two dimensions (X–Y) along a plane or surface. This plane is often the surface of the earth and defines many critical environments, including agricultural and range lands, wetlands, forests, or lake and ocean surfaces. The third dimension is the Z axis away from the X–Y plane. Thus, the Z dimension comprises height or depth and incorporates environments such as the atmosphere, the earth's subsurface, and the ocean depths. Human beings live inside the atmosphere and walk on the X–Y plane defined by the earth's surface (Figure 2.2). Therefore environmental scientists spend much time trying to quantify what happens at or very near the earth crust–atmosphere interface.

The collection of samples at multiple depths or altitude intervals adds a third dimension (Z) to two-dimensional (2-D) sampling. It is possible to collect samples at random intervals down a soil/geological profile. However, most of the time, either discrete sampling (at fixed intervals) or stratified sampling (defined by geologic layers) is chosen. In the laboratory, cores are visually inspected and often separated in layers. Similarly, for atmospheric measurements a priori knowledge of possible temperature inversions, winds, and turbulent layers helps

atmospheric scientists define sampling locations, altitudes, and ranges.

TEMPORAL PROPERTIES

Usually sample collection or measurements over time are defined with natural cycles such as daytime; nighttime; or daily, seasonal, or yearly intervals. Additionally, more precise intervals are sometimes simply defined in convenient time units such as seconds (or fractions), minutes, hours, weeks, or months. Therefore most temporal sampling programs can be defined as systematic because they are usually carried out at regular intervals. For example, groundwater monitoring at landfill sites is often done once every 4 months over a year. Farmers collect soil samples for fertility evaluations usually once a year in the spring before the planting season.

REPRESENTATIVE UNITS

Environments do not always consist of clearly defined units. For example, although a forest is composed of easily recognizable discrete units (trees), a lake is not defined by a discrete group of water units. The lake in fact has a continuum of units that have no beginning or end in themselves. However, these "water" units (like the

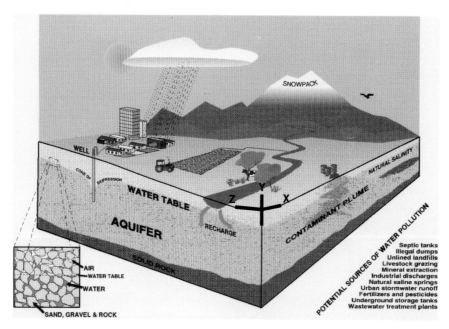

FIGURE 2.2 Three-dimensional section major environmental spatial patterns (landscape features, subsurface and, atmosphere) and potential sources water of pollution. Sources of air pollution (not included) require monitoring of the atmosphere. Most sampling/monitoring activities occur at or near the atmosphere-soil-water interphase. (From: Arizona Water, a poster published by the Arizona Water Resources Research Center [2002].)

forest units) occupy specific volumes in space at any given time. Furthermore, the water units in total reside within fixed boundaries defined by intersections with other components of the environment.

Within each environment we can now define a sample using an arbitrary unit of volume such as a liter or a gallon. No unique definition for a representative sample or unit exists. Each environment and scale has a different unit definition. Ideally, the sample support should be equal to the unit. However, this is not always the case. Because of this ambiguity, a few examples of this concept will be presented. Note that a unit is defined as the smallest sample or observation that has or is believed to have all the attributes of the targeted environment. In other words, it is considered to be "representative" of the target component. In reality this often translates to the smallest sample or observation that can be collected, handled, identified, or measured directly. Sampling protocols are also intimately related to how units are defined in an environment. Sampling protocols often bring an inherent bias to the process of sampling; this bias is discussed later in this chapter. Box 2.3 lists some examples of samples collected from different environments. From these examples, it is evident that sampling protocols are defined by the unique characteristics of each environment.

The size or dimensions of a sample are constrained by two important aspects. First, the sampling technique applied to the problem must be defined, with consideration to the physical limitations of the environment, which in turn limits the type of equipment available for use, sensor resolution, and mass of the material removed. For example, what is the smallest sample that can be recognized or identified visually? In some cases the sample can also be defined by the lowest common denominator such as an individual plant or animal in whole or in parts.

In many cases, the number of units comprising an environment may be so large as to be considered infinite. Because the collection of all units from a population is impossible, a few units that represent the environment (population) are selected. Thus, the second important aspect is the collection of an adequate and representative number of samples that are critical to the science of environmental monitoring. Sample error about the mean is related to the number of observations made and is defined in the classical statistics discussed in detail in Chapter 3.

SAMPLING LOCATIONS

Statistical-based monitoring plans require environmental scientists to collect samples from an environment at statistically determined locations. Ideally, each sampling location should be selected at random. Also, the number of samples must be defined with a maximum-accepted level of error in the results (see Chapter 3). In reality, sample location and number of samples must be considered in concert with several other important aspects unique to environmental science. For example, costs associated with

BOX 2.3 *Examples of Representative Units*

Example 1: A sample from a river taken from the middle and at one half the depth at a given position along the length of the river. The sample is collected from a location that has properties, such as velocity or chemical composition, representative of the mean properties of the river. In this case, sample size (volume) is defined by the selected analytical methodological requirements.

Example 2: A soil sample collected from an agricultural field. One or more locations are selected with no distinguishing features such as unique surface cover, depressions, or protrusions. In this case, the mass of soil sample(s) collected (usually between 300 and 1000 g) is determined by the sampling equipment used, as well as the analytical methodological requirements.

Example 3: An air sample collected from a street intersection during a certain time interval for analysis of particulates. This requires the passage of a fixed volume of air through a filter. The actual sample is the particulate matter collected from a known volume of air passed through the filter. However, the volume of air is limited by the mechanics of the filtering system, the minimum and maximum number of particles that needs to be collected for detection, and the sampling interval.

Example 4: A plant sampled to measure nutrient uptake or pollutant accumulations. Tissue from the same plant parts such as leaves and roots is chosen from plants at similar stages of growth. The sample support (typically 10–200 g weight/weight) is defined by plant genotype and morphology. With few exceptions, leaves can be collected in whole units. Nevertheless, plant roots and shoots can seldom be collected in their entirety and are often subsampled (Figure 2.3).

Example 5: A measure of surface ground cover with aerial photography. The number of shrubs per unit area may be counted. Therefore a resolution sufficient to resolve or distinguish individual shrubs of a minimum size must be chosen. The resolution of the picture in pixels will be determined by the minimum shrub size and the area of the coverage. The photography equipment and type of airplane, including minimum-maximum flight altitudes, must be considered (Figure 2.4).

sampling and analysis often limit the application of rigorous statistics in environmental monitoring. Also, the analysis of viruses in water samples or the analysis of Environmental Protection Agency (EPA)–designated priority pollutants in soil samples can cost in excess of $1000 per sample. Other factors such as accessibility and time may constrain statistical schemes and result in unintentional bias.

The degree of the bias varies with the type of knowledge available to the designer(s) of the sampling

FIGURE 2.3 Pecan tree leaf tissue sampling for nitrogen analysis; only the middle pair of leaflets from leafs on new growth must be collected. (Figure 49 from Doerge *et al.*, 1991. Reprinted from "Nitrogen Fertilizer Management in Arizona," T. Doerge, R. Roth, and B. Gardner, University of Arizona Cooperative Extension, copyright 1991 by the Arizona Board of Regents.)

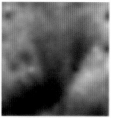

FIGURE 2.4 **(A)** Picture shows a digital picture of a landscape with partial vegetation. **(B)** Picture shows an enhanced section of one 20-cm-tall shrub (30 × 30 pixels). If **(A)** had a lower resolution (less pixels per unit area), the shrub could not be identified. (Source: J. Artiola.)

plan. Some bias is expected, acceptable, and even necessary to reduce costs. The use of previously acquired knowledge about an environment to select a specific location, soil depth, or plant species is acceptable. For example, if it is known that a type of plant found in an abandoned industrial site is a metal accumulator, it would make sense to sample this plant versus others found at the site, to estimate the potential impact of metals found in plants that may be consumed by grazing wildlife visiting the site. However, this process, if left unchecked, can quickly become judgment sampling, which has some inherent shortcomings. Judgment sampling assumes that the sampler "knows best" and that the location or time of the sample selected by the sampler is "representative." Often this approach produces biased data that have no defined relevance. Nonetheless, some forms of this approach are often used in environmental monitoring to reduce costs and save time. Box 2.4 lists common sample collection approaches and definitions used in environmental monitoring.

TYPES OF ENVIRONMENTAL SAMPLING

Environmental monitoring is paradoxical in that many measurements cannot be done without in some way affecting the environment itself. This paradox was recognized by Werner Heisenberg in relation to the position of subatomic particles in an atom and named the Heisenberg Uncertainty Principle. Nonetheless, varying degrees of disturbance are imposed on the environment with measurements. Destructive sampling usually has a long-lasting and often permanent impact on the environment. An example is drilling a deep well to collect groundwater samples. Although here the groundwater environment itself suffers little disturbance, the overlaying geological profile is irreversibly damaged. Also, soil cores collected in the vadose zone disrupt soil profiles and can create preferential flow paths. When biological samples are collected, the specimen must often be sacrificed. Thus, sampling affects an environment when it damages its integrity

---

BOX 2.4 *Sampling Definitions*

Random: Sampling location selected at random. All units have the same chance of being selected. Also, the original environment can be subdivided into smaller domains by visual observations and a priori information. This approach still yields random data sets, if data are also collected randomly, as they were in the original domain (see Figure 2.5A,B).

Systematic: This approach is a subset of random sampling if the initial sampling locations are selected randomly (see Figure 2.5C). This type of sampling is very useful to map out pollutant distributions and develop contour maps (see Chapter 3). Systematic sampling is also very useful to find hot spots, subsurface leaks, and hidden objects (see Figure 2.5B). Systematic sampling can be called *search sampling* when grid spacing and target size are optimized to enhance the chance of finding an object or leak (see Gilbert, 1987).

Grab, Search, or Exploratory: Typically used in pollution monitoring and may include the collection of one or two samples to try to identify the type of pollution or presence/absence of a pollutant. This haphazard approach of sampling is highly suspect and should be accepted only for the purposes previously stated. Exploratory sampling includes, for example, the measurement of total volatile hydrocarbons at the soil surface to identify sources of pollution. Faint hydrocarbon vapor traces, emanating from the soil, can be detected with a portable hydrocarbon gas detector (see Figure 2.5D).

Surrogate: Done in cases where the substitution of one measurement is possible for another at a reduced cost. For example, if we are trying to map the distribution of a brine spill in a soil, we know that the cost of analysis of Na^+ and Cl ions is much more expensive than measuring electrical conductivity (EC). Therefore a cost-effective approach may be to collect

BOX 2.4 *(Continued)*

samples in a grid pattern and measure EC in a soil-water extract.

Example: total dissolved solids in water can be estimated with EC measurements (see Chapter 9).

Composite (bulking): Commonly done to reduce analytical costs in sampling schemes where the spatial or temporal variances are not needed. This approach is common in soil and plant fertility sampling where only the average concentration of a nutrient is needed to determine fertilizer application rates. Composite sampling is usually limited to environmental parameters that are well above the quantifiable detection limits; common examples for soils are total dissolved solids, organic carbon, and macronutrients.

Path Integrated: Used in open path infrared (IR) and ultraviolet (UV) spectroscopy air chemical analysis (see Chapter 10).

Time Integrated: Commonly used in weather stations that measure ambient air properties such as temperature and wind speed, but report time-averaged hourly and daily values (see Chapter 14).

Remote sensing: Commonly used to collect two-dimensional photographs of the earth surface passive radiation using IR, UV, and Vis light sensors (see Chapters 10 and 11).

Quality Control:

a) Blanks are collected to make sure that containers or the preservation techniques are not contaminating the samples.

b) Trip samples are blank samples carried during a sampling trip.

c) Sample replicates are collected to check the precision of the sampling procedure: preservation and contamination.

d) Split samples are usually collected for archival purposes.

---

or removes some of its units. When samples are physically removed from an environment, it is called *destructive sampling*. Table 2.1 lists some forms of destructive sampling and their relative impact to the environment.

Nondestructive sampling, often called *noninvasive sampling*, is becoming increasingly important as new sensors and technologies are developed. Two major techniques are remote sensing, which records electromag-netic radiation with sensors, and liquid-solid or gas-solid sensors, which provide an electrical response to changes in parameter activity at the interface. The first sampling technique is best illustrated by satellite remote sensing that uses reflected visible, IR, and UV light measurements of the earth's surface. The second technique is commonly used in the direct measurement of water quality parameters such as E.C. or pH with electrical conductivity and H^+ activity–sensitive electrodes. Box 2.5 lists common methods that use nondestructive sampling. It is important to note that even "noninvasive" sampling can alter the environment. For example, inserting an instrument probe into the subsurface can alter the soil properties.

These methods are discussed at length in subsequent chapters.

SAMPLING PLAN

Several objectives must be defined in a good sampling plan. The most obvious objective is what is the objective of the study? What needs to be accomplished with the sampling plan? Who will be using the results? Examples include what is needed to quantify the daily amount of a pollutant being discharged into a river, to determine the percent of vegetative cover in a watershed area, or to measure the seasonal changes in water quality in a reservoir. Each of these objectives requires different sampling approaches in terms of location, number or samples, and sampling intensity. Therefore it is important that the

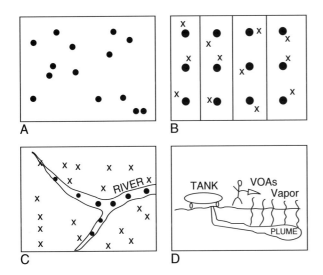

FIGURE 2.5   **(A)** Simple random sampling. **(B)** Systematic grid sampling *(dots)* and random sampling within each grid block *(x)*. **(C)** Stratified random sampling (soil, plants, etc.) *(x)* within each section of the watershed, stratified systematic sampling of the water in tributaries and river *(dots)*. **(D)** Search sampling of volatile gases *(VOAs)* associated with a subsurface plume of volatile contaminants with a vapor detector above surface.

TABLE 2.1
Forms of Destructive Sampling

Sample Type	Damage/Duration
Subsurface cores (geologic)	Major, permanent
Soil cores	Minor, permanent
Plants and plant tissue samples	Minor, may be reversible
Animals and animal tissue samples	Variable, may be reversible
Water samples	Insignificant, reversible
Air samples	Insignificant, reversible

BOX 2.5 *Nondestructive Sampling*

–Satellite-based optical (passive) and radar (active) sensors to measure topography, plant cover, or temperature (see Chapter 11).

–Portable sensors for water quality to measure pH, electrical conductivity (EC), or dissolved oxygen (DO) (see Chapter 9).

–Neutron probes with access tube to measure soil water content (see Chapter 12).

–Time domain reflectometry (TDR) to measure soil water content and salinity (see Chapter 12).

–Fourier transform infrared spectroscopy (open path) to measure greenhouse gases and hydrocarbon pollutants in air (see Chapter 10).

–Ground-penetrating radar and EC electrodes to measure subsurface geology, particle density, and salinity distributions (see Chapter 13).

---

objective be clearly stated, that it be attainable, and that its successful completion produce data that are usable and transferable. However, there are three issues that often limit efficacy and objectiveness in environmental monitoring:

- Number of samples (n), which is usually limited by sample analysis and/or collection costs.
- Amount of sample, which is often limited by the technique used.
- Sample location, which is often limited by accessibility.

Box 2.6 lists and describes the basic elements of a complete sampling plan.

ANALYTICAL DATA QUALITY REQUIREMENTS

A critical component of environmental monitoring is the type of analytical equipment used to analyze the samples. The choice of methods is usually dictated by the environment monitored, the parameter of interest, and the data quality requirements. Typically, we must select a scientifically sound method, approved by a regulatory agency. For example, drinking water quality methods require a specific laboratory technique. For example, in the case of the analysis of total soluble lead in drinking water, U.S. EPA Method 239.2 should be used. The method requires

---

BOX 2.6 *Elements of a Sampling Plan with Data Quality Objectives*

–**Number and types of samples collected in space and time.** This section should discuss the statistical basis for the number of samples and sampling patterns selected. These issues are discussed further in Chapter 3, which is devoted to statistics and geostatistics.

–**Actual costs** of the plan, including sample collection, analysis and interpretation. A cost analysis that provides a measure of the cost versus effectiveness of the plan. Alternate approaches can also be included. Sampling costs are determined by the precision and accuracy of the results.

–**Data quality control and objectives** are also needed in a sampling plan. Although the following requirements are borrowed from U.S. EPA pollution monitoring guidelines, these are generic enough that they should be included in any type of environmental sampling plan.

Quality: Discuss statistical measures of:
   Accuracy (bias): How data will be compared with reference values when known. Estimate overall bias of the project based on criteria and assumptions made.
   Precision: Discuss the specific (sampling methods, instruments, measurements) variances and overall variances of the data or data sets when possible using relative standard deviations or percent coefficient of variation (%CV).
Defensible: Ensure that sufficient documentation is available after the project is complete to trace the origins of all data.
Reproducible: Ensure that the data can be duplicated by following accepted sampling protocols, methods of analyses, and sound statistical evaluations.
Representative: Discuss the statistical principles used to ensure that the data collected represents the environment targeted in the study.

BOX 2.6 *(Continued)*

Useful: Ensure that the data generated meets regulatory criteria and sound scientific principles.

Comparable: Show similarities or differences between this and other data sets, if any.

Complete: Address any incomplete data and how this might affect decisions derived from these data.

–**Implementation:** A detailed discussion on how to implement the plan should be provided. Discuss in detail the following issues, when applicable.

Site location: Provide a physical description using maps to scale (photos, U.S. Department of Agriculture, topographic).

Site accessibility: Show maps of physical and legal boundaries.

Equipment needed: Down to the last pencil.

Timetable: List/graph dates (seasons) and times to complete.

Personnel involved: All personnel, chain-of-command, qualifications.

Personnel training: Any specialized training needed, certifications.

Safety: List any safety equipment/training needed. Type and level of protective equipment.

Sample containers: List types and numbers of containers.

Sample storage and preservation: Describe methods and container used to store and transport samples.

Sample transportation: Describe methods and equipment for sample transportation.

Forms: Provide copies of all the forms to be filled out in the field, including sample labels and seals, and chain-of-custody forms.

---

the use of graphite furnace atomic absorption spectroscopy. Additionally, the method provides a detailed laboratory operation procedure and quality control requirements for use with water samples. Many analytical methods are available for the analysis of air, water, soil, wastes, plants, and animal samples. These methods can be found in standard references for the analysis of soil, water, and wastes (Box 2.7).

These books provide comprehensive lists of methods, including laboratory operating procedures also known as *standard operating procedures (SOPs)*. Field and laboratory methods are not usually interchangeable, although they are often complementary. As shown in subsequent chapters, the choice of field analytical methods is often limited. When no direct methods of analysis exists, then sampling and analysis become two separate tasks. Field analysis procedures are often adapted from laboratory methods.

The reader is encouraged to review standard laboratory analysis references to understand basic analytical methods. In upcoming chapters, discussions of methods of analysis concentrate on field sampling and analysis procedures. Standard laboratory methods are only introduced when needed to complement a field protocol. As previously indicated, several national and international agencies provide guidelines and approval of methods. Because samples are collected in the field but analyzed in the laboratory, these standards may be applicable only to laboratory procedures. Box 2.8 lists of some agencies that provide methods and guidelines related to environmental monitoring.

PRECISION AND ACCURACY

Measurements are limited by the intrinsic ability of each method to detect a given parameter. These limitations are dependent on the instrument(s) and the method used, as well as the characteristics of the sample (type, size, matrix) and the human element.

Precision

Observations are made with instruments that are a collection of moving parts and electronic components

---

BOX 2.7 *Examples of Reference Books*

Soils:
Soil Science Society of America, *Agronomy No. 9 and No. 5 series.*

Wastes:
U.S. Environmental Protection Agency, *SW–486* and subsequent revisions. American Society of Testing Materials Methods.

Waters and wastewaters:
American Waterworks Association, *Standard Methods.* Editions. U.S. EPA Standard Methods, 500+600 Series & Contract Laboratory Program and subsequent revisions.

Microbiology:
American Waterworks Association, *Standard Methods.* Editions.

subject to changes. It is therefore impossible to guarantee that the same signal will produce the same response repeatedly. Precision is a measure of the reproducibility of a measurement done several times on the same sample or identical samples. A measure of the closeness of measurements is given by the distribution and its standard deviation. In most chemical measurements, instrument/method precision is computed under controlled conditions with no fewer than 30 replicate measurements. These measurements are done with standards near the detection limits of the instrument. Analytical measurements are usually assumed to have a normal distribution. The concept of normal distribution is discussed in Chapter 3.

Resolution is a term sometimes used interchangeably with precision and is applicable to modern measuring devices that convert a continuous analog (A) signal into a discrete digital (D) response (see Chapter 4). Thus, all instruments, including cameras; volt, amp, and resistance meters; and photometers, have an intrinsic resolving power (see Chapter 4). Resolution is the smallest unit that provokes a measurable and reproducible instrument response. How we define this response determines the instrument's detection limits discussed in the next section.

Accuracy-point of reference

The instruments used in environmental monitoring and analysis are often extremely sophisticated, but without a proper calibration, their measurements have no meaning. Thus, most instruments require calibration with a point of reference because measurements are essentially instrument response comparisons. A reference is usually a standard such as a fixed point, a length, a mass, a cycle in time, or a space that we trust does not change. Field and laboratory instruments must be calibrated using "certified" standards. Calibration is a process that requires repeated measurements to obtain a series of instrument responses. If the instrument produces a similar response for a given amount of standard, then we trust the instrument to be calibrated. Box 2.9 provides a list of several suppliers that provide common reference materials.

DETECTION LIMITS

All techniques of measurement and measuring devices have limits of detection. Furthermore, most instruments can be calibrated to produce predictable responses within only a specified range or scale. At the low end of the range, a signal generated from a sample is indistinguishable from background noise. At the upper range, the sample signal generates a response that exceeds the measuring ability of the instrument. When measurements are made at or near the detection limits, the chances of falsely reporting either the presence or absence of a signal increase.

Lower detection limits are very important in environmental monitoring and must be determined for each method-instrument-procedure combination before field use. These detection limits should be determined under controlled laboratory conditions. There is no consensus on how to measure detection limits, and they are still a subject of debate. The most common method is based on the standard deviation(s) of the lowest signal that can be observed or measured generated from the lowest standard available. Note that blank, instead of standard, readings can be used, but this is not recommended because blank and standard values often do not have the same distribution. It is also important to remember that detection limits are unique for each environment (matrix), method, and analyte. Consecutive standard readings should be made to determine detection limits no less than 30s. From these values the mean and standard deviation(s) should be computed. We can then proceed with the analysis of an unknown sample and set a reliable detection limit (RDL) to be equal to the method or minimum detection limit (MDL) of no less than 3s, (Figure 2.6A, α and β areas). However, even setting an RDL at 3s has problems in that 50% of the time, data that are the same as the MDL will be discarded (Figure 2.6A). This equates to a 50% chance of making a Type II error

(false negative), as compared with a less than 0.15% chance of making a Type I error (false positive).

If the RDL is increased to 6s units, then the chances of having either a Type I or Type II error are now both equal to or less than 0.15% (Figure 2.6B, α and β areas). If the

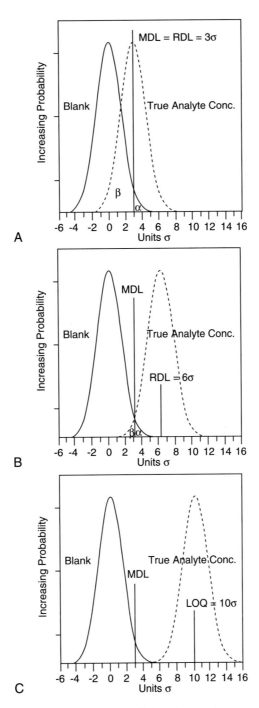

FIGURE 2.6 Instrument blank and real sample values can be misinterpreted when these are close to each other. (**A**) When reliable detection limits are set to equal minimum or method detection limits = 3 sigma (σ) units, blank and sample values overlap. (**B**) If reliable detection limits are set at 6 σ units at least, blank and sample reading overlap minimally. (**C**) Quantifiable detection limit should be set at least at 10 σ units above average blank to prevent overlap between blank and sample values.

RDL is increased to 10s units, no chance exists of making a Type I or Type II error (Figure 2.6C). This is also called the *limit of quantification (LOQ)*, as defined by the American Chemical Society Committee on Environmental Improvement.

Single measurements have large uncertainties. For example, if we assume that the relative uncertainty of a measurement at the 95% percent confidence level is defined by the following equation (Taylor, 1987):

$$\text{Relative uncertainty} = 2\sqrt{2}(s/\underline{x})100 \qquad \text{(Eq. 2.1)}$$

changing $\underline{x}$ (the measured value) for multiples *(N)* of the standard deviation(s), then Equation 2.2 becomes:

$$\text{Relative uncertainty} = 2\sqrt{2}(1/N)100 \qquad \text{(Eq. 2.2)}$$

This relationship is plotted in Figure 2.7. Therefore for a single measurement, if the limit of detection is set at *3s*, its relative uncertainty will be $\pm 70\%$ (at the 95% confidence level) about the true value. Whereas, if the limit of detection is set at 10s, the relative uncertainty will be $\pm 20\%$ (at the 95% confidence level) about the true value (see Figure 2.7) (Keith, 1991; Taylor, 1987). Box 2.10 presents a summary of the method detection limits. Other references on this subject include Funk *et al.* (1995) and McBean and Rovers (1998).

It is important that data be reported with the specified detection limits as qualifiers so that data can be censored (reported as "less than") if they fall below the specified detection limit. Failure to attach this information to data sets may lead to inappropriate use of data.

TYPES OF ERRORS

Field instruments with poor precision and accuracy produce biased measurements. Three types of errors can occur when making measurements:

BOX 2.10 *Quick Summary on How to Establish Method Detection Limits*

1—Measure a lowest standard signal several times (n ≥ 30)
2—Compute sample std (s), but assume population std (σ)
3—Set method detection limit (MDL) as $3s \simeq 3\sigma$
4—Set reliable detection limit (RDL) as $6s \simeq 6\sigma$
5—Set quantifiable detection limit (QDL) as $10s \simeq 10\sigma$

Detection limits should not be estimated with simple regression extrapolation.

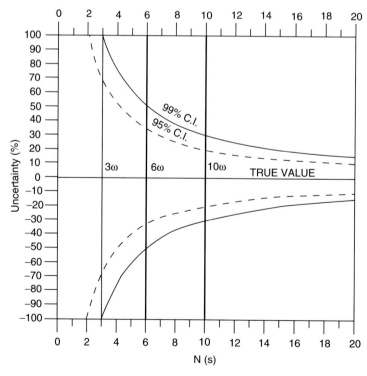

FIGURE 2.7 Single measurements have an increasing uncertainty when they approach the instrument limit of detection. Reporting data values that are less than 10 sigma units is not recommended.

Random errors are usually due to an inherent dispersion of samples collected from a population, defined statistically by variance or standard deviation about a "true" value. As the number of replicate measurements increases, this type of error is reduced. Precision increases as n (number of measurements) increases. Random errors include Type I and Type II errors, which were discussed previously.

Instrument calibration errors are associated with the range of detection of each instrument. Uncertainty about the calibration range varies. Typically, as the analyte concentration increases, so does the standard deviation. Also, if extrapolation is used, at either end of the calibration curve, the standard deviation of the confidence intervals increases quickly. For example, the common linear regression used to interpolate instrument response versus analyte concentration assumes that all the standards used have the same standard deviation. Because this is not true, often modern instruments incorporate the weighted regressions to optimize calibration curves.

Systematic errors or *constant errors* are due to a variety of reasons that include the following:

- Biased calibration-expired standards
- Contaminated blank; tainted sample containers
- Interference: complex sample matrix
- Inadequate method: does not detect all analyte species
- Unrepresentative subsampling: sample solids/sizes segregate, settle
- Analyte instability: analyte degrades due to inadequate sample preservation

Field measurements are very prone to large random and systematic errors because operators do not make enough replicate measurements and do not calibrate instruments regularly. Additionally, environmental conditions (such as heat, moisture, altitude) can change quickly in the field, which increases the magnitude and chance of occurrence of the preceding list of errors. Instrument environmental operating ranges should be carefully noted in sampling plans.

COMBINED ASPECTS OF PRECISION AND ACCURACY

Environmental data include all the factors that affect precision and accuracy. The previous sections focused on the analytical aspects of data precision and accuracy, but there are also inherent field variations (spatial and temporal variabilities) in the samples collected. There are also inherent variations in the methods chosen for the sample preparation before analysis. Thus, it is useful to discuss in

a final report, the precision, accuracy, and detection limits associated with each step of the monitoring process. The following is a suggested sequence of data quality characterization steps:

1. Instrument precision and detection limits.
2. Type of sample and sample preparation (method) precision and detection limits.
3. Combined sample spatial and temporal or random variations. Note that if the goal is to measure field spatial and temporal variabilities, this step should be omitted (see Chapter 3).
4. Overall precision of the data based on sum of all or some errors from the steps 1–3.

The precision values presented in the these steps should be in the form coefficients of variation $CV = [s/\times]$ or $\%CV = [(s/\times) * 100]$ where $\times$ is mean and s is standard deviation. These coefficients can be added as needed to provide an overall precision associated with each data point. For example, if the combined instrument and sample preparation %CV is $\pm 15\%$, and the field sample variability is $\pm 20\%$, then the overall certainty of the data must be reported as $\pm 35\%$.

QUALITY CONTROL CHECKS

Routine instrument, method precision checks, or both can be done in the field by analyzing the same sample or standard twice or by analyzing two samples that are known to be identical. This process should be repeated at regular intervals every 10–20 samples (Csuros, 1994). In this case the percent absolute difference (PAD) is given by the following formula:

$$PAD_{AB} = [abs(A - B)/(A + B)] * 200 \qquad (Eq.\ 2.3)$$

where abs = absolute value

Similarly, we can check the accuracy of the method if one of the two sample values is known to be the true value. The accuracy as a percent relative difference (PRD) of the measurement can simply be defined as:

$$PRD_B = [(A - B/B] * 100 \qquad (Eq.\ 2.4)$$

where A is the unknown value and B is the true value. Conversely, accuracy could also be checked by measuring the percent recovery (%R) of an unknown value against a true value:

$$\%R_B = [A/B] * 100 \qquad (Eq.\ 2.5)$$

where A is the unknown value and B is the true value.

Analytical limits of precision and accuracy may be determined and updated in ongoing field projects that require numerous measurements over long periods. For example, plotting Equations 2.3 and 2.4 or 2.5 over time with preset upper and lower limits would provide a visual indication of the precision and accuracy of each measurement over time. More commonly, control charts are made by plotting individual values by date against an axis value scale (Figure 2.8). If the precision of each measurement is needed, then the center line is a running mean of all the QC measurements. The upper and lower control limits can be defined in terms of confidence limits (see Chapter 3) as mean $\pm 2s$ or warning limits (WLs) and mean $\pm 3s$ or control limits (CLs). If the accuracy of each measurement is needed, then the central line represents the true value and the %R values (Eq. 2.5) are plotted with WL and CL lines (see Figure 2.8). For more details on the use of control tables, see Eaton *et al.* (1995).

REPORTING DATA

Most chemical, physical, and biological measurements have inherent limitations that limit their precision and consequently their accuracy to four or five significant digits. However, there are exceptions worthy of discussion. In the digital age, methods often use highly precise processing algorithms with more than 128 bits (significant digits) of precision. This high level of precision has meaning only in the context of the computational power (speed) of a computer and also serves to reduce rounding-off errors that can become significant when performing a series of repetitive computations. Digital processing does not add more digits of precision than those imposed by sensor (analog or digital), human, and environmental factors. Therefore computer processing does not add digits of precision to external data such as values entered in spreadsheets and graphs. In digital photography the resolution of a picture is often reported in numbers of pixels per unit surface. For example, a picture may have $512 * 256$ pixels, which is exactly 131072. But, this six-digit number refers to the digital composition of the image, not its visual precision. Atomic clocks routinely achieve eleven digits of precision by measuring highly stable frequency energies from light-emitting gaseous molecules. These examples of high-precision data and data processing are the exception rather than the rule.

Data manipulation often combines numbers of different precision. For precision biases to be reduced during data manipulation, round-off rules should be always be followed. These rules are listed in Boxes 2.11 and 2.12.

UNITS OF MEASURE

Use of appropriate units or dimensions in the final results is important to have transferability and applicability. There are several systems of measurement units, the most common being the British/American system and the metric system. The *Système Internationale*

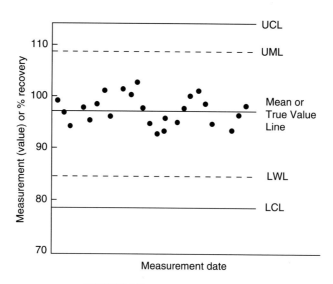

FIGURE 2.8   Quality control chart.

BOX 2.11 *Basic Rules for Determining Significant Digits*

1. Terminal zeros to the left do not count.

Examples: 0.012 has 2 significant digits
0.001030 has 4 significant digits
0.10234 has 5 significant digits
14.45 has 4 significant digits

Sometimes there is ambiguity about a final zero to the right; if it is significant, then leave it there. Note that many spreadsheets keep total number of digits constant and do not consider significance. Do not erase or add "0s" on the right unless you are certain about their significance.

2. Add up all nonzero digits to the left and all digits to the right.

Examples: 103.50 has 5 significant digits
02.309 has 4 significant digits

3. Special case. As values near the detection limit, the number of significant digits decreases. For example, if the mass detection limit of an instrument is $0.001 \, mg \, kg^{-1}$, and the precision is only good to 3 significant digits, then values below are correct and could be found reported in the same data set.

Examples:	Correct			Incorrect
sample 1	1.56	has 3 sig. digits		1.563
sample 2	0.125	has 3 sig. digits		0.1254
sample 3	0.013	has 2 sig. digits		0.0132
sample 4	0.009	has 1 sig. digit		0.00867

BOX 2.12 *Basic Rules for Rounding Off Numbers*

Often final data are the product of both a direct measurement and a multiplier or divisor factor, which is the result of another measurement. Sometimes the data are adjusted by some factor, again the result of another measurement. It is important to remember that when data are combined, there are some basic rules to follow to avoid biasing the final results.

1. Multiplying or dividing two numbers results in a value with the lowest number of significant figures.

Examples: 1. $0.34 * 4.349 = 1.5$
2. $23/2.35 = 9.8$
3. $1.50 \, ft * 30.48037$ (conversion factor) $=$ 45.7 cm (see the "Units of Measure" section)

2. Adding or subtracting two numbers results in a value with the fewest decimal place figures.

Examples: 1. $1.28 + 0.023 = 1.30$
2. $67 - 24.789 = 42$
3. $189 + 8.2 = 197$

3. If less than 0.5, round off to 0.

4. If more than 0.5, round to 1.

5. If 0.5 use odd/even rule: If preceding number is odd, then round off high; if preceding number is even, then round off low.

Examples: Applying rules 3, 4, and 5 (underlined number indicates application of odd/even rule) Note: "0" is assumed to be an even number.

1.   0.3̲5 to 0.4
2.   5.751 to 5.7̲5 to 5.8
3.   8.4̲5 to 8.4
4.   0.251 to 0.2̲5 to 0.2

---

d'Unité (SI) incorporates the metric system and combines the most important units and unit definitions used in reporting and processing environmental monitoring data. Scientists and engineers involved in monitoring and characterization activities must pay careful attention to units and often spend significant amounts of time converting data, from British/American to metric and SI, to/from non-SI units. This process undoubtedly adds transcription and rounding-off errors to data. It is common to add more significant figures to data values that have been converted, but because most con-

version factors are multiplications or divisions, the round-off rules spelled out in Box 2.12 still apply. It is unfortunate that there is no uniform or mandated use of the SI, and in particular the use of the metric system in the United States. Table 2.2 provides a summary of the most common units of length, mass, temperature, area volume concentration, density, and others used in environmental monitoring and characterization. Other important and special units and definitions used in environmental monitoring are discussed in each chapter.

TABLE 2.2

Common SI and Non-SI Units Used to Report Data in Environmental Monitoring and Characterization

Dimension	Unit	Abbreviation
Length	kilometer	km
	meter	m
	centimeter	cm
	micrometer	μm
	nanometer	nm
Mass	metric tonne	t (metric)
	kilogram	kg
	gram	g
	milligram	mg
	microgram	μg
	nanogram	ng
	pound	lb
Volume	cubic meter	m³
	liter	L
	milliliter	mL
	microliter	μL
	gallon	gal
Concentration	milligrams per L	mg kg⁻¹
	mg L⁻¹	mg L⁻¹
	parts per million	ppm
	part per billion	ppb
	milliequivalents per L	mEq L⁻¹
	moles of charge per L	mol_c L⁻¹
Plane angle	radian	rad
	degrees	°
Density	grams per cubic cm	g cm⁻³
Temperature	degrees Centigrade	°C
	degrees Kelvin	°K
Radioactivity	curie	Ci
Pressure	atmosphere	at
	pounds per square inch	psi
	pascal	Pa
Application rate	kilograms per hectare	kg ha⁻¹
	pounds per acre	lb acre⁻¹
Energy	joule	J
	British thermal unit	Btu
	calorie	cal

QUESTIONS

1. Define representative unit, sample support, bulking, and temporal pattern.

2. A wastewater treatment plant manager states in his report that a 25,000-gallon storage tank that contained 5 lb of ammonium (NH_4^+) was discharged into the local waterway. He knows that water with ammonium concentration above 5 mg L^{-1} may not be discharged into the local waterway, so he oxidized the ammonium to nitrate (NO_3^-) before discharging it into the waterway, but he apparently forgot (or did he?) that there is a regulatory limit on nitrate discharges of 20 mg L^{-1}, as NO_3^-N.

   (a) What data quality objective(s) may not have been met in the manager report?
   (b) What was the NO_3^-N concentration in the tank after oxidation?
   (c) Did the plant manager violate any discharge rules?

3. What is a standard? Explain in your own words using an example. Use Figure 2.8 and describe the use of a standard as related to precision and accuracy.

4. Why do the chances of making a false-positive error increase as data values near instrument detection limits? Explain your answer.

5. A portable x-ray fluorescence elemental analyzer (see Chapter 13) has generated the following data set with the same sample analyzed for chromium (mg kg^{-1}).

140	167
155	170
144	148
149	165
160	142

   Compute the reliable and quantifiable detection limits of this instrument. If you had a soil cleanup standard of 100 mg kg^{-1} for chromium, would you trust this instrument to give you the type of data needed to select areas for cleanup, or would you require that the samples be analyzed using another instrument? Or require more samples? Explain your answer.

6. Perform the following computations and report the correct number of significant digits.
   $(145.5 + 55) * 2.55 =$ _____
   $(34 - 1.3 + 44.55) * 12 =$ _____
   Round off these numbers to two significant digits.
   $1.553 =$ _____
   $244.2 =$ _____
   $0.5453 =$ _____

7. An instrument has a QDL of 10 μg L^{-1} and three digits of precision. Which numbers should not be reported in their present form (correct them as needed) or reported as censored data? Explain your answers.
   12.55, 455.1, 340, 1778, 1.34, 4.58. 9.917, 24.5.

REFERENCES AND ADDITIONAL READING

Csuros, M. (1994) *Environmental Sampling and Analysis for Technicians.* Lewis Publishers, Boca Raton, FL.

Eaton, A.D., Clesceri, L.S., and Greenberg, A.E. (1995) *Standard Methods for the Examination of Water and Wastewater.* 19th Edition. APHA/AWWA/WPCF. American Public Health Association.

Funk, W., Dammann, V., and Donnevert, G. (1995) *Quality Assurance in Analytical Chemistry.* VCH, Weinheim, NY.

Gilbert, R.O. (1987) *Statistical Methods for Environmental Pollution Monitoring.* Van Nostrand, Reinhold, NY.

Keith, L.H. (1991) *Environmental Sampling and Analysis: A Practical Guide.* American Chemical Society Publisher, Chelsea, MI.

Klute, A. (1986) *Methods of Soil Analysis—Part 1, Physical and Mineralogical Properties.* Second Ed. Agronomy No. 9. ASA, Inc., American Society of Agronomy, Inc. Publishers. Madison, WI.

McBean, E.A., and Rovers, F.A. (1998) *Statistical Procedures for Analysis of Environmental Monitoring Data & Risk Assessment.* Prentice Hall PTR Environmental Management & Engineering Series. Volume 3. Prentice Hall, Upper Saddle River, N.J.

Page, A.L., Miller, R.H., and Keeney, D.R. (1982) *Methods of Soil Analysis—Part 2, Chemical and Microbiological Properties.* Second Ed. Agronomy No. 9. ASA, Inc., SSSA, Inc. Madison, WI.

Sparks, D.L. (1996) *Methods of Soil Analysis-Part 3—Chemical Methods.* SSSA Book Series No 5. Soil Science Society of America, Inc., American Society of Agronomy, Inc., Madison, WI.

Taylor, J.K. (1987) *Quality Assurance of Chemical Measurements.* Lewis Publishers, Chelsea, MI.

Weaver, R.W., Angle, J.S., and Bottomley, P.S. (1994) *Methods of Soil Analysis—Part 2, Microbiological and Biochemical Properties.* SSSA Book Series No 5. Soil Science Society of America, Inc., American Society of Agronomy, Inc., Madison, WI.

3

STATISTICS AND GEOSTATISTICS IN ENVIRONMENTAL MONITORING

J.A. VARGAS-GUZMÁN, A.W. WARRICK, D.E. MYERS, S.A. MUSIL, AND J.F. ARTIOLA

Statistical methods are necessary for environmental monitoring and assessment because in general it is not possible to completely characterize a circumstance by direct observation. For example, one may wish to decide whether a plot of ground is contaminated. It would be both economically unrealistic and simply impractical to analyze all the soil in the plot (even for a fixed depth). Statistical methods allow using partial information to infer about the whole. With some number of locations in the plot having been selected, "data" will be collected in one of several possible ways. For example, soil cores may be extracted and taken to the laboratory for analysis. Alternatively, it may be possible to use an instrument to directly obtain a reading at each location. In the case of *in situ* instruments, it may be possible to obtain data at multiple times at the chosen locations. Both methods of gathering information at selected spatial locations or times are called *sampling*. The result of sampling then is a univariate data set or a multivariate data set. Multivariate means that several data values are generated for each location and time. For example, the soil core might be analyzed for the concentrations of several different contaminants. The data set is called a *sample* for either the univariate or multivariate case. It can be thought of as a subset of the possible values that could be generated by *sampling* the entire plot. This larger set of possible values is sometimes referred to as the *population*.

There are at least two general categories of statistical analysis of interest for environmental monitoring. The first, usually called *descriptive* or *exploratory,* consists of computing one or more *summary statistics* for the *sample.* A summary statistic is a single number that characterizes a data set in some way. Of course no one number can completely characterize a data set. Descriptive statistics often include the use of one or more graphical presentations of the data. The second kind of statistical analysis is *inferential,* using the data (which represents only partial information) to infer something about the *population.* It is a good practice to always consider descriptive or exploratory statistics before attempting to use inferential statistics.

It is important to mention that there are spatial statistical methods and nonspatial statistical methods; the former also incorporate information pertaining to the physical location of each sample, and the latter do not. Both types are important but in general do not answer the same kind of questions. A record of the "coordinates" for each data location is needed when applying spatial statistical methods. Particularly for environmental problems, it may be necessary to use spatial-temporal statistical methods, in that case the time coordinates are also needed.

Finally, it should be noted that in environmental monitoring and assessment, the objective is not merely to "characterize" a locale but rather to use the information to make decisions. For example, if it is concluded that a plot of ground is contaminated then the question is whether to remediate in some manner. Making decisions will nearly always incorporate some degree of risk. For example, there is the risk of making the wrong decision or the risk of choosing an inadequate remediation process.

SAMPLES AND POPULATION

Strictly speaking, a sample is the set of individual observations obtained from sampling. Each data measurement may represent multiple pieces of information, for example, the concentrations of one or more contaminants. These sets of numbers (one set for each contaminant) are called *data.* Before data are collected, it is necessary to consider how many and where the data are to be taken, as well as which methods of monitoring are going to be used. Accessibility, available technology, costs associated with the physical collection of samples, and the subsequent laboratory analyses may limit or constrain the amount and the quality of information that can be gained by sampling. This in turn can affect the reliability of any conclusions that are drawn from the statistical analysis of the data. It is always important to think about how the data will be used (e.g., what questions are to be answered and how reliable the answer must be) before the collection of any data.

One must recognize that what is important is not the individual numbers that are generated by sampling but rather the set of numbers (i.e., the data). In the case of the concentration of a chemical in a field, there is a concentration at each location in the field, but this set of numbers is not directly observable. This entire set of numbers is usually called the *population.* The set of numbers that will actually be generated (e.g., by laboratory analysis of soil samples or measured by some instrument) is called a *sample set* or simply a *sample* from the population. The count *(n)* on this set of numbers is called the *sample size,* not to be confused with sample support (see the "Sample Support" section). The population will nearly always be infinite, whereas the sample always will be finite. Multiple attributes may be measured at the same location (or time), and in this case the data are multivariate. In many cases, the values for different components (sample attributes) may be statistically interdependent.

RANDOM SAMPLING

The validity of conclusions drawn by the use of statistical methods depends on whether certain underlying assumptions are satisfied. "Random sampling" is an example of such an assumption. In the preceding section, a sample was defined as a set of numbers selected from a larger set of numbers (the population). The question is, how is the selection made? Random sampling means that the selection is made in such a way that every subset (with fixed sample size n) of the population is equally likely to be selected. Designing the sampling process to ensure random sampling is not always easy. Note that random sampling is not the same as "erratic" selection, such as that based purely on convenience.

SAMPLE SUPPORT

Data values often represent a volume of material or an area of measurement. This volume or area is called the *support of the sample.* For example, the porosity of a soil is the fraction of pores within the volume. Larger pores and fractures in soil and rocks cannot be detected from soil cores with small support. Hydraulic conductivity and chemical concentrations are average values over a volume, and the monitored volume is the sample support. Thus, the support depends on the equipment or devices used for monitoring. For example, the size of the thermalization sphere is the support of water content measured with a neutron probe (see Chapter 12). In remote sensing, support is the size (area) of pixels (picture elements). In a spatial context, the physical size of the soil or support is different from the sample size (n) as explained in the "Samples and Population" section.

RANDOM VARIABLES

Consider all the possible values for the concentration of a contaminant at locations in a field. If no information exists about these values for a specific field, it may only be possible to specify a range of possible values. Until a specific location is chosen in the field, we do not have a single number but rather a population of values. The concept of a random variable helps deal with the uncertainty. As a simple example, consider a die with six faces, each face having some number of dots. Usually the numbers are designated as 1, 2, 3, 4, 5, and 6. If we have not "tossed" the die, we cannot predict the number of dots that will show on the uppermost face. However, we may know the likelihood of occurrence of each number showing for a given toss. In the case of a "fair" die, each of these six numbers is equally likely to show. A random variable is then characterized by two things: the set of possible values and the associated set of relative likelihoods (the latter is called the *probability distribution*). Random variables are usually classified as one of two types: discrete or continuous. A discrete random variable is one such that when the possible values are plotted on the real line, there is always a space between two consecutive points. A random variable is continuous when the set of possible values is an interval (including the possibility of the entire real line) or the union of several intervals.

In some applications, attributes are measured without considering the location. A target population is identified, and systematic or random samples are drawn. An example of this is the analysis of errors from a laboratory instrument. Note that the limited precision of instruments with digital outputs may show a continuous random variable as discrete.

In environmental applications it may be useful to consider the location in the field. For example, it is obvious that one cannot average the annual rainfall from a desert region with the rainfall of a tropical region and postulate that the average represents the rain in both regions. Environmental problems are spatial, and the global monitoring of the earth, as well as large domains, needs careful considerations of the spatial variability. In those cases, one has to use more advanced methods with the spatial properties of the random variable (e.g., Chiles and Delfiner, 1999; Journel and Huijbregts, 1978).

FREQUENCY DISTRIBUTION AND PROBABILITY DENSITY FUNCTION

For a sample or a finite population, the number of times a specified value occurs is called the *frequency*. The *relative frequency* is the frequency divided by the sample size (or the population size); the relative frequency is also an estimate of the probability or chance that some event may happen. First, consider a case where the random variable has only a finite number of possible values. For example, in a parking lot you counted 100 cars (population) of which 30 were red, 40 were white, and 30 were other colors. Thus, 0.3 are red, 0.4 are white, and 0.3 are other colors. These fractions are the probabilities that a sample will contain cars of a certain color. In environmental science, the sample but not the population values are known. However, we attempt to represent the population through a random variable that may approximately follow a known discrete probability distribution model. Given a sample, if we compute the relative frequencies for each possible value of the random variable, we have an estimate of the probability distribution of the random variable. This can be shown in graphical form. In the case of a discrete random variable, we can construct a bar graph with the abscissa showing the values of the random variable and the ordinate, their relative frequencies.

In the continuous case, we can construct a similar graph by grouping members of the population within classes or intervals of values for the attribute. For the relative frequency to be obtained, the number of counted specimens falling within a given interval of values is divided by the sample size. Each relative frequency is divided by the width of the interval to give an ordinate $\hat{f}(y)$. A plot of that ordinate versus the attribute values in the continuous case is a histogram. The total area of the bars of the histogram must equal one. Sample histograms can be sensitive to the number of class intervals.

EXAMPLE 3.1. A data set for clay content in a soil horizon is detailed in Table 3.1. With these data a sample histogram is computed and shown in Figure 3.1A. In this case, the number of observations from 20–25, 25–30, etc., are each shown by the height of a bar (in the case

TABLE 3.1

Thirty-Six Values of Clay (%) Taken Over a 90-Hectare Area at a 30-cm Depth[a]

% Clay	% Clay	% Clay	% Clay
34.7	38.8	45.5	36.1
29.5	38.5	27.0	40.1
43.8	42.6	42.5	37.4
33.3	36.3	27.2	45.3
37.5	32.1	27.2	28.5
33.3	53.2	24.9	30.2
29.5	39.9	33.6	33.6
36.1	34.7	37.4	39.9
25.4	30.5	31.3	32.3

[a] The sample mean ($\bar{x}$) and standard deviation (*s*) are 35.3 and 6.38, respectively. (Data: Coelho, 1974.)

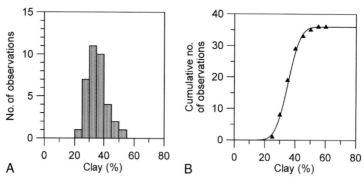

FIGURE 3.1   Measured frequency histogram (**A**) and cumulative frequency distribution (**B**) for the clay from Table 3.1. The solid line is the theoretical curve based on a normal distribution with the same mean and variance as those estimated from the data. (From Warrick *et al.*, 1996.)

of a "tie" such as at 25, the observation is included with the lower-valued observations). It is easy to see from the histogram that the largest frequencies are associated with the ranges of 30% to 35% and 35% to 40%. This gives a snapshot of the observed frequencies and can be used to infer the distribution of the assumed population of clay percentages.

A histogram can be constructed either for a sample or for a finite population but not for an infinite population. In particular, one cannot construct a histogram for a continuous random variable or a discrete random variable with an infinite number of possible values.

The alternative representation is the probability density function (pdf), which might be thought of as a continuous version of the histogram. The pdf is a function such that the area under the curve between two points is the probability that the random variable takes a value between those two points. A pdf fully characterizes a random variable. Most random variables have two important numerical characteristics called the *mean* (μ) and the *variance* (σ^2). The square root of the variance, σ, is called the *standard deviation*. The mean is also called the *expected value of the random variable* and might be denoted as E(X), where X is the random variable. The mean can be thought of as representing the balance point on the graph of the pdf. The variance quantifies how much the possible values are dispersed away from the mean. One very important and widely used random variable is called the *normal* or *Gaussian variable*. The pdf for a normal random variable Y is

$$f(x) = \frac{1}{(2\pi)^{0.5}\sigma} \exp\left[\frac{-(x-\mu)^2}{2\sigma^2}\right] \qquad \text{(Eq. 3.1)}$$

where μ is the population mean and σ the population standard deviation (see Figure 3.2A). (A list of symbols

and terms used is given as Table 3.2.) The graph of this function is bell shaped. There is very little area in the tails, in fact the area outside of the interval $\mu - 4\sigma, \mu + 4\sigma$, is less than 0.001. The mean can be estimated by the sample mean

$$\bar{x} = \frac{1}{n} \sum_{i=1}^{n} x_i \qquad \text{(Eq. 3.2)}$$

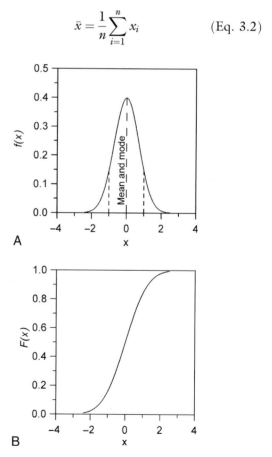

FIGURE 3.2   Frequency distribution *f(x)* for a normal distribution (**A**) and the associated cumulative distribution function for $\mu = 0$ and $\sigma = 1$ (**B**). In the first graph, the dashed lines show the center and $\pm$ standard deviation from the mean. (From Warrick *et al.*, 1996.)

TABLE 3.2
Statistical Symbols and Their Definitions

Term	Symbol	Definition
Sample size	n	The number of samples taken to represent the population.
Sample mean	$\bar{x} = \dfrac{\sum_{i=1}^{n} x_i}{n} = \dfrac{x_1 + x_2 + \ldots x_n}{n}$	Mean estimated from the n values $x_1, x_2, \ldots, x_n$.
Mean	μ	Population or true mean.
Sample variance	$s^2 = \dfrac{\sum_{i=1}^{n}(x_i - \bar{x})^2}{n-1}$	Based on n values given by $x_1, x_2, \ldots, x_n$. (If including all possible values of n, use n in place of $n-1$ for denominator.)
Variance	σ^2	Population or true variance for the population.
Standard deviation	s, σ	Defined for s^2 and σ^2 above. (Specific cases are identified by subscripts, such as s_x and s_y.)
Probability	$P(X \leq x)$	The probability that a random value X is less than or equal to a specified value x.
Probability density function (pdf)	$f(x)$	Function that gives probability density.
Cumulative density function (cdf)	$F(x)$	Probability of $X \leq x$.
Geometric mean	$\bar{G} = \sqrt[n]{\prod_{i=1}^{n} x_i} = \exp\left[\frac{1}{n}\sum_{i=1}^{n} \ln x_i\right]$	Antilog of the mean of the transformed variable $y = \ln x$ where x is the original measurement.
Maximum allowable error or tolerance	d	Specified error used in estimating sample size.
Coefficient of variation	$CV = \dfrac{\sigma}{\mu}$	A relative standard deviation. Can also be expressed as a percentage. (The estimator is $s/\bar{x}$.)
Covariance	$s_{xy} = \dfrac{1}{n-1}\sum_{i=1}^{n}(x_i - \bar{x})(y_i - \bar{y})$	Average of cross products of attributes of x and y.
Covariance matrix	S	Array constructed with covariances in the off-diagonal terms and variances in the major diagonal.
Slope	$b = \dfrac{\sum_{i=1}^{n}(x_i - \bar{x})(y_i - \bar{y})}{(n-1)s_x^2}$	Slope for linear regression for n data pairs. The data pairs are $(x_1, y_1), (x_2, y_2), \ldots (x_n, y_n)$.
Intercept	$a = \bar{y} - b\bar{x}$	y-Intercept for linear regression for n data pairs.
Sample correlation coefficient	$r = \dfrac{s_{xy}}{s_x s_y}$	Estimate of the sample correlation coefficient.
Coefficient of determination	r^2	Square of r (above) for linear correlation. Range is 0 to 1.
Predicted value	$y = a + bx$	Predicted value of dependent variable y.
Inverse distance estimator	$\hat{z}_0 = \dfrac{\sum_{j=1}^{m} k_j z_j}{\sum_{j=1}^{m} k_j}$	Interpolation found by weighing nearby measured values according to the inverse distance from the point of interest.
Sample variogram	$\hat{\gamma}(h) = \dfrac{1}{2N(h)}\sum_{N(h)}[z(x) - z(x+h)]^2$	An expression of the spatial interdependence of values; is similar to a variance but is a function of distance.
Kriging estimator	$\hat{z}_0 = \sum_{j=1}^{m} \lambda_j z_j$	Interpolation based on optimization following the assumed spatial interdependence.

The standard deviation σ can be estimated by the sample standard deviation:

$$s = \sqrt{\frac{\sum_{i=1}^{n} (x_i - \bar{x})^2}{n-1}} \qquad \text{(Eq. 3.3)}$$

The variance is the square of the standard deviation. An estimator is called unbiased if the expected value of the error is zero. That is why the divisor is $n-1$ in Equation 3.3. In the case of a random variable with a finite number of equally likely possible values, Equation 3.2 can be used to compute the mean, and Equation 3.3 can be used to compute the standard deviation, except that one should substitute n for $n-1$ in Equation 3.3.

When the probability density function is integrated from the lowest possible value to an arbitrary value x, the result is the cumulative distribution function (cdf) given by:

$$F(x) = \int_{-\infty}^{x} f(y)dy \qquad \text{(Eq. 3.4)}$$

The value of F is identical to the probability that a randomly chosen attribute is less than or equal to x. Figure 3.2B is a plot of $F(x)$ for the normal distribution. The value of F increases from 0 to 1 as x goes from the lowest to highest value of all possible values.

Figure 3.1B shows the cumulative number of observations for the clay percentages of Table 3.1. The triangles show the sum of the number of observed values that are less than the percentage shown on the abcissa. These were found by adding appropriate values represented by the bars in the frequency histogram of Figure 3.1A. Also shown by the solid line in Figure. 3.1B is the theoretical result found by calculating the $F(x)$ and multiplying by the total number of observations. The value of $F(x)$ was calculated based on the estimates of the mean (35.3) and standard deviation (6.78) and a normal distribution.

We can make use of $F(x)$ to evaluate the probability that a random value will be less than a specified amount (a "cutoff value") or that it will be between two specified amounts, which may be useful for establishing a "confidence interval." The probability that a specimen or sample randomly drawn is less than a cutoff value c is:

$$P(x \leq c) = F(c) = \int_{-\infty}^{c} f(y)dy \qquad \text{(Eq. 3.5)}$$

The probability that a random sample would take the value of the attribute between two specified values x_1 and x_2 is

$$P(x_1 < x \leq x_2) = F(x_2) - F(x_1) \qquad \text{(Eq. 3.6)}$$

One way to determine whether the data were obtained as a random sample from a normal population is to compare computed probabilities with relative frequencies (this is sometimes known as using a chi-square test).

Sometimes histograms exhibit strong asymmetry. In such cases, the distribution is skewed. This asymmetry is very common in chemical concentrations and other earth science attributes. When the random variable cannot take negative values (for example, concentrations cannot be negative), strictly speaking, the pdf cannot be normal because each large positive value should correspond to another lower value to the left of the mean. If the mean value is small, the distribution tends to be skewed. It is important to mention the estimators of Equations 3.2 and 3.3 are very sensitive to high values and are not good for skewed data. However, the distribution may become close to the normal when a natural log transformation is applied to the nonnormal random variable x, that is:

$$y = \ln x \qquad \text{(Eq. 3.7)}$$

In that case, we say the random variable x is log-normally distributed. Because of normalization, several calculations are facilitated. Special care should be taken into account to relate the mean μ_y and standard deviation σ_y of the transformed data to the mean μ and standard deviation σ of the nontransformed data:

$$\mu = \exp(\mu_y + 0.5\sigma_y^2) \qquad \text{(Eq. 3.8)}$$

$$\sigma^2 = \exp(2\mu_y + \sigma_y^2)\left[\exp\left(\sigma_y^2\right) - 1\right] \qquad \text{(Eq. 3.9)}$$

The exponential of the arithmetic mean of the log transformed data $\ln z$ is the geometric mean:

$$\bar{G} = \sqrt[n]{\prod_{i=1}^{n} x_i} = \exp\left[\frac{1}{n}\sum_{i=1}^{n} \ln x_i\right] \qquad \text{(Eq. 3.10)}$$

EXAMPLE 3.2. Table 3.3 gives 60 values of soil lead concentration (Englund and Sparks, 1988). The values are plotted in Figure 3.3A with a wide range of values. A histogram (Figure 3.3B) formed

TABLE 3.3
Soil Lead Data Showing Untransformed and Log-Transformed Values

IDᵃ	Lead (mg kg⁻¹)	ln Lead	IDᵃ	Lead (mg kg⁻¹)	ln Lead
01	18.25	2.904	31	19.75	2.983
02	30.25	3.410	32	4.50	1.504
03	20.00	2.996	33	14.50	2.674
04	19.25	2.958	34	25.50	3.239
05	151.5	5.020	35	36.25	3.590
06	37.50	3.624	36	37.50	3.624
07	80.00	4.382	37	36.00	3.584
08	46.00	3.829	38	32.25	3.474
09	10.00	2.302	39	16.50	2.803
10	13.00	2.565	40	48.50	3.882
11	21.25	3.056	41	49.75	3.907
12	16.75	2.818	42	14.25	2.657
13	55.00	4.007	43	23.50	3.157
14	122.2	4.806	44	302.50	5.712
15	127.7	4.850	45	42.50	3.750
16	25.75	3.248	46	56.50	4.034
17	21.50	3.068	47	12.25	2.506
18	4.00	1.386	48	33.25	3.504
19	4.25	1.447	49	59.00	4.078
20	9.50	2.251	50	147.00	4.988
21	24.00	3.178	51	268.00	5.591
22	9.50	2.251	52	98.00	4.585
23	3.50	1.253	53	44.00	3.784
24	16.25	2.788	54	94.25	4.546
25	18.00	2.890	55	68.00	4.220
26	56.50	4.034	56	60.75	4.107
27	118.00	4.771	57	70.00	4.248
28	31.00	3.434	58	25.00	3.219
29	12.25	2.506	59	33.00	3.496
30	1.00	0.000	60	40.75	3.707

(Data: Englund and Sparks, 1988.)
ᵃSample identification number.

using class intervals of 20 (mg kg⁻¹) shows a preponderance of small values in the 0–20, 20–40, and 40–60 range. However, there are a number of values larger than 100, and at least two values above 250. When the natural logarithm of the values is taken ($y = \ln z$), the range of values is somewhat more uniformly distributed, which is observed in the transformed histogram (Figure 3.3D). In fact, the last plot is somewhat like the normal distribution depicted in Figure 3.2A. This is not to say that the underlying distribution of lead values is log-normally distributed; further tests would be necessary to address this question.

Notice that in environmental monitoring the variance for the sample depends on the support of the sample. For example, if a contaminated soil is sampled and the variance of the sample is computed with the square of Equation 3.3, one may observe

that a larger variance is for smaller sample support. For this reason one has to take care when comparing data with different sample support. This is an important problem when remote sensing data are calibrated with data collected from the ground surface.

SAMPLE SIZE AND CONFIDENCE INTERVALS

A normal random variable is fully characterized if its mean and standard deviation are known. A common question in environmental monitoring and sampling is what sample size is required to adequately estimate either or both of these parameters. The sample size or number of locations (n) needed to estimate the mean depends on the tolerance or error d one is willing to accept in the estimation and also on the degree of confidence desired that the error is actually less than d.

Consider first the problem of estimating the mean of a normal random variable assuming that the standard deviation σ is known. It can be shown that the sample mean, Equation 3.2, is also normally distributed with the same mean and with variance

$$\sigma_{\bar{x}}^2 = \frac{\sigma^2}{n} \qquad \text{(Eq. 3.11)}$$

Then, using the normal table, we obtain

$$P\left(\mu - \frac{z_{\alpha/2}\sigma}{\sqrt{n}} \leq \bar{x} \leq \mu + \frac{z_{\alpha/2}\sigma}{\sqrt{n}}\right) = 1 - \alpha \qquad \text{(Eq. 3.12)}$$

where $z_{\alpha/2}$ is the value from a standard normal table or spreadsheet function (see Question 5) corresponding to probability $\alpha/2$. That is

$$P(z > z_{\alpha/2}) = 1 - \alpha/2 \qquad \text{(Eq. 3.13)}$$

Thus we can predict the likelihood that the sample mean will be close to the true mean. This is not quite the question we want to answer. But Equation 3.12 can be rewritten in the form

$$P\left(\bar{x} - \frac{z_{\alpha/2}\sigma}{\sqrt{n}} \leq \mu \leq \bar{x} + \frac{z_{\alpha/2}\sigma}{\sqrt{n}}\right) = 1 - \alpha \qquad \text{(Eq. 3.14)}$$

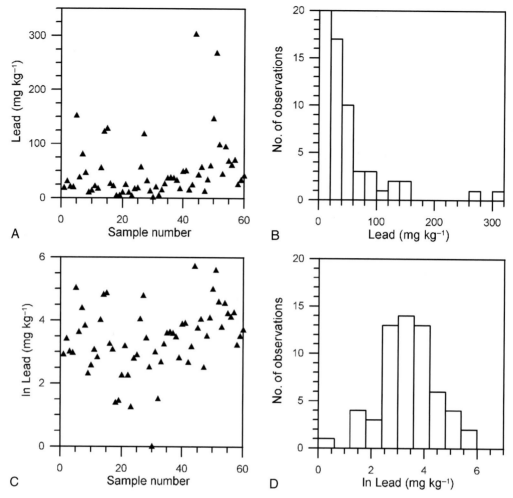

FIGURE 3.3   Scatterplot (**A**), frequency distribution (**B**), transformed distribution (**C**), and frequency distribution for the transformed distribution (**D**) for lead in soil as given in Table 3.3. Data: Englund and Sparks, 1988. (From Warrick *et al.*, 1996.)

It is important to remember that this is still a statement about the behavior of the sample mean. However, we can reinterpret this statement as follows:

$$\bar{x} - \frac{z_{\alpha/2}\sigma}{\sqrt{n}} \leq \mu \leq \bar{x} + \frac{z_{\alpha/2}\sigma}{\sqrt{n}} \qquad \text{(Eq. 3.15)}$$

which is a $(1 - \alpha)100\%$ *confidence interval* for μ. This means that we are really making a statement about the reliability of this method for estimating μ. If we use this method 100 times, we expect that our conclusion will be correct $(1 - \alpha)100$ times (but unfortunately we will not know which times are correct and which times are incorrect). If we take $(1 - \alpha)100 = 99$, the chance that the one time we are doing it incorrectly is not very great. If we take $(1 - \alpha)100 = 80$, the chance that the one time we are doing it incorrectly is larger. Using $(1 - \alpha)100 = 50$ results in essentially worthless information for most applications. The problem with being "incorrect" is

that all you know is that the population mean is "outside" the interval, and "outside" is a very big place. Now let the tolerance error d be:

$$d = \frac{z_{\alpha/2}\sigma}{\sqrt{n}} \qquad \text{(Eq. 3.16)}$$

or

$$n = \left(\frac{z_{\alpha/2}\sigma}{d}\right)^2$$

Thus for a given confidence level $(1 - \alpha)100\%$, and a given tolerance d, we can predict the required sample size.

Note that both the confidence interval and the sample size computations depend on two very important assumptions; that the random variable is normal and that the standard deviation is known. In Equation 3.15 a mean and standard deviation are used but these are never known. After the data are collected, then an interval similar to Equation 3.15 can be calculated:

$$\bar{x} - \frac{st_{\alpha/2,\,n-1}}{\sqrt{n}} \leq \mu \leq \bar{x} + \frac{st_{\alpha/2,\,n-1}}{\sqrt{n}} \qquad \text{(Eq. 3.17)}$$

where $t_{\alpha/2,\,n-1}$ is from a "t-table" or spreadsheet function (see Question 6) and will approach $z_{\alpha/2}$ when the sample size n becomes large. The value of $t_{\alpha/2,\,n-1}$ will be larger than $z_{\alpha/2}$. (Note the sample mean and sample standard deviation will not be known until the data are available.)

In applying these results to determine an adequate sample size for planning data collection, there is a circular problem (i.e., we are using a mean and standard deviation to determine the sample size to find the mean and standard deviation within chosen tolerances and confidence intervals). Despite the circularity, a necessary decision in collecting any data is the sample size and the cost of the study or monitoring program is likely to be strongly influenced by the choice of n. The best that can be done at the outset is to estimate values of s from previous studies. This previous estimate can be inserted in Equation 3.16 with the realization that it is only a rough approximation. After the data are collected, Equation 3.17 can be applied and a meaningful confidence interval established.

The following examples illustrate the construction of confidence intervals and prediction of sample size to obtain a confidence interval with a fixed tolerance and fixed confidence level.

EXAMPLE 3.3. Suppose that a random sample of size 16 has been selected from a normally distributed population with a sample mean of 22.4. Assume that the population standard deviation, σ, is 3.2. We wish to obtain 95% and 99% confidence intervals for the unknown population mean μ.

Because the population standard deviation is known, we use Equation 3.17. For $1 - \alpha = 0.95$, α is 0.05, and $\alpha/2$ is 0.025. From a standard normal table or spreadsheet function we find $z_{0.025} = 1.96$ (see Question 5). Thus the 95% confidence interval is

$$22.4 - \frac{(1.96)(3.2)}{4} \leq \mu \leq 22.4 + \frac{(1.96)(3.2)}{4}$$

or

$$22.4 - 1.568 \leq \mu \leq 22.4 + 1.568$$

If we believe that we should only express our results to one decimal place, then the 95% confidence interval would be

$$22.4 - 1.6 \leq \mu \leq 22.4 + 1.6$$

Note that rounding off should result in a wider confidence interval (otherwise the confidence level decreases by an unknown amount). For $1 - \alpha = 0.99$, corresponding values of $\alpha = 0.01$, $\alpha/2 = 0.005$, $z_{0.005} = 2.576$ follow. Then the 99% confidence interval is

$$22.4 - \frac{(2.576)(3.2)}{4} \leq \mu \leq 22.4 + \frac{(2.576)(3.2)}{4}$$

EXAMPLE 3.4. Suppose that we wish to obtain 95% and 99% confidence intervals for the mean of a normally distributed population where the population standard deviation is 3.2. If we want the maximum allowable error (tolerance) to be d = 0.75, how large a sample is necessary? The value of d either can be specified directly as done here or can be computed by multiplying a relative error (such as 15%) by an estimate of the mean value (resulting in 0.15 $\bar{x}$). From Equation 3.16 we find that

$$n \geq \left[\frac{(1.96)(3.2)}{0.75} \right]^2 = 69.93, \ \text{use } n = 70$$

for a 95% confidence interval and

$$n \geq \left[\frac{(2.576)(3.2)}{0.75} \right]^2 = 120.80, \ \text{use } n = 121$$

for the 99% confidence interval.

Note three things about the above results. First, if the sample size is fixed (and the population standard deviation is known), then the greater the confidence level, the wider the confidence interval. That is, if we are more certain in one way (greater confidence level), then we are less certain in another way (wider confidence level). Second, given the information (sample mean, sample size and population standard deviation), there is a confidence interval for each choice of the confidence level. Third, with everything else the same, a higher confidence level means a greater sample size.

EXAMPLE 3.5. Again suppose that a random sample of size 16 has been selected from a normally distribution population with a sample mean of 22.4 and a sample standard deviation of 3.2. We still wish to obtain 95% and 99% confidence intervals for the population mean. The difference is simply that we must use the t-table instead of the normal table. For $1 - \alpha = 00.95$, corresponding values of $\alpha = 0.05$, $\alpha/2 = 0.025$, $t_{0.025, 15} = 2.131$ (see Question 6) are found and hence the 95% confidence interval is

$$22.4 - \frac{(2.131)(3.2)}{4} \leq \mu \leq 22.4 + \frac{(2.131)(3.2)}{4}$$

For $1 - \alpha = 0.99$, corresponding values are $\alpha = 0.01$ and $\alpha/2 = 0.005$. From the table $t_{0.005, 15} = 2.947$, resulting in the 99% confidence interval of

$$22.4 - \frac{(2.947)(3.2)}{4} \leq \mu \leq 22.4 + \frac{(2.947)(3.2)}{4}$$

Again we see that the 99% confidence interval is wider than the 95% confidence interval, but each of these two are wider than their counterparts when the population standard deviation is known (due to greater uncertainty). As the sample size becomes much larger than 30, the values in the t-table will be much closer to the values found in the normal table.

EXAMPLE 3.6. Of course in real applications we may not know the population standard deviation, and we may still wish to predict a sample size for a confidence interval for the population mean. That is, we want to fix the confidence level and the tolerance. We will have to do the problem in steps (perhaps several). We first pick a sample size based on other considerations (cost of sampling, ease of sampling, etc.). We select a sample and compute the sample standard deviation. We use it as though it were the population standard deviation and predict a sample size. If the predicted sample size is larger than the original sample size, we must collect a new sample (caution that in general we cannot simply add to the original sample, it would not be a random sample). We would continue this process until the predicted sample size is about the same as our last sample size, then we use the sample mean from that sample and the sample standard deviation from that sample together with a value from the t-table to generate our desired confidence interval. There is no assurance that this process will stop very quickly; it turns out that estimating the standard deviation is more difficult than estimating the mean.

EXAMPLE 3.7. From the data in Table 3.1 (clay percentages), we find that the sample mean is 35.3 and the sample standard deviation is 6.38, the sample size is $n = 36$. First we specify 95% and 99% confidence intervals for the population mean (using the information given). For $1 - \alpha = 0.95$, we find $\alpha = 0.05$, $\alpha/2 = 0.025$, $t_{0.025, 35} = 2.030$ (see Question 6 on how to find the t-value) and hence the 95% confidence interval is

$$35.3 - \frac{(2.030)(6.38)}{6} \leq \mu \leq 35.3 + \frac{(2.030)(6.38)}{6}$$

Similarly consider $1 - \alpha = 0.99$, we find $\alpha = 0.01$ and $\alpha/2 = 0.005$. From the table $t_{0.005, 35} = 2.738$ and the 99% confidence interval is

$$35.3 - \frac{(2.738)(6.38)}{6} \leq \mu \leq 35.3 + \frac{(2.738)(6.38)}{6}$$

We also see that a larger standard deviation means either a wider confidence interval or a larger sample size.

EXAMPLE 3.8. We may think that these confidence intervals are too wide (the tolerance is too large) and are willing to obtain more samples. How large a sample is necessary to obtain a tolerance of 0.8, with a confidence level of 95%? Using Equation 3.16 but with the z value, we obtain

$$n \geq \left[\frac{(2.030)(6.38)}{0.8} \right]^2 = 262.09, \text{ use } n = 263$$

for a 95% confidence interval and

$$n \geq \left[\frac{(2.738)(6.38)}{0.8} \right]^2 \approx 477$$

for a 99% confidence interval. To complete the analysis we would have to select a random sample of the specified size, compute the sample mean and sample standard deviation, then compute the tolerance (to compare with the choice of $d = 0.8$). Of course the sample standard deviation may be larger than before. If the predicted sample size had been larger, the t-value would be smaller and the divisor in Equation 3.16 would be larger (the quotient smaller). In that case these may offset the larger sample standard deviation.

In general, for a normal population (and random variables in general) there is no direct relationship between the population mean and the population standard deviation (except of course, the standard deviation is always positive). The lognormal distribution is one important exception to this statement. However, the ratio of these two, μ/σ, called the *coefficient of variation*, is an important distribution characteristic. It can be estimated by the sample coefficient of variation. Finally, note the importance of the assumption that the population has normal distribution. There are ways to test this, but they are beyond the scope of this book. The assumption that the sample is random (i.e., that it was collected in the right way) is critical to the validity of the previous results.

In spatial environmental monitoring, data may be correlated and the number of locations for a good estimation at an unsampled location is a more complicated problem that needs spatial variability considerations (see Cressie, 1993).

COVARIANCE AND CORRELATION

There are two ways (which are somewhat related) to determine whether two random variables X,Y are interdependent. One is given directly in terms of probabilities. X,Y are said to be independent if

$$P(X \leq x \text{ and } Y \leq y) = P(X \leq x)P(Y \leq y) \quad \text{(Eq. 3.18)}$$

otherwise they are dependent. Unfortunately we do not usually know these probabilities, and they are hard to estimate. Thus another scheme is more useful. The covariance of X,Y is Cov(X,Y) defined by

$$Cov(X, Y) = E(XY) - E(X)E(Y) \quad \text{(Eq. 3.19)}$$

One of the consequences of independence is that if the X,Y are independent, Cov(X,Y) = 0 (the converse is also true if X,Y are bivariate normal). (A caveat is that the Cov[X,Y] may not exist, which would be the case if one or both random variables does not have a finite variance.)

A normalized version of the covariance is more useful; this is the Pearson correlation coefficient ρ

$$\rho = \frac{Cov(X, Y)}{\sigma_X \sigma_Y} \quad \text{(Eq. 3.20)}$$

Unlike the covariance, the correlation coefficient is bounded, $-1 \leq \rho \leq 1$. If $|\rho| = 1$, then X,Y are linearly dependent. Note that while the Pearson correlation coefficient is a measure of a linear relationship between the variables, nonlinear relationships will not show up in this coefficient. The covariance can be estimated by the sample covariance

$$s_{xy} = \frac{1}{n-1} \sum_{i=1}^{n} (x_i - \bar{x})(y_i - \bar{y}) \quad \text{(Eq. 3.21)}$$

The sample correlation coefficient is a normalized form of the sample covariance:

$$r = \frac{s_{xy}}{s_x s_y} \quad \text{(Eq. 3.22)}$$

where the univariate standard deviations are computed from Equation 3.3. The sample correlation coefficient is also bounded in the interval $-1 \leq r \leq 1$. The correlation coefficient is also interpreted as the slope of the line formed by the scatter plot of two random variables when they are standardized.

Recall that a normal distribution is completely determined by its mean and its variance. A multivariate normal distribution is completely determined by the vector of means and the matrix of variances and covariances. Of course, all of these can be estimated by their sample counterparts.

For example, for three random variables the sample covariance matrix might be written in the form

$$\mathbf{S} = \begin{bmatrix} s_1^2 & s_{12} & s_{13} \\ s_{21} & s_2^2 & s_{23} \\ s_{31} & s_{32} & s_3^2 \end{bmatrix} \quad \text{(Eq. 3.23)}$$

EXAMPLE 3.9. Using the data from Table 3.4 for lettuce and ozone concentration one can compute sample variances and covariances. In this case, s_{xy} is -00867, resulting in the sample covariance matrix S:

$$\mathbf{S} = \begin{pmatrix} 0.05320 & -0.00867 \\ -0.00867 & 0.00147 \end{pmatrix}$$

For the example above, the sample correlation coefficient is r = −0.98.

LINEAR REGRESSION

When the attribute of interest *y* is difficult to measure (this may be due to technological or financial reasons), one may measure another correlated attribute *x* in place of the original attribute. The proxy attribute may be related in some systematic or quantifiable manner to the original attribute. Such relationships may be characterized with regression equations. A linear model may exist that relates both attributes. A simple regression equation is when one attribute is proportional to the second attribute plus a constant as

$$y = \alpha + \beta x + \varepsilon \quad \text{(Eq. 3.24)}$$

where ε denotes an error term. The regression equation is then

$$y = a + bx \quad \text{(Eq. 3.25)}$$

TABLE 3.4
Ozone Concentrations and Corresponding Lettuce Yields

Average Ozone (μL L^{-1})	Lettuce Yield (kg)
0.106	0.414
0.043	0.857
0.060	0.751
0.068	0.657
0.098	0.437
0.149	0.251

Statistical Summary	
n	6
$\bar{x}$	0.087
$\bar{y}$	0.561
s_x	0.038
s_y	0.231
s_{yx}	−0.0086
	7
b	−5.90
a	1.08
r^2	0.96
r	−0.98

The lettuce yield is fresh head weight per plant in kilograms per plant. (Source: Heck *et al.*, 1982.)

where a is the estimated value of α and b the estimated value of β. Note that the data will consist of pairs of values. One way to estimate these coefficients is to use least squares. The objective is to minimize the sum of square deviations d^2 from the model of Equation 3.24 and real n data for the dependent attribute y. This is

$$\sum_{i=1}^{n} d^2 = \sum_{i=1}^{n} (y_i - a - bx_i)^2 \qquad \text{(Eq. 3.26)}$$

One form of these equations for a and b are given in Table 3.2; other forms are possible. Note that b is the slope of the line, and it is simply the sample covariance between x and y divided by the sample variance of y. Thus, if the variables have the same sample variance, b is the sample correlation coefficient. The a coefficient is the intercept on the ordinate and a function of the means of y and x; therefore a vanishes if both sample means are zero. In practice, both a and b are conveniently calculated in most spreadsheet and statistical software.

In some cases, the scatter plot of the data may clearly show that another more complicated polynomial model is preferable to fit the data. Estimators for the coefficients in a polynomial regression can also be obtained with the least squares method. For more details see Draper and Smith (1981).

EXAMPLE 3.10. Consider the results in Table 3.4 for lettuce yield (kg) for several ozone (O$_3$) concentrations (μL L^{-1}) in the atmosphere. We need to develop an

equation that relates the yield to ozone concentration. A first step in the data analysis is to plot one variable versus another as shown in Figure 3.4. This scatter plot shows that the higher yields tend to be for the lower ozone concentrations; conversely, the lower yields are for the higher ozone levels. As shown, a straight line can be drawn by placing a ruler across the points such that they are all reasonably close to the line. To develop an unbiased method, we apply the regression model of Equation 3.24.

The values of the slope and intercept are $b = -5.90 \, \text{kg}/(\mu\text{L L}^{-1})$ and $a = 1.08 \, \text{kg}$. The negative sign for the slope indicates the yield decreases as ozone concentration increases. The value of the intercept can be verified by extending the line up and to the left until the level of ozone would reach zero (with a corresponding yield of $a = 1.08 \, \text{kg}$).

For the lettuce yield plot, r is -0.98. The sign of r will always be the same as for b and also indicates whether the two variables increase together or are inversely related as in Figure 3.4. Common pitfalls to avoid with regard to linear regression include extrapolation of results beyond the measurements (rather than interpolation within the range of measured values) and the false identification of casual relationships.

There are many data sets that have more than two variables. In fact, for the preceding study, lettuce yields were measured for mixtures of ozone, sulfur dioxide (SO$_2$), and nitrogen dioxide (NO$_2$). Whereas linear regression is used for one variable, multiple regression deals with more than one variable, such as ozone and the two other attributes. A multiple regression analysis follows the

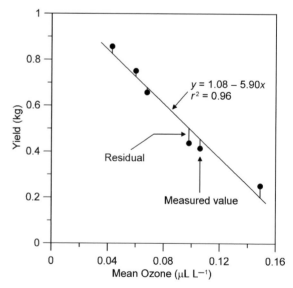

FIGURE 3.4 Lettuce yield (fresh head weight per plant) vs. ozone concentration and deviations of experimental points from regression line. Data: Heck *et al.*, 1982. (From Warrick *et al.*, 1996.)

same general steps as the linear analysis previously mentioned.

INTERPOLATION AND SPATIAL DISTRIBUTIONS

It is almost never possible to collect data at all possible locations in an area of interest. Although sampling will provide information at some locations, it is still necessary to use some method to "interpolate" the data, that is, to estimate or predict the values of an attribute at nondata locations. There are a number of different methods that have been and are used. All are based on the intuitive idea that values of an attribute at locations that are close together are more "similar" than when the locations are far apart. Typically estimates are made at each node on a regular grid. These estimates might then be graphically represented by a contour or relief map over a surface (an example of a contour map is a topological map showing changes in elevation). Interpolation and generation of a contour map are two separate steps although some software packages may make them appear as one. In environmental monitoring and assessment applications, the data locations may be in the three-dimensional (3D) space and hence a contour map may not be sufficient. We consider three commonly used methods for interpolation: (1) nearest neighbor estimates, (2) inverse distance weighting, and (3) kriging. For more information on interpolation and estimation, see Chiles and Delfiner (1999), Kane *et al.* (1982), and Myers (1991). There are other interpolation methods, but these three are widely used. "Similarity" is not a very precise term and there is still the question of how to quantify similarity and how to relate it to the distance (and possibly direction) between two locations. Finally how can we use this quantification to actually make estimates? We begin with a very simple approach to answering these questions. For all three methods it is assumed that there is a set of data locations and at each data location a value for the attribute of interest.

NEAREST NEIGHBOR ESTIMATES

The simplest notion of "similarity" is to estimate the value of an attribute at a nondata location by using the value at the nearest data location. This corresponds to partitioning the region of interest into polygons (polyhedrons in three space) such that each data location is the centroid of a polygon. At each location inside a polygon, the estimated value is the same as at the centroid, this "tiles" the region.

EXAMPLE 3.11. Textural information of porous media is of interest in environmental monitoring particularly in terms of water balance and in the evaluation of leaching potential. The x,y-coordinates and values of percent sand in a soil are given (Table 3.5) for 50 points in an 800×1600–m field (El-Haris, 1987). The sample mean and variance are 61.8 and 62.3, respectively. First, the nearest neighbor estimates were generated at each node of a grid (4700 points) and then the estimates were contoured. Both the estimates and the contour plot were obtained with Surfer software (2002) with the result shown in Figure 3.5B.* The contour plot clearly shows a tendency for lower values towards the lower left corner (on the order of 35% or less). There are a few regions identified with high values (on the order of 70%) in the lower center and to the left of the lower center. Note that the divisions between the different ranges of values are jagged and tend to be polygonal. This is a consequence of the way the figure is constructed; typically, contouring packages estimate the values on a fine grid throughout the region of interest, but still

TABLE 3.5
Fifty Values of Sand (%) Taken from an 800×1600–m Field at 0–25 cm Depth

x (m)	y (m)	% Sand	x (m)	y (m)	% Sand
152	251	35.53	446	351	68.12
936	148	65.01	292	301	59.31
782	98	62.38	1034	48	70.19
390	677	56.2	1272	496	63.72
544	148	72.77	1230	351	58.77
446	446	68.65	586	301	66.3
1426	351	61.9	1426	48	57.76
978	677	61.89	936	546	61.89
1230	251	61.13	1174	496	61.12
1230	546	61.88	684	301	66.57
740	727	57.78	1132	727	56.69
838	627	61.39	642	727	62.42
348	627	55.13	54	446	69.15
586	98	68.93	194	301	52.62
1230	48	56.43	54	351	28.78
54	546	69.15	782	677	62.42
292	677	58.79	1034	627	58.57
1230	627	63.98	1468	496	54.09
684	496	62.93	1034	251	66.32
1524	446	60.88	642	251	63.72
642	48	71.54	1132	148	68.93
390	496	63.46	488	677	62.93
1468	301	63.97	446	48	76.65
740	351	65.53	1426	251	60.29
1174	301	60.33	978	98	65.01

Data: El-Haris, 1987.

---

*This and other references to Surfer are not an endorsement but are included for the reader's information.

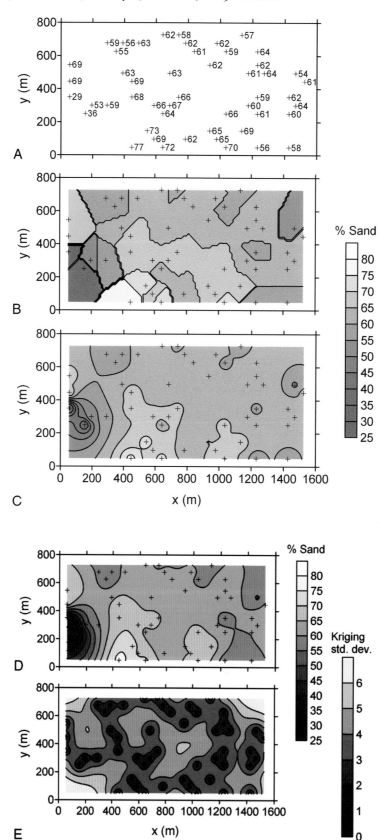

FIGURE 3.5 (**A**) Distribution of sand percentage measured values; (**B**) estimated with nearest neighbor approach; (**C**) estimated with inverse distance weighting approach; (**D**) estimated with kriging; (**E**) the distribution of kriging standard deviations.

must draw curves on a small scale to delineate regions. For a "perfect" graphical representation, there would be one polygon about each point and each delineation on the contour map would be a set of those individual polygons.

The advantage of the nearest neighbor method is that it is easy to use and it requires very little computation. It is also an exact interpolator; this means that the estimate at a data location is the same as the data value. The method is not based on any theory, and thus there are no assumptions to verify. It simply incorporates the intuitive idea that the values at locations close together are more similar than values for locations far apart.

The disadvantage of this method is that the set of estimates is strongly affected by the spatial pattern of the data locations. If the pattern of data locations is very dense, then the nearest data location will be close, but if there are few data locations, the nearest data location can be far away and the estimate may be very suspect. The size of the polygons is determined by the pattern of data locations.

INVERSE DISTANCE WEIGHTING

The nearest neighbor method uses only the datum location that is closest to the point where an estimate is desired. This means that for any one estimate, only a part of the information contained in the data set will be used. Inverse distance weighting extends the basic idea of using similarity and attempts to quantify it. Instead of only the value at the nearest data location being used, values at other data locations are used but are weighted. The weight for each data location is assumed to be inversely related to the distance between that location and the location where an estimate is desired. The estimator is of the form:

$$\hat{z}_0 = \frac{\sum\limits_{j=1}^{m} k_j z_j}{\sum\limits_{j=1}^{m} k_j} \qquad \text{(Eq. 3.27)}$$

where the z_j are from the "m" closest positions and the weights k_j are chosen larger for data values near to where the estimate $\hat{z}_0$ is to be made. The denominator is a normalizing factor. Note that this reduces to the previous nearest neighbor estimate when $m = 1$.

There are two possibilities for fixing m: one is to use all the data locations for generating all estimates, and a second is to use a *search neighborhood* for each location where an estimate is desired. This is typically taken to be a circle or an ellipse centered at the estimation location. Then only the data locations inside the search neighborhood are used in this equation. This search neighborhood might be specified by fixing the radius of the circle (lengths of the major and minor axes for the ellipse as well as the angle of orientation) or by specifying a maximum number of data locations to be used. In fact both conditions could be used. If the search neighborhood is taken to be too small, less "information" is used in generating an estimate; if the neighborhood is too large, irrelevant information might be incorporated.

There are different ways to relate the weights to distance, but one simple way is to choose

$$k_j = 1/\left[(x_0 - x_j)^2 + (y_0 - y_j)^2 \right] \qquad \text{(Eq. 3.28)}$$

where x_0, y_0 are the coordinates for $\hat{z}_0$. Other "powers" of the distance can be chosen, all of which result in greater weights for values from nearby positions and smaller weights for values at locations farther away. The user must choose a maximum distance for each location where an estimate is desired. If this distance is too small, then less "information" is used; if the distance is too large then irrelevant information may be incorporated.

Like the nearest neighbor estimator, this method is ad hoc and not based on any theory. There are several disadvantages, the first being that the weighting scheme is entirely dependent on the pattern of data locations and does not differentiate between different attributes. A second disadvantage is that this method assumes isotropy; that is, it is only the distance between two locations that is important in determining the weights and not the direction of the line segment connecting them.

An advantage of the inverse distance weighting method is that the resulting estimates are continuous over the area of interest. The method is easy to use and is a common choice in software contouring packages.

EXAMPLE 3.12. From the sand percentage data of Table 3.5, a map showing estimated values was prepared based on the inverse distance weighting method. Again Surfer (2002) was used to prepare the contour maps. The estimated values were based on all of the points within 800 m of each of the 4700 grid nodes used to draw the contours. The results (Figure 3.5C) show the same overall tendencies as when the nearest neighbor estimates were used. However, the contours are smoother, and the divisions are somewhat more complex.

KRIGING

In addition to the limitations already noted above for the nearest neighbor and inverse distance weighting interpolation methods, there is another disadvantage. Neither method provides any measure of the reliability of the estimates generated. Unfortunately, the only way to determine the true error(s) is to collect data at the locations where estimates have been produced. Alternatively, the error or reliability might be quantified in a statistical manner. Kriging, and more generally geostatistics, was introduced by G. Matheron (1971). The kriging estimator is very similar to a multiple regression estimator, and Matheron chose the name to recognize the work of D.G. Krige, who proposed the use of regression after reporting that the polygon method produces overestimation or underestimation of the proportion of high concentrations of metals in a field. Matheron (1971) introduced the variogram and used the theory of random functions to develop kriging. The theory and applications are extensive and much broader than simply interpolation. Numerous books are available on the subject, including those by Chiles and Delfiner (1999), Goovaerts (1997), and Journel and Huijbregts (1978).

The form of the estimate $\hat{z}_0$ of the attribute at an unmeasured location is essentially the same as before:

$$\hat{z}_0 = \sum_{j=1}^{m} \lambda_j z_j \qquad \text{(Eq. 3.29)}$$

Note that we could have used this same form for the inverse distance weighting by replacing the k_j with its normalized form and renaming it λ_j. For a guaranteed unbiased estimator, the sum of the λ_j is required to be 1:

$$\sum_{j=1}^{m} \lambda_j = 1 \qquad \text{(Eq. 3.30)}$$

This last condition is for ordinary kriging and assumes the mean in the domain is constant but unknown. Note that this condition is not necessary if the mean is known. In that case, the approach is termed *simple kriging*. There are important theoretical considerations that one can explore in more specialized literature (e.g., Chiles and Delfiner, 1999; Cressie 1993).

The kriging equations and the variogram

The origin and computations of the λ_j values are much more complex. For ordinary kriging the weights λ_j are found by minimizing the estimation variance, that is, the minimum of

$$\text{Var}[\hat{z}(x_0) - z(x_0)] \qquad \text{(Eq. 3.31)}$$

where for simplicity a single x represents a location given by a set of two coordinates in two-dimensional problems and three coordinates in three-dimensional problems. But how is it possible to formulate the minimization of this variance? Neither the estimate $\hat{z}_0$ nor the true value z_0 is known at x_0, which means more information needs to be known. The missing link is the variogram function $\gamma(h)$ defined by

$$\gamma(h) = 0.5 \, \text{Var}[z(x+h) - z(x)] \qquad \text{(Eq. 3.32)}$$

The parameter h is a separation vector, and x is a position. The variogram is a measure of dissimilarity and under some theoretical considerations is also independent of location (e.g., Journel and Huijbregts, 1978). If h is small, then the difference between the attribute at location $x + h$ and at x is expected to be small, and the variance of all such differences between positions separated by a vector h is expected to be small. The opposite is true for larger h, in that the differences between the attribute at $x + h$ and at x tend to be larger and the corresponding variance would also be expected to be large. Thus, one can estimate $\gamma(h)$ from data separated a distance h using the sample variogram as follows:

$$\hat{\gamma}(h) = \frac{1}{2\,N(h)} \sum_{N(h)} [z(x) - z(x+h)]^2 \qquad \text{(Eq. 3.33)}$$

Pairs of measured values of the attribute at all locations x that are separated by a distance h are collected together to form the estimate. There are $N(h)$ such pairs, and these provide one estimate of $\hat{\gamma}(h)$, which is for that specific value of h. This is repeated for all such pairs of measured attributes for which the locations are separated by a common distance. From a practical standpoint, it is necessary to group the $N(h)$ pairs based on separations that are close but not generally identical (except for one-dimensional transects or, to some extent, other regular sampling patterns).

Sample variogram values are fitted to theoretical variograms. For example, Figure 3.6 shows an example where the dots correspond to sample variogram values for a sand percentage based on 188 data locations from El-Haris (1987). The variogram was modeled with a spherical model. The spherical variogram is

$$\gamma(h) = s\left(\frac{3}{2}\frac{h}{(a)} - \frac{1}{2}\left(\frac{h}{a}\right)^3\right) \qquad 0 \le h \le a$$
$$= s \qquad\qquad\qquad h > a \qquad \text{(Eq. 3.34)}$$

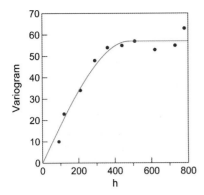

FIGURE 3.6 Sample and model variogram.

The parameters in the model describe the variability, a is the range at which samples are still correlated, and s is the sill that is close to the value of the variance. In Figure 3.6 the parameters are $a = 480$, $s = 57$.

Several models exist for modeling of different shapes of variograms, for example, the exponential and gaussian models are widely used (see Chiles and Delfiner [1999] and Cressie [1993] for details and the theory related to valid models). Sometimes modeling uses two or more valid models that are added together to fit to the sample variogram values.

If the sample variogram could fit into a horizontal line for all lag distances $h > 0$, the attribute is said to respond to a pure nugget effect, and in that case kriging would estimate just the global mean. The nugget effect is the model for uncorrelated variability; this is when the spatial covariance is zero. The nugget model is

$$\gamma(h) = g \qquad h > 0 \qquad \text{(Eq. 3.35)}$$

The choice of an appropriate variogram model and evaluation of the coefficients has a degree of subjectivity, and even the "experts" may not agree on a "best" choice. Ideally, the sample variogram will appear similar to a common model and the coefficients can be chosen by validation of estimates. Generally, fitting routines based strictly on least squares (or mechanical) criteria are viewed with skepticism.

Once the variogram function is chosen, a system of linear equations (ordinary kriging system) is set up as follows

$$\sum_{j=1}^{m} \lambda_j \gamma_{ij} + \mu = \gamma_{i0} \quad i = 1, 2, \dots m \qquad \text{(Eq. 3.36)}$$

where i and j are data locations and m is the number of data used in the estimation. An additional requirement is

that the sum of the λ_j is "1" (Equation 3.30). The solution of the $m + 1$ linear equations minimizes the variance of the estimation error.

Note that μ in the kriging system is a Lagrange multiplier that is part of the solution; γ_{ij} are known variogram values from the variogram function evaluated between data points i and j:

$$\gamma_{ij} = \gamma \left[\left\{ \left(x_i - x_j \right)^2 + \left(y_i - y_j \right)^2 \right\}^{0.5} \right] \qquad \text{(Eq. 3.37)}$$

and γ_{i0} are known variogram values between data points i and the estimated location x_0, y_0.

In addition to the estimate $\hat{y}_0$, the kriging variance σ_E^2 is found from

$$\sigma_E^2 = \sum_{i=1}^{m} \lambda_i \gamma_{i0} + \mu \qquad \text{(Eq. 3.38)}$$

For interpolation of several attributes one has to use more advanced cokriging methods, which has lead to a complete multivariate geostatistics (e.g., Chiles and Delfiner, 1999; Myers, 1982; Wackernagel, 1995). The geostatistical literature is commonly linked to mathematical geology. There are several available public domain software packages for geostatistics (e.g., Deutsch and Journel, 1998).

EXAMPLE 3.13. From the sand percentage data of Table 3.5, a map was prepared with kriging. The equations are solved using the variogram of Figure 3.6. Again Surfer (2002) was used to prepare the contour maps. The estimated values were for all of the points within 800 m of each of the 4700 grid points used to draw the contours. The results (Figure 3.5D) show the same overall tendencies as when the nearest neighbor and inverse distance weighting estimates. However, the contours are smoother. Also shown in Figure 3.5E is the kriging standard deviation (square root of the kriging variance based on Equation 3.38). Generally, the highest kriging variance occurs where there are fewer data points (this may be observed by comparing the locations from Figure 3.6A and the contours of Figure 3.6E).

Provisions for estimating the quality of the estimates are big advantages of kriging and are generally lacking in the other estimation methods. Although the application of kriging is somewhat more involved, user-friendly software packages are widely available, including popular contouring and mapping packages. Proper application of kriging requires knowledge of both properties and limitations of the kriging model and good understanding of the sampled reality. For this reason multidisciplinary assessments are always encouraged.

QUESTIONS AND PROBLEMS

1. Consider the first two rows of the clay data in Table 3.1. Calculate $\bar{x}$ and s^2. What will be the difference in the value of s^2 if n rather than $n - 1$ is used in the denominator of Equation 3.3. Would the difference be more significant for a large or a small number of samples? If you have a handheld calculator that calculates $\bar{x}$ and s^2, which form does it use for s^2?

2. Enter the values for clay percentage in Table 3.1 in a spreadsheet and calculate $\bar{x}$, s, and CV. Separate the data into classes, calculate their frequency, and then plot a frequency histogram.

3. Consider the first two rows of lead data in Table 3.3. Calculate $\bar{x}$ using Equation 3.2. Take the transformation $y = \ln(x)$ for each value and find the mean and variance for y. How does the estimate of the population mean μ based on Equation 3.8 compare with $\bar{x}$? Why are they different?

4. Use the mean and standard deviation for each of the variables shown in Table 3.6. Plot the mean and include "error bars" based on $\pm s$ for each.

5. In establishing the confidence interval (Equation 3.13), a value $z_{\alpha/2}$ was used along with a known sample mean $\bar{x}$ and standard deviation s. The definition of $z_{\alpha/2}$ is from

$$P\left(z > z_{\alpha/2}\right) = 1 - \alpha/2$$

Look up the value of $z_{\alpha/2}$ for $\alpha = 0.05$. (Hint: The value can be found directly in most spreadsheet programs, for example, in Microsoft EXCEL use "NORMSINV $[1 - \alpha/2]$.")[*]

6. A value from the Student's t-distribution was used to establish a confidence interval on the population mean in Equation 3.17. Find the value of $t_{\alpha/2,\,n-1}$ corresponding to probability level α and $n - 1$ degrees of freedom. (Hint: The value can be found directly in most spreadsheet programs, for example, in Microsoft EXCEL use "TINV$[\alpha, n - 1]$.")[†]

TABLE 3.6

Amount of Ammonium (mg kg^{-1}) in Soil Samples from an Agricultural Field Near Maricopa, AZ.

Replicate	1	2	3
x	4.8150	5.1760	5.6317
s	0.3775	2.0263	1.9767
n	4	15	12

(Source: G. Sower and T. Thompson, personal communication, May 20, 2002.)

7. The normal distribution assumes values from $x = -\infty$ to $x = \infty$. Suppose the variable of interest goes from 0 to 100. Discuss the appropriateness of a normal distribution for modeling the data.

8. a. Estimate the sample number to estimate the mean value of lead concentrations. Assume you want the estimate to be within 15% of the mean with a "95% probability level" and believe the estimates of s from Table 3.3 are a reasonable approximation.

   b. How would n change if the estimate is to be within 50% of the mean and with a "50% probability level"? The mean and standard deviation are 20.58 and 33.427, respectively.

   c. After the data are collected, how can you test whether your objective was likely met?

9. Plot the peanut yield vs. ozone concentration shown in Table 3.7. Calculate the mean and standard deviation for each variable. Calculate the slope and y-intercept of a line fitted through the points. Draw the line on your plot.

10. Fit a line to the data for the first 10 days only, using the data shown in Table 3.8. Report the calculated slope and y-intercept. Use the fitted line to predict the amount of carbon dioxide (CO_2) (milligrams) after 35 days. Discuss the difference, if any, between the predicted and the measured values. Is the predicted value reasonable? Why or why not?

TABLE 3.7

Peanut Yield[a]

Peanut Yield (g)	Ozone (μL L^{-1})
157.8	0.056
142.3	0.025
122.4	0.056
92.0	0.076
68.9	0.101
40.0	0.125

[a]Weight of marketable pods/plant (in grams) and ozone concentration (μL L^{-1}). (Source: Heck *et al.*, 1982.)

---

[*]This is not an endorsement for EXCEL or Microsoft in behalf of the authors or publisher. The value can also be found in tables of standard normal distributions found in most statistics books.

[†]This is not an endorsement for EXCEL or Microsoft in behalf of the authors or publisher. The value can also be found in tables of Student's t-distributions found in most statistics books.

TABLE 3.8
Partial Data Set for Carbon Dioxide (CO_2) Evolved and Time for Mine Tailings Water Plus Nutrients

CO_2 (mg)	Time (days)
8.702	0
7.631	0
9.072	0
11.67	3
10.54	3
12.26	3
11.55	6
13.96	6
15.14	6
14.93	9
12.39	9
15.59	9
17.46	35
24.08	35
19.30	35

(Source: Herman *et al.*, 1994.)

TABLE 3.9
Fifteen Pairs of Highly Correlated Data[a]

X^a	Y^b
295	73
339	78
343	85
344	91
357	100
359	109
368	119
395	125
414	129
406	135
385	142
394	139
404	140
420	147
446	156

(Source: Little and Hills, 1978.)
[a]The number of cigarettes (billions) used annually in the United States from 1944 to 1958.
[b]Index number of production per man-hour for hay and forage crops from the same time.

11. Plot the data in Table 3.9. Find the linear regression and draw it on the plot. Calculate the coefficient of regression. Does this indicate that X is a good predictor of Y? X is the number of cigarettes (billions) used annually in the United States from 1944 to 1958. Y is an index number of production per man-hour for hay and forage crops from the same time period. Do you still think that X is a reasonable predictor for Y? Why or why not?

12. Discuss why you think in practice the population of sample means of a nonnormal random variable should approach a normal pdf. (Hint: In theory the populations of means approach the normal due to the central limit theorem.)

13. Compute the number of samples for two confidences and one tolerance for an attribute with coefficient of variation 20% and another with coefficient of variation 40%. (Hint: Watch the units.)

14. What is the probability that a sample randomly selected will have a value lower than the mean? Consider a normal random variable and explain why? A cutoff value is 1.96 times the standard deviation at the right of the mean. What is the probability of a random sample to be larger that such a cutoff? (Hint: May see a normal scores statistical table.)

15. The nugget effect corresponds to a variogram defined by a horizontal line model. Describe the difference between this and the spherical variogram.

16. What can you explain from the spherical variogram of Figure 3.6?

17. We are estimating an attribute at one unsampled point with samples at several distances from the point. If the kriging weights are all equal, what kind of variogram model do you expect?

REFERENCES AND ADDITIONAL READING

Chiles, J.P. and Delfiner, P. (1999) *Geostatistics: Modeling Spatial Uncertainty.* John Wiley and Sons, New York, 695 p.

Coelho, M.A. (1974) Spatial variability of water related soil physical properties. Ph.D. Thesis, The University of Arizona.

Cressie, N. (1993) *Statistics for Spatial Data,* John Wiley & Sons, Inc., New York, 900 p.

Davis, J. C. (1986) *Statistics and Data Analysis in Geology,* John Wiley. New York, 646 p.

Deutsch, C. V. and Journel, A. G. (1998) *GSLIB Geostatistical Software Library and User's Guide.* Oxford University Press. New York, 369 p.

Draper, N.R. and Smith, H. (1981) *Applied Regression Analysis,* John Wiley. New York, 709 p.

El-Haris, M.K. (1987) Soil spatial variability: Areal interpolations of physical and chemical parameters. Ph.D. dissertation. University of Arizona. Tucson, 147 p.

Englund E. and Sparks A. (1988) Geo-Eas (Geostatistical Environmental Software) User's Guide. EPA600/4-88/033. U.S. Environmental Protection Agency, Las Vegas, NV.

Golden Software, Inc. (2002) *Surfer user's guide.* Golden, CO. 640 p.

Goovaerts, P. (1997) *Geostatistics for Natural Resources Evaluation,* Oxford Univ. Press. New York.

Heck, W.W., O.C. Taylor, R. Adams, G. Bingham, J. Miller, E. Preston, and L. Weinstein. (1982) Assessment of crop loss from ozone. *J. Air-Poll. Control Association.* **32,** 353–361.

Herman, D.C., Fedorak, P.M., MacKinnon, M.D., and J.W. Costerton. (1994) Biodegradation of naphthenic acids by microbial populations indigenous to oil sands tailings. *Can. J. Microbiol.* **40,** 467–477.

Journel, A.G., Huijbregts, Ch. J. (1978) *Mining Geostatistics,* Academic Press Inc. New York.

Kane, V., C. Begovich, T. Butz and D.E. Myers. (1982) Interpretation of regional geochemistry. *Comput. Geosci.* **8,** 117–136.

Little, T.M. and F.J. Hills. (1978) *Agricultural Experimentation.* John Wiley and Sons, New York.

Matheron, G. (1971) The theory of regionalized variables and its applications. English translations of Les Cashiers du Centre de Morphologie Mathematique, Fact. 5 Paris: ENSMP. 212 p.

Myers, D.E. (1982) Matrix formulation of co-kriging: Math. Geology, v. 14, no. 3, p. 249–257.

Myers, D.E. (1991) Interpolation and estimation with spatially located data. *Chemometrics and Intelligent. Lab. Syst.* **11,** 209–218.

Wackernagel, H. (1995) *Multivariate geostatistics.* Springer, Berlin, 256 p.

Warrick, A.W., S.A. Musil and J.F. Artiola. 1996. Statistics in pollution science. In: *Pollution Science,* Eds. I.L. Pepper, G.P. Gerba, and M.L. Brusseau. Academic Press, Inc., San Diego.

4

AUTOMATED DATA ACQUISITION AND PROCESSING

P. BROWN AND S.A. MUSIL

Collection of data serves as the foundation of environmental monitoring. To effectively monitor any environment, we must collect data for the environmental parameter or parameters of interest, then process these data into an appropriate format for the intended end use. These uses can range from issuing warnings related to severe weather (e.g., tornadoes, floods, and blizzards) to simply archiving data to assess longer-term trends such as changes in global temperature. Today, data collected for environmental monitoring are often acquired with some form of microprocessor-controlled automated data acquisition system. Mastering automated data acquisition systems requires an in-depth understanding of physics, electronics, and measurement theory. This chapter presents an overview of the physics and electronics necessary to understand the basic concepts of automated data acquisition. In addition, data management and quality control issues are discussed. Students seeking additional detail are encouraged to read the references listed at the end of this chapter.

MEASUREMENT CONCEPTS

Measurement is required for all sciences, fair trade (e.g., standard weights and measures), communication (e.g., telephone, satellite), transportation (e.g., speed, location), health (e.g., diagnostic tests), and public safety (e.g., air quality, weather). Metrology is the branch of science dedicated to measurement, which is defined as the process of comparing an unknown quantity, referred to as the *measurand*, with a standard of a known quantity. Simple examples of measurement include (1) measuring ambient temperature against a standard temperature scale such as the Celsius or Kelvin scale or (2) measuring the length of an object against a standard measure of length such as the meter.

Because all measurements are compared with a standard, these must be expressed in terms of units. Today's scientists use a set of standard units referred to as the Systéme Internationale d'Unites (SI) units. Just seven SI units are required to express or derive all other units used to define scientific measurements. Table 4.1 lists these seven units and their affiliated measurement parameters (see also Table 2.2). Standards may be either material or protocol. A *material standard* is a physical representation of the unit in question, whereas a *protocol standard* describes a series of procedures required to recreate the standard. Most industrial countries maintain a standards laboratory, such as the National Institute of Standards and Technology (NIST) in the United States (see also Chapter 2).

AUTOMATED MEASUREMENT

Today, most measurements are made with automated equipment. Figure 4.1 shows in very simple terms the process required to make an automated measurement of some environmental parameter. Two fundamental components are required to automate measurements: (1) an instrument to make the measurement called the sensor or transducer and (2) a data acquisition system (DAS). The role of the sensor/transducer is to "sense" or measure the parameter of interest and generate an electrical output signal. The role of the DAS is to monitor the output signal of the transducer and process the resulting data into a form that can be understood by the end user.

These two fundamental components of measurement are discussed in more detail in the following sections.

SENSORS AND TRANSDUCERS

Sensing devices, also called instruments, are commonly referred to as *sensors* or *transducers*—terms often used interchangeably. However, a sensor is a device that detects and responds to a signal or stimulus, whereas a transducer is a device that converts input energy of one form into output energy of another form. When making automated measurements, we generally require an instrument that incorporates both a sensor and a transducer. For example, temperature can be measured with a sensor known as a *thermistor*, which is a ceramic material that exhibits a change in electrical resistance in response to a change in temperature. The output of this sensor is of no value unless we can measure changes in electrical resistance. By placing the thermistor in an electrical bridge circuit, we can measure a voltage output that is related to the resistance of the thermistor. Thus, the thermistor sensor, together with its associated electrical circuitry, forms an instrument that converts thermal energy into electrical energy. We call this instrument a *temperature transducer*.

Some sensors also serve as transducers because they convert energy directly (Figure 4.2). Examples include silicon solar cells, which convert radiant light energy into electrical current and thermocouples, which convert thermal energy into electrical energy as voltage. However, many sensors require additional electrical circuitry to generate a measurable output signal.

CLASSIFICATION OF TRANSDUCERS

Transducers can be classified into two categories: active and passive. Active transducers require some form of external energy to generate output signal that can be measured with a DAS. In other words, active sensors must be activated. Examples of active transducers include

TABLE 4.1

The Seven SI Units Required to Address All Scientific Measurement Possibilities

Unit	Parameter
Meter	Length
Kilogram	Mass
Second	Time
Ampere	Electric current
Candela	Luminous intensity
Kelvin	Temperature
Mole	Amount of a substance

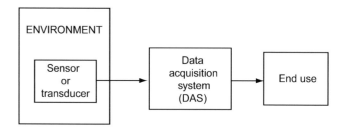

FIGURE 4.1 The process of making an environmental measurement involves placing a sensor or transducer in the environment of interest, connecting the transducer to a DAS, and then linking the resulting data to some form of end use.

thermistors, resistive displacement transducers, and variable capacitance humidity sensors. Active transducers typically generate large output signals that are easier to monitor with a DAS. Passive transducers generate an output signal without an external energy source. Examples of passive transducers include thermocouples, thermopiles, silicon cell pyranometers (Figure 4.2), and photovoltaic sensors. The simplicity and low cost of these sensors are often offset because they generate small output signals that require additional circuitry such as a signal amplifier prior to measurement with the DAS.

The output signal from a transducer can be in an analog or digital format. Analog signals are continuous signals that change in a way that is analogous to the change in the measurand (signal that is being measured) (Figure 4.3). In contrast, digital signals (Figure 4.4) have two discrete states that can be described as "on/off," "yes/no," or "high/low." Most physical and environmental variables change in a continuous fashion; thus, the bulk of the transducers generate analog output signals. These output signals are electrical in nature and typically appear as either a voltage (electrical potential) or current (flow of electrical charge). Voltage output signals are classified as alternating current (AC) or direct current (DC). AC signal outputs oscillate in a periodic (often sinusoidal) fashion about some mean voltage because the current in the transducer circuit is alternating in direction. For DC the transducer signal always moves in one direction. Most transducers generate low-voltage DC outputs, which are relatively simple to measure. Ohm's law, which describes the relationship between electrical potential energy and the flow of current in a circuit, applies to most simple DC circuits. Understanding the simple concept of Ohm's law can prove valuable when making measurements. A review of Ohm's law is provided in the next section.

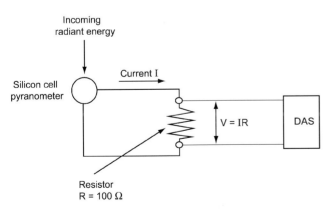

FIGURE 4.2 Electrical circuit consisting of a silicon cell pyranometer and a 100-Ω resistor. The pyranometer generates an electrical current proportional to the radiant energy flux. A change in electrical potential *(V)* results as the current flows through the resistor.

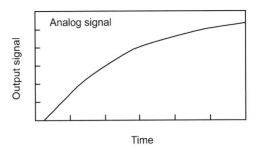

FIGURE 4.3 Analog signals are continuous over time.

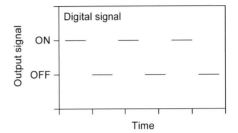

FIGURE 4.4 Digital output signals fluctuate between two discrete levels, such as "on/off" or "high/low."

OHM'S LAW

Ohm's law states that the flow of electrical current (I [amperes {Å}]) in a circuit is proportional to the difference in electrical potential (V; volts[v]) in the circuit:

$$\Delta V = R * I \qquad \text{(Eq. 4.1)}$$

The constant of proportionality, *R*, is known as the resistance and carries the units of ohms (Ω). Figure 4.5 shows a simple DC electrical circuit consisting of a 5-V battery and a 100-Ω resistor. The arrow shows the direction of current flow in the circuit. After some algebraic rearrangement of Equation 4-1, we can use Ohm's law to determine the current flow in the circuit:

$$I = \Delta V/R$$
$$I = 5\,V/100\,\Omega \qquad \text{(Eq. 4.2)}$$
$$I = 0.05A$$

What happens if we double the electrical potential in the circuit to 10 V? The answer is simple; we double the current in the circuit.

$$I = \Delta V/R$$
$$I = 10\,V/100\,\Omega$$
$$I = 0.10\ A$$

What happens if we double the resistance in the circuit to 200 Ω?

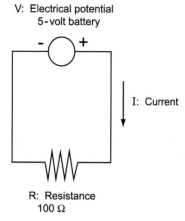

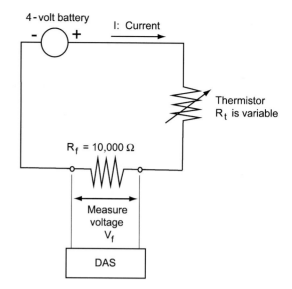

FIGURE 4.5 Simple electrical circuit consisting of a 5-V battery and a 100-Ω resistor. Ohm's law is used to compute the electrical current (I), which flows from high (+) to low (−) potential.

$$I = \Delta V/R$$
$$I = 5\,V/200\,\Omega$$
$$I = 0.025\,A$$

FIGURE 4.6 Simple electrical circuit consisting of a 4-V battery, a 10,000-Ω fixed resistor, and a variable resistor known as a thermistor. Changes in temperature change total circuit resistance, which alters current flow and the voltage drops across both resistors. Ohm's law is used to compute the current flow in the circuit and the resulting voltage drops across both resistors.

In this case, the current of the original circuit is reduced by half.

Ohm's law can help us understand how some very common transducers work. For example, many common sensing elements exhibit a change in electrical resistance in response to a change in measurand. When these resistive sensing elements are properly placed in simple DC circuits, changes in the measurand will alter the resistance, current flow, and electrical potential in the circuit.

Figure 4.6 shows a simple electrical circuit containing a 4-V battery and two resistors. One resistor has a constant resistance of 10,000 Ω, whereas the second resistor consists of a thermistor that is a piece of ceramic material that exhibits a repeatable change in resistance with temperature. The first and second columns of Table 4-2 provide

data showing how the resistance of the thermistor changes with temperature. Ohm's law can be used to determine the current flow in the circuit and the change in voltage that develops across each resistor as the current flows through the circuit. For current flow to be determined, the total circuit resistance must be calculated. This is obtained by the sum of the fixed resistor, R_f, and the variable thermistor, R_t. Thus, for a temperature of 0° C, R_f and R_t equal 10,000 and 9800 Ω, respectively.

$$I = \Delta V/(R_f + R_p)$$
$$I = 4\,V/(10,000\,\Omega + 9800\,\Omega) \qquad \text{(Eq. 4.3)}$$
$$I = 0.000202\,A = 0.202\,mA$$

TABLE 4.2

Example Calculations of the Current Flow and Voltage Values in the Simple Thermistor Circuit (Figure 4.5) at Various Temperatures

Temperature (°C)	Thermistor Resistance (R_t) (Ohms)	Fixed Resistance (R_f) (Ohms)	Circuit Current (mA)	Voltage Across R_f (Volts)	Voltage Across R_t (Volts)
0	9800	10,000	0.202	2.02	1.98
5	7600	10,000	0.227	2.27	1.73
10	5900	10,000	0.252	2.52	1.48
15	4700	10,000	0.272	2.72	1.28
20	3750	10,000	0.291	2.91	1.09
25	3000	10,000	0.308	3.08	0.92

Column 2 provides the resistance of the thermistor element at the temperatures presented in Column 1.

Note that the current flow is quite small. In this case milliamperes (mA) rather than amperes are the preferred unit for current (1 mA = 0.001 A). Once the flow of current is known, one can solve for the voltage drop across each resistance in the circuit. Again, with Ohm's law, the voltage drop across the fixed resistor is:

$$\Delta V = I * R$$
$$\Delta V = 0.202 \text{ mA} * 10,000\,\Omega$$
$$\Delta V = 0.000202 \text{ A} * 10,000\,\Omega \qquad \text{(Eq. 4.4)}$$
$$\Delta V = 2.02 \text{ V}$$

And the voltage drop across the thermistor is:

$$\Delta V = I * R$$
$$\Delta V = 0.000202 \text{ A} * 9800\,\Omega$$
$$\Delta V = 1.98 \text{ V}$$

The fourth, fifth, and sixth columns of Table 4.2 show the results of similar current flow and voltage drop computations for the other temperatures listed in the table. Note how a change in temperature of the thermistor element causes a change in current flow, which in turn causes a change in the voltage drop across both the fixed and thermistor resistance elements. Operationally, temperature is measured by attaching one side of the lead wires across the fixed resistor and the other side of the leads to the input terminals of a DAS system, preconfigured to accept a DC input signal of 0–4 V. Figure 4.7 shows graphically how the output signal of the thermistor circuit would vary with temperature. Note that we chose to connect the DAS across the fixed resistor (sometimes referred to as the *pick off resistor*) because the voltage

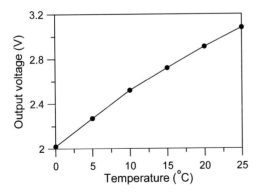

FIGURE 4.7 Voltage output of the thermistor circuit in Figure 4.5 as a function of temperature.

increases with temperature. In contrast, the voltage drop across the thermistor decreases with increases in temperature.

TRANSDUCER SPECIFICATIONS

Manufacturers subject transducers to a series of tests to calibrate the transducer and to assess transducer performance under various conditions. The results of these tests are summarized in a document referred to as a *specification sheet*, or *spec sheet* for short. An understanding of information contained on the spec sheet can greatly improve the chances of making proper measurements. A complete review of the information listed in the spec sheet is beyond the scope of this chapter; however, we review some of the important information commonly found on spec sheets later in the chapter.

Transducer specifications generally fall into three categories: static, dynamic, and environmental. Important static specifications include accuracy, precision, resolution and sensitivity (described in Chapter 2), linearity, and hysteresis. Each of these concepts, applied to transducers, is described in more detail in Box 4.1.

SELECTING MEASUREMENT TRANSDUCERS

The decision of which transducer to select will depend on a number of factors described in Box 4.1 including (1) required accuracy, resolution, and precision of the measurements, (2) environmental conditions, (3) the DAS used to monitor the transducers (Box 4.2), and (4) the cost of the transducer. The first step is to determine the required accuracy, resolution, and precision of the proposed measurement because higher accuracy, precision, and resolution typically result in increased cost. If the goal is to measure temperature with an accuracy of 0.5° C, it would be wasteful to purchase a more expensive transducer that is accurate to within 0.01° C. The next step is to determine the environment where the transducer will be located. Environmental conditions that can have an affect on the measurement of transducers, such as temperature, humidity, vibration, and contamination should be assessed, and a transducer selected that can withstand the environment. A review of the transducer environmental specifications can assist with questions pertaining to environmental tolerance.

Some transducers may need to be plugged into a conventional, high-voltage power source (e.g., 120 or 240 V AC) to function correctly. If a transducer is to be placed in

BOX 4.1 *Transducer Specifications*

Static Specifications. Static characteristics relate to transducer performance when the sensor is operating in a steady state situation (measurand is not changing).

Accuracy is defined as the difference between the measured value and true value of a parameter and is generally presented as the maximum error one can expect in the measurement. Accuracy is commonly expressed as a percentage of the full scale output (highest output) of the transducer. For example, if one obtains a temperature transducer with a rated accuracy of 1% of the full scale output and the full scale output of the transducer is 50° C, then the transducer should measure temperature to within 0.5° C (1% of 50° C) (see also Chapter 2).

Precision is defined as the variation in transducer output when the measured values are repeated. A high-precision instrument generates a small variation in output when a static measurand is repeatedly measured, whereas a low-precision instrument generates larger variations in output under the same circumstances. Precision is often confused with accuracy, but the two parameters are distinctly different (see also Chapter 2). Precision refers only to the spread or variation of instrument outputs and does not indicate whether the readings are correct (accurate).

Resolution is defined as the smallest change in the measurand that can be detected by the transducer. Resolution may be presented either as an absolute value (e.g., 0.1° C) or a percentage of full-scale reading. Again, with our temperature transducer with a full scale output of 50° C, a resolution of 1% would mean the smallest change that a transducer can detect is 1% of 50° C or 0.5° C (see also Chapter 2).

Sensitivity is defined as the change in transducer output that occurs in response to a given change in the measurand. For example, the sensitivity of a copper-constantan thermocouple is ~40 µV/°C when used to measure temperatures in the ambient range (see also Chapter 2).

Linearity. It is desirable to have a sensor generate a linear output in response to changes in the measurand (Figure 4.8). Most transducers are engineered to provide a linear response to changes in the measurand, but all deviate slightly from this desired linearity. This deviation is commonly presented as a percentage of full scale value. Suppose for a temperature transducer the maximum deviation from expected linear output is 0.2° C and the full-scale output is 50° C. The

linearity would therefore be computed as follows: (100%*[0.2° C / 50° C]), which is equal to 0.4% of the full-scale value.

Hysteresis is defined as the difference in output values corresponding to the same measurand depending on whether the measurand is increasing or decreasing. Figure 4.9 provides a graphical representation of hysteresis. The dashed curve in Figure 4.10 shows transducer output as the measurand is increased and then decreased over the operating range of the transducer. The maximum difference in output resulting from hysteresis is generally expressed as a percentage of full scale value.

Dynamic specifications pertain to sensor performance when the measurand is changing (not at steady state). These are two dynamic specifications most commonly encountered on a spec sheet.

Response time is defined as the time required for a sensor to fully respond to a step change in the measurand and is generally presented as the time required for transducer output to attain 95%–98% of imposed change in measurand.

Time constant is the time required for the sensor output to equal 63.2% (1–[1/e]) of the imposed change. Often this value is reported in lieu of response time (see Figure 4.10).

Environmental specifications relate to performance of a transducer when subjected to harsh and difficult environments. The environmental specifications are designed to assist the user in determining whether a particular transducer will work for a particular measurement task. A discussion of four of the more important environmental specifications—temperature, humidity, pressure, and vibration—follows.

Temperature. The impact of temperature on transducer performance may be expressed in several ways. A common temperature specification is the operating range, which is defined as the range of temperatures over which the transducer will perform. For example, the listed operating range may be from −40° C to +50° C. However, even within the operating range, the transducer may exhibit errors because of temperature changes. Temperature errors may be expressed as a percentage of full scale value per degree or as a bias or change in sensitivity per degree.

Humidity. Like temperature, it is common to have a humidity operating range for a transducer. If a transducer will be placed outdoors or in a saturated indoor environment, it is especially important to note whether the sensor can tolerate a condensing

BOX 4.1 *(Continued)*

environment. Normally, humidity specifications will include both an operating range and the label "condensing" or "noncondensing" to indicate whether the sensor is designed to withstand liquid water that results in a saturated environment. For example, a humidity specification might be presented as "0–95% relative humidity, noncondensing."

Pressure may have an impact on sensor performance, especially if transducers are to be

submerged in water or installed in low-pressure environments such as at high elevation or high altitude. A common pressure specification would be the ambient pressure error, which is defined as the maximum change in output resulting from a change in ambient pressure.

Vibration may have a serious impact on transducer performance on aircraft, vehicles, and machinery. Vibration specifications generally indicate the amplitude and frequency of vibration that can be tolerated or the error resulting from the application of a vibration to the transducer.

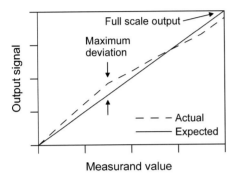

FIGURE 4.8 All transducers exhibit some deviation from the ideal or expected linear response. The linearity specification is reported as the maximum deviation from linearity expressed as a percent of full scale output.

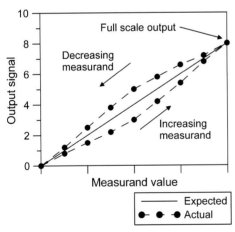

FIGURE 4.9 Hysteresis results when output values differ depending on whether the measurand is increasing or decreasing. The hysteresis specification is reported as the maximum deviation cause by hysteresis expressed as a percent of full scale output.

a remote environment with limited power supplies, then transducers with low power requirements should be selected. For example, if the spec sheet lists 5-V DC and a maximum current draw of 5 mA, the transducer power requirement (*P*) is:

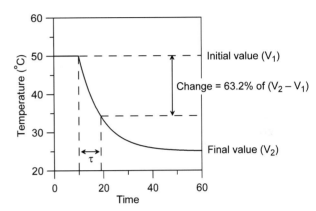

where τ = time required = time constant

FIGURE 4.10 The time constant is a measure of how quickly a transducer responds to a step change in the measurand. In this example, temperature is changed in a stepwise fashion from 50° C to 25° C. The time required for transducer output to attain $1/(1-[1/e])$ of the required temperature change is defined as the time constant.

$$P = 5\,\text{V} * 0.005\text{A} = 0.025\,W^* = 25\,mW \quad \text{(Eq. 4.5)}$$

THE DATA ACQUISITION SYSTEM

The DAS occupies the second block of the general diagram depicting the measurement process (see Figure 4.1). For many years, outputs from environmental monitoring transducers were monitored with analog chart recorders that provided a continuous record of the parameters in question. Today, the chart recorders and the tedious process of data extraction and analysis through human interaction are obsolete, thanks to the rapid

---

*Note: Electrical power is usually measured in watts (W) (W = V*A) or milliwatts (mW).

development of microprocessors and digital computers. Modern digital DASs can monitor and process the output of multiple environmental transducers with an accuracy and speed unimagined just a few decades ago and are now integral to environmental monitoring.

Signal Conditioning

The first functional component of a DAS is the signal conditioning unit (Figure 4.11). The primary role of the signal conditioning unit is to modify the output signal from transducers so the signal matches the required input characteristics of the DAS. Examples of modification include amplifying small magnitude signals from low-output devices, such as thermocouples and piezoelectric devices, and filtering to remove unwanted components or noise in the transducer output signal. As indicated earlier, active transducers require some form of power or excitation to function correctly. The signal conditioning unit also includes the circuitry required to generate these necessary activation signals.

Analog to Digital Conversion

Once the output from a transducer has been properly conditioned, the signal is directed to the analog to digital converter (ADC), which converts analog signals into digital (base 2) format necessary for processing. In digital or base 2 format, our normal base 10 numbers occur as a sequence of "1s" and "0s." An important characteristic of ADC process is the converter resolution that is expressed as a number of bits. The bit value of an ADC

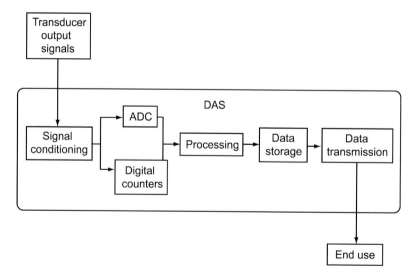

FIGURE 4.11   Block diagram showing the basics components of a data acquisition system *(DAS)*. *ADC*, Analog to digital converter.

represents the length of the binary output value that emerges from the ADC. Common ADC resolutions are 8 and 12 bits. Essentially, the ADC divides the maximum range of the analog input into 2^n bits, where n is the resolution in bits. The size of each bin or segment represents the resolution of the ADC or, in other words, the smallest change in input signal that will be detected by the converter. Consider the example of an analog input signal with a range of 0–5 V being processed with 8-bit ADC. The ADC divides the 5-V range into 2^8 or 256 segments with each segment equal to $5 V/256 = 0.195 V$. Thus, the ADC can only detect changes in the input signal that equal or exceed 0.195 V. In comparison, a 12-bit ADC converter would divide the 5-V range into 2^{12} or 4096 segments, which would increase resolution to 5 V/4096 or 0.00122 V. Clearly, the 12-bit ADC has a much higher resolution than the 8-bit ADC. Cost becomes a factor in selecting an ADC because price increases as the resolution increases. The time required for ADC processing represents a second factor. The process of converting analog signals to digital format involves comparing the input signal with known signals generated internally by the ADC (see references for more details). Higher ADC resolution requires more comparisons and thus a longer processing period. If the sampling rate or frequency of data acquisition exceeds the ADC processing time, sample points will be lost.

DIGITAL COUNTERS

Intermittent on/off signals, for example a magnet placed on a rotating shaft, can trigger a switch each time the shaft rotates. The number of switch closures recorded over a set time interval would measure the rotation rate of the shaft. The signal of interest—the closure of the switch—is clearly a digital output; however, the number of switch closures over time is the measurement of interest. Many digital DAS systems can count digital inputs ranging from switch closures to voltage outputs that change between two discrete "digital" levels. These digital counters, like ADC, are rated based on the bit size (n) of the maximum base 2 number they can accommodate (2^n).

TYPES OF DATA ACQUISITION SYSTEMS

There are two types of DASs: turnkey or general purpose. Turnkey DASs are designed to monitor the output of a specific transducer and typically are provided by the manufacturer of the transducers. Turnkey systems are generally easier to use because they are designed to interface with a specific set of transducers. Often, turnkey DASs are designed to be "plug and play" devices, mean-

ing the user simply connects the transducer and then configures the DAS using software designed for the DAS/transducer combination. Common weaknesses of turnkey systems include (1) their lack of compatibility with transducers from other manufacturers; (2) limited DAS flexibility, which can limit the types of measurements one can make; and (3) future compatibility with new transducers.

The general-purpose DAS offers much more flexibility because it can be configured to handle nearly any measurement, including analog and digital inputs, voltages, and currents. However, the general-purpose DAS is usually more difficult to use because the user must understand more about the transducer, its output signal, and the connection and interface of the transducer to the DAS. General-purpose DASs come in three general forms: (1) stand alone data loggers; (2) modular component systems that must be assembled to create a DAS; and (3) simple computer boards. Stand-alone data loggers are typically portable, low-power DASs that operate off standard AC power and batteries. Modular component systems allow the user to customize the DAS to fit specific needs. One selects key components, including the signal conditioning units, ADC, processing modules, storage devices, and output interfaces. The obvious advantage of a modular DAS is that the system can be changed or expanded by simply changing components. Modular systems do require considerable knowledge/experience with instruments and DAS equipment, which may be a disadvantage to the novice user.

The most common DAS today is one that uses a desk or laptop computer in combination with a plug-in computer DAS board. The DAS is created by installing the DAS board in a computer card slot and then loading attendant DAS software onto the computer hard drive. The software allows the user to program the DAS board to make specific measurements at particular intervals and store the resulting values on various output devices (disks, magnetic tape, or CD-ROM). DAS boards are available for most measurement needs and often the costs of the board and personal computer (PC) together are much lower than a stand-alone or modular DAS. However, DAS board–PC combinations are usually not rugged enough for continuous field use.

SELECTING A DATA ACQUISITION SYSTEM

A number of factors influence the selection of a DAS.

Compatibility of the DAS with measurement and transducer. It is important from the outset to identify the required accuracy, resolution, and frequency of the intended measurement, and make sure that the DAS is compatible and correctly configured to produce a correct measurement. For example, suppose you have a temperature transducer that operates over the range from 0 to 50° C with stated accuracy of 0.1% of full-scale output and sensitivity of 0.2 V/°C. The accuracy of the transducer is 0.1% of 50° C or 0.05° C. With a stated sensitivity of 0.2 V/° C, the DAS system must be able to resolve sensor output voltages of 0.01 V (0.05° C * 0.2 V/°C = 0.01 V) to obtain the full benefit of transducer accuracy. If for example, the ADC of the DAS has 8-bit resolution over the 0- to 10-V output range of the transducer, the DAS will be able to resolve changes in transducer voltage equal to $10 V/2^8 = 10 V/256 = 0.039$ V. In this situation, the resolution of the DAS limits the accuracy of the measured temperatures to 0.195° C (0.039 V/0.2 V/°C). For the obtainment of temperature measurements with sufficient resolution to fully realize the rated accuracy of the temperature transducer, the DAS would need to have a 12-bit ADC. A 12-bit ADC would resolve changes in voltage equal to $10 V/2^{12} = 10 V/4096 = 0.002$ V, which is less than the 0.01 V change associated with the rated accuracy of the transducer. This example shows that the accuracy of a given measurement is limited by the component of the measurement process that is most limiting.

Modern DASs do not continuously monitor analog signals, but rather make a series of discrete measurements over time. The desired objective is to be able to faithfully recreate the actual analog signal from these discrete measurements. Problems develop when the sampling rate is similar to the signal frequency change, resulting in data aliasing. When aliasing occurs, the recreated signal from the DAS contains false, low frequency oscillations (Fig. 4.12). Data aliasing can be overcome if the DAS sampling rate is at or above the Nyquist Frequency, which is defined as equal to two times the highest frequency fluctuation in the output signal.

Measurement environment. The DAS must be compatible with the measurement environment. Harsh environments (extreme temperatures, vibration, saturated conditions) require specialized transducers and DASs designed for extreme conditions. All DASs come with environmental specifications that should be reviewed to ensure the equipment can withstand the measurement environment. The DAS equipment should be in weather-tight boxes and may require desiccants to regulate humidity. Inadequate power is a common cause of failed data acquisition. Transducers and DASs that use minimal power or can operate off low voltage battery power are advantageous when making measurements in remote areas away from conventional sources of electrical power. Solar panels and rechargeable batteries are commonly used to supply power in remote environments and must be engineered to handle the extremes of temperature and available solar energy (e.g., winter vs. summer).

On-site data processing. Data that emerge from the ADC and the digital counters in digital format are subsequently processed. The processing section typically uses a microprocessor that is controlled through some form of programming. It also has an affiliated memory area that is used to store so-called intermediate data, which are data awaiting processing or the results of intermediate calculations. For example, if a DAS is sampling an analog signal every minute and is programmed to compute an average value of the signal every 15 minutes, processing memory might be used to store a running sum of the sampled analog signals until the required 15 minute averaging time is reached. The next space saving step may be to erase the 15 data points acquired and store only the average 15-minute value. On-site processing is especially important if the intended measurements are to be used in real-time to generate a warning or control a process. Other positive aspects of on-site processing include reduced costs associated with transfer, reduction and storage of data. The negative side of storing and transferring

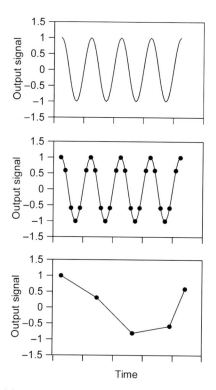

FIGURE 4.12 The sampling rate of the DAS can have an effect on the quality of the resulting measurements. For a periodic signal to be properly recreated from a transducer, the DAS must sample at a frequency equal to 2 or more times the frequency of the signal. A sinusoidal analog output signal is presented in the top figure. This signal is recreated from data collected by a DAS set to sample the signal at 6 times (*middle graph*) and 0.8 times (*bottom graph*) the frequency of the signal. The original signal can be recreated with considerable accuracy at the high sampling rate. At the low sampling rate, aliasing occurs whereby the recreated signal contains a low frequency variation that does not resemble the original output signal.

only processed data is discussed in the "Data Storage Considerations" section.

Data Storage and Retrieval

A DAS must provide adequate on-site data storage to avoid loss of data. On-site storage serves as a temporary repository for data collected between retrieval events. Both internal memory and peripheral devices such as disk or CD-ROM drives can serve as on-site storage. An external interface can be used transfer the data. This can be internal memory, any number of internal or external data storage devices such as tape drives, external memory modules, floppy and hard disks, or CD-ROMs. The size of the on-site storage depends on the amount of data acquired and the frequency with which data are removed from the system. It is therefore important to properly estimate the expected data load when selecting a DAS. If data are to be stored on-site for extended periods (e.g., remote location), then it is important that the DAS have an abundance of memory for data storage. So-called nonvolatile memory is recommended because this memory does not depend on the DAS power supply to provide power to the memory. Nonvolatile memory has its own power supply in the form of a battery.

Retrieval of data from DASs can proceed through direct hardwire links to a computer, telephone connections (both land lines and cellular telephones), and radio frequency (RF) links. Direct hardwire links are common when the DAS and computer system are located in proximity to one another. Telephone or RF links can be used to retrieve data from remote areas. Each type of data retrieval will require special communication equipment and software. For example, data retrieval by telephone will require the use of a modem, whereas a transceiver will be required to transfer data via RF links. With telephone networks and RF links, it is important to determine (1) if service is available at both end of the proposed transmission path, (2) the rate at which data may be transferred, and (3) the cost of using the service. Telephone service (both cellular and land lines) may prove limiting in remote areas. The establishment of a line of sight path between the DAS and the RF network is required when using RF for data transfer. Use of land-based RF may require the installation of signal repeaters to connect the DAS to the RF network. In very remote areas, satellite-based RF systems may be the only available means of communicating with DAS equipment. The rate of data transmission must match or exceed the rate of data collection to avoid loss of data. Some DASs will simply "write over" existing data if on-site data storage reaches its capacity. Other DASs may terminate data collection when on-site storage is exhausted. On-site storage cap-acity should be sufficient to handle periods when the communication link is inoperable.

DATA MANAGEMENT AND QUALITY CONTROL

DASs are excellent tools for automating data collection, but eventually the researcher must evaluate and process the data. Novice users often naively expect the data to be ready to use and to be completely accurate. This section discusses the steps needed to develop a usable data set.

There are three major aspects to processing research data. For the purposes of this chapter we call these aspects (1) management, (2) quality evaluation, and (3) analysis. Management involves manipulating the acquired data to transform it into a more workable form and then storing the results. Examples include reorganization of the data stream into a more convenient format, conversion of the data with calibration equations, and data storage. The next step, quality evaluation, involves inspecting the quality/reliability of the entire data set and is often neglected with large data sets because it tends to be time and labor intensive. Analysis is the final stage whereby the data are analyzed and evaluated with regard to the objectives of the monitoring program. The first two aspects of data processing are discussed from the viewpoint of working with the large data sets that commonly occur with automated data acquisition. Data analysis is not discussed in this chapter.

Text Versus Binary Files

Data acquisition systems generate files full of information, but the information is often in a proprietary format. A discussion of computer file formats may be helpful in understanding some of the implications of working with a variety of file formats.

TEXT FILES

Most of the information found in a user's computer file is "text," which can be very loosely thought of as the characters available from standard keys on a keyboard. Note that by this definition, text includes numbers and punctuation marks. Text files have a minimum of special embedded codes, including characters such as a carriage return, line feed (both are needed to indicate the end of a line), and an end of file marker. When a user looks at the information in a text file, there will be no special formatting such as bolding, equations, and graphics available in the file. These unformatted files are simple enough that most programs can read them because they use a system called the American Standard Code for Information Interchange (ASCII, pronounced "as-key") to represent text. Text files are sometimes referred to as ASCII files and often have a .TXT or .DAT file extension (Sidebar 1).

SIDEBAR 1 *More about ASCII files.* The American Standard Code for Information Interchange is a standard code that was established to provide compatibility between various types of data processing equipment and has become a common code for microcomputers. The standard ASCII character set consists of 128 decimal numbers ranging from 0 to 127, each of which is assigned to a particular letter, number, punctuation mark or the most common special characters (such as %). The extended ASCII character set ranges from 128 to 255, representing additional math, graphic and foreign characters. There is a lot to learn about ASCII files, but from the data processing point of view, the important concept is that ASCII code can be considered the lingua franca code. All software should be able to read ASCII (text) files correctly.

BINARY FILES

There is another type of file, commonly referred to as *binary,* that can be thought of as a "coded" file. These files, which include spreadsheets and word processing documents, include numerous embedded codes to ease formatting, math operations, etc. Embedded codes can be thought of as secret instructions for the program so that it remembers necessary steps. Binary files often include special instructions at the top of the file that contain information about what software created the file or for what printer it is formatted. These codes are usually specific to the computer program that created the file. For instance, a spreadsheet will save information to a file that can be read only by that same program, unless another program has a special decoder ability to convert the file to its own format. As an example, WordPerfect has to convert a Word document to the WordPerfect format before the user can see the file contents.

DATA IMPORT AND EXPORT

Modern spreadsheets are able to read a text file directly by opening the file in the same manner that other spreadsheet files would be opened. Ideally, the first cell (A1) in the spreadsheet would hold the first data value, the second cell (B1) would be the second data value, for example. However, the spreadsheet program may read in the data as text (where letters and numbers are treated the same), rather than numbers, and store it all in a column format (Figure 4.13). Fortunately, there is an easy solution. Most programs have standardized their terminology and refer to importing and exporting data. Importing data means the program

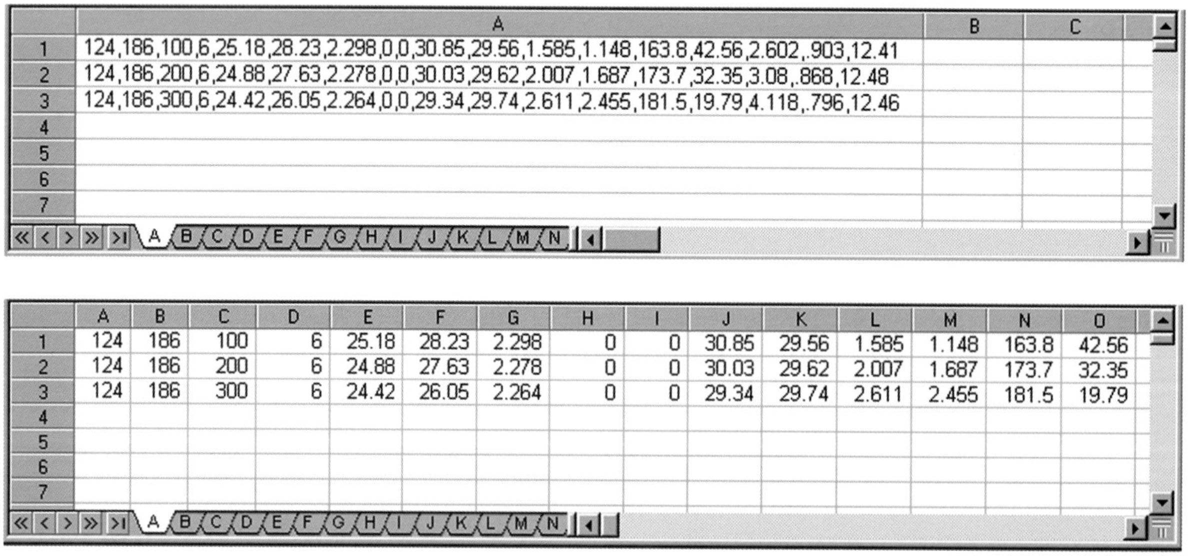

A

	A	B	C
1	124,186,100,6,25.18,28.23,2.298,0,0,30.85,29.56,1.585,1.148,163.8,42.56,2.602,.903,12.41		
2	124,186,200,6,24.88,27.63,2.278,0,0,30.03,29.62,2.007,1.687,173.7,32.35,3.08,.868,12.48		
3	124,186,300,6,24.42,26.05,2.264,0,0,29.34,29.74,2.611,2.455,181.5,19.79,4.118,.796,12.46		
4			
5			
6			
7			

B

	A	B	C	D	E	F	G	H	I	J	K	L	M	N	O
1	124	186	100	6	25.18	28.23	2.298	0	0	30.85	29.56	1.585	1.148	163.8	42.56
2	124	186	200	6	24.88	27.63	2.278	0	0	30.03	29.62	2.007	1.687	173.7	32.35
3	124	186	300	6	24.42	26.05	2.264	0	0	29.34	29.74	2.611	2.455	181.5	19.79
4															
5															
6															
7															

FIGURE 4.13 Examples of how text data is read into a spreadsheet. **A,** Results of opening a text file as if it were a spreadsheet file. All information is treated as text and is stored in the first column. **B,** Results of importing a text file. Each number is placed in a single cell, allowing the data to be further manipulated by the user.

can decode files in formats that are different from its own proprietary format. Exporting data means the program can convert its data format to a specific list of formats that are used by other programs. Almost all software programs can import and export ASCII (text) files, although a considerable amount of formatting information such as equations or bolding may be lost in the process of exporting to ASCII. This feature is useful when moving data from one software program to another (e.g., exporting data from a spreadsheet to a statistics program).

TEXT FILE FORMATS

Downloaded data can be in a variety of formats, which refer to the way the individual data points are stored in the file. Some common examples are comma delimited (Fig. 4.14), tab delimited, and space delimited. Comma delimited text is usually the easiest format to import into spreadsheets and databases because there is no ambiguity about where each "column" of data begins and ends. The easiest file formats to read use either space or tab delimiters, particularly when the delimiters are used to line up the data in columns. Space-delimited data appear to be unambiguous but can cause problems when importing if the number of digits for numbers in a given column changes within the file. The reason is that the program reading the data file often decides on where the data column starts and ends, by looking at the first line or two of data. Errors may occur if data columns in later lines do not line up precisely with the columns in the first few lines. Tab delimited data eliminates that problem, but in practice, some software does not work well

with tabs, again leading to errors. Some programs are more versatile than other programs with respect to importing data, but typically have quirks, such as interpreting tabs incorrectly.

DATA MANAGEMENT

The first step in data processing is usually to rearrange the data set because most files from DASs have extra information or are in a format that is difficult for the average user to read. Figure 4.14 shows an example data set that is difficult to read because of the comma-delimited format. The data set includes information that probably does not need to be included in the final data storage file, such as the DAS program code and the battery voltage. The DAS program code indicates the type of data being recorded. In this example, 124 indicates hourly data, and the 202 on the final line indicates daily (24-hour) values. These program codes can be useful while manipulating the data into the final format but do not need to be retained in the final storage file.

COMPLEX DATA SETS

A challenge in working with DASs is that data sets can be extremely complicated, with multiple lines of data for one period. Often data acquisition devices give the user a minimum of control over the format of stored data. Consider Figure 4.15, which shows a subset of data for approximately 150 sensors. Note that the DAS has split the data for one time period into multiple data lines, resulting in 62 records for one sampling period.

124,186,100,6,25.18,28.23,2.298,0,0,30.85,29.56,1.585,1.148,163.8,42.56,2.602,.903,12.41
124,186,200,6,24.88,27.63,2.278,0,0,30.03,29.62,2.007,1.687,173.7,32.35,3.08,.868,12.48
124,186,300,6,24.42,26.05,2.264,0,0,29.34,29.74,2.611,2.455,181.5,19.79,4.118,.796,12.46
124,186,400,6,23.61,27.26,2.122,0,0,28.58,29.65,2.507,2.423,161.9,14.81,3.4,.793,12.45
124,186,500,6,21.77,35.99,1.672,0,0,27.92,29.61,1.85,1.331,167.6,42.89,3.08,.936,12.44
.
.

.
.
124,186,2300,6,25.62,24.59,2.491,0,0,32.15,29.26,1.626,1.311,171.5,35.67,3.24,.801,12.51
124,187,0,6,24.09,27.96,2.164,0,0,31.15,29.36,1.538,1.472,181.3,16.83,2.761,.837,12.48
202,187,0,39.87,20.56,30.87,43.99,5.253,19.1,3.98,32.12,0,39.23,27.23,32.87,30,28.94,29.44,2.335,
1.025,226.6,60.67,6.193,13.56,12.41,12.67,1.198,.378,.746

Data columns for hourly data (lines beginning with 124 are, from left to right, data logger program code, day of year, hour, station location number, air temperature, relative humidity, vapor pressure deficit, solar radiation, precipitation, 2-inch soil temperature, 4-inch soil temperature, average wind speed, wind vector magnitude, wind vector direction, wind direction standard deviation, maximum wind speed, actual vapor pressure, and battery voltage.

FIGURE 4.14 Sample of one day's data downloaded from a DAS. Data columns (lines beginning with 124), from left to right, are DAS program code, day of year, hour, station location number, air temperature, etc., and ending with battery voltage. The final line (beginning with 202) holds 24-hour means, maximums, and minimums for a variety of sensors. Note that some values, such as wind direction standard deviation, are calculated by the DAS before the data are downloaded.

```
110,1997,121,1200
112,13.78
117,32.25,30.75,30.15,30.66,29.85,26.97,36.77,31.41,30.42,29.98,26.71
124,24.114,-3.0603,2.7285,2.3373,2.4265,2.476,2.4958,2.5255,2.5503,2.5404,2.5998,2.58
126,24.517,-1.8768,3.0752,2.6938,2.5998,2.5205,2.4859,2.4512,2.4512,2.4165,2.3967,2.4215
128,24.517,-1.5103,1.0449,1.1885,1.3172,1.3568,1.3767,1.4311,1.4559,1.4856,1.4559,1.5302
130,25.092,-.3714,.80222,.03467,.07429,.1139,.11886,.13371,.1139,.13866,.15351,.12876
132,25.385,-.91611,2.0352,1.956,1.9411,1.9164,1.9263,1.8569,1.9065,1.9114,1.8768,1.8768
137,24.933,-.83191,2.6739,.41101,.36644,.34663,.29216,.30207,.2773,.27236,.28721,.24761
139,25.581,-99999,-99999,-99999,-99999,-99999,-99999,-99999,-99999,-99999,-99999,-99999
141,24.444,-.03466,2.4462,.17826,.08913,.07923,.05447,.02971,.04457,.02971,.00495,.04952
143,25.263,-1.0746,.79724,.2377,.24264,.23274,.22779,.23276,.20302,.22284,.21293,.24762
145,25.178,-.49023,1.2478,.72297,.62889,.55461,.51004,.51994,.48528,.45062,.41596,.45062
150,25.643,-1.842,1.8817,1.3023,1.3617,1.4014,1.3915,1.4063,1.4558,1.4509,1.441,1.4212
152,25.399,-.91607,3.1246,1.1438,.90122,.75267,.64869,.59422,.55956,.5249,.50509,.47043
154,25.155,-1.2627,1.3667,1.2577,1.2528,1.2528,1.2577,1.2627,1.2676,1.2429,1.2676,1.2379
158,3.4158,4.987,-2.4442
315,3.7308,1.1911,5.8627,1.1249,5.7587,2.1236,5.2429,3.8312,2.2362,5.5967,-99999,6.1382,4.2879,1.7812,-99999,6.269
326,32.428,31.012,22.216,22.993,24.826
326,32.428,31.012,22.696,24.011,25.571
326,32.428,31.012,22.721,23.653,24.972
326,32.428,31.012,23.227,24.253,25.744
326,32.428,31.012,23.569,24.475,25.809
                  .
                  .
                  .
```

Continues for another 39 lines for just one sampling period.

FIGURE 4.15 Part of a complex data set from a research project involving over 150 sensors and holding 62 lines of data for each sampling period. The first number on each line is the data logger table number, indicating the type of data. Note the −99999s on line 10 indicating sensor failure.

Some lines of data hold multiple data points and calculated values for one sensor, whereas other lines hold data for multiple sensors on one line. This complicates cleaning up the data stream for later use. It may be advantageous to split such a complicated data set into multiple tables or spreadsheets, depending on the software used to store the data and on the researchers needs. For example, the data from Figure 4.15 was separated into three data groups, which eased data analysis and storage problems unique to this project. Information was grouped by location and measurement type. Decisions on how to store data should be based on making the final data files easier to work with and also on ease of understanding.

SOFTWARE CHOICES FOR MANIPULATING DATA

The most common software choices for rearranging or cleaning up data sets include a spreadsheet, a database, or a programming language such as BASIC and Java. There are advantages and disadvantages to each choice, although the usual practice for beginners is to use what they already know, which typically is a spreadsheet program.

Spreadsheets generally have three functions: (1) worksheets, which allow the user to enter, calculate, and store data, (2) graphics, and (3) database techniques that allow the user to sort data, search for specific data, and select data that meet certain criteria. Spreadsheets are easy to learn, powerful in scope, and may be the only software tool needed to manage data for some projects. However, they are restricted as to how many data points they can handle at one time, which is potentially a fatal limitation for large data sets. It can also be difficult and time-consuming to sort through a large spreadsheet to find the one data record needed. Conversely, calculations are stored with the data set, so there is an automatic record of how values were computed. Spreadsheet data can be manipulated with macros (Sidebar 2) for repetitive work. For example, the data in Figure 4.14 is fairly simple in organization, with 24 records of hourly data and 1 record of 24-hour summary data. This data pattern would then be repeated for each day of data. In this case the hourly data might be in one worksheet, and the 24-hour summary data might be moved into a second worksheet within the same spreadsheet. Moving each summary line for a month of data could be a tediously repetitive and mistake-prone process, but a spreadsheet macro could accomplish the task almost automatically.

The term *database* refers to a collection of information organized in some manner that allows the user to retrieve the data as needed. Strictly speaking, an organized collection of information in a file cabinet could be called a database. However, common usage of the term *database* refers to a relational database, which is a computer program that allows the user to add, change, and sort data, as

SIDEBAR 2 *What are macros?*   A macro is a small program that can be created in spreadsheet and word processing software. A macro is a series of software commands that can be stored and run whenever the user needs to perform the task. Macros can perform keystrokes, mouse actions, and menu commands. They are excellent for automating repetitive tasks. Software usually includes macro "recorders," which capture the command sequence as the user works through the task. The recorder automatically records the actions taken and then stores the information for later use. It is also possible to edit macros manually.

well as create forms for entering data and reports for extracting data.

The strength of relational databases lies in the way data is stored in multiple tables of information, with each line of data in a table called a record. Information in various tables can be cross linked. As an example, consider a database that holds information on students studying environmental science at a university. One data table might hold home addresses and phone numbers, a second table could list each student's advisor and year of study, whereas a third table might hold information on scholarships. A relational database can search all the tables simultaneously, and then display the desired merged information. This process is very powerful, particularly when working with tens of thousands of records.

Consider the relatively simple data set in Figure 4.14, where 1 year of data generates 9125 records of data. That much data are relatively slow and difficult to search in a spreadsheet, whereas in a database, a search would be nearly instantaneous. The data set shown in Figure 4.15 is very large. Sampling occurred every 4 hours, which resulted in 372 lines of data per day, or 135,780 lines per year. Unlike spreadsheets, a relational database would be able to store all the data in one table, if desired, or facilitate breaking the data into multiple, linked tables. A database could also execute a search of multiple tables in a split second. This sorting capability is the primary reason databases are so useful. Most databases do not store equations directly with the data in an effort to minimize storage requirements, so more effort is required to keep a permanent record of how the data were calculated. Most databases also offer a programming option in the background for advanced users, which can ease repetitive data analysis operations and can also provide a record of how data is processed. A database is also an integral part of Geographical Information Systems (GISs), which will undoubtably become an increasingly important data analysis tool for environmental science investigators (see Chapter 6).

Programming languages such as BASIC and Java offer the most flexibility for manipulating data and can offer the added advantage of doing data quality evaluation at the same time. Programming languages can be difficult and time-consuming to learn, and often the program writer is the only person who can use the programs effectively. Searching for a particular data subset may be problematic, requiring special programming for each search.

DATA STORAGE CONSIDERATIONS

No matter what choice is made for manipulating data, we still need to carefully determine the format for storing data. This becomes more critical when working with large data sets because the volume of material makes them difficult to work with. Normally, it is important to always keep track of units and to always be able to work backwards, to show exactly when, where, and how a particular piece of data was collected. There should also be a way to document instrument damage, repair, or replacement.

A typical choice would be to import the unprocessed data into a spreadsheet, and then eliminate the extraneous information, leaving a data set as shown in Figure 4.16. At this point, all data columns should be clearly labeled as to information type, including units. Unfortunately, full labeling is often omitted. Units in particular can be forgotten very quickly, or in the case of group work projects, coworkers may not be as familiar with the data set, so this step should not be overlooked. Data sets should be labeled clearly so that a person new to the research project would be able to understand the contents. The same clarity should be applied to data file names. Part of data management is the process of converting data values into recognizable units. For example, a heat dissipation sensor measures changes in soil temperature in °C, which is then plugged into a calibration equation to give the desired result of soil water potential (cm of water). Most data acquisition equipment can do this conversion to save a step in data analysis after data is downloaded. However, if at all possible, it is better to output just the raw (original or unprocessed) values (in this case, °C), or both the raw value and the final result (°C and cm of water). This is done because if there is a problem with the instrument, it is usually much easier to determine the problem when the raw data are available.

If bad data are found during the data quality evaluation, the data points could be replaced with a standard value that indicates bad data. This standard value could be a blank cell, which would make statistical or other numerical analysis easy (blank cells are ignored in most spreadsheets and databases). Another common choice is −999,

	A	B	C	D	E	F	G	H	I	J
1	Day of	Hour	Air	Relative	Vapor pressure	Solar	Precipitation	2" soil	4" Soil	Wind speed
2	year		temperature	humidity	deficit	radiation		temperature	temperature	(mean)
3			C	%	kPa	MJ-2	mm	C	C	m/s
4	186	100	25.18	28.23	2.298	0	0	30.85	29.56	1.585
5	186	200	24.88	27.63	2.278	0	0	30.03	29.62	2.007
6	186	300	24.42	26.05	2.264	0	0	29.34	29.74	2.611
7										
8										

FIGURE 4.16 Sample final data storage file with the first three lines of data from Figure 4.1. The example does not include all the data to save space. The DAS program code and station location number have been removed.

or something similar, where care is taken to choose a value that could never be mistaken for real data. The advantage to this approach is that different values can be used to represent different problems. For instance, −999 could indicate bad data and −888 could indicate missing data. A paper trail for this type of data substitution is a necessary part of a quality assurance/quality control program. A better option is to use a blank cell for bad data, plus include a column for comments about data problems, changes of equipment, special conditions, etc.

Choosing a data format is also affected by how the data is managed. If blanks are chosen to indicate bad or missing data, then a space (blank) could not be used as a data point delimiter in the text file. If commas are chosen, then a line of data could include a blank value by entering a space and then a comma as shown below:

$$203, 1408, 28.25, \quad , 26.2, 0.447, 0$$

This illustrates the need to carefully consider the methods used to indicate bad data for final data storage, etc. Modern spreadsheets and databases interpret blank cells as empty (i.e., no data), rather than zeros. This is a critical distinction when trying to find an average value for a group of values that includes one or more blank cells. On the other hand, programming languages are likely to interpret blanks as zeros.

Truncated values. There can be serious problems with the way numbers are truncated in computer programs. Assume a sensor returns a reading of 1.435 mV, which is then put through a calibration equation to yield a value of 25.67834710. In a spreadsheet, the cell can be set to show a certain number of digits after the decimal place as determined to be appropriate by the researcher. In this case it might be appropriate to show three digits (this usually is determined by the precision of the instrument regardless of the number of digits stored by the DAS). Spreadsheets will continue to store the unseen additional digits, so the user can change the setting to four digits later if desired. Databases usually clip the data where indicated, permanently losing the extra digits of data, with the goal of decreasing the amount of storage space.

If the number in the example is stored as 25.678 in a database, all the numbers after the 8 are permanently lost. This is another reason to store the raw, unprocessed data.

NUMERICAL PRECISION

There is a more serious aspect to this storage issue that is independent of the type of software being used. Computers do not recognize the difference between the numbers 12.720 and 12.72, yet a researcher may want to indicate that the first number has a precision to three decimal values, whereas the second value is precise only to the two digits after the decimal. The same problem can occur with numbers without a zero on the end—occasionally you may have a value that is precise to a different number of digits than the rest of the data set. The awkward solution involves saving the numbers as text rather than numerical values, which will be stored precisely as written. The text values need to be converted to numbers when using the data for calculations or for plotting.

DATA QUALITY EVALUATION

One of the hidden challenges of working with automated DASs is the sheer volume of data that can be collected. Numerous data points are often desirable, but there is a trade-off in terms of the amount of time it will take to evaluate the data quality. This is particularly important with "near real-time" systems, where data points are expected to be downloaded to a computer, and then made available for use within 24 hours or less. Most near real-time systems use programs that can evaluate data points by comparing them to predetermined acceptable ranges, and then flag questionable data for manual evaluation.

Data quality evaluation requires a comprehensive understanding of how the sensor works, and what data can be reasonably expected. Some data sets can be simplistically checked by determining whether the data is outside expected sensor ranges (i.e., does the data fall between predetermined minimum and maximum values?). As an example, consider a thermocouple measuring air temperature at a desert site. Suppose experience suggests setting the range from −5 to 50°C. A reading of −10°C would be identified for further investigation. A value of −10°C in summer is probably wrong, whereas in winter it might be correct due to an unusually cold weather front. A more sophisticated approach would narrow the expected data range to better match temperatures as they change during the year.

All sensors have upper and lower measurement limits. Some will give meaningless data outside of those limits, usually due to extrapolation beyond the calibration curve limits. Sometimes the sensors will limit the possible data values. As an example, a Met-One anemometer (measures wind speed) is listed as having a data range of 0 to 60 m s^{-1} with a threshold value of 0.5 m s^{-1}. The threshold speed is the speed at which the anemometer cups begin to turn, with the result that speeds between 0 and 0.5 m s^{-1} are treated the same. In this particular case the calibration curve that calculates the wind speed has a y-intercept of 0.447 m s^{-1}, so the lowest speed recorded is 0.447 m s^{-1}, even if the air is completely still. In this case, a value below 0.447 would probably be bad data and require further investigation. Some sensor data can be evaluated with a more rigorous statistical analysis to determine whether data are outside of acceptable limits. For example, some sensors might yield data that are expected to fall within two standard deviations from the mean. Using statistics is often the best possible approach and is discussed in Chapter 3.

PLOTTING DATA TO FIND OUTLIERS

Checking data ranges or performing basic statistical analysis is an excellent beginning, but may not be sufficient to identify all potential problems. With small data sets, the next step would be scanning the data by eye to find data outliers or unexpected data trends. Quick scanning is impractical with large data sets, because it is too easy to miss problem data points. The best choice is to plot all the data. Figure 4.17 shows soil water tension versus time for a tensiometer installed 5 m below the surface. Water was continuously applied at the soil surface from days 0 to 23, with the curve showing water movement past the tensiometer over time. A smooth curve is expected, with the tension values dropping as the wetting front moves past the tensiometer. The plot clearly shows several seemingly random, aberrant data points. The behavior suggests an electrical short of some type. When precipitation data for the same period is added to the plot, it suggests that precipitation is correlated to the bad data. The instruments were checked at the field site, and it was

determined that some wire connectors were not watertight, permitting electrical shorts when it rained.

A similar approach for finding outliers is to plot related data sets together. For example, relative humidity can

---

CASE STUDY 4.1 *New Sensor to Measure Soil Water Tension*

It can be useful to have multiple ways of measuring the same parameter, particularly if you are working with a new sensor or a sensor that tends to have problems. As an example, Figure 4.18 shows data from a relatively new sensor called a *heat dissipation sensor*. It measures soil water potential by creating a pulse of heat and then determining how quickly the heat dissipates. A calibration curve can convert the change in temperature over time to soil water potential. In this case, the calibration is a power curve, with the result that at high tensions, small changes in temperature can cause large changes in tension. As the soil becomes wetter, decreasing the tension, the same small temperature changes cause smaller changes in tension. In other words, we expect to see larger fluctuations in tension because of a given change in temperature at high tensions than at low tensions. This can be seen on the graph, and the results seem reasonable. Unexpectedly, the data are consistently offset from the results from a tensiometer placed in the same location. In evaluation of the reasons for different results, it is important to consider potential problems with the calibration curve and the raw data stream. In this case, the raw data comprised the initial soil temperature, the final temperature after heating, and the temperature from a thermocouple that was installed in the same location, but attached to a second data acquisition device. The temperature provided by the thermocouple was significantly different than that provided by the heat dissipation sensor. After a great deal of investigation, it was determined that the voltage regulator on the data acquisition device was faulty, resulting in faulty temperature data for the heat dissipation sensor, whereas the tensiometer and thermocouple data was unaffected. The result is that the heat dissipation sensor data is incorrect in terms of actual data values, yet the overall response (curve shape) of the sensor is correct. Without either the tensiometer or the thermocouple data, this problem may not have been identified. This illustrates the need to be inherently skeptical of acquired data, particularly with sensors that are new to the user.

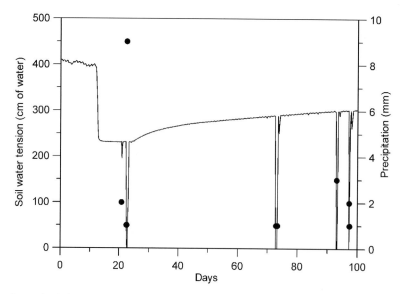

FIGURE 4.17 Tensiometer data at 5 m below the soil surface showing movement of the wetting front and the effect of precipitation causing electrical shorts. Circles indicate precipitation events.

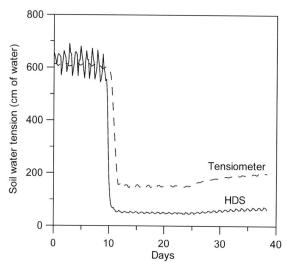

FIGURE 4.18 Heat dissipation sensor *(HDS)* and tensiometer data at 1.5 m depth below the soil surface showing water movement past the sensors due to continuous irrigation at the surface. Note the offset between the sensors at lower tensions.

increase or decrease without precipitation, yet if it does rain, the relative humidity would increase in most situations. If the humidity does not increase with rain, there may be a problem with the sensor. Surface soil temperature could also be linked to precipitation, with a summer rain cooling the soil, and a winter rain possibly increasing or decreasing the temperature, depending on conditions. This approach requires some careful consideration about the trends portrayed in the plot.

REASONABLE CAUSE

What if there is not a clear and reasonable cause for a bad data point? The conservative response would be to leave the data as they are, or, in other words, to leave them in place without changes. The example in Figure 4.17 is clear cut, but often it is difficult to know whether the unusual data are usable. It is unacceptable to make the decision based only on instinct; there needs to be a clear, demonstrable reason to exclude data. In some cases, the decision may be reasonably based on a researcher's experience or on published values in scientific literature. If there is no clear rationale for excluding data, the data could still be excluded from further analysis if the researcher makes a note of the exclusion in the final reports. This area can be fuzzy and is full of potential pitfalls. The best approach is to keep apparently anomalous data if there is any doubt about their validity.

DOCUMENTATION

Quality assurance programs often require that the procedure and criteria used to determine acceptable/unacceptable data be documented. This helps maintain consistency when there is more than one researcher working on a project. Documentation is highly recommended even when it is not specifically required because the process clarifies procedures and also provides a historical record for use years later when the data set is revisited.

FIELD VISITS AND OBSERVATIONAL SKILLS

Unexpected events can cause problems at field sites. Animal damage can include large birds landing on and bending wind vanes; horses rubbing on and toppling over instrument towers; rodents biting wires; and spiderwebs causing electrical shorts. Human vandalism occurs occasionally, whether intentional or not, including shooting guns at equipment or damaging sensors with golf balls. Extreme weather events can damage sensors or blow over instrument towers. Regular inspection of sensors at a research site is necessary to find these problems. It is just as important to develop keen observational skills. Small discrepancies/observations in the field can sometimes make all the difference in the final interpretation of the data.

QUESTIONS

1. Describe in general terms how sensors work. List and describe the two major types of sensors.

2. With Ohm's law, electric power (energy) consumption per unit time can be calculated by the following general relationship: watt-hour $= V * I$. You have a solar panel that generates 1 watt-hour of power at full sun exposure. Could you run a sensor and data logger that require a combined 0.070 amps at 12 V of power with it 24 hours a day? Explain your answer and discuss any other parts (if any) you would need to power this remote station 24 hours a day.

3. Figure 4.2 shows a simple circuit one can make to convert the current output to a voltage output. An electrical resistor is placed in series with the silicon cell the resulting voltage drop that occurs as current flows through the resistor is measured. Suppose for example, the pyranometer sensor generates an output signal of 10^{-4} A KW^{-1} m^{-2}. Assuming a midday summer solar radiation flux density of 1.0 KW m^{-2}, the sensor will generate a current flow of 10^{-4}A. When a 100-Ω resistor is placed in series with the silicon cell, what is the voltage drop across the resistor?

4. What is a sensor time constant? Explain why it is important to know the time constant of a sensor before using it.

5. Explain analog to digital signal conversion (ADC) and its relation to modern environmental monitoring. Can a 10-bit ADC detect a 0.01-V signal from a sensor? Explain your answer.

6. Is a word processor file a text or a binary file? Discuss your answer.

7. Discuss one advantage and one disadvantage to using a spreadsheet instead of a database for managing large data sets.

8. You are in charge of automating data quality evaluation from two sites in the United States. Site 1 has a mean annual precipitation of 42 inches, evenly spread out throughout the year. The temperature ranges from 0° to 100° F. Site 2 has a mean annual precipitation of 11 inches, most of it during the summer months. The temperature ranges from 40° to 105° F. Discuss some reasonable criteria that you would use to spot outlier precipitation and temperature data for each of the two sites.

REFERENCES AND ADDITIONAL READING

Dyer, S.A. (2001) *Survey of Instrumentation and Measurement.* John Wiley & Sons, New York.

Heaslip, G. (1975) *Environmental Data Handling.* John Wiley & Sons. New York.

Minor, G.F., and Comer, D.J. (1992) *Physical Data Acquisition for Digital Processing: Components, Parameters, and Specifications.* Prentice Hall, Englewood Cliffs, NJ.

Morris, A.S. (1996). *The Essence of Measurement.* Prentice Hall, London.

Omega Technologies. (1998) *Transactions in Measurement and Control.* Vol. 2. Data acquisition. Putnam & Omega Press LLC, Stamford, CT.

Taylor, H. R. (1996) *Data Acquisition for Sensor Systems.* Chapman & Hall, London.

Turner, J.D. (1988) *Instrumentation for Engineers.* Macmillan, London.

Usher, M.J., and Keating, D.A. (1996) *Sensors and Transducers: Characteristics, Applications, Instrumentation, Interfacing,* 2nd Ed., Macmillan, London.

5

MAPS IN ENVIRONMENTAL MONITORING

D.M. HENDRICKS

Maps are scale models of a portion of the earth's surface that show details of the size, shape, and spatial relations between features. Maps provide a method of reducing large amounts of information to a manageable size.

Maps are usually drawn on a flat surface to a known scale. The science of map making, known as *cartography,* is now intimately related to environmental monitoring because maps are generated from remote sensing, including aerial photography and satellites, as well as from field surveying and observations. Therefore maps build on themselves by facilitating the location and study of land features that in turn can be used to generate new or more precise maps.

The focus of this chapter is the use of maps by environmental scientists. Basic concepts of scale and systems of map coordinates such as latitude and longitude, the Public Land Survey, and the Universal Transverse Mercator grid system are considered. Particular emphasis is given to the use and application of topographic and soil maps.

PRINCIPLES OF MAPPING

A variety of information about many types of features found on the earth's surface are shown on all maps. Those maps that do not attempt to show the relief features in measurable form are called *planimetric maps.* Planimetric base maps are often used to provide the background framework for *thematic maps,* which present information about some special subject, as for example with human population maps. Maps that show the shape and elevation of the terrain are generally called *topographic maps* to distinguish them from the planimetric type. Within the categories of either planimetric or topographic maps, several types can be distinguished. *Engineering maps* are detailed maps, sometimes called *plans,* used for guiding engineering projects, such as the construction of bridges or dams. Of particular interest to environmental scientists are *flood-control maps,* which are used to provide information about areas subject to flooding; *vegetation maps,* which show the distribution of plant communities; *climate maps,* which show the distribution of climatic parameters such as mean annual precipitation and temperature; *geological maps,* which show surface geologic features; and *soil maps,* which show the distribution of soils.

In the United States, Canada, and other English-speaking countries, most maps produced to date have used the English system of measurement. Distances are measured in feet, yards, or miles; elevations are shown in feet; and water depths are recorded in feet or fathoms (1 fathom = 6 ft). In 1977, in accordance with national policy, the U.S. Geological Survey (USGS) formally announced its intent to convert all its maps to the metric system. The conversion from English to metric units will take many decades in the United States. The linear units used in the English system and the metric system and means of conversion from one to another are given in Table 5.1.

Map Scale

The scale of a map is the ratio between map distances and earth distances. This relationship is expressed in one of three different, but equivalent ways: (1) as a word statement, (2) as an arithmetic ratio, or (3) as a graphic symbol.

A *word statement* of a map scale gives a quick idea of size relationships, but it is in a form that is awkward to use for many applications. Typical word statements might say, for example, that a given map is at a scale of "1 inch to the mile," which means that 1 inch on the map represents 1 mile on the earth's surface.

A *representative fraction (R.F.) scale* is the ratio between the map distance and the ground distance between equivalent points. It is expressed as a true ratio, such as 1:100,000, or as fraction, such as 1/100,000. Both notations have the same meaning. In this example, one unit of distance on the map represents 100,000 of the same unit on the earth's surface. A representative fraction is "unit free." That is, the ratio between map-distance and

TABLE 5.1

Units of Measurement in the English and Metric Systems and the Means of Converting One to the Other

---

1. English units of linear measurement

12 inches	=	1 foot
3 feet	=	1 yard
1 mile	=	1760 yards, 5280 feet, 63,310 inches

---

2. Metric units of linear measurement

10 millimeters	=	1 centimeter
100 centimeters	=	1 meter
1000 meters	=	1 kilometer

---

3. Conversion of English units to metric units

Symbol	When you know	Multiply by	To find	Symbol
in.	inches	2.54	centimeters	cm
ft.	feet	30.48	centimeters	cm
yd.	yards	0.91	meters	m
mi.	miles	1.61	kilometers	km

---

4. Conversion of metric units to English units

Symbol	When you know	Multiply by	To find	Symbol
mm	millimeters	0.04	inches	in.
cm	centimeters	0.4	inches	in.
m	meters	3.28	feet	ft.
m	meters	1.09	yards	yd.
km	kilometers	0.62	miles	mi.

earth-distance values remain the same, regardless of whether the measurements are expressed in feet, miles, meters, kilometers, or any other unit of distance. Another way of saying this is that both portions of the ratio (or the numerator and denominator of the fraction) are the same, always expressed in the same units of measurement.

A *graphic scale* is a line drawn on a map and subdivided into units appropriate to the scale of the map (Figure 5.1). A graphic scale allows a distance measured on the map to be translated directly into earth distance by comparison with the scale. Alternatively, the graphic scale allows the transfer of a desired earth distance directly onto the map. Graphic scales on a single map often include a variety of scale units such as kilometers, miles, or feet. This arrangement allows measurements to be made directly in the desired units, without the necessity of converting from one unit to another. Graphic scales are particularly useful because they change size in proportion to map size. If a map is photographically enlarged or reduced, the scale is altered proportionately. For this reason, neither the original word statement nor the R.F. scale relationship is valid on the reproduced map. If a graphic scale is included with the rescaled map, its length changes in proportion as the map, and it continues to be correct.

It is important to distinguish between small scale and large-scale maps. Small-scale maps are so named because the fraction represents a small numerical value (e.g., 1/100,000), whereas a large-scale map indicates a fraction with a larger numerical value (e.g., 1/10,000). Thus, small-scale maps cover a larger area and show less detail than large-scale maps.

It is often helpful to convert a map scale from one to another. For example, conversions are made from a fractional scale to a verbal or graphic scale or from verbal to fractional. Conversions involve simple problems in arithmetic.

Examples of Typical Conversions:

1. Convert an R.F. of 1:62,500 to a verbal scale in terms of inches per mile. In other words, how many miles on the ground are represented by 1 inch on the map? One unit on the map equals 62,500 to express the R.F. as an equation of what it actually means. Because the interest is in 1 inch on the map, we can substitute inches for units in the above equation. Thus:

1 in. on the map = 62,500 in. on the ground   (Eq. 5.1)

The left side of the equation is now complete because we initially wanted to know the ground distance represented by 1 inch on the map. We can next convert the right side of the equation into miles. Because there are 5280 feet in a mile and 12 inches in a foot:

$$5280 \times 12 = 63,360 \text{ in. in 1 mile}$$

If we divide 62,500 inches by the number of inches in a mile, we arrive at:

$$\frac{62,500}{63,000} = 0.986 \text{ miles}$$

Thus, the verbal scale can be expressed as 1 in. = 0.986 mile or approximately 1.0 mile.

2. On a certain map, 5.6 in. is equal to 2.12 miles on the ground. Express the verbal scale as an R.F. By definition:

$$\text{R.F.} = \frac{\text{distance on the map}}{\text{distance on the ground}} \quad \text{(Eq. 5.2)}$$

Substituting the appropriate values we get:

$$\text{R.F.} = \frac{5.6 \text{ in. on the map}}{2.12 \text{ miles on the ground}}$$

Next, change the denominator of the equation to inches:

$$2.12 \times 63,360 = 134,323 \text{ inches}$$

Then

$$\text{R.F.} = \frac{5.6 \text{ in. on the map}}{134,323 \text{ in.}}$$

To express the numerator as unity:

$$\text{R.F.} = \frac{5.6}{134,323} = \frac{1}{23,986}$$

or rounded off to $\frac{1}{24,000}$ or $\frac{1}{2.4 \times 10^4}$

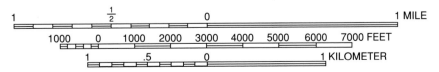

FIGURE 5.1 Bar scales from a recent U.S. Geological Survey topographic map.

TABLE 5.2
Scales Commonly Used with U.S. Geological Survey Topographic Maps

Series	Scale	1 inch represents	Standard quadrangle size (latitude-longitude)	Quadrangle area (square miles)
7.5 min	1:24,000	2000 ft	7.5 × 7.5 min	49 to 70
15 min	1:62,500	about 1 mile	15 × 15 min	197 to 282
U.S. 1:250,000	1:250,000	nearly 4 miles	1E × 2E or 3E	4580 × 8669
U.S. 1:1,000,000	1:1,000,000	nearly 16 miles	4E × 6E	73,734-102,759

Table 5.2 shows the commonly used scales of topographic maps published by the USGS. These maps, referred to as *quadrangle maps,* cover four-sided areas bounded by parallels of latitude and meridians of longitude (discussed later). *Quadrangle size is given in minutes or degrees.* The usual dimensions of quadrangles are 7.5 by 7.5 minutes, 15 by 15 minutes, and 1 degree by 2 or 3 degrees. Map series are groups of maps that conform to established specifications for size, scale, content, and other elements.

LOCATIONAL AND LAND-PARTITIONING SYSTEMS

LOCATION

Because topographic maps are drawn with great accuracy based on field surveying, they are most useful in determining the precise location of features on the earth's surface. For example, the precise location of a sampling or monitoring site can be readily obtained on large-scale maps. There are three coordinate systems used on USGS maps in the United States: the latitude and longitude system, the Public Land system, and the Mercator grid system.

LATITUDE AND LONGITUDE SYSTEM

The earth's surface is arbitrarily divided into a system of reference coordinates called *latitude* and *longitude*. This coordinate system consists of imaginary lines on the earth called *parallels* and *meridians*. Both of these can be described by assuming the earth to be represented by a globe with an axis of rotation passing through the north and south poles. Meridians are labeled according to their position, in *degrees,* from the zero meridian, which by international agreement passes through Greenwich, near London, England. The zero meridian is commonly referred to as the Greenwich or prime meridian. If meridianal lines are drawn for each 10 degrees in an easterly

direction (toward Asia) and in a westerly direction from Greenwich (toward North America), a family of great circles will be created. Each one of the great circles is labeled according to the number of degrees it lies east or west of the Greenwich, or zero, meridian. The 180 west meridian and the 180 east meridian are one and the same great circle, which constitutes the International Date Line.

Another great circle passing around the earth midway between the two poles is the equator. It divides the earth into the Northern and Southern Hemispheres. A family of lines drawn on the globe parallel to the equator constitute the second set of reference lines needed to locate a point on the earth accurately. These lines form circles that are called *parallels of latitude* and are labeled according to their distances in degrees north or south of the equator. The parallel that lies halfway between the equator and the North Pole is latitude 45 north.

The position of any point on the globe can be accurately designated by expressing the coordinates of intersected meridians and parallels in terms of *degrees, minutes, seconds,* and *fractions of seconds.* The exact location of the Washington Monument, for example, is 38° 53′ 21.681″ north latitude, 77° 02′ 07.955″ west longitude.

Meridianal lines converge toward the North or South Poles from the equator. Latitudinal lines, on the other hand, are always parallel to each other. Thus, the area bounded by parallels and meridians is not a true rectangle. USGS quadrangle maps are also bounded by meridians and parallels, but on the scale at which they are drawn, the convergence of the meridianal lines is so slight that maps appear to be true rectangles.

U.S. PUBLIC LAND SURVEY

For purposes of locating property lines and land descriptions on legal documents, the U.S. Public Land Survey (USPLS) is used throughout much of the central and western United States. In this system an initial point was first established for the particular region to be surveyed. Astronomical observations were used to determine the location of this initial point. A north-south principal meridian was first established, and an east west baseline was set at right angles to it, with both lines passing through the initial point. Thirty-four principal meridians and primary baselines have been established in the United States. In the latitude/longitude system, the earth's geographical grid is based on the Greenwich meridian (0 longitude) and equator (0 latitude). In the USPLS system each of the 32 survey areas is based on its own 0 principal meridian. For example, the San Bernardino Principal Meridian and the Gila and Salt River Principal Meridian were established for southern California and Arizona, respectively.

Starting with these meridians and base lines, the land is divided into congressional townships, six miles square. Townships are further divided into 36 sections, each one mile square (Figure 5.2). Sections are numbered in consecutive order, from one to 36, beginning with 1 in the *northeast* corner (Figure 5.2). Each section contains 640 acres.

Rows or tiers of townships running east and west are numbered consecutively north and south of the baseline (e.g., T2N), whereas rows, called *ranges,* running north and south are numbered east and west of the principal meridians (e.g., R16E, R4W).

For purposes of locating either cultural or natural features in a given section, an additional convention is used. This consists of dividing the section into quarters

called northeast one-quarter (NE ¼), the southwest one-quarter (SW ¼), and so on. Sections may also be divided into halves such as the north half (N ½) or west half (W ½). The quarter sections are further subdivided into four more quarters or two halves, depending on how refined one wants to make the description of a feature on the ground. For example, an exact description of the 40 acres of land in the extreme southeast of section 24 in Figure 5.2 would be as follows: SE ¼ of the SE ¼ of Section 14, T1S, R2W.

UNIVERSAL MERCATOR GRID SYSTEM

Although latitude and longitude and the Public Land Survey are the most often used system for determining locations on the earth, a third method, which is somewhat more complicated, called the *Universal Tranverse Mercator (UTM) Grid System,* is also shown on most topographic maps published by the USGS. This is based on the transverse Mercator projection and covers the earth's surface between 80 south and 84 north latitudes.

The UTM system involves establishing 60 north-south zones, each of which is six degrees longitude wide. In addition, each zone overlaps by two degrees into adjoining zones. This allows easy reference to points that are near a zone boundary, regardless of which zone

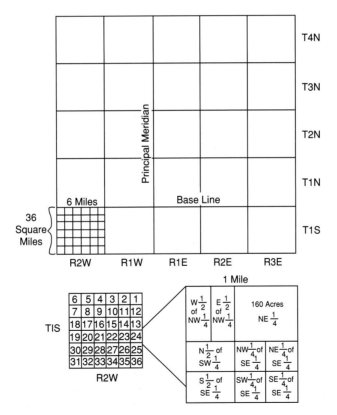

FIGURE 5.2 Townships, ranges, and further subdivisions of the U.S. Public Land Survey.

is in use for a particular project. Each UTM zone is assigned a number. Grid zone 1 is assigned to the 180 meridian with zones numbered consecutively eastward. A false origin is established 500,000 m west of the central meridian of each UTM zone. In the Northern Hemisphere, this origin is on the equator; in the Southern Hemisphere it is 10,000,000 m south of the equator. A square grid, with lines extended north and east from the origin, provides a basic locational framework. With this framework, any point on the earth's surface, within each zone, has a unique coordinate. UTM coordinates are shown on the edges of many USGS topographic maps.

Although the latitude and longitude system is used for locating points on a sphere, the UTM system is a location grid that uses straight lines that intersect at right angles on the flat (plane) map sheet. Each UTM grid area is divided into true squares, each consisting of 100,000 m². Because meridians of longitude converge at the poles, the straight lines of UTM do not follow the lines of longitude on a map.

TOPOGRAPHIC MAPS

Topographic maps have long been an important tool in various disciplines within the earth and biological sciences. A topographic map is a graphical representation providing a three-dimensional configuration of the earth's surface. Thus, these maps provide means for determining the nature of landforms, hydrology, and in some cases the vegetation of an area. Such maps are extremely useful in environmental monitoring; not only can sampling sites be located and selected, but also the physical features of an area can be interepreted and the area's impact or potential impact on the system monitored.

GENERAL NATURE

The USGS, a unit of the Department of the Interior, is the main agency that makes topographic maps in the United States. These maps follow a similar or standard format.

Features shown on topographic maps may be divided into three groups: (1) relief, which includes hills, valleys, mountains, and plains, for example; (2) water features, including lakes, rivers, canals, swamps, and the like; and (3) culture (i.e., human works such as roads, railroads, land boundaries, or similar features). Relief is printed in brown, water in blue, and culture, including geographical names, in black. On maps that are more recent, major roads are shown in red. In addition, urban areas are often shown in salmon, and forests and woodlands in green. A number of symbols are used to represent roads, trails, railroads, bridges, dams, buildings, transmission lines, mines, boundaries, elevations, contours, water features, and types of vegetation, among others.

The direction of a topographic map sheet is indicated by a diagram (Figure 5.3) located within the lower margin of the map sheet. Several "norths" are indicated: (1) true north with a star shows the direction of the earth's north pole and (2) magnetic north, indicated with an MN, is the direction in which a compass needle points. Magnetic north is located northwest of Hudson Bay, about 500 miles from the geographical (true) North Pole. Because the earth's geographical North Pole is different from its magnetic, only in a few areas do the two north directions align. One of these areas is located in the central United States and runs from Wisconsin south through Mississippi. The magnetic north arrows show both the direction that magnetic north is from true north, and the number of degrees it varies from true north. The difference between magnetic north and true north is called the *magnetic declination*. Because

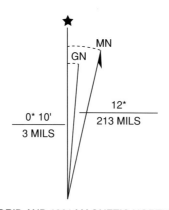

UTM GRID AND 1981 MAGNETIC NORTH
DECLINATION AT CENTER OF SHEET

FIGURE 5.3 Declination diagram. The star indicates true north. GN stands for grid north, and MN and the arrow designate magnetic north. (Empire Ranch Quadrangle, Arizona, 1:24,000.)

the earth's magnetic pole moves slowly but constantly over a small area, the magnetic declination of a place can vary from year to year. Consequently, the year of the declination is also given. On newer topographic maps a third north, called *grid north (GN)*, is also indicated. Grid north shows the relation between true north and the north-south lines of the UTM grid system used for location.

CONTOUR LINES

Topography is the configuration of the land surface and is shown by means of contour lines. A *contour line* is an imaginary line on the earth connecting points of equal elevation above an accepted base level, usually mean sea level. Visually, a contour line would look like a line running around a hill or up a valley. It is actually the line formed by the intersection of an imaginary horizontal plane and the surface of the earth (Figure 5.4). The first contour line (the line formed by the intersection of the land and sea) is zero feet above sea level. The contour interval is the difference in elevation of any two adjacent contour lines. Contour intervals are given in the lower margin of a map sheet below the graphic scale. Two basic contour intervals can be used on one map when an area contains abrupt contrasts in relief. In such cases, a closer interval is used in the areas that have relatively steep terrain, and a wider one is used in the flat areas.

Contour lines are brown on the standard USGS maps. The contour interval is usually constant for a given map and may range from 5 ft for flat terrain to 50 or 100 ft for a mountainous region. Usually, every fifth contour line, called the *index contour*, is printed in a heavier line than the others and has numbers inserted at intervals along its length to indicate the elevation it represents (Figure 5.5). Regular contours or intermediate contours spaced at the normal interval and printed in a finer line weight lie between the index contours. Supplementary contours are additional contours, usually drawn at intervals that are some fraction of the basic contour interval, which is often one half. Supplementary contours are appropriate in areas of flat terrain, where contours drawn at the basic interval would be spaced relatively far apart. They are usually drawn as dashed or dotted lines to distinguish them from contours drawn at the basic interval.

When an area lies at a lower elevation than all the surrounding terrain, it forms a depression. A depression that fills with water forms a lake. The land surface of a depression that is not filled with water, however, is mapped with contours on its surface. These depression contours are distinguished from the regular contours by adding short ticks at right angles to the contour line known as *hachures*. The ticks point downslope toward the bottom of the depression (Figure 5.6).

When contours are drawn to represent the bottom configuration of a water body, they are called *depth curves*, or *isobaths*. Depth curves are measured down from the specific water surface and therefore are not directly related to the overall map datum. Depth curves in ocean areas are frequently the exception because they are normally related to the mean low-tide level.

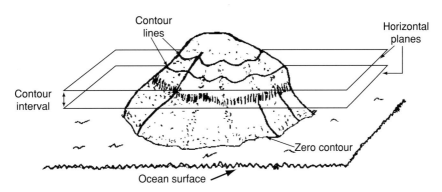

FIGURE 5.4   Schematic showing the relation between horizontal planes and contour lines.

FIGURE 5.5   Contour lines are shown on most topographic maps as brown lines. Heavier lines (every fifth contour) are used to designate major units of elevation.

FIGURE 5.6   Enclosed depressions (e.g., sink holes, kettles) are shown by hachure lines on contours. Hachures point towards the center of the depression.

The following principles may be useful in the use and interpretation of contour lines:

1. All points on a contour line have the same elevation.
2. Contours separate all points of higher elevation from all points of lower elevation.
3. Contour lines never intersect or divide; they merge at a vertical or overhanging cliff.
4. Every contour closes on itself either within or beyond the limits of the map.
5. Contours that close within a relatively small area on a map represent hills or knobs.
6. Steep slopes are shown by closely spaced contours, gentle slopes by widely spaced contours.
7. Uniformly spaced contour lines represent a uniform slope.
8. When contour lines cross streams, they bend upstream; that is, the segment of the contour line near the stream forms a V with the apex pointing in an upstream direction.
9. The difference in elevation between the highest and the lowest point of a given area is the maximum relief of that area.

Refer to Figure 5.7 for the relationship of contour lines to topography.

The ability to read and interpret contours is largely a matter of practice and of recognition of certain common formations. Perhaps the simplest formations are those of hill slopes. In addition to the steepness or gentleness of a slope being determined by the spacing of the contour lines (principle 6 in the list) variations in steepness can be shown. For example, Figure 5.8 shows an even slope *(A)* a slope steeper at the top and gentler at the bottom representing a concave slope *(B),* and a slope gentler at the top and steeper at the bottom representing a convex slope *(C).* A valley is a low area of ground, usually cut by a river or stream between areas of higher ground. As shown in Figure 5.9A from point "O," located in a valley bottom, there will be one downhill direction (downstream), and three uphill directions, steeply up from the stream on each side of the valley and more gently towards the source of the stream. The characteristic shape for a valley, as shown in Figure 5.9B, is a V, but there are many variations in valley form. Figure 5.9C through H are some of the main ones. In all of these examples, natural streams would flow in the bottoms of the valleys, bisecting the angle formed by the V of the contours, except in the case of the through valley.

THE USE OF SYMBOLS IN MAPPING

These illustrations show how various features are depicted on a topographic map. The upper illustration is a perspective view of a river valley and the adjoining hills. The river flows into a bay which is partly enclosed by a hooked sandbar. On either side of the valley are terraces through which streams have cut gullies. The hill on the right has a smoothly eroded form and gradual slopes, whereas the one on the left rises abruptly in a sharp precipice from which it slopes gently, and forms an inclined tableland traversed by a few shallow gullies. A road provides access to a church and two houses situated across the river from a highway which follows the seacoast and curves up the river valley.

The lower illustration shows the same features represented by symbols on a topographic map. The contour interval (the vertical distance between adjacent contours) is 20 feet.

FIGURE 5.7   Topography expressed by contour lines. (Source: USGS.)

Contours in a V shape may also mean a projection from a hill, known as a *spur*. This is shown in Figure 5.10A where from point "P" there is only one uphill direction (to the top of the hill or ridge) and three downhill directions. An isolated hill is illustrated in Figure 5.10B. A long narrow hill or range of small hills is known as a ridge (Figure 5.10C). A high break in a ridge is called a *saddle* (Figure 5.10D). If the break is low enough to allow easy communication between the lowlands on each side, it is referred to as a *pass*.

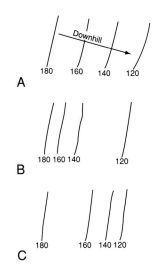

FIGURE 5.8 Use of contours to show types of slopes and gradients. (**A**) Even slope. (**B**) A slope steeper at the top and gentler at the bottom, a concave slope. (**C**) A slope gentler at the top and steeper at the bottom, a convex slope.

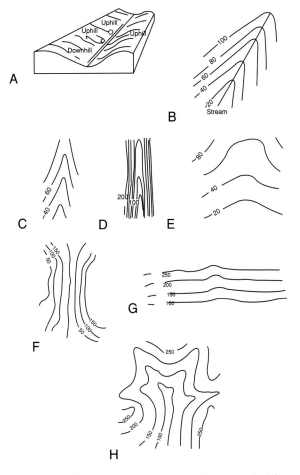

FIGURE 5.9 Use of contours to represent valleys. (**A**) Theoretical example in a block diagram. (**B**) Theoretical example in contours. (**C**) A narrow valley. (**D**) A gorge. (**E**) A wide valley. (**F**) A through valley. (**G**) A wide valley. (**H**) A meeting of valleys.

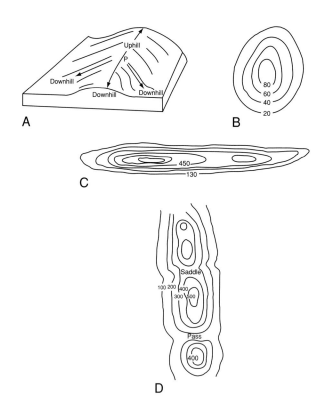

FIGURE 5.10 Use of contours to represent hills. **(A)** Theoretical example in a block diagram. **(B)** An isolated hill. **(C)** A ridge. **(D)** A ridge with a saddle and a pass.

In addition to contour lines, the heights of many points on a map, such as road intersections, summits of hills, and lake levels, are shown to the nearest foot or meter. More precisely located and more accurate in elevation (as determined by surveying) are benchmarks. Benchmarks are indicated by brass plates permanently fixed on the ground. On a map a special symbol is usually used to distinguish it from other types of spot elevations and includes the elevation (e.g., BM 5653).

The Mapping Agency of the Department of Defense also produces topographic maps, as well as aeronautical charts, nautical charts, and precise positioning data and digital data for military purposes. The topographic maps are similar to those produced by the USGS, except that the UTM grid system, with a few modifications, is used for locations and distances are measured with metric units.

APPLICATION

DETERMINATION OF ELEVATION

On contour maps the elevations of benchmarks and other spot locations are often given exactly. Elevations at other locations must be estimated. If the point whose elevation is to be estimated falls on a contour line, the elevation of that contour is the best estimate of the point. If the point falls between contours, however, a technique called *linear interpolation* may be used to arrive at the estimated elevation.

Linear interpolation is based on the assumption that the slope between contours is constant, an assumption that may not actually be correct, and that introduces potential errors into the estimation. In the absence of additional information, linear interpolation is a logical procedure that involves using rules of proportionality for the estimate. If, for example, a point is located halfway between two contour lines, the point's elevation is estimated as being halfway between the elevation values of those lines. The procedure for estimating the elevation at any location is to first draw a line through the point whose elevation is to be determined in the direction of maximum slopes, that is, at right angles to the contour lines. Next, determine the distance from the point to the lower contour, as a proportion of the total distance between the two contours. Then apply that proportion to the contour interval. Finally, add the resulting fraction of the contour interval to the lower contour elevation to obtain the point's estimated elevation. As an example, on a map with a contour interval of 20 ft, the distance of a point from the lower contour line is 10 cm, and the distance between the lower contour and the next higher contour line is 30 cm. If the elevation of the lower contour line is 820 ft, then the estimated elevation is:

$$\text{Elevation} = 820 + (10/30)(20) = 826.7 \text{ ft} \quad (\text{Eq. 5.3})$$

SLOPE MEASUREMENT

Slopes are expressed in three alternate ways: (1) as a gradient, (2) as a percent, or (3) in degrees. The same slope can be expressed in any of these three forms.

The gradient of a slope is the fraction that represents the ratio between the vertical distance and the horizontal distance. Both distances must be stated in the same units. For this ratio to be computed, a starting point and ending point are located on the map. The elevation of each point and the horizontal distance between them are determined. The slope is then written as the ratio between the vertical and horizontal distances. If the vertical distance is 35 ft, for example, and the horizontal distance is 350 ft, the slope is 35/350, which reduces to 1/10 or 10 units (feet, meters, etc.) of horizontal per unit increase in elevation.

Slope is expressed as a percent by simply determining the increase in elevation per 100 units of horizontal distance. This is determined by dividing the elevational increase by the distance and multiplying by 100. In the previous example:

$$\text{Percent slope} = 35/350 \times 100 = 10\% \quad (\text{Eq. 5.4})$$

The third means of expressing the slope is as an angle in degrees from the horizontal. This is obtained by dividing the increase in elevation by the horizontal distance. The numerical value is the tangent of the slope angle from which the degrees of slope can be obtained.

TOPOGRAPHIC PROFILE

Topographic maps represent a view of the landscape from above, and even though contour lines show the relief of this landscape, it is often desirable to obtain a better picture of the actual shape of the land in an area. This can be achieved by constructing a topographic cross section (i.e., a cross section of the earth's surface along a given line). Profiles may be constructed quickly and accurately from topographic maps with graph paper.

The procedure for constructing a profile is as follows:

1. Examine the line of profile on the map and note the elevations of the highest and lowest contour lines crossed by it. This will determine the number of horizontal lines needed by the profile land and will help in determining a convenient and meaningful vertical scale. Select the vertical scale to be used. Label on graph paper the horizontal lines that correspond to the elevation of each contour line (use only alternate or heavy contour lines if they are closely spaced).
2. Place the edge of the graph along the profile line. Opposite each intersection of a contour line with the profile, place a short pencil mark on the graph paper and indicate the elevation of the contour. Also, mark the intersections of all streams and depressions.
3. Project each marked elevation perpendicularly to the horizontal lines of the same elevation on the graph paper.
4. Connect the points with a smooth line and label prominent features.
5. State the vertical exaggeration. Profiles are drawn with equal vertical and horizontal scales to avoid distortions. However, it is usually necessary to use a vertical scale several times larger than the horizontal scale. This is done to bring out topographic details in areas of low relief. The vertical exaggeration is calculated by dividing the horizontal scale by the vertical scale. Thus a map with horizontal scale of 1:24,000 and a vertical scale of 1 in = 400 ft (1:4800) = 5X.

An example of a profile is given in Figure 5.11.

DELINEATION OF A WATERSHED

The delineation of a watershed of a stream may be useful if the stream is being monitored when the pollution or potential pollution of the stream is from either point or nonpoint sources. The extent of the watershed can be delineated by drawing a line along the ridge tops or drainage divides adjacent to the stream. The closed ends of contours that represent ridges point in the downslope direction. An example is given in Figure 5.12.

DISTANCE MEASUREMENT

The simplest method of determining the straight-line distance between two points on a map is to measure the map distance directly with a ruler, a pair of dividers, or marks on the edge of a strip of paper. The earth distance is then determined by checking the measurement against the appropriate bar scale. If the measured distance exceeds the length of the bar scale, several partial measurements are added together.

The advantage of the paper strip method is that distances other than straight-line distances are easily accommodated. The distance along an irregular route is measured by individually marking sections of the route,

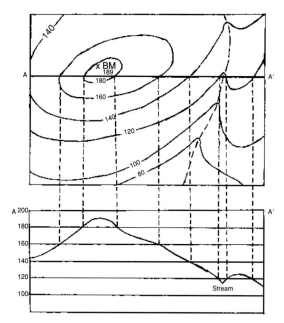

FIGURE 5.11   A profile generated from a topographic map.

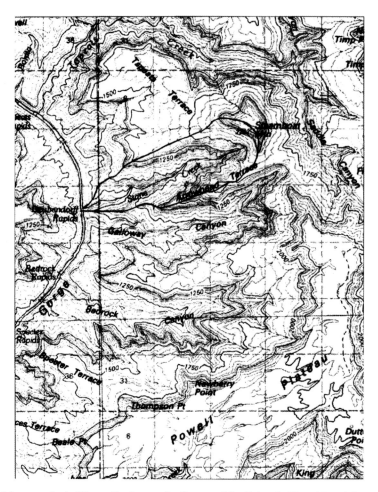

FIGURE 5.12   Delineation of the watershed of Stone Creek, a small tributary of the Colorado River (Grand Canyon, Arizona, 1:100,000, 1984). (Source: USGS topographic maps.)

end to end, on the paper strip. These sections are kept as short as necessary to ensure that the length of each one closely approximates the curved-line distance between the marked points. Once the total length of the route is marked on the strip, the distance is scaled by the same technique as is used in straight-line distances.

Another method of measuring distances on a map is to use a special device called a *map measurer*. This device consists of a tracing wheel connected to scaled dial. The tracing wheel rotates as the device is guided along the route being measured. The number of inches or centimeters that the device travels is indicated on the dial.

The method just described and other applications can be carried out with computer digitizing techniques and the appropriate graphical software. USGS topographic maps are currently available for every state, together with supporting software that enables various operations to be performed. For example, the latitude and longitude coordinates and elevation of any point on the map can be easily determined, distances and slopes can be measured, areas can be calculated, and profiles can be easily drawn and displayed. In addition, scales and slopes can be easily changed by zooming in or out, routes can be traced, custom text added and a custom photo-quality map can be printed on ink-jet or laser printer or plotter.

SOIL SURVEY MAPS

Soil maps are the result of soil surveys. Soil surveys classify, map, and describe soils as they appear in the field. In the United States soil surveys are carried out under the auspices of the National Cooperative Soil Survey Program (NCSS), a joint effort of the U.S. Department of Agriculture (USDA), National Resource Conservation Service (NRCS), formerly known as the Soil Conservation Service and other federal agencies such as the U.S. Forest and the State Agricultural Experiment Stations. Most of the actual surveying is done by soil scientists of the NRCS. Published soil maps are accompanied by soil survey reports, which provide scientific information useful to all who interact with soil resources.

A soil survey report lists the kinds of soils found in a county (or other area covered by the map and report). The soils are described in terms of location, profile characteristics, relation to each other, suitability for various uses, and needs for particular kinds of management. Each report contains information about soil morphology, soil genesis, soil conservation, and soil productivity. This allows the user to determine the suitability of a soil for such uses as crop production, pasture or rangeland, growth of trees, various engineering interpretations or wildlife management. Such information can be of direct or indirect interest to those concerned with environmental problems and monitoring. Of direct relevance is information about the suitability of soils for septic tank installation, for example. Basic soil characteristics can be used to assess an area for its susceptibility of being affected by hazardous materials and for location of monitoring sites. For example, it would be very undesirable to dispose of hazardous material on coarse-textured, particularly gravelly or sandy soils, that have very rapid permeabilities. Components of such materials could readily reach and contaminate groundwater.

A detailed soil map, based on the original soil survey, usually consists of numbered sheets in the back of the soil survey report, printed on aerial photographs. The objective of soil mapping is to separate areas of the landscape into segments that have similar use and management requirements. The maps delineate areas occupied by different kinds of soil, each of which has a unique set of interrelated properties characteristic of the material from which it formed, its environment, and its history. The soils mapped by the NRCS are identified by names that serve as references to the national system of soil taxonomy. Basic to the soil map is the map unit, which is a collection of areas, defined and named the same in terms of their soil components. Each map unit differs in some respects from all others in a survey area. Each unit is uniquely identified on a soil map. Because soils are natural objects, however, a mapping unit will typically include a range of properties characteristic of all natural-occurring entities.

Soil survey reports also contain a small-scale map called a *general soil map*. The general soil map contains soil map units or soil associations that are generated by combining the map units of the original detail soil maps. A *soil association* is a group of soils or miscellaneous areas geographically associated in a characteristic repeating pattern and defined as a single map unit. A general soil map shows broad areas that have a distinctive pattern of soils, relief, and drainage. The general soil map can be used to provide an overview of the soil distribution of an area and to compare the suitability of large areas for general land uses. Descriptions of the soil associations of the general soil maps are given in the text of the soil survey report.

A user interested in the soils of a particular area or site must first locate the area or site on the map and determine what soils are there. Index sheets are included to help one find the right portion of the map. The map legend gives the soil name for each symbol. Once the soils of a given area or site are determined, the user can then refer to the descriptions and appropriate tables to learn about the properties of the soils. Additional tables can be referred to by the user to obtain more information about the uses, limitations, and potential hazards associated with the soils.

Refer to Figure 5.13 for an example of the use of a soil map. This figure shows a portion of the detailed soil map from the *Soil Survey of Gila Bend-Ajo Area, Arizona, Parts of Maricopa and Gila Counties*. The area shown is northwest of Gila Bend. Oatman Flat contains soil mapping unit 41, which is the Indio silt loam. This unit is dominated by the Indio silt loam, which is described in the soil survey report as consisting of deep, well-drained soils on floodplains of the Gila River. The permeability is moderate, the water holding capacity is high, and the soils are generally subject to occasional periods of flooding. The soils have severe limitations for septic tank absorption fields, sewage lagoon areas, and areas for landfills because of potential for flooding.

AERIAL PHOTOGRAPHY

In addition to providing a base for soil and other applied maps, aerial photography allows the rapid acquisition of accurate, up-to-date information about the size, location, and character of objects on the earth's surface. It can also reveal the configuration of the surface itself. For these reasons, aerial photography is essential to modern topographic mapping. Vertical aerial photographs for mapping purposes are taken in a series, with each photo overlapping the preceding photo. Two adjacent overlapping photos are called *stereo pairs*. In use, stereo pairs are placed in a simple instrument, called a *stereoscope*, that allows one eye to see one view and the other eye the other. Parallax (the difference in two aspects of the same object) is thus obtained, and the result is an apparent three-dimensional views of the earth's surface, a stereoscopic image. Aerial photos used in this manner can be valuable in themselves for environmental monitoring.

When stereo pairs are used for preparing topographic maps, they are viewed in a more sophisticated instrument, called a *stereo plotter*. In conjunction with aerial photography, ground control points are established that are easily identifiable, both on the ground and in aerial photos. A survey establishes the precise locations of the control points, and a map, called a *control diagram*, is prepared on the basis of the survey. This provides the base for a map, which is prepared in conjunction with the data obtained from aerial photo stereo pairs with a stereo plotter.

Aerial photographs obtained from aircraft have been used for the most part in map preparation. In addition, images are being obtained with earth-orbiting satellite

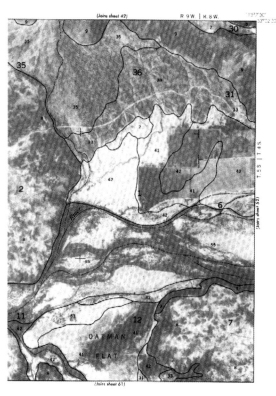

FIGURE 5.13 A portion of a detailed soil map showing an area along the Gila River northwest of Gila Bend, Arizona. (From U.S. Department of Agriculture and Natural Resources Conservation Service, 1997. Soil Survey of Gila Bend-Ajo Area, Arizona, Parts of Maricopa and Pima Counties. Issued May, 1997.)

images from space. These images, used in conjunction with computer technology, are increasingly being used to supplement the use of the more conventional maps. This topic is discussed in Chapter 11.

GLOBAL POSITIONING SYSTEM

The use and application of coordinate systems as described in the "Principles of Mapping" section have been based on geodetic and related field-surveying methods. With the development of new technologies, including computer technology, a new system of determining location has evolved known as the Global Positioning System (GPS). GPS is now used by researchers and field workers to obtain more accurate and real-time positional information. The GPS is an all-weather, space-based navigation system developed by the Department of Defense (DOD) to satisfy the requirements for the military forces to accurately determine their position, velocity, and time in a common reference system, anywhere on or near the earth on continuous basis. Because the DOD was the initiator of GPS, the primary goals were military ones. It is now widely available for civilian uses and is an important tool to conduct all type of land and geodetic control surveys. Relatively inexpensive GPS receivers are commercially available, with which the user can determine his or her position and track locations on a moving vehicle.

The GPS receiver can mathematically determine a precise geographical position by processing signals from four satellites simultaneously. Five basic steps are essential for the GPS receiver to obtain accurate coordinates. They include (1) trilateration, (2) distance measurements by radio signals, (3) the use of clocks and computers, (4) satellite locations, and (5) atmospheric correction.

Trilateration is the method by which the GPS receiver triangulates its location by simultaneously processing signals from four or more satellites. The first satellite establishes the receiver's position with a single sphere, which is the distance the radio signal travels. A second satellite refines the receiver's position to the intersection of two spheres. The third satellite corrects the receiver's position to three spheres, and finally, the fourth satellite resolves the receiver's position to a single point. A single point is acquired by solving mathematical equations with four variables (latitude, longitude, elevation, and time). Four satellites, one for each variable, are needed to triangulate a three-dimensional point.

The distance to the satellites must be known to obtain a GPS location. GPS determines the distance to the satellites with the travel time of the radio signal emitted from the satellites and the speed the radio signal travels. Clocks are needed to measure the travel time of the radio signal

transmitted from the satellite to ensure the locations recorded by the GPS receiver are accurate. Digital clocks in the receivers synchronize themselves with the atomic clocks in the satellites.

DOD computers at five monitoring stations around the world validate the accuracy of the clocks in all the satellites. If satellite clock errors are detected, the computers make orbital calculations and corrections as needed. A simplified GPS system in Figure 5.14 shows schematically the relationship between the satellites, the monitoring stations, and the receivers held by the users.

The satellites are in geosynchronous orbit with an orbital period of 12 hours (actually 11 hours and 58 minutes because it is based on the observation of the stars in sidereal time) in designated planar orbits at an altitude of approximately 12,600 miles. Together the GPS satellite orbits are so arranged that at least four satellites are in view above the horizon at any time. Although 24 satellites consisting of 21 working units and 3 spares were considered optimum, other satellites are being added. Figure 5.15 illustrates and describes the arrangement of the orbital satellites. Additional information about satellites and types of orbits can be found in Chapter 11. Knowing the exact satellite position is critical. The four DOD monitoring stations function as uplink installations, which means they are capable of transmitting data to the satellites, including positions as a function of time, clock

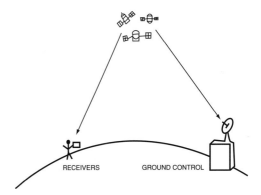

FIGURE 5.14 Simplified GPS system.

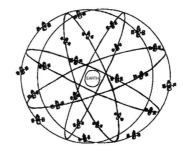

FIGURE 5.15 The satellites and spares are arranged in six nearly circular orbits. The orbits are tilted at an angle of 55 degrees relative the axis of the earth.

correction, and other broadcast message data. A fifth station monitors satellite performance. Atmospheric interference results when charged particles and water vapor in the earth's atmosphere affect the travel time of radio signals. The monitoring station computers can correct for these factors.

QUESTIONS

1. Compare a small-scale map with a large-scale map with regard to:

    a. Area covered
    b. Amount of detail
    c. Usefulness for environmental monitoring

2. In the public land coordinate system it is common for the ranges to be offset in an east-west direction as one moves north or south on a map. What is the reason for this?

3. Describe how a topographic map can be used for interpreting the physical features of an area that might be useful in an environmental monitoring system.

4. Describe the usefulness of delineating a watershed on a topographic map in evaluating potential pollution hazards.

5. Describe how a soil survey report and accompanying soil maps can be used to establish an environmental monitoring site

6. How can global positioning systems (GPSs) be used to supplement topographic and soil maps in environmental monitoring?

REFERENCES AND ADDITIONAL READING

Anonymous. (1981) *Topographc maps: a guide to map reading.* Australia Division of National Mapping, Canberra.

Anonymous. (1996) *Understanding GPS: principles and applications.* Artech House, Boston.

French, G.T. (1996) *Understanding the GPS: an introduction to the Global Positioning System: what it is and how it works.* Geo-Research, Bethesda, MD.

Gold, R.L. (1988) Procedures for determining drainage areas using a digitizer, digital computer, and topographic maps. *U.S. Geological Survey, Water Resources Investigations Report* 86–4083, 624–B.

Larijane, L.C. (1998) *GPS for everyone: how the global positioning system works for you.* American Interface Corp, New York.

Miller, V.C., and Westerback, M.E. (1989) *Interpretation of topographic maps.* Merrill Publishing, Columbus, OH.

Olson, G.W. (1981) *Soils and the environment: A guide to soil surveys and their applications.* Chapman and Hall, New York.

Pinzke, K. (1994) *Introduction to topographic maps.* (CD–ROM), TASA Graphic Arts, Inc. Albuquerque, NM.

Soil Survey Division Staff. (1993) Soil survey manual. U.S. Department of Agriculture Handbook No. 18.

Weiss, M.P. (1960) *Topographic maps and how to use them.* Earth Science Curriculum Reference Series: RS–7. Prentice-Hall, Englewood Cliffs, N.J.

6

GEOGRAPHIC INFORMATION SYSTEMS AND THEIR USE FOR ENVIRONMENTAL MONITORING

R. MACARTHUR

Until recently, geographic information systems (GISs) and related technologies such as global positioning systems (GPSs) and remote sensing were largely the instruments of a select group of researchers. Now, more environmental professionals have a better understanding and use of these tools. These professionals are building applications to help their own decision making and to facilitate interaction with the public. There are a few major categories of GISs applications, as shown in Box 6.1.

GISs are especially useful in environmental applications for several reasons. First, environmental issues are the subject of widespread interest and heated debate. As the public becomes more involved with environmental discussions, they also demand greater access to information. GISs focus on the use of maps that are understandable. Second, GISs can handle large amounts of disparate data and organize these data into topics or themes that represent the multiple aspects of complex environmental issues. Third, GISs serve as collaborative tools that promote interaction.

This chapter introduces GISs and attendant technologies. The maps are generated with ArcView GIS software from Environmental Systems Research Institute, the larg-

est manufacturer of GIS software for natural resource use. Important GIS concepts are illustrated with graphic examples with data from the Santa Rita Experiment Range south of Tucson, Arizona.

DESCRIPTION OF GEOGRAPHIC INFORMATION SYSTEMS

GEOGRAPHIC INFORMATION SYSTEMS COMPONENTS

GIS has two major components. One component is a database of information about a given location. The other component is a computer with cartographic (geocoordinates) display capabilities that is used to create maps from the database. The two components are linked so that when data are added to or changed in the database, the maps change too. In the example shown in Figure 6.1 the properties of soil type and vegetation are related to elevation ranges and represented as locations A, B, and C. For simplicity the spatial area (X,Y data) occupied by each location is not included in Figure 6.1. GIS tables usually contain multiple data sets, which are called *themes* or *layers*. These data are combined as shown in Figure 6.2.

GIS is built on the logic of common database programs (see Chapter 4). Typical databases are used to perform queries against tables of data. A typical query is, "Tell me all the people in this city that own their home, are at least 50 years old and make at least $30,000 per year." GIS performs this same kind of query, but adds a spatial component. Location is involved (e.g., "Show me [on a map] all the private residences owned by persons who are

---

BOX 6.1 *General Geographic Information Systems Applications*

- Building digital maps
- Inventorying resources, such as a forestlands or crops
- Spatial data modeling for land management and environmental decision making.

LOCATION	Area (x,y)	ELEVATION (z)	SOIL TYPE	VEGETATION
$A(x_a, y_a)$	--	300-500	Clay	Grass
$B(x_b, y_b)$	--	400-600	Loam	Pines
$C(x_c, y_c)$	--	100-300	Sand	Shrubs

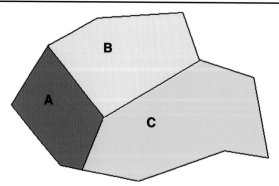

FIGURE 6.1 Graphical representation of soil data and vegetation locations.

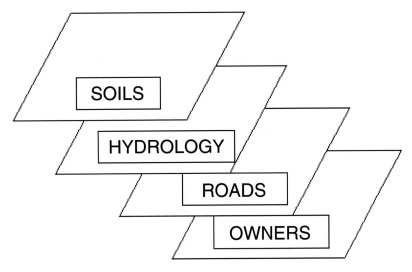

FIGURE 6.2 Combining data layers in a geographic information system.

over 50 and make at least $30,000 per year"). The added dimension, once again, is *where*. In GIS, all data are geo-referenced to some location.

HOW GEOGRAPHIC INFORMATION SYSTEMS REPRESENT THE REAL WORLD

There are two types of GIS programs—they are distinguished by how each represents real-world objects. *Raster-based* GISs represent data with dots (bits) on a map (Figure 6.3). Locations are represented on a grid of cells. In the figure, the surface of a lake is mapped to a raster GIS and results in the blue cluster of cells on the right. The cells do not exactly match the outlines of the lake. The "squaring off" of edges is called *aliasing* and is caused by imperfect sampling of smooth curves. Taking more samples (i.e., using smaller grids with more bits per unit area) results in less aliasing and a better representation of physical reality. This is called higher *resolution*, and it is a term used throughout the domain of spatial tech-

nology to measure error. Better resolution means less error, but it costs more to acquire. The accuracy of GIS maps depends on how many data samples are taken per unit area.

Vector-based GISs represent the physical features in nature as points, lines, or polygons (Figure 6.4). Note that there is also aliasing in vector GISs. In this case, the curvilinear boundaries of features like lakes or roads are measured with bits of straight line. The more little lines we use, the better we will capture the true nature of the feature.

BASIC GEOGRAPHIC INFORMATION SYSTEMS OPERATIONS

There are many GIS operations, but they can be grouped into three categories (Box 6.2).

These operations are explained in more depth later in this chapter, after the process of adding data into a GIS is explained.

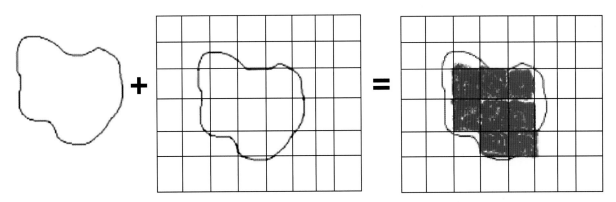

FIGURE 6.3 Raster or grid geographic information systems representation of a lake.

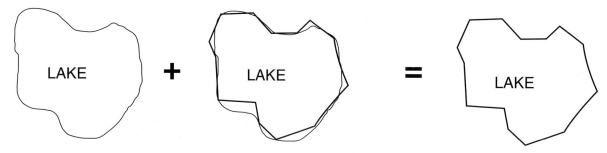

FIGURE 6.4   Vector geographic information systems representation of a lake.

GEOGRAPHIC INFORMATION SYSTEMS DATA

TYPES OF GEOGRAPHIC INFORMATION SYSTEMS DATA

GISs require two types of data: location data, including location and elevation, and thematic data or the properties at each location. GISs use the location data as a base map to tie together the layers of thematic data. Techniques for collecting and managing each type of data are discussed in the subsequent sections.

LOCATION DATA

Location data can come from paper maps, from field surveys with GPS and surveying instruments, or, more recently, from airborne photogrammetric technology such as orthophotos and LiDAR. These location data must be converted into computer maps. Therefore it is useful to talk about some considerations regarding maps and cartography in general.

BOX 6.2 *Geographic Information Systems Operations*

Reclassification: this means working with just one map, usually to perform an inventory or a count of some feature (e.g., how much of a given area is vegetated or what is the extent of this oil spill).

Overlay: combining two or more maps (e.g., lay a soils map on top of a vegetation map and see how many places there is a coincidence of a certain soil with a certain vegetative type).

Measuring distance and connectivity: (e.g., how far will owls have to travel over this clear-cut to get between stands of forest).

MAP CONSIDERATIONS

Maps are useful when they represent the world in standardized ways and use common terminology so that they can be related to one another. Three important terms are scale, projection, and datum (Box 6.3). See also Chapter 5 for a complete description of maps.

COLLECTING LOCATION DATA FROM THE AIR

As previously noted and as discussed in detail in Chapter 8, location data can be collected with GPS. Location data, including elevation, can also be collected from airborne sources. The latter includes stereoscopic aerial photographs, sonar (measures ocean floor topography), and airborne radar. These technologies send out a continuous burst of signals, measure the time it takes for the signals to bounce back, and use that information to map out terrain or buildings. Orthophotos are another means of collecting airborne location data (Fig. 6.5). Orthophotos are actually a series of overlapping photos that take pictures of the same ground features from several angles. These photos are then combined and the features are "orthogonalized" or straightened (like an orthodontist straightens teeth) so that everything in the photograph looks as though it is directly below you.

REPRESENTING ELEVATION WITH DIGITAL ELEVATION MODELS

Areal surface (x,y) locations are easily represented in two-dimensional maps but the elevation (z) value is not. A location map is incomplete without the elevation data that are often required for thematic data overlays. In GIS, elevation data are collected into what are called *digital elevation models (DEMs)*. DEMs are constructed by *interpolating* continuous elevations from data points taken all across a surface. Interpolation is a process of using algorithms to build a surface based on averages or best-guess estimates of points between the data that are actually taken in the field.

BOX 6.3 *Map Consideration Terms*

Scale: Maps are a reduced representation of the earth's surface. The extent to which they are reduced is the map's scale, which is expressed as a fraction. A common scale is 1/24,000, which means that 1 inch on the map represents 24,000 inches on earth. The larger the number in the denominator, the smaller the scale (a confusing use of terminology.) Large scale means close up, higher resolution, or more detail (see Chapter 5). Maps must be the same scale if they are going to be overlaid on one another.

Projection: The earth is not flat as seen on most maps, but a spheroid. Maps try to imitate the curvature of this spheroid with a projection. There are many different projections and their value varies depending on which part of the earth you are studying. The most widely used projection is Universal Transverse Mercator, or UTM, invented by the Flemmish geographer Gerhardus Mercator in the 16th century,

and standardized by the U.S. Army in the late 1940s. This projection most accurately portrays the earth's curvature in the middle, between 84° N and 80° S. It divides the world into 20 zones of latitude, north-south, each 8° high (except in the most northerly and southerly, which are 12°) and 16° wide (see Chapter 5).

Datum: Datums are a measurement of the earth's curvature. The earth is not perfectly spherical but slightly egg-shaped, a spheroid. Datums are estimates of this spheroid, and they represent complex coordinate systems generated from a set of control points, or known locations. These control points, called *monuments,* are embedded in sidewalks or streets across the country, and surveyors use them to get origin points for their measurements. They are a bright, brass color. The current standard that is used in the United States is the North American Datum of 1983 (NAD83).

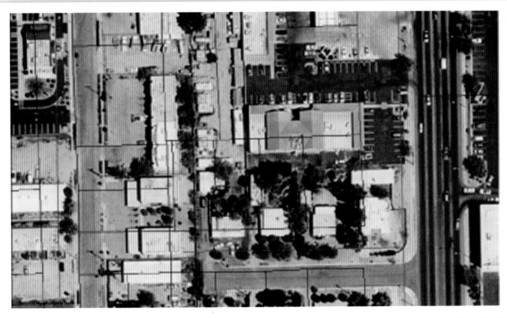

FIGURE 6.5 Orthophoto of Tucson, Arizona. Note that the parcel boundaries do not match up to the lots and will need to be corrected with "rubber-sheeting."

There are many different forms and names of DEMs. The most simple one is the altitude matrix DEM, which is just numbers on a chart. In Figure 6.6, different cells are numbered to represent different elevations, in this case seven categories, like a raster map. DEMs are all essentially altitude matrices like this, but they can be made to look like continuous surfaces with different interpolation techniques. Figure 6.7 illustrate a DEM shown in ArcView: if you click on any of the points of the map, you will get elevation data for a single cell (DEMs are raster-type GIS).

Another common DEM is the contour model, in which interlapping curved lines describe the ridges and valleys that make up terrain relief (Figure 6.8). Resolution is defined by the distance between these lines (e.g., 100 m, 200 ft, 10 ft). These lines are normally constructed from data points with interpolation. Below, the DEM for the Santa Rita Experiment Range has been turned into a contour map.

As was stated, DEMs are raster data, or grids, but they can also be converted to vector maps. The most common

	1	1	2	2	2	2	2	2
1	1	2	2	2	2	3	3	3
2	2	2	3	3	3	3	3	3
3	3	3	3	4	4	4	4	4
3	3	3	4	4	4	4	5	5
3	4	4	4	5	5	5	6	6
4	4	5	5	6	6	7	7	7

FIGURE 6.6   Altitude matrix showing gradual rise in elevation from northwest top to southeast bottom.

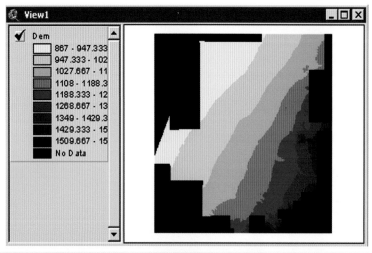

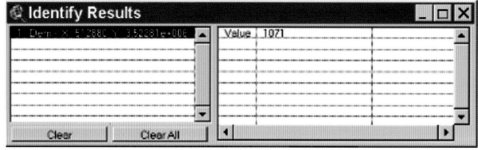

FIGURE 6.7   (A) Digital elevation map of the Santa Rita Experiment Range, south of Tucson. **(B)** Using the Identify tool to click on a point on the DEM. Notice x and y coordinates on the left and the elevation, or z-value, on the right: 1071 m.

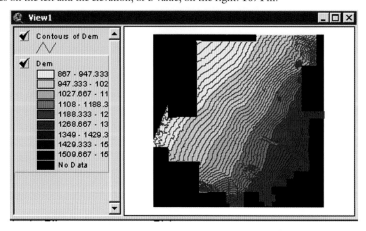

FIGURE 6.8   Digital elevation map of Santa Rita converted to contours.

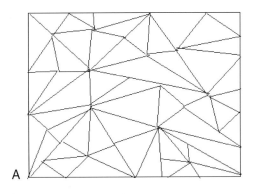

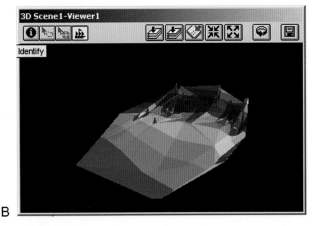

FIGURE 6.9 **(A)** Example of a triangulated irregular network (TIN). **(B)** Santa Rita Experiment Range shown as a TIN.

example of a DEM vector map is the triangulated irregular networks (TINs) elevation model. TINs are made up of continuous, connected triangles (Figure 6.9). Because TINs are vector based rather than raster based, they are easier to manipulate mathematically. TINs also better represent the direction a surface faces (e.g., northwest or southwest) than do raster DEMs. The surface direction is called *aspect*. TINs belong to a class of surface representations known as *tessellations*. TINS represent a surface with patches, much the way a soccer ball does, only with TINs the triangle-shaped patch size varies to match a different elevation. For TINs, more and smaller triangles indicate more relief (greater changes in elevation).

THEMATIC DATA

As previously stated, thematic data (sometimes called attribute data) are the data that GISs combine with location data to describe real-world features. An example is soil type. Soils types vary across a landscape. The combined map of these variances is the soils theme (Figure 6.10).

Theme data can come from field surveys, such as soil sampling, from reports (for example, growers reporting

their field crops combined into a crops theme), or from remote-sensing technology. The latter is an increasingly important source for data and, because of its complexity bears more explanation.

COLLECTING THEMATIC DATA WITH REMOTE SENSING

Remote-sensing tools include everything from handheld cameras and sensing devices, to aerial photography and satellite imagery. Remote sensors read the light of an object or natural feature and can discern information from its "spectral signature." For example, the brightness value can indicate whether the location is vegetated or barren and the stress status of the vegetation. A field with crops gives off a different signal than one that is plowed. A paved street reflects differently than an unpaved one. These sensors are *multipsectral*, meaning they can see beyond the visible image that our eyes give to us. They can measure infrared signals, for example, which are very good at detecting vegetation (see Chapter 11). The full range of radiation sources is categorized in the table of electromagnetic spectrum (see Chapter 11, Figure 11.2).

A

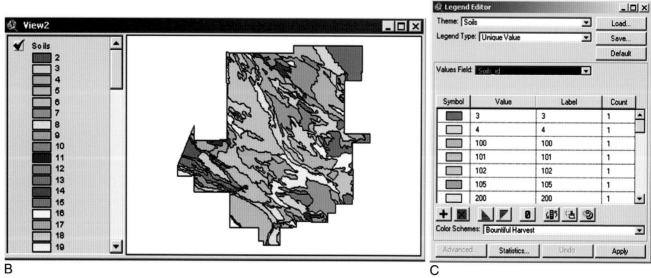

B   C

FIGURE 6.10 (**A**) Landsat image with the outline of the Santa Rita Experiment Range superimposed. (**B**) Soils map for the Santa Rita Experiment Range. (**C**) Legend for the soils theme. Note that the soil types have been grouped into *classifications* with a number assigned to them.

PROCESSING DATA FOR USE IN GEOGRAPHIC INFORMATION SYSTEMS

GIS data must be *classified* into themes or covers—recall the soils theme map of the Santa Rita Range (see Fig. 6.10). Notice in the legend down the side that there are a number of soil *classes* listed—this is what a classification does—it places similar features in the same category. In a similar manner, satellite data are classified by grouping the measured reflectance values in a given sensor's band. Those values are then assigned some meaning. For example, in the image of the High School Wash area of Tucson, AZ (Fig. 6.11), various reflectance values have been grouped and false colored to represent grass, buildings, streets, and so on, as shown in the legend. This classification has turned this image into a land use theme. Note from the following figures that as the resolution becomes coarser, so do the classification categories.

MODELS OF REAL-WORLD PROBLEMS WITH GEOGRAPHIC INFORMATION SYSTEMS

In the "Map Considerations" section, three basic GIS operations are mentioned (reclassification, overlay, and measuring distance). This section gives examples of each to demonstrate how a GIS can model real-world problems. The real-world example will be an endangered species application to find and isolate a certain endangered plant species on the Santa Rita Experiment Range.

RECLASSIFICATION

Reclassification is rearranging the classes of data in a theme to isolate a certain variable. In the example, it is determined that the endangered plant species is more likely to be located at elevations over 1100 m. Figure 6.12 shows the DEM theme, which has been classified by ArcView into nine classes (A) and then in to two classes (B). A reclassification operation will change that into two classes—above and below 1100 m. Reclassifications are usually done to inventory themes. Figure 6.12, shows the location and size of areas exceeding 1100 m in elevation.

OVERLAY

GIS operations that involve two or more themes are called *overlay*. In the example, it is determined that fur-

ther information can help reveal more likely places to find the endangered plants. Soil type is another factor. Collating certain soils types and elevations exceeding 11 m will narrow the search. As with the DEM, the soils map is reclassified to reflect the sought-after soil types. Overlay will combine these two binary maps to find the coincidence of the soils types and elevations over 1100 m (Figure 6.13). This is the equivalent of a Boolean "AND" operation in database terms (more on Boolean operations below).

DISTANCE MAPPING

Distance mapping can supplement the applications illustrated in Figure 16.13. The plant species used in the example is often found near washes. GIS has a buffer operation that will find the cells that are a specific distance from the washes on the Santa Rita Experiment Range (Figure 6.14). Now this map can be overlaid on the prior map to yield those locations that are a coincidence of elevations over 1100 m, the favoring soil types, and areas near the washes.

MORE COMPLEX SPATIAL MODELING

The preceding example provides a hint of the more complex modeling that can be done with GIS by combining Boolean operations with binary mathematics. The reclassified binary themes for elevations over 1100 m and soils of a certain type can be used in various ways. In Figure 6.14, a Boolean AND operation was used. Boolean operations, named after their inventor, George Boole, are the basis of logic circuits and database programs. In GIS Boolean operations are used to join themes or maps. The Boolean AND operation is used to disclose the locations where both conditions exist—in this instance, where elevations over 1100 m and soils of a certain type occur.

In Boolean operations, all the numbers are binary—either 0 or 1. A 0 is used for locations where the plant species will not be found—either cells less than 1100 m in elevation or the wrong soil types. A 1 is used for cells that have the selected properties. To yield the cells where *both* conditions occur, we use a Boolean AND because Boolean ANDs work like a multiplication:

$$0^*0 = 0, 1^*0 = 0, 0^*1 = 0, 1^*1 = 1 \qquad \text{(Eq. 6.1)}$$

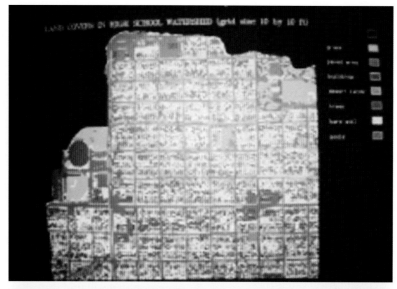

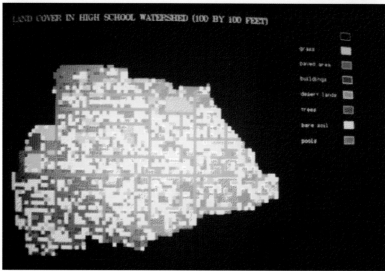

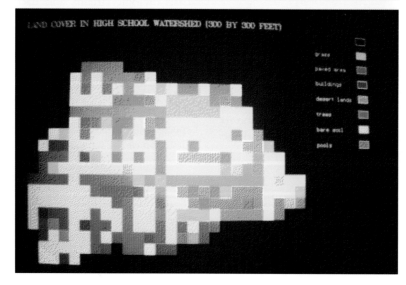

FIGURE 6.11   Images of the High School Wash in different resolutions. As in any statistical sampling program, the larger the sample (pixel) size, the coarser the classification system.

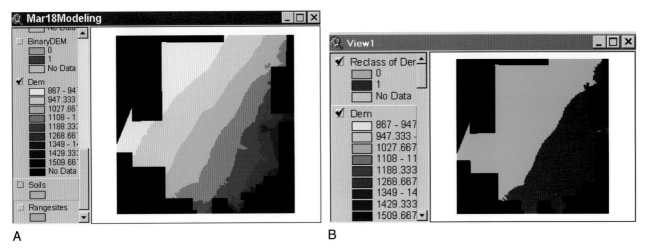

FIGURE 6.12 (**A-B**) Reclassification. Nine classes have become two.

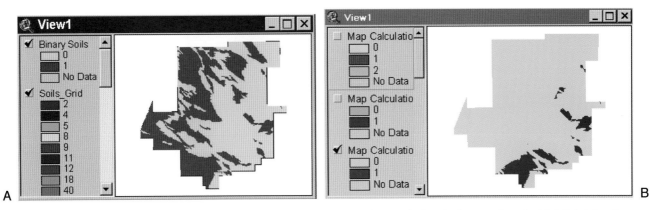

FIGURE 6.13 (**A**) Binary map of soil types where plant species is likely to be found. (**B**) Result of overlay between the reclassified digital elevation map and soils maps. The dark area at the bottom of the map is the coincidence of those two themes.

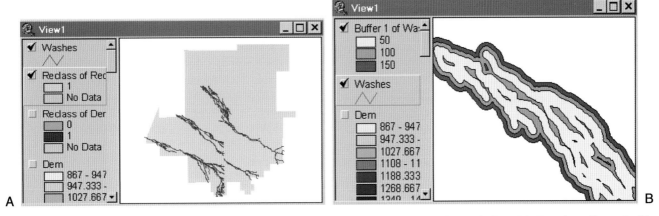

FIGURE 6.14 (**A**) Washes on the Santa Rita Experiment Range. (**B**) Close-up of buffers on the washes on the Santa Rita Experiment Range. In this case there are three layers of buffers to rank those areas that are closest as the highest priority and work outward from there.

where the last case, both conditions being 1, is the only one that comes out with a 1 (Figure 6.15A).

However, suppose the plant species is more likely to be found where both conditions are met, but that it still may occur when either of the two conditions (>1100 m elevation or appropriate soil) are met. In this case,

a Boolean *OR* will show where any condition will be met:

$$0 \text{ OR } 0 = 0, 1 \text{ OR } 0 = 1, 0 \text{ OR } 1 = 1, \\ 1 \text{ OR } 1 = 1 \quad \text{(Eq. 6.2)}$$

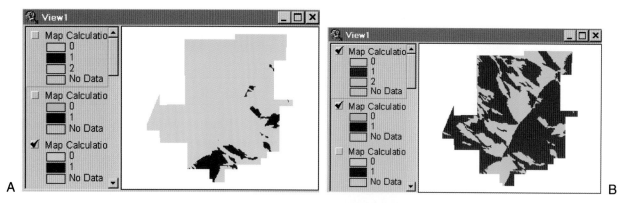

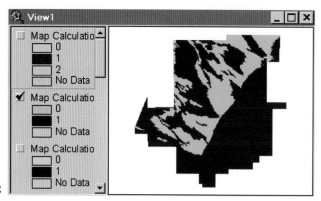

FIGURE 6.15   (A) Result of Boolean AND. **(B)** Boolean OR results. All the dark areas have at least one condition met. **(C)** Simple ranking model ends with three classes.

Clearly, this will result in many more "solutions" (Fig. 6.15B).

Of course, the areas that meet both conditions are more likely to have the plant species. A ranking model will show them as a preference. Ranking models are not Boolean. Their logic works like a mathematical plus:

$$0 + 0 = 0, 0 + 1 = 1, 1 + 0 = 1, 1 + 1 = 2 \quad \text{(Eq. 6.3)}$$

The result will be that those areas that meet neither condition for finding this plant species—those with a 0—will still result in a 0 when combined. Those areas that have one, but not both conditions will form a large middle group of cells, and those with both conditions will form a third category that is a priority area (see Fig. 6.15C).

MORE COMPLEXITY-WEIGHTED RANKING MODELS

The ranking model demonstrated how to bring preference into spatial modeling. The same preference can be introduced by an expert to weight the model. For example, if elevation is a more important factor than soil type, a multiplier can be applied to that part of the equation. This is called *weighting the model* and would make a ranking model a weighted ranking model. Scientists may apply very complex polynomials weights to parts of an equation to reflect the information they have discovered in their research. This underscores a very important point: GISs are only tools; they do not do the modeling—the researcher does. It is tempting to think of GISs and related technologies as a black box, spewing forth answers to problems with certainty and precision. The old expression "Garbage in/garbage out" applies to GIS as well as any other computer technology. GIS will not correct poor science. As previously shown, GIS is also only an approximation of physical reality. Resolution is always an issue and affects the use of this technology at many levels.

GEOGRAPHIC INFORMATION SYSTEM ONLINE TOOLS AND VISUALIZATION

GISs and other environmental modeling tools have long been off-limits to most users because of their cost and their complexity. Recently, these tools have become much more affordable, and the technology has become more accessible through the introduction of user-friendly online tools that reduce the steep learning curve traditionally associated with GIS and make it an accessible device for nontechnical practitioners.

ONLINE TOOLS

GIS has become a part of the products offered by companies dealing in location-based services (LBS). These services include online mapping such as is found at Yahoo and other Web-based search tools or the directional technology available in some upscale automobiles. Soon many devices, such as cellular phones, will come equipped with GPS. Other companies are becoming more GIS-invested in their marketing strategies and at a small business level.

VISUALIZATION

GISs are visualization-based by design. Integrating the GIS visualizations with digital photography is another useful way to make GIS more appealing. Figure 6.16A shows some panoramas created by participants in a community-mapping project. These participants capture 360-degree pictures of their environment and then map them to an online resource (Fig. 6.16B) where they can be shared. The panoramas are constructed by combining several overlapping digital photographs. They can be tied to data, and they can be strung together so that you

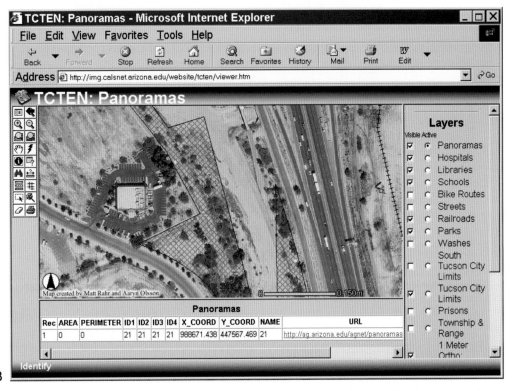

FIGURE 6.16 (A) Photo panoramas. (B) Panoramas mapped to online global information systems with other themes using an orthophoto as base map.

can "travel" through an entire region. They help users orient themselves to the study terrain.

VIRTUAL REALITY MODELING

Panoramas simulate three-dimensional (3-D), but they are not true 3-D modeling. GIS can do true 3-D modeling, which is very useful in applications where there is a visual impact assessment involved, such as designing scenic look outs on trails and highways. Figure 6.17A illustrates how a 3-D model of the Santa Rita Experiment Range is generated from a DEM.

Other 3D effects can also enhance GIS queries. GIS and Virtual Reality Modeling Language (VRML) have been successfully integrated so that users can "fly" through a landscape and then query data behind the model. Figure 6.17B shows an example of a VRML created with a satellite photo draped over the DEM of the Santa Rita Experiment Range south of Tucson integrated with a GIS, and the layer for Pastures turned on.

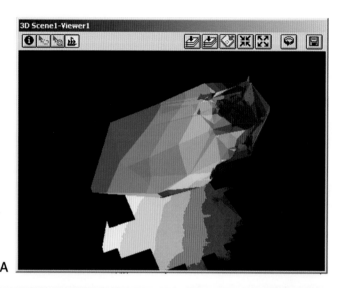

FIGURE 6.17   **(A)** DEM-generated 3-D model of Santa Rita Experiment Range. **(B)** Virtual reality model of the Santa Rita Experiment Range.

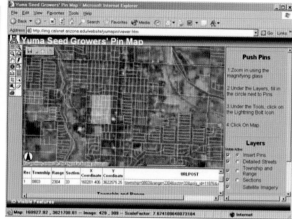

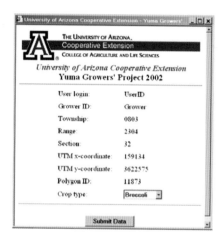

FIGURE 6.18 Yuma, AZ, seed growers project Website. Open a password-protected page, zoom in, click on a field, and insert a "pin" with information about the crop being planted.

THE FUTURE: GEOGRAPHIC INFORMATION SYSTEM AS A COMMUNITY RESOURCE

Online data and visualization are making GIS a community-accessible tool. GIS can also serve as a community collaboration vehicle. Figure 6.18 shows screens from a Yuma, AZ, seed growers, application Website. These growers have to work with each other at planting time to prevent cross-pollination. They had been using a physical bulletin board in the Yuma County Extension office to communicate with each other, which meant several trips a day to the office, and the danger that the pins would fall out or be disturbed in some way, and some growers would not get a message. In this case, with the application by ArcIMS, a Web-based GIS tool, the base map is a satellite photo of the Yuma Valley, and the data on the back end is managed by an Oracle database program. This project demonstrates two principles of community GISs: interactivity and peer-to-peer communication. Interactivity is an important component of public service GIS.

QUESTIONS

1. What spatial considerations are there in enviromental decision making, and how can GIS help address these considerations?

2. How does GIS tie data to location?

3. What is the difference between raster and vector GIS?

4. Define these basic GIS operations: reclassification, overlay, distance measurement, and connectivity.

5. What are the different ways to collect data for GIS?

6. Describe these spatial models: simple binary AND and OR models; ranking models; weighted ranking models.

7. Solve the following problem using the modeling techniques above:

   You are trying to locate the areas in a given region that are most susceptible to mosquito breeding. You are looking for these factors: low, near water, like a lake or a stream, and grassy. Show how you would solve that problem using a simple Boolean AND. Now weigh the model so that the low factor gets twice as much weight as the others.

REFERENCES AND ADDITIONAL READING

Bernhardsen, Tor. (1992) *Geographic Information Systems*, Viak IT, Longum Park, Arenoal, Norway.

Berry, Joseph K. (1993) *Beyond Mapping: Concepts Algorithms and Issues in GIS*, GIS World Books, Fort Collins, CO.

Berry, Joseph K. (1995) *Spatial Reasoning for Effective GIS*, GIS World Books, Fort Collins, CO.

Burrough, Peter A., and McDonnell, Rachaell A. (1998) *Principles of Geographic Information Systems*, 2nd Ed, Oxford University Press, Oxford, England.

Lein, James K. (1997) Environmental Decision Making: *An Information Technology Approach*, Blackwell Science Inc., Malden, MA.

Longley, Paul, and Batty, Michael, Eds. (1996) *Spatial Analysis: Modeling in a GIS Environment*, John Wiley and Sons, New York.

Luarini, Robert, and Thompson, Derek. (1992) *Fundamentals of Spatial Information Systems*, Academic Press, London.

7

SOIL AND VADOSE ZONE SAMPLING

L.G. WILSON AND J.F. ARTIOLA

The vadose zone is defined as the geological zone between a regional groundwater body and the land surface and is composed of unsaturated material. The vadose zone includes weathered and unweathered minerals and geological deposits, which are sedimentary, metamorphic, or igneous in nature. The soil environment is a subset of the vadose zone defined more precisely as weathered geological materials and biological residues that come into contact with the earth's atmosphere. The soil environment includes viable plants and animals (roots, residues) and microorganisms that reside in the pore spaces of and are attached to the geological materials. The pore spaces of this environment also contain water with dissolved minerals and air enriched with carbon dioxide and other trace gases. The pore water and soil gas sustain plants, animals, and microorganisms that live in this environment. When excess water percolates below the root zone, soluble constituents (nutrients and pollutants) can contaminate groundwater.

The challenge facing environmental scientists when sampling the soil or vadose zone environment is formidable because this environment is heterogeneous at all scales (Figure 7.1). Often, anthropogenic activities enhance the complexity of the soil environment. These activities can mask the true nature of the soil environment in ways that can constrain accurate characterization. For example, many agricultural activities often modify the physical, chemical, and biological nature of soils. Field preparations include surface leveling, which may cut across and mix several soil horizons, and the removal of all plant species to facilitate crop production. Over time this "modified" soil environment develops its own set of unique characteristics. Regular surface manipulation of agricultural soil may homogenize the first 0.5 m of soil,

but wide variations in soil types can occur even over short distances.

In this chapter we emphasize the methods and equipment needed to collect and store soil, soil-water, and soil-gas samples for the following purposes: conducting soil surveys, assessing the fertility status for agricultural production, and monitoring the vadose zones for pollution

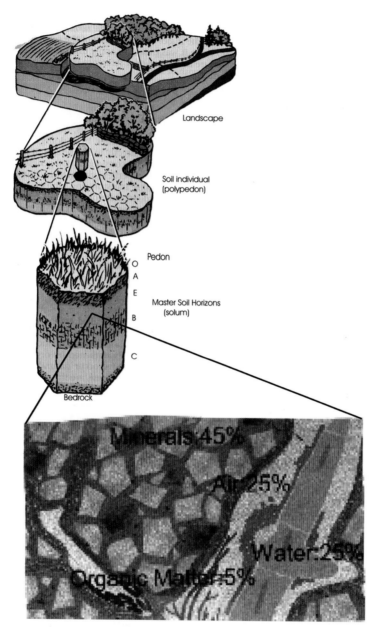

FIGURE 7.1   Soil scales, from landscape to a soil core sample. Topography and vegetation are visible (external) soil macroheterogeneities. Soil landscape and polypedons and soil horizons held define intermediate and field scale heterogeneities of the soil environment. Inset shows the microheterogeneity of soils. (Adapted from Brady and Weil, 1996, Fig. 3.1, and Pepper *et al.*, 1996, Fig. 2.3.)

characterization. Inasmuch as the distinction between soil and vadose zones is often blurred, the sampling techniques discussed in this chapter apply to both environments.

SOIL SAMPLING STRATEGIES

Chapter 2 introduced sampling units and strategies in two and three dimensions, with examples that are applicable to the soil environment. In classic soil science a soil "unit" or pedon is defined as "the smallest volume of soil than can be called a soil" (Brady and Weil, 1996). This definition has particular relevance for soil classifiers who must rely on observations made from volumes of soil that can vary from 1 to 10 m² in surface area and from 1 to 3 m in depth (see Figure 7.1). However, a more generic definition is often adopted in soil environmental monitoring. This may in fact be based in the definition of scales that are discussed later. Modern soil environmental monitoring methods often do not consider classic soil classification to optimize sampling schemes. In subsequent sections we briefly discuss classic soil sampling for classification applications.

In terms of surface, a *homogeneous soil unit* may be defined as an area that has similar physical features, smaller or larger, as a pedon. Thus, a soil unit may be considered as a mass of soil with no distinguishing profiles; that is, there are no visible changes in characteristics such as texture, color, density, plant (root distributions), moisture content, and organic-matter content. This definition of a soil unit is used in the remainder of this chapter.

If other changes occur within a "unit," in one or more of the categories previously listed, it is subdivided into smaller units. Thus, the process of subdividing the soil environment continues by defining another unit of soil within these units. Soil scientists recognize that intermediate scale units such as a landscape (see Figure 7.1) are far from being homogeneous. However, intermediate scales are important in the management of soil processes associated with agricultural production, land treatment of waste, and remediation of contaminated soils. In the practical sense, intermediate-scale soil units must be subdivided into smaller macroscale and microscale units to facilitate the sampling and analysis processes.

CLASSIC SOIL SAMPLING

Soil surveying methods, originally developed by the Soil Conservation Service of the U.S. Department of Agriculture (USDA), have their origins in classic soil science, specifically soil genesis, pedology, and morphology. Soil morphology involves observational analysis of soil formations, structures, and the like. Soil morphology requires direct field observations, which, when combined with laboratory measurement of field samples, allow final classification of the soil pedon. For example, field observations include the semiquantitative measurements shown in Table 7.1.

Other important soil classification information that can be obtained from field samples and analyzed in the laboratory are shown in Table 7.2.

Classic soil classification, with field observations, requires extensive field experience. Soil classifiers must be certified and have several years of supervised field experience in soil surveying and classification. The purpose of soil mapping is not to define exact areas or volumes of soil with unique or exclusive properties, but to define the locations of soil series (the most specific unit within the U.S. Soil Classification System) within soil map units or landscapes (Figure 7.2). Most USDA Soil Conservation Service Soil Surveys (see also Chapter 5) have been conducted by trained soil scientists relying almost exclusively on field topographical and morphological observations.

EQUIPMENT FOR SOIL CLASSIFICATION

A subset of equipment listed in Box 7.1 may be used to inspect and collect soil samples. A more complete list of field equipment definitions, charts, and guidelines is provided in a practical field guide by Boulding (1994) or *Manual Agricultural Handbook 18* (1993) by Soil Survey Staff of the U.S. Conservation Service. Typically, the soil

TABLE 7.1

Semiquantitative Measurements Obtained from Field Observations

Soil horizon definitions and boundaries: O, A, E, B, C
Color-predominant: red, black, gray, bluish related to moisture, and soil chemical composition
Mottling: color distributions by size and contrast
Texture-particle size distribution: gravel, sand, silt clays (field estimates)
Porosity-density-compaction: voids
Structure: ranges from loose (single grain) to massive (prismatic)
Moisture regime: saturated (water table location), unsaturated (duration)
Water drainage: infiltration rate

TABLE 7.2

Quantitative Measurements from Laboratory Analyses of Field Samples

Particle size distribution: in percent sand, silt, and clay
Hydraulic conductivity: saturated, unsaturated
Saturated pH and electrical conductivity: from water saturated paste
Organic carbon and organic nitrogen: organic matter
Cation exchange capacity: base saturation, sodium exchange ratio

BOX 7.1 *Field Equipment Necessary for Soil/Vadose Zone Sampling*

Auger (hand-mechanical), drill rig
Shovel, scoop, mechanics tool box
Charts: soil morphology and classification,
   Munsell soil color, mineral and texture
   classification
Bottles: hydrochloric acid (tenth: normal) solution,
   deionized water
Recording devices: log book, camera, waterproof
   pen

Maps: Soil survey, topographic, geologic,
Forms: Soil profile, borehole forms, chain-of-
   custody, sample labels

Compass, global positioning system (GPS) handheld unit
Soil sample collection paper or plastic bags, auger
   sleeves
Sample storage container (cooler)

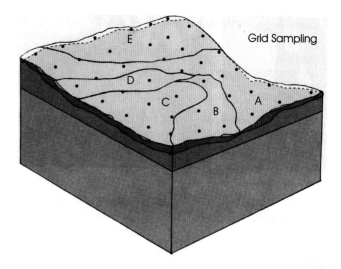

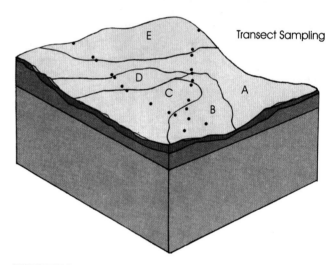

FIGURE 7.2   Soil sampling approaches used for the delineation of soil boundary series *(A–E)*. Grid sampling points *(top)* are more costly and ignore topographic features. Transect sampling makes use of soil topographic information to find boundaries, reducing sampling costs. (Adapted from Brady and Weil, 1996, Fig. 19.6.)

classifier walks across the landscape in either a grid or transverse pattern and verifies the soil properties by collecting and examining samples using a screw type hand auger (see Figure 7.2). The soil classifier may also collect soil samples for laboratory chemical and physical analyses using a truck-mounted drill rig. Soil classifiers also may dig trenches to view soil profiles. This practice, although expensive and disruptive, is effective for visually delineating soil profiles and microfeatures. Finally, soil scientists may also rely on topographical features, vegetation distribution, and unique landmarks for the final delineation of soil series on an aerial photograph (see Chapter 5).

SOIL SAMPLING IN AGRICULTURE

Agricultural environments are unique, intensively managed soil systems defined and modified by human beings to sustain food production. These changes require large energy inputs like grading and plowing and regular additions of fertilizers, pesticides, and, in arid and semi-arid areas, significant amounts of irrigation water. Monitoring the status of agricultural soils is primarily done to assess soil fertility. However, soil samples are sometimes collected to determine major chemical and physical properties such as salinity, water infiltration, or organic matter content. These soil properties can change significantly over time with repeated fertilizer and waste additions and tillage practices.

As a general guideline, one single or composite (bulk) sample per four hectares ($\simeq$ 10 acres) should be collected from the top 30 cm of the soil profile with a bucket auger or a spade. However, obvious changes in topography should also be taken into consideration. Bulking samples to save on analytical costs is also frequently done under these conditions. This procedure is acceptable if each sample contributes the same amount of soil to the final sample and only if an estimate of the mean concentrations is needed without regard to potential spatial variability. The collection of these types of samples is typically accomplished with a simple random sampling scheme (Figure 7.3). The scale intervals must be selected before sampling and can be approximated by steps or measured with a tape measure (see also Klute, 1986). Alternatively, a

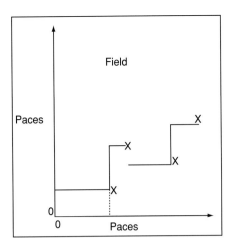

FIGURE 7.3 Random walk sampling used to collect soil samples from agricultural soils. A randomly selected starting location in the field and other subsequent sampling locations, based on randomly selected directions and paces. (Adapted from Klute, A. [1986] *Methods of Soil Analysis: Part 1—Physical and Mineralogical Methods,* 2nd edition. American Society of Agronomy, Soil Science Society of America. Madison, WI.)

field can be subdivided into units with a grid to facilitate sample location. Samples can then be collected within randomly selected grids. Here the number of grid divisions should be at least 10 times the number of samples collected. However, with global positioning system (GPS) devices, grids may no longer be necessary to collect samples randomly (see Figure 2.5B).

When a given soil parameter is suspected to have a strong spatial dependency, a systematic or grid sampling pattern can be chosen. A grid sampling pattern may be selected by dividing the field into squares or triangles (Figure 2.5B) and collecting samples at the nodes or within each grid. Here the number of squares or triangles is equal to the number of samples needed. This technique provides a set of samples with X–Y coordinates that can be used to develop contour maps with geostatistical methods (see Chapter 3). Modern precision farming techniques that make use of detailed spatial information on the soil texture of agricultural fields is critical to optimize fertilizer and water applications.

EQUIPMENT FOR SOIL SAMPLING IN AGRICULTURE

The equipment necessary to collect soil samples is limited to a subset of the list presented in Box 7.1 Unsaturated soil samples for soil fertility analyses should be collected with a bucket auger and kept at room temperature. Microbial activity in soil samples is minimized by air-drying (1 to 3 days) in a dry environment with daily mixing. Once dried, the average loam soil should not have more than 5% soil moisture content. After drying,

the samples should be stored in sealed plastic bags, pending analysis. Properly dried and sealed soil samples may be stored indefinitely. If microbial analyses are needed, the samples should not be air-dried; they should be sealed and stored at 4° C before analysis (see Chapter 20 for details on soil sampling for microbial analyses).

VADOSE ZONE SAMPLING FOR POLLUTION CHARACTERIZATION

In the last 25 years, a new emphasis on more precise methods for sample collection has emerged, following the need to quantify vadose-zone pollution. Industrial and municipal waste treatment and disposal, uncontrolled chemical spills, atmospheric depositions, and physical disturbances (mining, reclamation) add new dimensions to traditional soil sampling. Accordingly, statistical methods to define the number of samples are necessary to meet environmental pollution and cleanup standards, as well as control the costs associated with soil sampling and analyses. However, soil sampling still requires a combination of systematic and random sampling and knowledge of site characteristics, as well as its history.

In some cases, the source of the pollutant is known, and the objective is to characterize the extent of contamination. In such cases, sampling density is usually greatest at the origin of contamination and decreases radially outward (Figure 7.4). Systematic sampling may be done along possible migration pathways, defined by topography and wind patterns. In other cases, the source of pollution is not known, and the objective is to locate the source. Grid sampling is then used to locate pollutant "hot spots." In such cases, a statistical analysis is required to define the grid size, considering the area and probability of occurrence of hot spots (see also Chapter 3). For agricultural applications, soil profiles are used to assist in sampling. Soil profiles typically extend from 0 to 3 m below the surface, depending on the number and thickness of each profile. For pollutant characterization, the goal of soil sampling is to determine the depth of pollution, and soil-profile data may or may not be useful. Typically, a more important consideration is the depth to groundwater. Therefore sampling is often extended beyond the soil environment to include the vadose zone.

SOIL SAMPLERS

Soil samplers are grouped into manually operated and power-driven samplers (Dorrance *et al.*, 1995). Factors affecting the selection of samplers between and within these two categories include required sampling depth, soil conditions (e.g., presence of caliche, gravel layers),

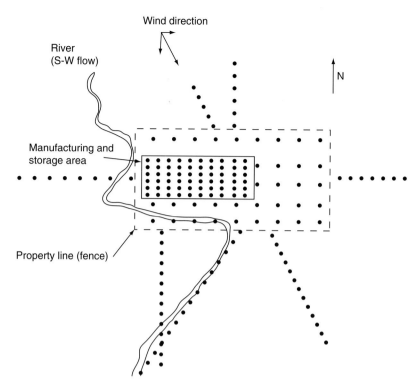

FIGURE 7.4   Grid and directional (exploratory) soil sampling patterns assume a pollution point of origin. Note that sampling can be done in the direction of potential pathways of pollution (transport via air and water). Dots represent sampling points.

sample size, moisture conditions, site accessibility, costs, and personnel availability. Manually operated samplers are primarily used for surface sampling (e.g., less than 5 m deep, depending on conditions). Mechanically driven devices are more suited for sampling deeper regions of the vadose zone or in conditions that are unsuitable for manual samplers. For additional details on samplers described in this chapter, see Dorrance *et al.* (1995).

HANDHELD (MANUALLY OPERATED) SAMPLING EQUIPMENT

The most common manually operated samplers are shown in Table 7.3, and their descriptions follow.

BULK SAMPLERS

Simple tools for obtaining soil samples include stainless steel spoons, sampling scoops, shovels, trowels, and spatulas. Stainless steel construction ensures that these samplers will be inert to typical pollutants. These devices are also used to obtain soil samples from excavated trenches. Samples are stored in appropriate containers. Bulk samplers expose the sample to the atmosphere, enhancing loss of volatile constituents including volatile organic compounds (VOCs). Because they are disturbed samples, bulk samples are not suitable for geo-technical studies.

AUGERS

Many soil environments have developed profiles (A and B horizons) with depths that seldom exceed 2 m. There fore most soils can be sampled with handheld equipment similar in design to that shown in Figure 7.5. There are several types of augers and tubes designed for sampling under many soil textural and moisture conditions (see Figure 7.5). Auger samplers do not prevent

TABLE 7.3
Manual Sampling Equipment

1. Bulk samplers:
     Shovels
     Scoops
2. Auger samplers:
     Screw-type augers
     Posthole augers
     Regular or general-purpose barrel augers
     Sand augers
     Mud augers
     Dutch-type augers
3. Tube-type samplers:
     Open-sided soil sampling tubes
     Veihmeyer tubes
     Thin-walled tube samplers (Shelby tubes)
     Split-barrel drive samplers (split-spoon samplers)
     Ring-lined barrel samplers
     Piston samplers
     Maccauley sampler

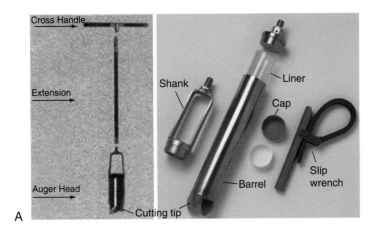

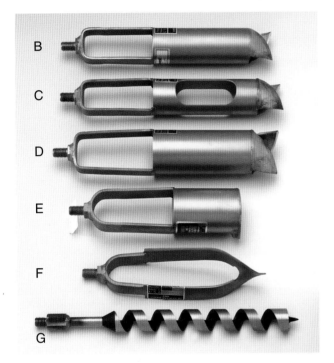

FIGURE 7.5 **A,** *Left,* basic hand-operated soil auger showing the auger head, an extension rod, and a handle. *Right,* details on the auger head showing cutting tip, barrel, slip wrench, and shank for attachment to extension rods. This illustrated design allows insertion of a replaceable plastic liner and caps for collection and retention of an intact sample suitable for geotechnical studies. Alternative auger head designs are available depending on soil conditions. **B,** Sand. This sampler is designed for use in extremely dry, sandy soils. The cutting bits touch so that they can hold dry sand samples. **C,** Mud. This head has an opening cut out of the barrel for easy removal of heavy, wet soil and clay samples. Cutting bits are similar to regular augers but are spaced further apart. **D,** Regular. This head is for sampling under ordinary soil conditions. **E,** Planer. This head cleans out and flattens the bottom of predrilled holes. It removes the loose dirt left in the borehole by other augers. **F,** Dutch. This head is excellent for drilling excessively wet, boggy soil and fibrous, heavily rooted swampy areas. It is forged from high-carbon spring steel and is available only in a 3-inch diameter. **G,** Screw. Extra-heavy-duty screw-type auger is used for taking smaller samples. Flighting is 6 inches in length. (From Ben Meadows Company, a division of Lab Safety Supply, Inc.)

cross-contamination of soil profiles and are not recommended for sampling soils for trace chemical analyses. Also, core integrity cannot be maintained because in most cases the soil sample must be dislodged by tapping the outside shell.

SCREW OR SPIRAL AUGERS

These augers are similar to wood augers (Figure 7.6) modified for attachment to extension rod and tee-type handle (Dorrance *et al.* 1995). A commercial spiral auger consists of a steel strip spiraled to 25-cm units. These samplers are best suited for use in moist, cohesive soils, free of gravel. They are frequently used to bore through very dense layers of soil. Because the samples are exposed to the atmosphere, they are not

suitable for analyses of VOCs. Screw-type augers are usually 1 in (2.5 cm) to 2 in (5 cm) in diameter, work best in clay soils, and are designed to collect small amounts of soil.

BARREL OR BUCKET AUGERS

The basic design of a barrel auger includes a bit with cutting edges, a short tube or barrel within which the soil sample is retained, and two shanks (see Figures 7.5 and 7.6). Barrel augers generally provide larger samples than screw-type augers. Some designs permit insertion of plastic, brass, stainless steel, polytetrafluoroethylene (PTFE), or aluminum liners (see Figure 7.5). The auger head is attached to extension rods and a tee-handle (see Figure 7.5). Additional extension rods are added

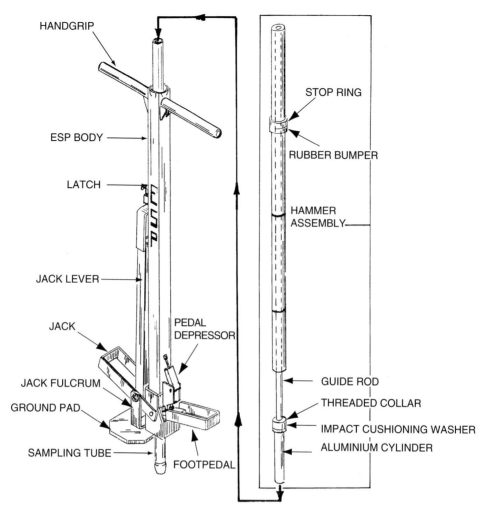

FIGURE 7.6   Veihmeyer tube example. Manual JMC subsoil probe, hammer driven, with plastic lined sampling tube that can be removed and capped to minimize sample contamination. Hammer option and extensions allow for manual sampling below 1 m depth, depending on soil type and moisture conditions. (From JMC manual.)

as needed. Samplers are available in various diameters. Bucket augers vary in diameter from about 2 in (5 cm) to 4 in (10 cm); the most common are 3.5 in (8.9 cm) in diameter. The basic variations in design allow sampling in sandy, muddy, or stony conditions (see Figure 7.5) and collect up to 1 kg of soil. Core catchers are often installed within the tip to prevent loss of very dry or loose samples.

TUBES

Soil sampling tubes (Figure 7.6) provide an alternate method of sample collection that limits cross-contamination and maintains core integrity. Tube-type samplers generally have smaller diameters and larger body lengths than barrel augers. The basic components of tube-type samplers include a hardened cutting tip, body tube or barrel, and a threaded end. In some designs, extension rods are screwed onto the body tube as required to reach

the total sampling depth. The tubes are constructed of hardened steel, and some units are chrome plated. Tube samplers permit the insertion of liners for the collection and direct storage of an intact soil core. Liners are used to obtain undisturbed samples. This reduces sample handling in the field and minimizes sample contamination. Because these samplers have a diameter that seldom exceeds 2 in (5 cm), they can be driven into the ground by hand, with a foot support adapter, or with a slide hammer. Tube samplers fitted with a slide hammer can be driven several meters into the ground, depending on soil textures. Transparent plastic liners are very useful for collecting soil samples for trace chemical analysis and for recording and separating soil textural layers. Nonetheless, soil-sampling tubes have one serious sampling problem: they can compress soil profiles in heavy-textured soils. In light-textured soils these sampling devices often exclude or displace coarse soil material.

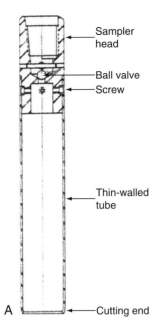

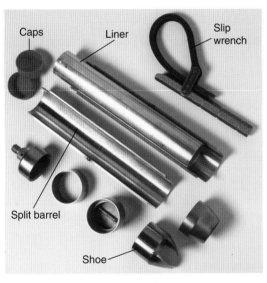

FIGURE 7.7 (**A**) Shelby tube, comprising a single piece, thin-gauge metal tube with a sharpened point. Sampler head connects tube to extension rod for manual or power sampling. Ball valve releases air during sampling. Ends are capped following sampling, and the tube is stored in an ice cooler for laboratory analyses of sample. (**B**) Split spoon sampler. (**A**, After Foremost Mobile Co., Inc. **B**, (From Ben Meadows Company, a division of Lab Safety Supply, Inc.)

Common sampling tubes include open-sided soil-sampling tubes, Veihmeyer tubes (See Figure 7.6), thin-walled tube samplers (Shelby tubes) (Figure 7.7A), split-barrel drive sampling (split-spoon) (Figure 7.7B), ring-lined barrel samplers, and piston samplers. Some of these tubes can also be used with power-driven equipment. For a description of these tubes see Dorrance *et al.* (1995).

Other specialized types of soil sampling equipment work only in organic soils that are soft or saturated. Other samplers and sampling equipment adapted for sampling in peat soils and even in frozen (permafrost) soils are discussed by Carter (1993) (Boxes 7.2 and 7.3).

MECHANICAL SOIL AND VADOSE ZONE SAMPLING EQUIPMENT

The manual samplers described in the previous section are suitable when relatively shallow depths are involved and where drilling conditions are favorable. Otherwise,

---

BOX 7.2 *Advantages of Manual Sampling*

1. In contrast to mechanically operated samplers, there are no major mobilization and demobilization requirements (i.e., the technician walks to the sample site and can begin sampling almost immediately).

2. Samples can be taken at locations not easily accessible by power equipment (e.g., hillsides).

3. Hand-operated equipment may be safer to operate than mechanically operated samplers (e.g., no moving parts). However, there is always the risk of muscle or back injury to the technician when withdrawing the tool and samples from the hole.

4. Hand-operated devices do not generate as much frictional heat compared with mechanically operated samplers. Therefore the associated loss of volatile constituents is reduced.

5. The cost of manual sampling is generally much lower than that for mechanically operated samplers.

BOX 7.3 *Disadvantages of Manual Sampling*

1. Sampling is generally limited to shallow depths.

2. Collecting intact or complete samples may be difficult or impossible in coarse-grained soils.

3. Samples obtained with bucket and screw type augers are disturbed and not suitable for geotechnical observations. Using tube-type samplers or augers with liners reduces this problem.

power-driven equipment is necessary. The units described include handheld power augers, solid-stem flight augers, hollow-stem augers, sonic drilling, and direct push soil probes. These methods are preferred over drilling methods commonly used to install water supply wells, such as cable tool and rotary units, because drilling fluids are generally not required. For additional details on these samplers see Lewis and Wilson (1995).

HANDHELD POWER AUGERS

These devices consist of a spiral flight auger driven by either a small air-cooled engine or an electric motor. They are available with handles for either one or two operators. Additional flights are attached as required. Flights are available commercially in diameters ranging from 2 in (5 cm) to 12 in (30.5 cm) and lengths of 30 in (76.2 cm) or 36 in (91.4 cm). Unpainted flights are available to avoid sample contamination. As the auger advances into the profile, soil cuttings are brought to the surface. The samples are placed in containers with spatulas or trowels. A problem with this method is that samples are disturbed, making it difficult to relate them to specific depths. An alternative approach is to withdraw the auger flights at the depth of interest and drive a probe (e.g., Shelby tube) into undisturbed soil. Handheld power augers are limited to shallow depths and unconsolidated soils free of gravel or cobbles. If sampling tubes are not used, the disturbed samples are not useful for soil profile characterization or geo-technical studies (e.g., determination of layered conditions). Volatile constituents are lost by exposure to the air and by heat generated during drilling.

SOLID-STEM AUGERS

Mounted on drill rigs, these augers are capable of sampling to a greater depth than the hand-operated units. Single auger flights consist of spiral flanges welded to a steel pipe. Multiple sections are joined together as required to produce continuous flighting. The bottom flight contains a cutting head with replaceable carbide teeth. Single flights are commonly 5 ft long. Outside diameters are commercially available in sizes ranging from 4 in (10 cm) to 24 in (61 cm). New single flights should be sandblasted before use to remove paint that may interfere with soil pore-solution analyses. An engine-driven system turns the auger sections, and as the auger column is rotated into soil, cuttings are retained on the flights. The augers are then removed from the hole and samples are taken from the retained soil. Alternatively, samples are brought to the surface and collected. Unfortunately, with either method it may not be possible to precisely relate samples and depths. The reason is that soil

moves up the flights in an uneven fashion as the auger column is advanced. A preferred approach is to remove the auger flight at the desired sampling depth, lower a tube sampler (e.g., Shelby tube) in the open hole, and drive the sampler into undisturbed soil. Both methods are subject to cross contamination by caving and sloughing of the borehole wall as the auger flights are removed and reinserted. This is of particular concern when the objective of sampling is to obtain depth-specific contaminant profiles.

Several factors affect maximum sampling depth, including soil texture, excess soil moisture, available power, and auger diameter. A typical depth under ideal conditions may range up to 37 m. Greater drilling depths are attained in firm, fine-textured soils. Unless core samplers are used, this methods is not suitable for characterizing VOCs because of volatilization as samples are exposed to air and because of heat generated by the drilling process.

HOLLOW-STEM AUGERS

Hollow-stem augers are similar to continuous flight, solid-stem augers except their hollow interior allows access of sampling tools (Figure 7.8). A center rod with a pilot bit is inserted into the continuous flights to help advance the cutting bits. Alternatively, a knockout plug is located at the end of the lowermost flight to keep fine material from rising into the auger head. A common length of an auger flight is 5 ft (1.5 m). The inside diameters of commercially available auger flights range from 2.5 in (6.3 cm) to 10.25 in (26 cm). The corresponding actual hole sizes range from 6.25 in (15.9 cm) to 18 in (45.7 cm). Maximum drilling depths depend on the texture of the vadose zone, size of rig, and hole diameter.

Soil sampling is possible by collecting samples brought to the surface, as is done with solid-stem augers. A preferable approach is to lower sampling tubes through the hollow core of the flights and drive them into undisturbed soil. Because the continuous flights of a hollow-stem auger remain in place during sampling, soil from upper horizons is prevented from falling into the bottom of the hole. Thus, the cross contamination problem associated with solid-stem augers is avoided. When sampling at a given depth, drilling is stopped and the center rod or plug is removed. The sampler is then lowered inside the auger and driven by a hammer into undisturbed soil ahead of the auger flights. For a characterization of the pore-solution chemical properties throughout the vadose zone, samples are commonly collected at 5-ft (1.5-m) intervals or at distinct textural changes. It is also sometimes helpful to maintain a record of the number of hammer blows required to advance the sampler a given distance.

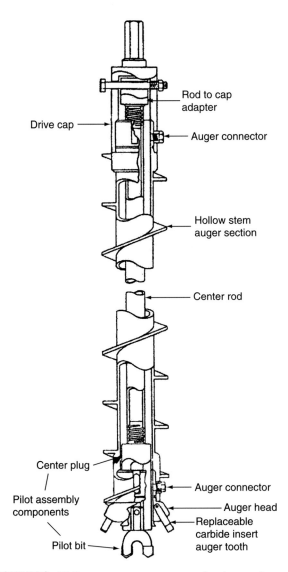

Rod to cap
adapter

Drive cap

Auger connector

Hollow stem
auger section

Center rod

Center plug

Pilot assembly
components

Auger connector

Auger head

Replaceable
carbide insert
auger tooth

Pilot bit

FIGURE 7.8 Hollow-stem auger components showing attachment to rotating drill rods. Primary use is with power equipment. Center rod is withdrawn for insertion of sampling equipment or suction lysimeters. (After Central Mine Equipment Co.)

The results can be related to textural changes. Samplers used during hollow-stem boring include Shelby tubes, split-spoon samplers, and ring-lined barrel samplers. Additionally, 5-ft (1.5-m) long, continuous samplers are commercially available for sampling soils free of cobbles.

Hollow-stem augering provides a rapid, economical means of obtaining undisturbed samples. Equipment is routinely available. Because hollow-stem boring uses no drilling fluid, the potential for chemical contamination or alteration of pore liquids is minimal. A major limitation is that the method is not suitable in extremely dry, fine materials; in saturated soils; or in soils containing cobbles or boulders. Heat that is generated during driving the samplers will promote VOC loss.

SONIC DRILLING

Sonic drilling is a relatively new technique for obtaining vadose zone soil samples for pore liquid analyses (www.sonic-drill.com). This technique is essentially a dual-cased drilling system that uses high-frequency mechanical vibrations to allow continuous core sampling and advancement into the profile. The drill head uses offset counterrotating weights to generate sinusoidal wave energy, which operates at frequencies close to the natural frequency of the steel drill column (up to 150 cycles per second) (Figure 7.9). The counterrotating balance weights are designed to direct 100% of the vibration at 0 degrees and 180 degrees. This action causes the column to vibrate elastically along its entire length. Resonance occurs when the vibrations coincide with the natural resonant frequency of the steel drill casing. This allows the rig to transfer timed vibrational energy to the top of the drill string. The high-energy vibrations are transmitted down to the face of the drill bit, producing the cutting action needed for penetration. Rotation and application of a downward force causes the drill string to advance through the profile. During drilling, the walls of the steel pipe expand and contract, causing the fluidization of soil particles along the drill string, which enhances drilling speed.

A dual-string assembly allows the use of an outer casing to hold the borehole open and an inner core barrel for collecting samples. Diameters of the outer casing range up to 12 in (30.5 cm). Diameters of the sampling core barrel range from 3 in (7.5 cm) to 10 in (25 cm). When the borehole is drilled, the core barrel is advanced ahead of the outer casing in 1 to 30 ft (95 m) increments, depending on physical conditions. After the core barrel is removed from the borehole, a plastic sheath is slipped over the barrel. The sample is extruded into the sheath. Subsamples are taken and stored in appropriate containers. Lined split-spoon samplers or Shelby tubes can also be used for sampling. In addition to the drilling of vertical holes, sonic rigs can angle-drill holes up to 75 degrees from horizontal. This is advantageous when sampling beneath existing waste disposal facilities such as landfills and impoundments.

The principal advantages of sonic drilling are that water is not required, perched water can be identified, and there is a relatively safe operating environment. Other advantages include the ability to collect continuous cores; drilling rates (up to 10 times faster than hollow-stem rigs [no drilling fluids needed]; fewer drill cuttings to dispose of (up to 80% less than other rigs); the ability to drill through bedrock, cobbles, and boulders; and the ability to drill to greater depths than hollow-stem augers (sonic drills can provide 3-in cores to a depth of 500 ft [153 m]). Disadvantages include a higher cost than that

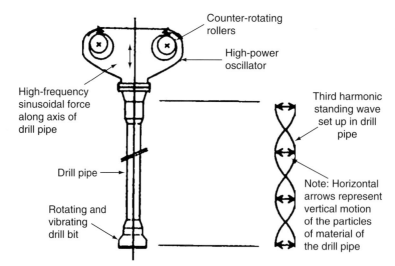

FIGURE 7.9 Basic components and principle of operation of sonic drilling. Besides usage for environmental sampling, the technique is used for mineral exploration, angle drilling, and constructing water supply wells. (After Sonic Drilling Ltd. and Boart Longyear Co.)

for hollow-stem augers, equipment breakdowns, and potential volatilization of VOCs due to the heat that is generated.

DIRECT-PUSH SOIL SAMPLING

Direct-push soil sampling is also known as direct drive, drive point, or push technology. This type of sampling involves forcing small-diameter sample probes into the subsurface and extracting samples at depths of interest (Lee and Connor, 1999). The samplers are advanced into the vadose zone by static hydraulic force coupled with percussion hammering and/or vibration. Under favorable conditions, this advancement system is capable of pushing sampling tools at a penetration rate of 5 ft (1.5 m) or more per minute to depths of 30 ft (9 m) to 60 ft (18 m). Some systems use a large rig capable of conventional drilling that has been modified with a hydraulic hammer. Other systems combine hollow-stem augering and direct push sampling to reach greater depths. Up to 14,500 kg of downward force may be applied with this system.

Samplers used with this method include nonsealed soil samplers (solid and split barrel samplers, thin-walled samplers), and sealed soil samplers (piston samplers). Liners can be used with the nonsealed samplers to minimize loss of volatile constituents. Although single probe sampling is possible, the raising and lowering of samplers in the borehole results in cross-contamination. Dual-tube systems reduce this problem but require greater force, limiting sample depth and sample diameter compared to conventional systems. Unlike the typical hollow-stem auger rigs where the raised masts may reach 30 ft (9 m) in the air, the smaller direct-push samplers have lower operating masts 12 ft (3.7 m). Thus, they can be used

under power lines or in other areas that would preclude hollow-stem augers. They can also be mounted on pickup trucks, increasing mobility and decreasing mobilization and demobilization times. The mass of waste soils brought to the surface is much less than with conventional rigs, reducing disposal costs. The requirement for grout to seal the abandoned hole is also reduced. The main limitation with this technique is the decreased sampling depth compared with other rigs, especially where cobbles and boulders are present in the profile. Ongoing improvements in the technique will undoubtedly minimize this problem in the future. Heat that is generated during sampling causing the loss of VOCs is once again a potential problem.

SOIL SAMPLE STORAGE AND PRESERVATION

Soil samples collected for issues related to contamination, public health, and risk and safety assessment usually require special procedures. Typically, samples are not allowed to dry and are collected and preserved "as is," meaning that soil moisture and chemical field conditions are maintained. Therefore these soil samples are usually collected in glass or plastic jars, sealed, and kept cool (4° C). Cooling the samples to near freezing is also necessary to reduce biological activity. Freezing soil samples, although sometimes done, is not recommended because freezing will change the biological and perhaps even the physical nature of the soil medium. No chemical preservatives are added to soil samples. Table 7.4 lists the major container types and holding times allowed for soil samples.

Intact soil cores collected using the sampling tubes, as described earlier, must also be quickly capped and sealed

TABLE 7.4
Soil Sample Storage Containers and Holding Times

Parameter[a]	Container (Plastic, Glass)	Maximum Holding Time (days)
pH, alkalinity	P,[b] G	14
Major anions: Cl, SO$_4$, Br, PO$_4$ ... Hg, CN total P	P,[b] G	28
Nitrate, nitrite, sulfide, sulfite, ammonia	P,[b] G	2
Cr(VI), oil and grease, organic carbon	P,[b] G	28
Metals	P,[b] G	6 mo
Extractable organics	G (Teflon-lined cap)	~7 (until extraction)
		30 (after extraction)
Volatile organics	G (Teflon-lined septum)	14
	Methanol immersion[c]	Unknown

(From U.S. Environmental Protection Agency.)
[a]All samples should be stored at 4° C.
[b]Plastic (polyethylene) containers preferred.
[c]Soil samples may be immersed in methanol in the field to prevent volatilization and degradation.
Br, Bromine; *Cl*, chloride; *CN*, cyanide; *Cr(VI)*, hexavalent chromium; *Hg*, mercury; *P*, phosphorus; *SO$_4$*, sulfate.

using tape, and stored at 4° C until dissection and analysis. Care must be taken not to invert or shake these cores to prevent mixing and disruption of core integrity.

Soil samples collected for volatile organics analysis can be collected in glass vials fitted with Teflon-lined septa. The use of jars or vials with septa reduces losses of volatile organic chemicals because the soil samples need not be exposed to the atmosphere before analysis. Another option is to transfer subsamples into glass jars to minimize headspace or into jars containing a known volume of methanol, usually in a 1:1 ratio (volume/volume). Methanol emersion effectively preserves the volatile components of the sample at the time of containerization.

SOIL-PORE WATER SAMPLING

All common pollutants dissolve in water to some extent and are prone to move with the water. Therefore soil-pore water sampling is often used to determine the degree and extent of vadose zone contamination. Table 7.5 shows the pollution sources for which pore-liquid sampling is appropriate. The table lists specific sources under three classes—industrial, municipal, and agricultural—and possible pollutants. Most of these sources are regulated by federal statutes designed to protect groundwater, for example, the Resource Conservation and Recovery Act (RCRA) and the Safe Drinking Water Act of 1974. For additional information on techniques discussed in this section, see Wilson *et al.* (1995).

SOIL SAMPLING VERSUS SOIL-PORE WATER SAMPLING

Choosing between soil sampling and soil-pore water sampling depends on the goals of the sampling program. If the purpose of sampling is to characterize the distribution of pore liquid constituents throughout the vadose zone on a one-time basis, soil sampling is appropriate. However, soil sampling is essentially a destructive process in that it does not allow for the measurement of changes in a profile over time. Soil-pore water sampling allows collection of multiple samples over time, thus providing a means to evaluate specific locations in the vadose zone. The device often used to collect soil-pore water samples is called a *suction lysimeter*. We define suction lysimeters as devices with porous segments facilitating the extraction of pore liquid samples from variable-saturated regions of the vadose zone. Although these devices can be used to sample both saturated and unsaturated regions, their major role is sampling from unsaturated regions. They are used to detect pollutant movement in the vadose zone underlying new waste disposal facilities, serving as an early warning system. They also have a role in compliance monitoring and postclosure monitoring at waste sites. Suction lysimeters are also widely used in research projects (e.g., in weighing lysimeters).

In contrast to soil sampling, where samples are obtained at incremental depths (e.g., every meter) throughout a profile, because of cost, suction lysimeters are generally positioned at greater intervals. Additionally, lysimeters may not be located in the same vertical profile. The trade-off is the ability to sample throughout time.

SOIL-PORE WATER SAMPLERS (SUCTION LYSIMETERS)

The basic components of a suction lysimeter are a porous segment or cup connected to a body tube of the same diameter (Figure 7.10) (Wilson *et al.*, 1995). Two small-diameter tubes are inserted through the top, one for applying pressure/vacuum, and the second for

TABLE 7.5
Pollution Sources Suitable for Pore-Liquid Sampling

Pollution Source Class	Possible Pollutants
Industrial	
Surface impoundments (pits, ponds, lagoons)	Salinity (brine), pesticides, nerve gas, chlorinated solvents, other organic and inorganic compounds
Landfills	Salinity, toxic and hazardous wastes, metals, incinerator ash
Land treatment sites	Oily wastes, nitrogen compounds, phosphorus, heavy metals, refractory organic compounds
Underground storage tanks	Gasoline, diesel fuel, fuel oil, liquid hazardous waste
Mine drainage	Heavy metals, radionuclides, sulfuric acid
Waste piles	Salinity, sulfuric acid, heavy metals
Municipal	
Sanitary landfills	Salinity, nitrogen compounds, trace metals, pathogenic microorganisms, possibly trace organic compounds, methane
Septic systems	Nitrogen compounds, phosphorus, BOD, pathogenic microorganisms, possible trace organic compounds (e.g., TCE)
Oxidation ponds	Nitrogen compounds, BOD, phosphorus, pathogenic microorganisms
Artificial recharge facilities (soil aquifer treatment)	Nitrogen compounds, phosphorus, BOD, TOC (THM precursors), pathogenic microorganisms
Wetlands	Nitrogen compounds, phosphorus, BOD, TOC (THM precursors), pathogenic microorganisms
Urban runoff drainage wells (AKA vadose zone wells, dry wells)	Salinity, oils, grease, gasoline, nitrogen compounds, pesticides, suspended solids, pathogenic microorganisms
Agricultural	
Irrigation return flows	Salinity, nitrogen compounds, phosphorus, pesticides
Feed lots	Salinity, nitrogen compounds, phosphorus, pathogenic microorganisms
Surface impoundments	Salinity, nitrogen compounds, phosphorus, pathogenic microorganisms

Adapted from Bedient *et al.* (1999) and Fetter (1999).
AKA, Also known as; *BOD,* biological oxygen demand; *TCE,* trichloroethylene; *THM,* trihalomethane; *TOC,* total organic carbon.

transmitting collected pore liquid to the surface (see Figure 7.10). When it is properly installed (i.e., when the saturated porous segment is placed in intimate contact with the surrounding soil), the liquid within the sampler pores forms a hydraulic connection with the liquid in the soil.

OPERATING PRINCIPLES

When pore spaces of the vadose zone are completely filled with water, pore liquid may be extracted with either wells or suction lysimeters. However, when the water content is less than saturation, pore liquids will not drain into a well. The simple analogy of a household sponge may help explain this phenomenon. When a sponge is placed in a basin of water for some time, the pores become full of water (i.e., the sponge is saturated). If the sponge is removed from the water and held vertically by one end, water will drain from the other end. In time, drainage stops although ample water remains within the pores. The sponge is said to be unsaturated, but not dry. Water is retained within the sponge against the force of gravity by capillarity and adsorption of water molecules to the sponge matrix. Squeezing the sponge slightly causes drainage from the large pores. When even more pressure is applied, the finer pores drain. Similarly, we have to supply energy by squeezing or applying a

suction to get water out of unsaturated soil. Suction lysimeters are a means of applying suction to the surrounding unsaturated soil.

When the air pressure is reduced within a suction lysimeter in a soil, a suction gradient is established across the porous segment into the soil-pore liquid (see Figure 7.10). This causes a flow of pore liquid through the porous segment into the interior of the sampler. Application of a small vacuum drains water from the larger pore spaces. Drainage from the smaller and smaller pores requires more energy than may be available. Eventually drainage slows and stops unless a greater suction is applied (i.e., by further reducing the air pressure inside the sampler). Depending on soil texture, the upper limit of water movement in most soils is between about 65 and 85 centibars (cb). This means that although there may be water remaining in unsaturated pores at higher soil water suctions (e.g., in fine-textured soils), suction lysimeters are incapable of sampling them. Box 7.4 shows the common units to describe pressure and vacuum.

CONSTRUCTION MATERIAL AND BUBBLING PRESSURE SELECTION FACTORS

The porous segments of commercially available suction lysimeters are made from ceramics, PTFE, or stainless steel (Wilson *et al.*, 1995). Manufactured ceramic

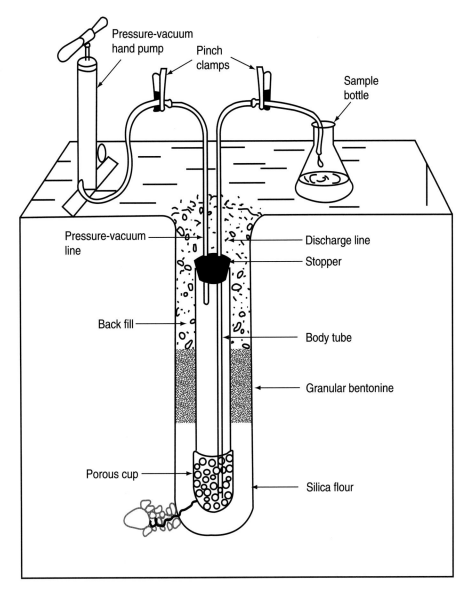

FIGURE 7.10 Design and operation of a pressure-vacuum lysimeter used to collect soil pore water samples at shallow depths. The ceramic cup is fragile and can crack or can be crushed under soil pressure. (After Fetter, 1999.)

segments are made from formulations of kaolin, talc, alumina, ball clay, and other feldspathic materials. The PTFE samplers are constructed from porous TEFLON7. Stainless steel porous segments are made from sintered stainless steel. Ceramic cups are epoxied onto the body tube. Alternative commercial ceramic and PTFE cups are threaded onto the body tube and sealed with an O-ring. The porous segments of stainless steel samplers are integrated into the body tube during construction.

The porous segments in ceramic and stainless steel samplers are hydrophilic, whereas PTFE segments are hydrophobic. The property of hydrophophilicity or hydrophobicity affects the ability of menisci in the pores to withstand suctions (i.e., to prevent air entry into the sampler). *Bubbling pressure* is defined as the air pressure at which bubbling occurs (air passes through) in the porous segment of a sampler submerged in water. The bubbling pressure is equivalent to the maximum suction that can be applied to the sampler before air entry occurs, which disrupts the flow of water into the device. Table 7.6 lists representative bubbling pressures for the three types of samplers and equivalent, maximum pore sizes. Table 7.6 also shows that the bubbling pressure for hydrophilic segments is greater than for hydrophobic samplers. Thus, the operating ranges of ceramic and stainless steel samplers are greater than for PTFE samplers. This means they will operate more effectively under drier soil conditions.

Other factors governing the selection of a porous segment type include expected contaminants, soil texture and expected pore-liquid suction range, and strength.

BOX 7.4 *Pressure/Vacuum Equivalents*

One (1) bar is the equivalent of:

100 centibars (cb)
100 kilopascals (kPa)
0.1 megapascals (MPa)
0.987 atmospheres (atm)
10^6 dynes cm^2
33.5 ft of water
401.6 in of water
1020 cm of water
29.5 in of mercury
75 cm of mercury
750 mm of mercury
14.5 pounds per square inch (psi)

Porous segments may attenuate certain pollutants during passage into the interior chamber. For example, ammonia-nitrogen, salts, and metals in pore liquids can sorb or precipitate onto ceramic and stainless steel suction lysimeters.

SAMPLER TYPES

The three basic types of suction lysimeters are single-chamber, vacuum operated; single-chamber, pressure vacuum (P/V) operated; and dual-chamber P/V operated (Wilson *et al.*, 1995). Selection among these alternatives depends largely on required sampling depths. These samplers are frequently laid horizontally. Table 7.7 lists appropriate depths for each of the common sampler types.

Single-chamber lysimeters

A simple vacuum-operated suction lysimeter consists of a porous segment integral with a body tube; a one-hole rubber stopper sealing the upper end of the body tube; small-diameter tubing, one end inserted through the stopper to the base of the porous segment, and the other end inserted into a collection flask; and a vacuum source connected to the collection flask. A vacuum ap-

propriate to the moisture conditions and the sampler type is applied (e.g., 30 cb for moist soils with stainless-steel samplers). The partial vacuum draws soil-pore liquid through the porous walls of the sampler and, via the small-diameter tubing, up into the collection flask.

A second type of chamber lysimeter called the P/V-operated sampler is designed to sample at depths beyond the reach of simple vacuum-operated units (see Figure 7.10). Although this design shows the porous segment at the base of the sampler, units are available with the porous segment integrated into the body tube at the top of the sampler. Two small-diameter lines are located within the sampler; one extends to the base of the sampler while the other terminates near the top. The longer inside tube is the sample collection line. The shorter tube is connected to a surface P/V source. The tubes are clamped off during sampling. When soil-pore water is sampled, a partial vacuum (e.g., 60 to 80 cb) is applied to the P/V line. This draws pore liquid through the porous segment into the base of the sampler. For the sample to be recovered, the vacuum is released on the sample line, pressure is applied to the P/V line, and the liquid is forced through the sample line into a collection bottle.

Dual-chamber, pressure-vacuum–operated samplers.

A concern with single-chamber samplers is that the pressure applied to retrieve the sample may force the sample out through the cup. Sample ejection may not be a problem in units containing the porous segment on top of the body tube, although applying pressures greater than the bubbling pressure may force air outside the sampler. Dual-chamber samplers avoid this problem by separating the lower, porous segment from the upper collection chamber. The line connecting the two chambers contains a check valve to prevent sample backflowing into the lower chamber. The sample collection procedure is the same as for the single-chamber units except the applied retrieval pressure cannot force the sample into the porous segment. Greater depths are possible, limited mainly by head losses in the tubes.

TABLE 7.6
Common Porous Materials Used in Suction Lysimeters

Porous Segment Material	Bubbling Pressures (cb)	Equivalent Maximum Pore Size (μm)
Ceramic	50 100 200	(6)(3)(1.5)
Stainless steel	20–60	5–15
Polytetrafluoroethylene	7–20	15–42

TABLE 7.7
Maximum Sampling Depths of Suction Lysimeters (meters)

Material	Vacuum	Pressure Vacuum	High-Pressure Vacuum
Ceramic	2	15	91
Stainless steel	1.75	3.3	>3
Polytetrafluoroethylene	1	1	1

INSTALLATION PROCEDURES

Cleaning, assembly, pressure testing, and prewetting new samplers is recommended before they are installed in the vadose zone. Cleaning is necessary to remove dust or other possible contaminants during the construction process. Ceramic and PTFE samplers are cleaned by passing dilute hydrochloric acid through the porous segment followed by distilled water. Stainless steel samplers are cleaned with isopropyl alcohol and 10% nitric acid. Samplers that can be dissembled are cleaned by placing the acid solution inside the sampler and forcing the solution through the cup by pressurization. Alternatively, the porous segments are placed in solution, which is drawn in by vacuum (Figure 7.11).

Suction lysimeters function equally well in vertical, slanted, and horizontal boreholes. The installation procedures are the same. Hand-operated bucket augers described in a previous section are suitable for constructing shallow boreholes for suction lysimeters provided soils are not too coarse grained. Power-driven units such as hollow-stem augers are required for greater depths or in coarse-grained deposits. The hollow auger flights also keep the borehole from caving during insertion of the sampler assembly. If possible, the soil removed from the borehole should be set aside in the order of excavation so that backfill soil will be replaced in the proper sequence. It is advisable to screen the excavated material (e.g., with a 3-in mesh screen) to remove gravel and cobbles that might interfere with a uniform backfilling of the hole.

Soil samples taken in regular intervals are useful for determining textural changes in the profile. When possible, the samplers should be terminated in soils above a textural change where water contents are likely to be elevated.

Soil slurry is placed in the bottom of the borehole to ensure an intimate contact between the porous segment and the surrounding soil. This placement is particularly important in coarse-textured soils. Soil slurry is prepared by mixing distilled water and sieved soil from the sampler depth to a consistency of cement mortar. An alternative approach in coarse textured soils is to prepare a slurry with 200 mesh silica flour. Silica flour slurry ensures good hydraulic contact and minimizes clogging of the porous cup by fine particulate matter transported by pore liquid. Silica flour slurry also extends the sampling range of PTFE suction cup samplers. In shallow holes, the slurry can be carefully poured to the bottom of the hole.

FREE DRAINAGE SAMPLERS

A *free drainage sampler* is a collection chamber placed in the soil to intercept liquid from overlying macropore

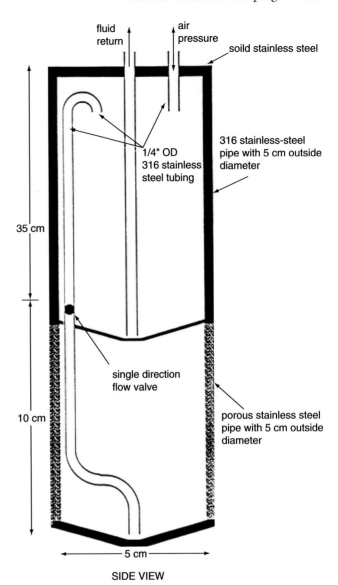

FIGURE 7.11 Diagram of a stainless steel dual chamber vacuum lysimeter used to collect soil pore water samples. The durable stainless-steel construction makes it less likely to fail when installed at depths below 3 m. (From Soil Measurement Systems.)

regions of the vadose zone that are intermittently saturated by infiltration of surface water (Wilson *et al.*, 1995). *Macropore flow* is defined as water movement through holes or channels that are more than 1 mm or more in width or diameter. These samplers are passive, collecting water by gravity rather than by vacuum. (However, some free drainage samplers are designed to apply a slight vacuum to the collecting surface.) A sand-filled funnel is a simple example of a free drainage sampler. The funnel is filled with clean sand and inserted into the roof of a cavity excavated into undisturbed soil. The funnel is connected through tubing to a collection bottle located at a lower elevation. Water percolating through macropores in the overlying soil is intercepted by the funnel and drains through the sand into the collection bottle. Water from the bottle is drawn up to the surface by vacuum. When macropore and micropore flow both occur, it is prudent to use a system of free drainage samplers and suction lysimeters. Free drainage samplers intercept saturated flow and suction lysimeters unsaturated flow.

PORE WATER SAMPLE STORAGE AND PRESERVATION

Pore water samples collected should be handled as any other water samples.

QUALITY CONTROL

EQUIPMENT CARE

All equipment used in soil sampling should be kept in working order, be cleaned promptly, and stored properly. Often, cutting edges of augers and tubes are damaged during use. Common maintenance of this equipment includes sharpening and tip replacement. It is helpful to have backup augers or tubes when conducting extensive field sampling.

Auger and tube cleaning should occur after each sampling when soil samples are collected for trace contaminant analyses. This is particularly important when textural discontinuities are encountered. For example, wet clay layers or lenses stick to the equipment and leave residues that can be dislodged in the next sampling event and mixed in with the new soil sample.

Prescribed cleaning protocols should be followed, which include brushing off all soil residues and washing the augers, tubes, and extensions with deionized water. Special detergents can be used when the equipment is contaminated with oily residues. Steam cleaning is also a very fast and effective method to decontaminate the auger flights between boreholes. This is especially important when samples are used to obtain pore liquids for chemical analyses. Although many augers are constructed of stainless steel, some are coated with nickel, cadmium, or zinc metal to reduce oxidation. These augers should not be used to collect soil samples for trace metal analysis. Most sampling tubes are constructed of stainless steel and are therefore rust resistant. Usually plastic liners are the best choice to sample and store soil samples with sampling tubes. New liners should be used for each sample because recycling them is not recommended.

SAMPLE RECORDS

Most soil, water, and gas samples collected in the field are subsequently analyzed in the laboratory. Therefore to avoid expensive resampling, all data should be logged and all samples tagged for identification. All containers must be labeled with waterproof labels. Additionally, custody seals should also be used on each container cooler used for sample storage and transportation. Chain-of-custody forms should follow each group of samples and container(s) from the point of sampling to the point of delivery, which is usually the laboratory. Finally, note that the holding time listed in Table 7.4 begins from the time of field collection.

QUESTIONS

1. Soil scientists define the smallest soil unit as a pedon. Explain why it may be impractical to collect a soil sample support the size of a pedon.

2. Explain why it may be advantageous to sample and characterize each soil horizon in a soil profile. Hint: consider how soil characteristics affect the pollutants.

3. Explain under what conditions soil sample bulking is (a) necessary and (b) not advised.

4. Figure 7.4 shows an industrial site slated for closure. Why would systematic grid soil sampling be necessary in the areas within the property fence? What environmental factors would determine the depth of soil and vadose zone sampling at each location?

5. Discuss the advantages and applications of hand tube samplers, hollow-stem auger samplers, and sonic drilling.

6. Soil pore water samples may not collect any water when the soil matrix potential is higher than their bubbling pressure. Explain.

7. Explain the potential use of soil pore water samplers to monitor the following:

   (a) Soil salinity
   (b) Total soil phosphorus

(c) Nitrate-nitrogen
(d) Total soil iron
(e) Soluble zinc
(f) pH
(g) Soluble hydrocarbons

REFERENCES AND ADDITIONAL READING

Aller, L., Bennett, T.W., Hackett, G., *et. al.* (1989) *Handbook of Suggested Practices for the Design and Installation of Ground-Water Monitoring Wells*, EPA 600/444–89/034, Published by National Water Well Association, 6375 Riverside Drive, Dublin, OH.

American Society of Testing Materials. ASTM D1452, D1586, D1587, D3550, D4700–91, D6001. American Society of Testing materials, 1916 Race Street, Philadelphia, PA.

Barcelona, M.J. (1989) Overview of the Sampling Process, in Keith, L.H., ed. *Principles of Environmental Sampling*, American Chemical Society, Washington, DC.

Barth, D.S., Mason, B.J., Starks, T.H., and Brown, K.W. (1989) *Soil Sampling Quality Assurance User's Guide*, 2nd Edition. EPA 600/8B89/04,Tl U.S. EPA, EKSL–LV, Las Vegas, NV.

Bedient, P.B., Rifai, H.S., and Newell, C.J. (1999) *Ground Water Contamination Transport and Remediation*. Prentice Hall PTR, Upper Saddle River, NJ.

Boulding, J.R. (1994) *Description and Sampling of Contaminated Soils: A Field Guide*. 2nd Edition. Lewis Publishers, Boca Raton, FL.

Brady, N.C., and Weil, R.R. (1996) *The Nature and Properties of Soils*. 11th Edition. Prentice Hall, Upper Saddle River, NJ.

Buol, S.W., Hole, F.D, and McCracken, R.J. (1973) *Soil Genesis and Classification*. The Iowa State University Press, Ames, IA.

Carter, R.M. (1993) *Soil Sampling and Methods of Analysis*. Canadian Society of Soil Science. Lewis Publishers, Boca Raton, FL.

Dorrance, D.W., Wilson, L.G., Everett, L.G., and Cullen, S.J. (1995) A compendium of soil samplers for the vadose zone, in Wilson, L.G. L.G. Everett, and S.J. Cullen, Eds., *Handbook of Vadose Zone Characterization and Monitoring*, Lewis Press, Chelsea, MI.

Fetter, C.W. (1999) *Contaminant Hydrogeology*, 2nd Edition. Prentice Hall PTR. Upper Saddle River, NJ.

Hackett, G. (1987) Drilling and constructing monitoring wells with hollow-stem augers; Part 1: Drilling considerations, *GWMR*, 7, 51–62.

Klute, A. (1986) *Methods of Soil Analysis: Part 1—Physical and Mineralogical Methods*. (2nd Edition). American Society of Agronomy, Soil Science Society of America. Madison, WI.

Lee, R.S., and Connor, J.A. (1999) Hydrogeologic site investigations. In Bedient, P.B., Rifai, H.S., and Newell, C.J., *Ground water contamination and remediation, Transport and Remediation*, 2nd Edition. Prentice Hall, Upper Saddle River, NJ.

Lewis, T.E., Crockett, A.B., Siegrist, R.L., and Zarrabi, K. (1991) *Soil Sampling and Analysis for Volatile Organic Compounds*, EPA/540/4–91/001, Ground Water Forum Issue, U.S. EPA, Las Vegas, NV.

Lewis, T.E., and Wilson, L.G. (1995) Soil sampling for volatile organic compounds, in Wilson, L.G. L.G. Everett, and S.J. Cullen, Eds., *Handbook of Vadose Zone Characterization and Monitoring*, Lewis Press, Chelsea, MI.

Norris, F.A., Jones, R.L., Kirkland, S.D., and Marquardt, T.E. (1991) Techniques for collecting soil samples in field research studies, in Nash, R.G. and Leslie, A.R., Eds. *Groundwater Residue Sampling Design*, ACS Symposium Series 465, American Chemical Society, Washington, DC.

Pepper, I.L., Gerba, C.P., and Brusseau, M.L., eds. (1996) *Pollution Science*, Academic Press, San Diego, CA.

Soil Survey Staff. (1993) *Soil Survey Manual Agricultural Handbook* 18. U.S. Soil Conservation Service.

Wilson, L.G., Everett, L.G., and Cullen, S.J., eds. (1995) *Handbook of Vadose Zone Characterization and Monitoring*, Lewis Press, Chelsea, MI.

8

GROUNDWATER SAMPLING

M. BARACKMAN AND M.L. BRUSSEAU

The topic of groundwater sampling encompasses a broad and still expanding range of disciplines and methods. The field has evolved rapidly throughout the 1980s and 1990s as increased attention has been focused on problems of groundwater contamination by various organic and inorganic compounds. As advances in laboratory analytical methods allow detection of ever smaller concentrations of increasing numbers of compounds in groundwater, the need for accuracy and precision in sampling techniques continues to grow.

The purpose of this chapter is to provide an overview of the diverse and evolving field of groundwater sampling. It is not possible to give a complete treatment to this topic within the confines of a single chapter. References are provided to direct the reader to more complete and detailed coverage of specific topics.

GROUNDWATER SAMPLING OBJECTIVES

To conduct an effective groundwater sampling effort, one must understand the specific objectives of the sampling program. Often, one or more of the objectives will dictate certain sampling methods, locations, or protocols. Some of the more common objectives include satisfying requirements of state or federal regulatory programs, providing data needed for management or cleanup of a contaminated site, and assessing baseline conditions before site development.

Groundwater sampling is often conducted under the requirements of state or federal regulatory programs. Federal regulatory programs that include requirements or guidelines for groundwater monitoring include the following:

- Comprehensive Environmental Response Compensation and Liability Act (CERCLA), commonly known as the "Superfund" program that regulates characterization and cleanup at uncontrolled hazardous waste sites
- Resource Conservation and Recovery Act (RCRA), which regulates management and remediation of waste storage and disposal sites
- Safe Drinking Water Act (SDWA), which regulates drinking water quality for public water supplies and also includes the Underground Injection Control (UIC) Program and the Well Head Protection Program
- Surface Mining Control and Reclamation Act (SMCRA), which regulates permitting for open pit mining operations

In addition to the federal regulatory programs, many states have regulatory programs that are concerned with groundwater monitoring. Some regulatory programs specify only the general requirement for groundwater monitoring. Others, such as RCRA, provide detailed guidelines for most aspects of the sampling program, including well design and placement, sample collection methods, target analytes, and statistical techniques used to evaluate the results. A more complete discussion of the federal regulatory programs mandating groundwater sampling is provided by Makeig (1991).

Much of the groundwater sampling conducted during the past two decades has been focused on characterization and cleanup of sites where groundwater has become contaminated through spills, leaks, or land disposal of wastes. Groundwater sampling programs at contaminated sites are usually conducted to obtain data necessary for making decisions on site management or cleanup in addition to satisfying regulatory requirements. The specific objectives of the groundwater sampling program may change as the site becomes better characterized and as site remediation progresses. The frequency of sampling may also be adjusted depending on how rapidly groundwater conditions are changing at the site.

Groundwater sampling is sometimes conducted to establish baseline conditions before development of a new facility or to characterize ambient conditions across a large area such as in basin-wide studies. Baseline sampling is usually meant to provide a "snapshot" of groundwater conditions at a particular time. The analytical suite for baseline sampling usually includes common anions and cations and, depending on the objectives of the study,

may include trace metals, pesticides, and a range of organic compounds. If baseline sampling is being conducted before industrial development of a site, the analytical suite will typically include potential contaminants that may be present on the site once the facility is in operation.

Leak detection monitoring, sometimes referred to as *compliance monitoring,* is commonly conducted at landfills, hazardous waste storage facilities, and chemical storage and manufacturing facilities. This type of monitoring is designed to provide early warning of contamination associated with releases from these facilities. To be effective, leak detection monitoring must be conducted at wells that are properly located, generally in areas downgradient from the facility of interest. The sampling frequency and list of analytes must be adequate to detect a release before the contaminants have migrated to any sensitive downgradient receptors. Typically, upgradient monitor wells are also included in leak detection monitoring to provide baseline water quality data for groundwater moving to the site to allow for comparative evaluations.

A common objective for all types of groundwater sampling is to obtain samples that are representative of the groundwater conditions in the subsurface where the sample originated. Many factors can affect the properties of a groundwater sample during the process of sampling. The following sections provide a discussion of some of the more important factors that must be considered to design and conduct a successful groundwater sampling program.

LOCATION OF MONITOR WELLS

Groundwater monitoring is typically conducted by collecting samples from a network of monitor wells. There are many site-specific variables that must be considered in the design of a monitor well network. If the monitor wells are not located or constructed appropriately, the data collected may be misleading. For example, if monitor wells are too widely spaced or improperly screened, the zones of contaminated groundwater may be poorly defined or even completely undetected. The task of designing a monitor well network is generally more difficult for sites with more complex or heterogeneous subsurface conditions compared with sites with more homogeneous subsurface conditions.

For initial investigations of sites where existing hydrogeological data are sparse, it is usually most cost-effective to implement a phased program of monitor well construction. In the phased approach, a group of wells is installed, and lithological data (e.g., rock/sediment type), groundwater levels, and groundwater quality data from the initial set of wells are collected and interpreted before deciding on placement and design of the second phase of monitor wells. This approach allows for more informed decision making, and generally results in a more cost-effective monitor well installation program. The following sections describe some of the factors that affect the design of monitor well networks.

HYDRAULIC GRADIENT

Groundwater flows from regions of higher hydraulic head to regions of lower hydraulic head. The change in hydraulic head along a groundwater flow path is termed the *hydraulic gradient*. The hydraulic gradient has both a magnitude and direction. The velocity of groundwater flow is proportional to the magnitude of the hydraulic gradient and the hydraulic conductivity of the aquifer (see Chapter 12). Groundwater flows faster where the hydraulic gradient and/or hydraulic conductivity are larger. Groundwater flow velocities are much slower than surface water flow velocities, except in limestone karst formations, where groundwater flows through caves and large solution channels. The range of groundwater flow velocity varies greatly, but does not commonly exceed a few meters per day.

During initial investigation of a site where there are no existing wells nearby, it is sometimes possible to infer the direction of groundwater flow in shallow, unconfined aquifers by observing the surface water drainage patterns. On the regional scale, groundwater flow patterns often follow surface water flow patterns. In basins where extensive pumping of groundwater is occurring, groundwater flow directions may be primarily controlled by pumping wells. In lowland areas near surface water bodies, groundwater flow directions may fluctuate in response to changes in surface water levels. In basins where groundwater is considered an important resource, there are often published reports available that provide information on groundwater levels and flow directions. If the direction of regional groundwater flow can be determined, this information can be used to help select the location of the initial monitor wells at a site. A series of groundwater level measurements should be collected after the initial wells are installed to confirm the groundwater flow patterns at the site. The hydraulic gradient is often depicted with contours of equal groundwater level elevation, similar to the way terrain contours are drawn on a topographical map. This is a useful means of displaying the complex nature of groundwater flow patterns, particularly on a regional scale.

All of the factors mentioned above may influence decisions about placement and design of monitor wells. At sites where groundwater contamination is being investigated, it is usually desirable to have wells located both upgradient and downgradient of the location where the contaminants were suspected of being released to the groundwater. This placement of wells allows for an assessment of the impact of the particular contaminant source area being investigated.

GEOLOGICAL CONDITIONS

Groundwater will tend to follow the path of least resistance through the aquifer. Groundwater flow paths will change where there are abrupt contrasts in hydraulic properties of the aquifer, such as where a low permeability bedrock outcrop intrudes into a sedimentary basin. In alluvial basins, groundwater flow may tend to preferentially follow zones of coarse-grained sediments such as buried stream channels or point bar deposits. In fractured rock aquifers, groundwater generally exhibits significant preferential flow, following the open, conductive fractures and bypassing the dense matrix. Such geological properties are often an important factor in the interpretation of groundwater levels and groundwater quality data and in the siting of monitor wells. This section provides an overview of a few of the more common types of geological conditions that complicate the task of design and placement of monitor wells.

Heterogeneous aquifers. Heterogeneity refers to the variation of hydrogeological properties from place to place in the subsurface. The degree of heterogeneity varies from site to site and is related to the environment in which the sediments were deposited. Glacial and fluvial sediments typically have higher degrees of heterogeneity than marine or lacustrine sediments. Discontinuous lenses of different sediment types are common in glacial and fluvial environments (Figure 8.1).

Groundwater flow and contaminant transport can be strongly affected by heterogeneity. In highly heterogeneous systems, groundwater flow paths can be very complex and convoluted. Groundwater will tend to flow more rapidly through the coarse-grained lenses and channels and very slowly through fine-grained lenses. Selection of well locations and screened interval depths is difficult in these environments.

Selection of screened interval depths is particularly difficult in these environments. Wells that are screened over large intervals of the aquifer have a better chance of intercepting a contaminated zone but may result in a diluted sample because of the contribution of uncontaminated groundwater from surrounding zones. Conversely, wells that are screened over short intervals may not access those zones where contaminants are present. Because of these uncertainties, the number of monitor wells required to adequately characterize a heterogeneous site is larger than that for a relatively homogeneous site.

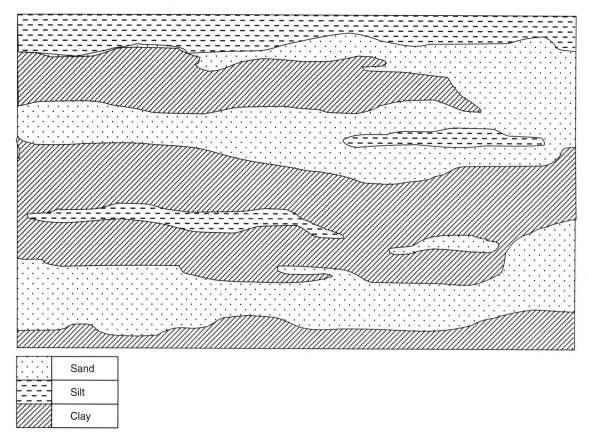

FIGURE 8.1   Vertical cross section illustrating the physical heterogeneity typical to subsurface environments (drafted by Colleen McColl).

Multiple aquifer systems: It is not uncommon to encounter multiple aquifer systems in sedimentary basins. Multiple aquifer systems are formed where laterally extensive low permeability silt or clay layers separate higher permeability layers of more coarsely grained sediments. The low permeability layers separating the aquifer zones are termed *aquitards*. A schematic drawing of a multiple aquifer system is presented in Figure 8.2. Groundwater flow directions and quality may vary considerably between the different aquifer zones in a multiple aquifer system.

Monitor wells in multiple aquifer systems must be located and constructed with care to avoid cross contamination between aquifer zones. Wells that are improperly sealed or constructed can provide conduits for contaminants to move between aquifer zones and result in greatly increased cleanup costs. Typically, multiple aquifer systems are monitored separately, with different sets of monitor wells constructed in each aquifer zone. Multiple completion monitor wells may be constructed by installing two or more well casings in a single borehole, as shown in Figure 8.3. However, care must be taken to insure correct placement and integrity of the seals between the aquifer zones.

Fractured rock aquifers: In fractured rock aquifers, groundwater moves almost exclusively through the network of open fractures. The locations of and connections between the fractures are very difficult to map. In some fractured rock environments, wells constructed a few meters apart may penetrate different fractures that are hydraulically disconnected. Because the groundwater flow paths are circuitous, movement of contaminants through fracture networks is often unpredictable. Where bedrock is exposed, it is sometimes possible to map the orientation of the fracture sets where they are visible on the surface. Monitor wells that are located along the principal axis of the fracture sets that intersect a contaminant source area may have a higher probability of encountering fractures through which groundwater is flowing preferentially.

WELL CONSTRUCTION

Many of the current monitor well drilling and construction methods evolved during the 1970s. Techniques have been adapted from both the water well and petroleum drilling industries to meet the special challenges of

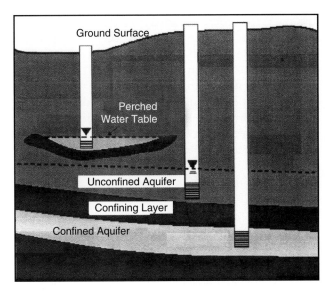

FIGURE 8.2 Vertical cross section illustrating a three-aquifer system: perched aquifer; unconfined upper aquifer; and lower, confined aquifer (drafted by Gale Famisan).

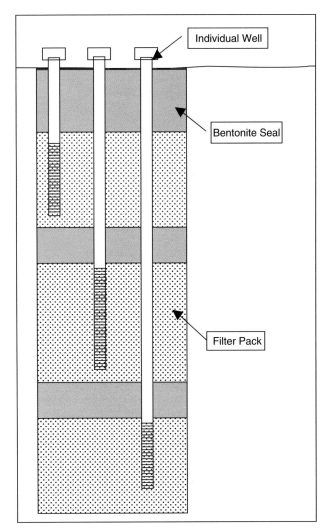

FIGURE 8.3 Vertical cross section showing a well cluster (multiple wells in a single borehole) (drafted by Colleen McColl).

monitor well drilling. As regulatory guidelines have grown more stringent and a better understanding about potential problems caused by improper materials or well construction has developed, monitor well installation methods have become more sophisticated. In the United States, many drilling contractors have developed special equipment and techniques specifically for monitor well installation.

Although many site-specific variables must be considered in the design of groundwater monitor wells, most monitor wells share some common design elements (Figure 8.4). Typically the upper section of the well is protected with a short section of steel casing called surface casing. The surface casing protects the well casing from physical damage by vehicles or from frost heaving in cold climates. Surface casing may also be grouted to provide a seal against surface water infiltration. The well casing extends downward through the surface casing. Grout materials are installed around the outside of the well casing to prevent movement of surface water down the borehole. Well screen is installed in the interval where the groundwater is to be monitored. In monitor wells, the well screen is typically either wire wrapped (also known as *continuous slot*) or slotted. Surrounding the well screen is a gravel pack, which serves to filter out fine sand and silt particles that would otherwise enter the well through the well screen.

A monitor well is designed to provide relatively small volume samples (e.g., tens of milliliters to a few liters) from the aquifer. It is desirable that these samples be as representative of the groundwater in the aquifer as pos-

sible. Small-diameter monitor wells are desirable (2.5 to 10 cm) to avoid pumping large quantities of contaminated water during well purging. Care must be taken during drilling of monitor wells to avoid contamination of the aquifer from overlying soils; contaminated drilling tools; or lubricants, adhesives, and drilling fluid additives used in the drilling operations. There may also be a need to collect discrete samples of soil or groundwater during the drilling of monitor wells.

MONITOR WELL DRILLING METHODS

Monitor wells are typically constructed by one of the following principal methods: (1) driving, (2) jetting, (3) auger drilling, (4) cable tool drilling, (5) rotary drilling, (6) casing hammer or Becker drilling, or (7) resonant sonic drilling. Each of these methods has

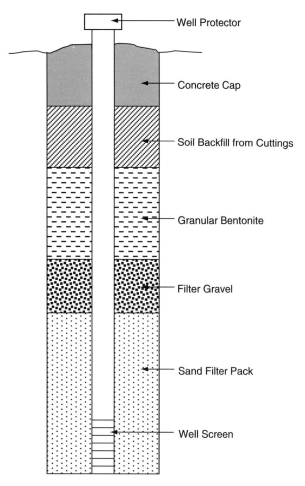

Well Protector

Concrete Cap

Soil Backfill from Cuttings

Granular Bentonite

Filter Gravel

Sand Filter Pack

Well Screen

FIGURE 8.4 Seal schematic of generic monitor well design elements (drafted by Colleen McColl).

advantages in different hydrogeological environments. For a more complete discussion of drilling methods, see Driscoll (1995). Brief descriptions of each of these methods are provided in Box 8.1.

Wells are generally drilled vertically downward. However, for certain conditions, so-called directional or angled drilling may be used. Directional drilling was originally developed for use in the petroleum industry and has recently been adapted for groundwater monitoring and reclamation wells. In environmental applications, horizontal drilling has been used where buildings or other obstructions prevent the installation of vertical wells or where the number of vertical wells needed to capture a contaminant plume would be impractical, such as in low permeability sediments. This method is more costly and time-consuming than vertical drilling methods. The horizontal well also presents challenges for the installation of pumps and other sampling equipment.

MONITOR WELL CONSTRUCTION MATERIALS

Materials that are used to construct monitor wells generally include a well casing and a gravel pack installed around the well screen, and grout plugs installed around the outside of the casing above or between the screened intervals. The following sections describe some of the considerations related to the choice of materials for monitor well construction.

Well casing and screen materials. The monitor well casing must be sturdy enough to withstand the stress of

---

BOX 8.1 *Comparison of Monitor Well Drilling Methods*

Driven Wells—Driven wells are used for shallow aquifers in unconsolidated formations. Driven wells consist of small-diameter steel pipe tipped with a specially constructed section of well screen that terminates in a tapered drive point. The drive point and screen are mounted on the bottom of the pipe, and the assembly is driven into the ground by hammering. Driven wells can be installed by hand in unconsolidated formations to depths of about 10 m. With the use of machine-driven hammers, depths of up to about 30 m can be reached. The inside diameter of the steel pipe used in drive point wells is typically 2 in (5 cm) or less.

Advantages—Driven wells offer low cost and rapid installation. There are no cuttings or spoil generated during the installation process. Driven wells can be

installed by hand in areas that are inaccessible to drilling equipment.

Disadvantages—Driven wells provide no opportunity to observe or sample the sediments as the well is installed. Unless a borehole is constructed before driving the well, there is no way to install grout seals in a driven well. The relatively small diameter of the drive pipe limits the types of sampling devices that can be installed.

Jetted Wells—The jetting method involves forcing a stream of high velocity water from the bottom of a section of well casing or screen. The water stream loosens the sediments at the bottom of the hole and transports the cuttings to the surface around the outside of the well casing. This action allows the casing to sink or be easily driven into the formation. The jetting method can be used to install small-diameter wells to depths of up to approximately 60 m in sand and fine gravel formations. Depending on the depth of the well and the characteristics of the formation, it may be necessary to increase the viscosity of the jetting fluid by adding bentonite or other additives. Jetting is sometimes used to place the well screen below the bottom of well casing that has been installed and grouted in place with other drilling methods.

Advantages—Jetting wells are low cost and can be rapidly installed.

Disadvantages—Jetting is generally not successful in formations containing layers of clay or cobbles. The jetting method does not allow for placement of a gravel pack around the outside of the well screen or grout seals around the outside of the well casing. It is not possible to obtain reliable lithologic information during jetting. Because the cuttings and water return to the surface, this is not a good choice for contaminated systems.

Auger Drilling—Auger drilling uses a rotating drilling string with spiral-shaped cutting edges called *flights*. There are several related drilling methods that are included under the category of auger drilling. Bucket augers and earth or construction augers consist of short sections of auger flights attached to a telescoping steel-drilling column. Bucket augers are operated by drilling a short distance until the flights become filled with cuttings, then raising the cutting head to the surface where the cuttings are spun or tipped off the cutting head. Continuous flight augers consist of interlocking sections of drilling column that are all equipped with auger flights. The cuttings from the bottom of the hole are continuously lifted to the surface as the auger spins. Hollow-stem augers are continuous flight augers constructed so that an open section of drill column extends through the center of the flights. Hollow-stem augers are available with center column diameters ranging from about 6.4 to 33 cm.

Advantages—With hollow-stem augers, it is possible to construct wells in unstable formations by installing the well casing, gravel pack, and well sealing materials through the hollow central column. During this time the auger remains in the borehole and is withdrawn in stages. In this way, the hollow-stem auger acts as a temporary casing. Hollow-stem augers also provide an excellent means of obtaining undisturbed soil samples during drilling. Various types of samplers can be inserted through the open central column and driven into the soil ahead of the auger. No drilling fluids are used with auger drilling methods. This can result in considerable cost savings when drilling at contaminated sites where spent drilling fluids must be handled as hazardous waste.

Disadvantages—Auger drilling is generally limited to use in unconsolidated formations at depths of less

than 60 m. For the construction of wells with bucket augers, earth augers, or solid-stem augers, the formation must be stable so that the borehole will remain open long enough to allow removal of the drilling tools and placement of the well casing and screen. Where clay is present, continuous flight augers may smear borehole walls by the rotation of the auger. This smearing can partially seal off the more permeable zones in the formation. This may result in unreliable estimates of aquifer hydraulic properties derived from pumping tests in auger-drilled wells.

Cable Tool Drilling—Also known as *percussion drilling*, cable tool drilling uses a string of heavy drilling tools that are attached to a cable and repeatedly lifted and then dropped into the borehole. The repeated action of the drill bit crushes or loosens the formation. Above the water table, a small amount of water is added to the borehole, which forms a slurry in the bottom of the hole. As drilling progresses, this slurry is periodically removed by means of a bailer. In unstable formations, a steel casing is driven into the borehole, behind the advance of the drill bit.

Advantages—Cable tool drilling is applicable to a wide variety of geological environments and can be used to install wells to depths of several hundreds of meters. When the casing is driven during drilling, excellent lithological information can be obtained because the cuttings bailed from the borehole are not mingled with sediments from higher elevations.

Disadvantages—There is typically a lot of splash as the bailer is emptied, making cable tool drilling a poor choice from the standpoint of worker health and safety at heavily contaminated sites. Cable tool drilling is also slow in comparison with rotary methods.

Rotary Drilling—Rotary drilling encompasses a number of similar drilling methods, all of which incorporate a rotating drilling column through which the drilling fluid, which can be liquid or air, is circulated to bring drill cuttings to the surface. There are several variations in the types of drill columns that are used and in the way that the liquid or air circulates in the borehole. In conventional rotary drilling, drilling fluid is pumped down through the hollow drill column through holes in the face of the drill bit. The circulating fluid cools and lubricates the drill bit and lifts cuttings from the hole as it returns up the borehole to the surface. At the surface, the cuttings are screened or allowed to settle, and the drilling fluid is recirculated back into the borehole. In reverse rotary drilling, the drilling fluid moves downward through the annular space in the borehole and back up through the drill column. Because the fluid is moving upward through the relatively small-diameter drill pipe, the fluid

(Continued)

BOX 8.1 *(Continued)*

velocities are higher, and cuttings are more effectively lifted from the well. Dual tube rotary methods use a drill string made of concentric steel pipe. The circulating fluid or air moves downward through the annular space between the inner and outer pipes and returns to the surface through the inner pipe. Some rotary drill rigs have the capability to drive the casing into the borehole during drilling. This is particularly useful where there are unstable formations or when there is a need to isolate shallow contaminated zones from deeper portions of the formation. The specific type of rotary method chosen will depend on the drilling conditions, the need for obtaining soil samples and lithological information during drilling, and the considerations for controlling the possible spread of contaminants during drilling.

Advantages—Rotary drilling methods are adaptable to most geological environments and can be used to install wells to depths of several hundreds of meters. In most geological formations, rotary drilling will progress faster than percussion or auger drilling. Air rotary drilling methods offer the added advantage of not having to manage potentially hazardous drilling fluids.

Disadvantages—Dust and vapors generated during air drilling are difficult to contain and may present health and safety risks for workers at contaminated sites. Rotary drilling is also generally more expensive than other methods.

Casing Hammer Drilling—Casing hammer drilling involves driving heavy walled casing into the ground with a steam or hydraulic-driven hammer similar to those used to drive bridge pilings. A double-walled pipe is used for the drive casing. A hardened drive shoe is attached to the bottom end of the drive casing. Compressed air is circulated down the annular space between the inner and outer casing pipes and returns up the inside of the inner pipe, lifting the cuttings from the borehole. The discharge airstream is routed through a cyclone separator to remove the cuttings from the airstream.

Advantages—The drill cuttings produced by casing hammer drilling arrive at the surface relatively intact and are representative of a discrete depth. This facilitates the preparation of detailed lithological logs. Sampling tools can easily be inserted into the open inner drill casing for collection of sediment, water, or vapor samples during drilling. When the borehole reaches the desired depth, the well casing and screen are installed through the open inner drill casing. Gravel pack and grout materials are placed around the well casing sequentially, as sections of the drill casing are withdrawn. The smooth inner walls of the drill casing and the uniform diameter of the borehole aid accurate placement of these materials. Casing hammer drilling is adaptable to a wide range of geological conditions and can drill to depths of more than 100 m. Casing hammer drilling can produce very rapid drilling rates in many geological formations.

Disadvantages—Casing hammer drilling is expensive and may be slower than rotary drilling methods in clays.

Resonant Sonic Drilling—The basic operation principles of sonic drilling are presented in Chapter 7 (see the "Sonic Drilling" section). Wells that are more than 100 m deep may be constructed with resonant sonic drilling methods.

Advantages—It is particularly suited to drilling in zones of cobbles, where other drilling methods may have difficulty. This method is well suited to construct monitoring wells more than 100 m deep. This method has several advantages for constructing monitor wells. Excellent lithological information can be obtained from the soil core produced with resonant sonic drilling. Subsamples of sediments can be obtained from any depth in the core and submitted for laboratory analysis. Many resonant sonic drilling rigs have the capability to drill angled boreholes, which can be useful if there are buildings, utilities, or other obstructions blocking the area where a monitor well is needed, or when attempting to intersect vertical fractures. Resonant sonic drilling produces minimal worker exposure to contaminated soil because the soil core remains intact and can be easily contained as it is extruded from the core barrel.

Disadvantages—This method is less suited to drilling in thick sequences of plastic clays. Depending on the drilling conditions and depth of the wells, resonant sonic drilling may be somewhat slower and more expensive than rotary or casing hammer methods. The sediment collected may experience heating and some size separation, which may affect their integrity for physical and chemical characterization.

emplacement in the borehole and the pressure exerted by the backfill materials. It must also be chemically and physically compatible with the subsurface environment where it will be installed. Ideally, the material should be chemically inert with respect to the type of contaminants being monitored. In most instances, cost is also a consideration in choosing well casing materials. The casing used in monitor wells is typically constructed of one of four general types of materials (Nielsen & Schalla, 1991, p. 249):

- Thermoplastic materials, including polyvinyl chloride (PVC) and acrylonitrile butadiene styrene (ABS)
- Metallic materials, including low carbon steel, galvanized steel, and stainless steel (particularly types 304 and 316)
- Polytetrafluoroethylene (PTFE), known by the trade name Teflon
- Fiberglass-reinforced epoxy (FRE)

Each of these materials has properties that may make it more suitable than others for a particular set of site conditions and groundwater sampling requirements.

Structural considerations: The stress on a monitor well casing occurs both during and after the well construction. During well construction the casing hangs suspended in the borehole. The casing joints must be strong enough to support the weight of the hanging casing string. During and after construction, the casing is subjected to the weight exerted by the backfill materials. Metallic materials generally provide the greatest structural strength. FRE is nearly as strong as stainless steel but much lighter in weight. Thermoplastic materials, particularly in the thicker wall sizes can also provide adequate structural strength for most monitor well installations. PTFE has the poorest structural strength and special care is needed to avoid distortion or bending of the well casing and screen when constructing monitor wells with PTFE. In addition to the casing material, the structural strength of the well casing also depends on the type of joints used. The most common jointing technique for monitor well construction is flush-threaded pipe. Other joining methods include welding for metallic casings, and the use of threaded couplings for plastic and metal pipe and groove locking couplings for FRE casing.

Physical/chemical compatibility considerations: At some sites, the physical or chemical conditions in the subsurface may be incompatible with certain casing materials. For example, at some highly contaminated sites the sediments may be saturated with various organic solvents. Although thermoplastic materials generally have good resistance to degradation from fuels and chlorinated solvents, some organic solvents such as methyl ethyl ketone, tetrahydrofuran, and methyl isobutyl ketone will soften PVC. Barcelona *et al.* (1983) provide a list of the groups of chemical compounds that may degrade thermoplastic well casing materials. Unfortunately, the threshold concentration at which the degradation occurs is not well documented. Metal, PTFE, and FRE casing materials are generally resistant to degradation from organic solvents.

Metal casing may be subject to rapid corrosion in certain geochemical environments. The factors contributing to corrosion are complex. In general, corrosive conditions are associated with high or low pH; with high concentrations of dissolved oxygen, dissolved carbon dioxide, or dissolved chloride; or with the presence of hydrogen sulfide. Corrosive conditions are commonly associated with landfills and sites affected by acid mine drainage. Stainless steel is much more resistant to corrosion than low carbon steels. Thermoplastic, PTFE, or FRE casing is generally not subject to corrosion and can be used in place of metal casings in corrosive environments.

For any bias to be avoided in the samples obtained from a monitor well, the casing materials should be inert with respect to the contaminants being monitored at the site. In 1986, the U.S. Environmental Protection Agency (EPA) (1986) issued a Groundwater Monitoring Technical Enforcement Guidance Document that established guidelines for groundwater monitoring at waste storage and disposal sites administered under the RCRA program. This guidance document specified that only fluoropolymer (PTFE) or stainless steel casings could be used for monitor wells because these were considered to be the most inert. However, since that time, several studies have shown that PVC may in some cases be used in place of these more expensive materials with no significant loss of sample quality.

DESIGN AND PLACEMENT OF MONITOR WELL SCREENS

Well screen describes any of several different designs of perforated casing used to allow water to enter a well. When a monitor well is designed, consideration must be given to the type of screen to be installed and the placement of the screened interval in the well. Proper screen design is necessary so that the well can provide adequate quantities of sediment-free groundwater for sampling. Placement and length of the screen will determine whether the monitor well provides discrete samples from a thin vertical section of the subsurface zone of interest or provides integrated samples from a larger vertical cross section of the subsurface domain.

Well screen design: Many different types of well screens are available for use in production wells. Most monitor wells use either machine slotted screen or continuous slot wire-wrapped screen. Machine slotted screen is manufactured by cutting slots of precise widths in ordinary well casing. The slots in slotted screen are a consistent width throughout the wall thickness of the casing. Wire-wrapped continuous slot screen is manufactured by spirally winding wire with a triangular cross section around an array of longitudinally oriented rods. At each point where the wire crosses a rod, the two are joined by welding. Continuous slot screen, given its larger slots, is more resistant to plugging from particles becoming trapped in the slots.

The well screen is designed to allow water to enter the well while preventing the entry of the surrounding gravel pack. For effectiveness, the slot size of the well screen must be chosen to be smaller than the majority of the grains in the gravel pack. The slot size is generally chosen to hold back between 85% and 100% of the grains in the gravel pack materials. The gravel pack grain size is chosen based on the character of the geological formation in which the well is screened. See Aller *et al.* (1989) for a more complete discussion of gravel pack and well screen design.

Length and placement of monitor well screens. The selection of the length of a monitoring well screen depends on the intended use of the well. In addition to providing groundwater samples, monitoring wells may be used as water level measuring points and as test wells for evaluating hydraulic properties of the aquifer. These different objectives sometimes present conflicting requirements for screened interval length.

For baseline monitoring in an aquifer used to supply potable water, it may be appropriate to construct monitor wells that are screened over a large portion of the aquifer thickness. However, contaminant concentrations may vary substantially across a vertical cross section of the aquifer. Depending on the sampling method, wells with long screened intervals may provide an integrated sample that is representative of some average concentration across the screened interval. It is generally more difficult to use monitor wells with long screened intervals to obtain information on the vertical distribution of contaminants in the aquifer.

At sites with stratified geological features, it is not uncommon for contaminants to be segregated into specific horizons or zones within the subsurface. Wells that are screened across a large thickness of the subsurface can provide conduits for movement of contaminants into previously uncontaminated zones. It is therefore important to understand something of the vertical distribution of contaminants in the subsurface before installing wells that are screened across multiple aquifer zones.

When it is necessary to obtain information on the vertical distribution of contaminants in the subsurface, monitor wells are constructed with short sections of screen open only to selected zones within the subsurface. Screen lengths as small as 0.3 m may be used in detailed investigation of contaminant plume geometry. More commonly, screen lengths ranging from 1 to 3 m are used for characterizing vertical contaminant distributions. A variation of this approach uses drive-point well devices, which in essence are pipes with a drive point and a screen located at the leading end (see Box 8.1). Wells with short screened intervals provide samples and water level data that are representative of a discrete subsurface horizon. With a sufficient number of selectively screened monitor wells, the three-dimensional distribution and direction of movement of contaminants can be characterized. Multiport sampling devices may be installed as an alternative to monitor wells. In place of well screen, these sampling devices use small sampling ports installed on the outside of the well casing. Multiport sampling devices provide a larger number of sampling ports in a single borehole than can be achieved with conventional screened wells. However, they are more costly and difficult to construct than conventional monitor wells.

Aquifers with changing groundwater levels complicate the decision on well screen placement. Groundwater levels may fluctuate seasonally in response to changes in recharge or to changes in pumping of production wells. In many areas, average groundwater levels are declining because of overpumping of the aquifer. Monitor wells that are installed with short screened sections located near the water table may be left dry and useless when groundwater levels decline. For an increase in the useful life of monitor wells in these areas, longer screened sections should be installed in those wells that are at risk of being left dry by declining regional groundwater levels.

MONITOR WELL SEALING MATERIALS AND PROCEDURES

The borehole that is drilled for well installation is by necessity larger than the well casing. The area between the well casing and the wall of the borehole is called the *annular space* or *annulus*. To ensure that the samples produced by the well are representative of the screened intervals, one must prevent groundwater flow through this annular space. Various sealing materials are installed in the annular space to accomplish this. The seals are used to prevent the infiltration of water from the surface and to isolate the screened intervals in the well. For a more complete discussion of monitor well sealing, see Aller *et al.* (1989).

Monitor wells are usually sealed with either a cement-based grout or one of several types of bentonite clay materials. The choice of grout materials may depend on the type of well casing materials used, the location of the sealing materials in relation to the water table, and local or state regulations. It is not uncommon for bentonite-based sealing materials to be used in the lower portions of the well, with cement-based grout placed near the ground surface.

Cement-based grout materials are formulated from a mixture of sand and cement, with no gravel. Cement grouts are mixed with water to form a slurry and placed in the annulus either by inserting a temporary pipe, known as a *tremie pipe*, that extends to the bottom of the grouted interval or by pouring the grout from the surface down the annulus. The tremie pipe method is preferred because it allows the grout to fill the annulus from below and minimizes the chance of voids being formed during grout placement. Cement grout shrinks slightly as it hardens. This shrinkage can result in the formation of cracks that can compromise the integrity of the grout seal. Special low-shrinkage cements may be used to mitigate this problem. Up to 10% powdered bentonite may be added to a standard cement grout mixture to reduce shrinkage and cracking. Cement-based grout materials are highly alkaline, with pH ranges from 10 to 12. Cement grout may alter the pH of groundwater with which it comes in contact. This effect can be severe and persistent. It is therefore necessary to ensure that no cement grout is allowed to enter the gravel pack or is placed in the vicinity of the screened section of the well.

Bentonite is a clay mineral that expands when hydrated with water. Bentonite does not harden like cement and is therefore not subject to cracking so long as it remains hydrated. Bentonite is available in several forms, including pellets, granules, and powder. It may be placed in the borehole as a slurry, using the same methods as for cement grout, or it may be placed in dry form and allowed to hydrate in place. Bentonite is less alkaline than cement grout and is therefore more suitable for use in proximity to the screened interval of the well. Like most clay minerals, bentonite has a high cation exchange capacity. This allows it to exchange cations such as sodium, calcium, aluminum, iron, and manganese with the groundwater. These reactions can alter the chemistry of the groundwater that comes in contact with the bentonite.

To protect the chemical integrity of the samples and the hydraulic properties of the well, one must prevent grout materials from infiltrating the gravel pack. If grout slurry is placed directly in contact with the permeable sand or gravel pack, the pressure of the overlying grout and back-fill materials can force the grout into the coarse sand or gravel. This can be prevented by placement of a 0.3- to 0.6-m–thick layer of fine sand between the top of the gravel pack and the overlying bentonite grout.

WELL DEVELOPMENT

During the drilling process the walls of the borehole are subject to plugging by fine particles in the drilling fluids or smearing by the movement of the drilling tools or temporary casing into and out of the borehole. For water from the aquifer to freely enter the well and to reduce the production of suspended sediment, it is neces-sary to remove the fine particles from the borehole wall. The process by which this is accomplished is referred to as *well development.*

Well development is accomplished by energetically moving water into or out of the well screen and gravel pack. For effective development, sufficient energy is required to physically disturb the gravel pack out to the radius of the borehole wall. During the process of development, silt and fine sand particles migrate through the gravel and enter the well screen. This migration of fine particles actually improves the ability of the gravel pack to provide effective filtration.

A variety of methods are used for well development. The choice of development method depends on the design of the well, the hydraulic characteristics of the aquifer, and the concentration of contaminants in the groundwater. The following sections discuss some of the more common methods of monitor well development.

Surging: One of the most common methods of monitor well development involves use of a plunger, called a *surge block*, that fits loosely inside the well casing and is forcibly moved up and down. As the plunger is moved down and then up, water is alternatively forced out of and drawn back into the well screen. This surging action provides the best means of effectively developing the gravel pack (Driscoll, 1995). Following a period of surging, the well is either pumped or bailed to remove the accumulated sediment from the bottom of the well. Development by surging results in a minimal amount of groundwater removed from the well. This can result in considerable cost savings at sites where groundwater is contaminated and therefore may need to be disposed of as hazardous waste when pumped to the surface.

Bailing: A bailer is constructed of a section of pipe with a diameter that is sufficiently small to fit inside the well casing and a check valve that allows it to fill from the bottom. The bailer is attached to a cable and repeatedly lowered into the well, allowed to fill with water, pulled to the surface, and emptied into a drum or tank. The action of quickly lowering the bailer forces some water to move outward through the well screen. As the bailer is filled and retrieved, water moves inward through the screen. Bailing is not as effective as surging, but may be sufficient for wells in formations that contain small amounts of silt and clay. Bailing produces larger quantities of water than surging.

Air lifting: Air lifting is accomplished by inserting a pipe into the well to a depth of at least 1 to 2 m beneath the water level. Compressed air is directed down the pipe and displaces the water in the well. When the displaced water reaches the surface, the air supply is turned off, and the water column is allowed to fall back down the well. Depending on the depth of the well and the height of the water column in the well, air lifting can produce surging action equal to or greater than that of a surge block. Air lift pumping is difficult to control, and water can be forcibly ejected from the top of the well. To help prevent this, a pipe or hose can be fitted to the top of the well to direct water to a storage tank. Nevertheless, this method may not be appropriate for the development of wells where the water discharged from the well must be contained and managed as hazardous waste.

Over pumping: Monitor wells in relatively permeable formations can be developed by intermittently pumping at high flow rates, then allowing the water level in the well to recover. This method is not as effective as the other methods mentioned previously but may be adequate for wells in coarse-grained formations. It is difficult to achieve the flow velocities necessary to remove fines from the borehole wall through pumping. Development by pumping also produces relatively large volumes of groundwater, which may result in higher waste disposal costs.

DESIGN AND EXECUTION OF GROUNDWATER SAMPLING PROGRAMS

The previous sections describe many of the factors that must be considered when constructing and installing monitor wells capable of delivering representative samples of groundwater. However, the most carefully designed and installed monitor well network can provide misleading or erroneous data through random or systematic errors introduced in the process of sample collection and analysis. Proper sampling methods are therefore equally as important as proper well placement and construction methods in the overall success of a groundwater sampling program. The following sections provide a discussion of the major aspects of designing and conducting a groundwater sampling program.

SAMPLING OBJECTIVES

The first step in development of a groundwater sampling program should be to define the objectives that are to be met and identify any factors that may constrain accomplishing those objectives. A clear understanding of the objectives will allow development of a focused and effective sampling program that will provide the necessary data in a cost effective manner. Some of the general objectives for groundwater sampling are discussed in the first section of this chapter. These objectives should be stated more specifically when applied to an individual site or project. Analytical and sampling protocols should reflect the general purpose and data quality objectives necessary to meet the project goals. The following examples illustrate some of the important aspects of sampling program design and execution.

EXAMPLE 1 *Objective: Determine the distribution of nitrate in private drinking water wells in a county-wide district.* This objective defines a regional-scale reconnaissance type of study focused on determining water quality characteristics for a large area. The study will be conducted with existing drinking water wells, so the investigator will have no control over well placement and design. Uncontrolled variability will undoubtedly be introduced because of variations in well construction and placement. It would be desirable to develop standard protocols for sample collection activities to minimize the variability introduced through sample collection. These protocols might define where samples could be collected (e.g., from the kitchen tap or from the closest outlet to the well), how long the well would be pumped before sampling; what water quality parameters would be measured in the field at the time of sampling; what data would be recorded about the sampling event; and how the samples would be preserved, packaged, and transported to the laboratory. With standardization of all the sample collection and handling procedures, the total amount of random variability in the data can be minimized.

EXAMPLE 2 *Objective: Identify the source of elevated chromium concentrations recently observed in samples collected from specific drinking water wells.* This objective defines a focused and probably localized study that may need to be conducted with some urgency. Of primary importance to the objective of this study is an understanding of the direction of groundwater flow in the vicinity of the affected wells. If no local measurements of water level data are available, regional water level contour maps may be available from state or federal resource management agencies. Once the direction of groundwater flow is determined, a search might be conducted to locate possible sources of chromium in the upgradient direction. Monitor wells could be installed between any suspected sources and the affected production wells.

A phased approach is often the most cost-effective way to conduct this type of investigation. The first phase may involve installation of monitor wells that will be located on the basis of very limited information. After the data from the initial investigation are available, the placement and design of the Phase-2 wells can be more focused and effective. Sampling and field data collection protocols should be developed before the initial investigation and modified as needed in subsequent phases.

Initial investigations of this nature can sometimes be conducted more effectively with groundwater sampling methods that do not require the installation of monitoring wells. These methods involve insertion of a specially designed probe into the subsurface using a drill or hammer rig, and retrieval of a groundwater sample. In shallow aquifers, the probe can be pushed from the surface. For deeper aquifers or aquifers overlain by consolidated geological formations, the probe is used in conjunction with a drilling rig and pushed into the bottom of the borehole ahead of the drilling tools. These sampling methods do not allow for repeat or duplicate samples from the same locations, but they can be useful as a screening tool for sampling many locations in a short period of time at minimal cost.

EXAMPLE 3 *Objective: Provide ongoing leak detection monitoring for a hazardous waste storage facility, in compliance with the Resource Conservation and Recovery Act (RCRA) permit requirements.* This objective describes a routine monitoring program that is conducted to meet a specific regulatory requirement. Most aspects of this sampling program, including sampling frequency, sampling methods, analyses to be performed, and statistical methods for data analysis will be prescribed by the regulatory agency that administers the permit. Quality control and quality assurance (QA/QC) is important in all groundwater sampling programs. However, QA/QC is of primary importance in compliance monitoring sampling. False-positive results can result in unnecessary public concern, as well as additional costs to verify that the facility is not leaking. False-negative results, or a negative bias in the data, can result in a leak remaining undetected. QC is maintained through standard practices being defined, followed, and documented. QA involves collection and analysis of duplicate and blank samples and audits of sampling procedures. Duplicate samples are samples that are collected at the same time from the same well. Analytical results from duplicate samples provide a measure of the variability inherent in the sampling and analysis process. Blank samples are samples that contain clean water. The blank samples accompany the actual samples through all phases of handling and analysis. Detection of contaminants in the blank samples may indicate contamination of the samples from extraneous sources. For large, ongoing sampling programs, QA/QC procedures are often documented in a Quality Assurance Project Plan (QAPP). The purpose of the QAPP is to define the data quality objectives for the program and detail all the procedures and personnel responsible for ensuring that those objectives are met.

EXAMPLE 4 *Objective: Monitor the changes in the trichloroethene (TCE) plume associated with a hazardous waste site.* This objective is typical of many ongoing monitoring programs at contaminated sites. The contaminant of concern has been defined and the source of the contaminant is known. The area of contaminated groundwater, often referred to as a *contaminant plume*, changes through time, either as a result of natural processes or in response to groundwater remediation efforts. Tracking the changes in the plume provides the data necessary for effective site management.

Ongoing monitoring programs benefit from consistency. Often, when changes are made in the way samples are collected, handled, or analyzed, random errors or new bias will be introduced in the data. This can lead to misinterpretation of results. Sampling protocols and methods should be established and followed. QA/QC should be in place so that spurious results can be identified. Periodically, it may be desirable to send a subset of duplicate samples to another analytical laboratory as a check on the possible bias in results from the primary laboratory. Audits of field procedures are also useful in ensuring that sampling protocols are being followed.

It is usually desirable to collect a set of samples over a short period of time. This provides a picture of the groundwater conditions during a single week or month. Comparison with conditions from previous or subsequent sampling rounds is used to determine where changes are occurring in the plume. Once sufficient data are available to develop an understanding of the plume dynamics, the monitoring intervals and set of wells to be monitored can be adjusted to match site conditions. In areas where concentrations are changing slowly, wells can be sampled less frequently than other wells. Wells in particularly sensitive areas, such as around the perimeter of the plume, may be sampled more frequently. When unexpected changes occur in groundwater levels or concentrations, the monitoring program may need to be modified so that appropriate data can be obtained to understand the causes and effects of these changes.

The preceding examples have demonstrated some of the differences in groundwater sampling programs designed to meet different objectives. There are common elements to most groundwater sampling programs. Such a list, presented by Herzog *et al.* (1991), is provided in Box 8.2. However, the relative importance of each element depends on the specific program objectives.

GROUNDWATER SAMPLING DEVICES

Many groundwater sampling devices are currently commercially available. A comparison of the more commonly used devices is presented in Box 8.3. The use of

BOX 8.2 *Common Elements of Groundwater Sampling Programs*

1. **Objectives** of the groundwater sampling and analysis program
2. **Site-specific parameters of concern** to be sampled and analyzed
3. **Location**, condition, and access to sampling points (e.g., wells, discharge points, surface water) to be included in the program
4. **Number and frequency of samples** to be collected
5. **Sampling protocol**—well purging procedure and equipment needs; field parameter monitoring/sample screening; sample collection; parameter specific techniques and equipment needs; field QA/QC controls
6. **Field sample pretreatment requirements**— filtration, preservation
7. **Sample handling, delivery method, and transport time**
8. **Chain-of-custody requirements**—sample documentation
9. **Sample chemical analysis**—identification of analytical methods; storage and holding times; laboratory QA/QC

From Herzog, B, Pennino, J, and Nielsen, G. (1991) Groundwater sampling. In Nielsen, DM. *Practical handbook of groundwater monitoring*, Lewis Publishers (an imprint of CRC Press), Chelsea, MI.

appropriate sampling equipment is critical for implementation of a successful sampling program. There is no one sampling device that is ideal for all circumstances. Barcelona *et al.* (1985, p. 98) list the following four important characteristics of groundwater sampling devices that should be considered for every sampling program:

1. The device should be simple to operate to minimize the possibility of operator error.
2. The device should be rugged, portable, cleanable, and repairable in the field.
3. The device should have good flow controllability to permit low flow rates ($<100 \, \text{ml min}^{-1}$) for sampling volatile chemical constituents, as well as high flow rates ($<1 \, \text{L min}^{-1}$) for large-volume samples and purging stored water from monitoring wells.
4. The device should minimize the physical and chemical disturbance of groundwater solution composition in order to avoid bias or imprecision in analytical results.

The choice of sampling equipment is dependent on several site-specific factors, including the monitor well design and diameter; the depth to groundwater; the constituents being monitored; the monitoring frequency; and the anticipated duration of the sampling program. Some of the major factors to consider are briefly discussed.

The first requirement for a sampling device is that it be able to be inserted in the monitoring wells. Although this seems intuitive, it is sometimes overlooked in the design

BOX 8.3 *Comparison of Common Groundwater Sampling Devices*

Bailers—Bailers are one of the oldest and simplest types of groundwater sampling devices. Conceptually, bailers are analogous to the bucket and rope type of system that has been used to lift water from wells for centuries. The major refinement for bailers used in groundwater sample collection is the use of a check valve to allow the bailer to fill from the bottom. This feature reduces the turbulence and aeration that would otherwise occur when water cascaded down into the open top of the bailer.

Advantages—Bailers used for groundwater sampling are typically fabricated from PVC, PTFE, or stainless steel. They are available in a variety of diameters and lengths. The length of line to which the bailer is attached limits the depth from which a bailer can lift. Electric winches are available that allow rapid retrieval of bailers from deep wells. Bailers are lightweight and simple to operate. They are inexpensive enough to be considered disposable,

thereby avoiding costs of decontamination and check sample analysis. Clear bailers are useful for obtaining an indication of the presence of floating hydrocarbon layers in wells.

Disadvantages—Purging wells with bailers is tedious and labor intensive. There is also a greater potential for worker exposure to contaminants with bailers than with most other sampling devices. Because the operator does not control the opening and closing of the check valve, it is difficult to know the depth in the well from which the sample was obtained. Bailing does not supply a continuous stream of water to the surface, so it may be necessary to fill sample bottles in stages, potentially allowing for the loss of volatile contaminants. The turbulence and aeration that occur when the bailer is splashed down into the water surface in the well can also affect volatile contaminant concentrations.

Syringe Samplers—Syringe samplers incorporate a movable plunger that fits tightly in a cylindrical sample chamber. The operator controls the movement of the

BOX 8.3 *(Continued)*

plunger. Initially, the plunger is forced to the end of the syringe by pressure applied through the air line. After the sampler is lowered to the desired depth in the well, the pressure is released from the air line and the hydrostatic pressure pushes the plunger back, allowing a sample into the syringe body. The sampler, with the sample inside, is then lifted to the surface. Syringe samplers that use an electric motor to move the plunger have also been constructed.

Advantages—The syringe sampler produces minimal negative pressure and virtually no aeration of the sample. The syringe can be made from inert materials, and samples can be obtained from discrete depths in the well. Syringe samplers are portable, easy to operate, and inexpensive. It is possible to use the syringe for a sample container, thereby avoiding the possible loss of volatile compounds when the sample is transferred to another container. There are no inherent depth limitations for syringe samplers. For pneumatically operated syringe samplers, the tubing used in the air line must be able to withstand pressures greater than the hydrostatic pressure at the depth where the sampler is to be used.

Disadvantages—The small volumes that can be obtained with syringe samplers make them ineffective for purging wells and inefficient for collecting large volume samples. Syringe samplers are also not suited to sampling water with high sediment content. The particles of sediment can score the syringe barrel, resulting in leakage around the plunger. Syringe samplers are not as widely used as other some other types of samplers and may not be readily available.

Inertial Pumps—Inertial pumps consist of a length of semirigid tubing with a check valve attached to the bottom end. The tubing is lowered into the well and water enters through the check valve. The tubing is then forcefully moved up and down in the well, either manually or through the use of a motor-driven pump jack. As the tubing moves upward, the check valve closes. The column of water is lifted along with the tubing and gains upward momentum. When the tubing changes direction and begins the downward stroke, the upward momentum in the water column results in low pressure at the check valve, causing the check valve to open and admit more water to the tubing. When the tubing reaches the bottom of the downstroke, the check valve closes. Through repetition of this process, water is lifted to the surface. The maximum flow rate is dependent on depth of the well and depth of submergence at the check valve.

Advantages—Inertial pumps are simple, portable, and easy to operate. They can lift water from depths as great as 75 m. The pumps are inexpensive and thus can be dedicated to each well, thereby eliminating problems with decontamination. They can be used with silt-laden water and can also be used for development of small-diameter wells. Finally, they can be used in wells as small-as 3 cm in diameter.

Disadvantages—Inertial pumps must be operated vigorously. Such operation results in turbulence and possible aeration of water in the upper portion of the well. There are also pressure fluctuations in the water column that could result in degassing or loss of volatile compounds. The surging action of the pump can agitate the water in the well and cause increased sediment in the samples.

Suction Lift Devices—Suction lift devices operate by drawing groundwater up from the well through a suction line. The maximum practical depth at which suction lift can operate is about 8 m. Therefore these sampling devices are limited to use in shallow wells. Flexible tubing is inserted to the desired depth in the well and connected to a flask. The flask is connected to a vacuum source. When a vacuum is applied, water is lifted from the well and passes through a sample delivery tube into the bottom of the flask. The flow rate can be controlled by limiting the amount of vacuum applied. After the desired sample volume is obtained, the sample is transferred from the flask to the appropriate sample containers. The sample is not allowed to enter the pump. A vacuum can be applied through use of a variety of hand-operated or electrically driven vacuum pumps.

Advantages—Suction lift sampling devices can be used in a well of any diameter and only a length of tubing need be inserted in the well. Suction lift devices are simple and easy to use. There are no moving parts that contact the water sample; thus these devices can pump sediment-laden water. Decontamination can be simplified by placing dedicated suction tubing in each well. With hand-operated vacuum pumps, suction lift devices are highly portable.

Disadvantages—Sampling is limited to wells where the depth to water is less than 8 m. The small volume pumped with suction lift devices makes them inefficient for purging wells. Samples are exposed to low pressures that can cause degassing and loss of volatile compounds. Additional losses of volatile compounds can occur when the sample is transferred from the flask into sample containers.

Gas-Drive Devices—Gas-drive devices typically consist of a cylindrical sample chamber connected to a gas-entry tube and a sample delivery tube. A check valve is installed in the bottom of the sample chamber.

(Continued)

BOX 8.3 *(Continued)*

Valves at the surface allow for the opening and closing of the sample delivery tube and admit and release pressure in the gas-entry tube. Before entry into the well, the sampler is pressurized with the sample delivery tube closed. This closes the check valve and prevents water from entering the sampler as it is lowered into the well. At the desired depth, the pressure is released to allow the sample chamber to fill with water. The pressure is then reapplied, and the check valve closes, trapping the sample in the sample chamber. For the sample to be delivered to the surface, the valve on the sample delivery tube is opened. The sample is then displaced by gas pressure and moves up the sample delivery tube to the surface.

Advantages—Gas-drive samplers are available in many different materials and diameters. They are portable and relatively inexpensive. The depth from which samples can be obtained is limited only by the burst strength of the tubing used. Gas-drive samplers can be used to obtain samples from discrete depths.

Disadvantages—The sample comes into contact with the gas used to drive the sampler and with the length of tubing that delivers the sample to the surface, either of which may result in changes in the chemical composition of the sample. If the driving gas is allowed to escape up the sample delivery tubing, it can strip volatile compounds from the water film left in this tubing from the previous sample. Volatile concentrations in the subsequent sample will then be reduced through dilution with the relatively contaminant free water film in the sample tubing. An air compressor or compressed-gas cylinders must be transported to the site where the sampler is to be used. If sediment partially jams the check valve, gas bubbles may be released into the well, possibly altering the chemistry of the water in subsequent samples.

Bladder Pumps—Bladder pumps are similar in concept to gas-drive pumps. However, they contain a flexible bladder that prevents contact between the sample and the driving gas. The interior of the bladder is connected to the inlet check valve and to the sample delivery tubing. The annular space between the sampler body and the exterior of the bladder is connected to the gas entry tube. When the sampler is pressurized, the gas pressure bears on the outer walls of the bladder, causing it to compress and displace the water on the interior of the bladder up the sample delivery tube. When the gas pressure is released, the hydrostatic pressure forces water to enter the interior

of the bladder, and the bladder expands. Automated control systems are available to optimize the cycling of the pump. With automated controls, bladder pumps can deliver a nearly continuous discharge stream.

Advantages—The bladder prevents contact between the driving gas and the sample, minimizing problems with gas stripping and aeration. Bladder pumps can produce high flow rates sufficient for purging wells and, by reducing the gas pressure or cycle timing, produce low flow rates suitable for sampling. Most bladder pumps are capable of lifts in excess of 60 m. The pumps themselves are portable and relatively easy to service and decontaminate.

Disadvantages—Operation at greater depths requires larger volumes of gas and longer cycle times. Check valves can be fouled by sediment in the water. The higher-quality pump units are relatively expensive.

Electric Submersible Centrifugal Pumps—Submersible centrifugal pumps operate by forcing water through a series of rotating impellers, powered by an electric motor. The motor section is located at the bottom of the pump. Water enters through a screen between the motor section and the impeller section. Different numbers of concentric impeller sections may be attached depending on the pumping volumes and pressures required. Electric submersible pumps are widely used in both domestic and public water supply wells. More recently, they have been adapted and become popular for the purposes of groundwater sampling.

Advantages—Electric submersible centrifugal pumps are available in inert materials (stainless steel, Viton, and PTFE). They can produce large flow rates that are sometimes needed for purging wells and, when equipped with variable frequency drives, be slowed down to produce small flow rates for sampling. At least two manufacturers produce models that will fit into a 2-in (5-cm) well casing. The pumps designed specifically for groundwater monitoring are relatively easy to repair and decontaminate in the field. They provide a continuous stream of water at the surface.

Disadvantages—The equipment cost is higher than for many other types of sampling devices. The lift for the smaller diameter pumps is limited to about 60 m. They require a generator or other source of electricity and a pump controller box for operation and are therefore not as portable as some other devices. The pumps are also sensitive to the presence of fine sediments, which may erode pump components at the high speed of operation.

of a sampling program. Almost all commonly used sampling devices will fit into 4-in (10-cm) diameter wells. Many devices are also available for 2-in (5-cm) diameter wells. Relatively few devices are available for wells smaller than 1 in (2.5 cm) in diameter. Wells that are crooked or do not have flush-fitting casing joints may prevent the use of samplers that would otherwise work in a straight, smooth well casing of the same diameter. PVC well casing has different internal diameters depending on the respective strength rating. A sampling device specifically designed for schedule 40 PVC casing may not fit in a well constructed with thicker wall schedule 80 PVC casing, even though the nominal diameters of the two casings are the same.

The depth to groundwater is another factor that must be considered in choosing a sampling device. The depth to groundwater determines the pressure that the sampling device must generate to lift water to the surface. Generally, suction lift devices will only operate where the lift is 8 m or less. Positive displacement devices can be designed to operate with lifts of several hundreds of meters. Generally, the selection of sampling devices becomes more limited when lifts of more than about 60 m are involved. Larger lifts also result in sampling activities becoming more time consuming, particularly if the sampling device must be inserted and removed from the well with each sample. For programs with large lifts, it is generally more cost effective to install dedicated sampling devices in each well.

After it has been determined which sampling devices will fit in the wells and are capable of lifting samples to the surface, the next consideration is generally whether the device is capable of delivering representative samples for the constituents being monitored. Monitoring programs typically include a list of constituents for which samples will be analyzed. Dissolved gases and volatile organic compounds are typically two of the more difficult classes of compounds to sample. These are generally sensitive to aeration, changes in temperature and pressure, and leaching or adsorption in the sampler tubing or components.

For ongoing monitoring programs, consideration should be given to equipping each well with a dedicated sampling device. The use of dedicated sampling devices reduces labor and avoids the potential errors associated with equipment decontamination. When a single sampling device is used for sampling contaminants from multiple wells, it must be decontaminated before each sampling event. To confirm that the decontamination process is effective, equipment blank samples are obtained with the sampling device to collect a sample of water that is known to be free of contaminants. The equipment blanks are analyzed at the analytical laboratory with the same methods as for the actual samples. In an ongoing monitoring program, the costs of analyses for equipment blanks and labor for decontamination can rapidly exceed the costs of equipping each well with a dedicated sampling device. Disposable bailers offer another alternative to decontamination; however, bailed samples may not be representative for all contaminants.

Ideally, the same device could be used for both purging and sampling the well. This requires a device capable of high pumping rates for purging and low pumping rates for sampling. When sampling for volatile compounds or compressed gases, it is generally not acceptable to use a valve for flow control. Valves may create large pressure drops in the sample delivery line that may result in turbulence and depressurization of the sample stream, and thus possibly alter the chemical quality of the sample (Neilsen and Yeates, 1985). It is better to be able to vary the pumping rate of the sampling device so that the flow can be regulated without extreme pressure changes in the sample delivery line. For larger wells, it may be necessary to install both a high-flow pump that is used only for purging the well and a dedicated sampling device used only for sampling after the well is purged.

PURGING MONITOR WELLS BEFORE SAMPLING

Under most sampling protocols, groundwater is removed from the monitor wells before sampling. This is referred to as *purging the wells*. The basis for purging is the assumption that the groundwater standing in the well casing is not representative of the groundwater in the surrounding aquifer. Processes that can affect groundwater in the well casing include exposure to the atmosphere and interaction with the well casing materials. Wells may also harbor microorganisms that can alter the chemistry of groundwater in the casing.

Three criteria have commonly been used to determine when a well has been purged sufficiently to yield representative samples. The most common "rule of thumb" is to purge a prescribed minimum number of casing volumes before sampling. Commonly three to five casing volumes are purged, although the number can range as high as 20. In this context, the casing volume is defined as the volume of water standing in the well. Another common criterion involves purging the well until a set of field-measured indicator values stabilize. Parameters commonly used as indicators include pH, temperature, specific conductance, and dissolved oxygen. A third common criterion involves calculation of the purge volume based on the aquifer transmissivity and the diameter of the well (Schuller *et al.*, 1981). This method provides an estimate of how long the well would need to be pumped at a specific flow rate to produce samples representative of the aquifer. It is generally recommended that field indicator parameters be measured to verify that the calculated purge volume is adequate (e.g., Barcelona *et al.*, 1991).

The rate at which a well is purged can also affect the quality of the samples. If a well is purged at a pumping rate higher than what the formation can deliver, fine particles may be drawn into the well, thereby increasing the turbidity of the samples. Overpumping a monitor well that is screened near the water table can result in dewatering of a portion of the formation, exposing the sediments in the vicinity of the well to the atmosphere and potentially changing the geochemical conditions.

The placement of the pump in relation to the well screen will also affect the volume of water that must be purged from the well. A pump placed within the screened interval and operated at a rate consistent with what the well can produce will draw water primarily from the well screen, leaving the standing column of water in the casing above the screen undisturbed. If a pump is placed above the well screen, the standing water in the well will be displaced upward toward the pump by groundwater entering the well screen. The standing water column will have to be completely removed from the well before the pump will deliver water that is representative of the aquifer. If a dedicated sampling device is not used, removal of the standing water column may be desirable. Recent studies have indicated that for small-diameter wells with relatively short screened intervals, a micropurge sampling technique can yield representative samples with a minimum volume of water purged.

Sample Collection and Processing

The actual process of collecting a groundwater sample and its handling (transport, storage process) after collection are important components of the groundwater sampling program. The methods used for sample collection and processing depend on the type of contaminants present or suspected of being present. The primary concern of sample collection and processing is to maintain the integrity of the samples so that the concentrations of analytes are the same as when the samples were collected. Some of the major considerations for sample collection and processing are briefly reviewed.

Sample aeration: It is often important for several reasons to minimize exposure of the sample to the atmosphere. For example, many organic contaminants of particular concern, such as chlorinated solvents and aromatic fractions of petroleum derivatives, are volatile. Thus, prolonged exposure to the atmosphere may result in loss of contaminant mass from the sample. The concentration of carbon dioxide in groundwater is often higher than that in the atmosphere. Thus, prolonged exposure of a sample

to the atmosphere can lead to loss of carbon dioxide, which can result in an increase in pH. A change in pH can in turn cause changes in contaminant properties, such as inducing precipitation of metals. For these concerns to be reduced, samples should be collected in a manner that minimizes their exposure to the atmosphere. In addition, the samples should be stored in vials that are airtight and have no air space (termed *headspace*).

Sample filtration: One question to address is whether the samples should be filtered. Samples are sometimes filtered to remove suspended particles and colloidal material. However, it is critical to determine if the filtering process may disturb sample integrity. For example, filtration should not induce sample aeration, for the reasons noted previously. Therefore, in-line filtration is generally preferred to vacuum filtratation. In addition, some contaminants may associate with suspended or colloidal particles. In such cases, filtration that removes these particles will also remove some of the contaminant mass.

Sample preservation and storage: A reagent is sometimes added to the samples to help preserve their integrity. For example, a buffer may be added to samples that are sensitive to changes in pH. A biocide may be added to prevent microbial activity from altering sample properties. Samples are often stored with ice packs during transit to the laboratory and are often refrigerated until processed. These procedures reduce the impact of processes such as biodegradation and volatilization on sample integrity. The length of time for which a set of samples can be stored while maintaining their integrity depends on the properties of the contaminant. For example, the storage time for samples containing volatile contaminants should generally be limited to a few weeks at most.

QUESTIONS

1. What are the main objectives of a groundwater sampling program?

2. What are some of the major factors that influence the design of monitoring well systems? How must each of these factors be taken into account in the design process?

3. What are the key elements for a monitoring well?

4. Why is it important to carefully evaluate well screen design for a monitor well?

5. What are the commonly used methods for well development?

6. Why are wells usually purged before sample collection? Are there any possible drawbacks to well purging?

7. Why are groundwater samples filtered in some cases? Are there any possible drawbacks to sample filtration?

8. For which type of constituents is it important to control/eliminate exposure of groundwater samples to the atmosphere? Why?

REFERENCES AND ADDITIONAL READING

Aller, L., Bennet, T.W., Hackett, G., Petty, R.J., Lehr, J.H., Sedoris, H., Nielsen, D.M. (1989) *Handbook of Suggested Practices for the Design and Installation of Ground-Water Monitoring Wells.* National Water Well Association, Dublin, Ohio. pp. 188–189.

Australian Drilling Industry Training Committee, Ltd. (1996) *Drilling, the Manual of Methods, Applications, and Management* Lewis Publishers, New York. p. 615.

Barcelona, M.J., Wehrmann, H.A., Varljen, M.D. (1994) Reproducible Well-Purging Procedures and VOC Stabilization Criteria for Ground-Water Sampling. *Ground Water*, Vol. 32, No. 1.

Barcelona, M.J., Gibb, J.P., Helfrich, J.A., and Garske, E.E. (1985) *Practical Guide for Ground-Water Sampling.* U.S. Environmental Protection Agency, Ada, OK; EPA/600/2–85/10.

Barcelona, M.J., Boulding, R., Heath, R.C., Pettyjohn, W.A., and Sims, R. (1991) *Handbook: Ground Water Volume II: Methodology.* EPA/625/6–90/016b. National Technical Information Service PB93–129740. Springfield, Virginia.

Driscoll, Fletcher G. (1995) *Groundwater and Wells*, 2nd Edition, Johnson Screens, St. Paul, Minnesota. p. 1089.

Herzog, B., Pennino, J., and Nielsen, G. (1991) *Ground-Water Sampling in Practical Handbook of Ground-Water Monitoring*, David M. Nielsen, (ed.), Lewis Publishers, Chelsea Michigan. pp.449–500.

Makeig, K.S. (1991) *Regulatory Mandates for Controls on Ground-Water Monitoring in Practical Handbook of Ground-Water Monitoring*, David M. Nielsen, (ed.), Lewis Publishers, Chelsea Michigan. pp. 1–17.

Nielsen, D.M., and Yeates, G.L. (1985) A Comparison of Sampling Mechanisms Available for Small-Diameter Ground Water Monitoring Wells. *Ground Water Monitoring Review*, Vol. 5, No. 2.

Nielsen, D.M., and Schalla, R. (1991) *Design and Installation of Ground-Water Monitoring Wells in Practical Handbook of Ground-Water Monitoring*, David M. Nielsen, (ed.), Lewis Publishers, Chelsea Michigan. pp. 239–331.

Schuller, R.M., Gibb, J.P., and Griffin, R.A. (1981) Recommended Sampling Procedures for Monitoring Wells. *Ground Water Monitoring Review*, Vol. 1 No. 1.

Sykes, A.L., McAllister, R.A., and Homolya, J.B. (1986) Sorption of Organics by Monitoring well Construction Materials. *Ground Water Monitoring Review*, Vol. 6 No. 4.

U. S. Environmental Protection Agency. (1986) *RCRA Ground-Water Monitoring Technical Enforcement Guidance Document.* U.S. Government Printing Office, Washington, DC.

9

Monitoring Surface Waters

J.F. ARTIOLA

Seven tenths of the world's surface is covered by water, and about 97% of the water is saline, residing in the oceans. The remaining 3% is fresh water and exists frozen as ice caps (2%), in groundwater or aquifers, or surface waters such as rivers, lakes, and soil moisture; or as vapor in the atmosphere (1%) (Figure 9.1). Although saltwater is not generally fit for direct consumption, most forms of life depend on the ocean directly or indirectly for their survival. For example, photosynthesis by algae living in the oceans accounts for most of the oxygen regeneration in the world, and the oceans store vast quantities of carbon dioxide and heat. However, much of human history is tied directly or indirectly to fresh water sources such as lakes, rivers, and estuaries. Thus, this chapter focuses on the monitoring of fresh water bodies.

Human beings, as well as other terrestrial and aquatic life forms, are sensitive to changes in the quality of the fresh water supply. Changes in properties such as total dissolved solids, pH, and dissolved oxygen in particular may affect the mortality of aquatic life. In addition, the presence of contaminants can be hazardous to organisms that drink the water and to those living in it. The dynam-

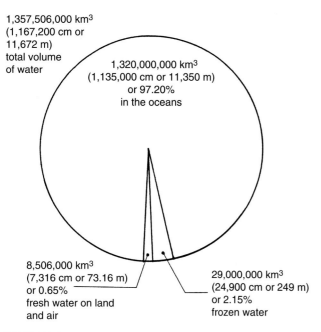

FIGURE 9.1   Water availability and distribution on earth. Over 97% of all water in the world is salt water found in the oceans. (Adapted from Van der Leeden, F., Troise, F.L., and Todd, D.K. [1990] *Water Encyclopedia*, 2nd edition. Lewis Publishers, Chelsea, MI. An imprint of CRC Press.)

ics of both natural weathering processes and anthropogenic activities can have a significant impact on water quality. Rainfall tends to dissolve and carry away minerals and contaminants found in the soil and the atmosphere. Water used for irrigated agriculture concentrates dissolved solids in the water as water undergoes evapotranspiration. Evaporation, on the other hand, purifies water as it changes from liquid to vapor. However, if consumptive uses of water exceed evaporation and precipitation cycles, the water supplies decline, and water quality typically degrades.

In Arizona, for example, continuous use of surface and groundwater for agricultural irrigation has not only lowered groundwater levels, but also caused degradation of the remaining water. The lowering of the groundwater table during the past 100 years has eliminated many perennial surface streams, causing the loss of more than 90% of the riparian habitats. As more and more surface water supplies are overused, groundwater supplies are also becoming exhausted with little hope of replenishment because only 10% of the precipitation may be recharged (Schlesinger, 1997) (Figure 9.2).

Monitoring fresh water quality and quantity has become a national and global concern. Fresh water supplies determine the very existence and survival not only of plants and animals, but also of human habitat locations. Diminished, degraded, and polluted water supplies are the immediate concern of all. The preservation and sustainability of fresh water resources can only come from careful monitoring of their sources and of their quality.

THE CHANGING STATES OF WATER

Water, the universal solvent, can dissolve to some extent all organic and inorganic components found in the environment. Thus, the nature of water changes through natural or anthropogenic impacts and as it moves from one environment to another (Figure 9.3). Water can also dissolve and suspend other liquids, gases, and fine particulates, resulting in a complex mix of components. Many natural water cycles contribute to this "mixing" process. Because water can exist in three states—gas, liquid, and solid—it is found virtually everywhere in one or more of these forms. Evaporation, solidification

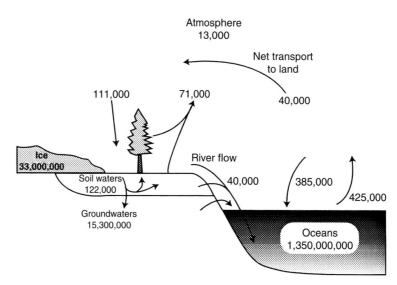

FIGURE 9.2   The global water cycle. River outflows equal net transport from oceans to land. (From Schlesinger, W.H. [1997] *Biogeochemistry: an analysis of global change*. 2nd ed. Academic Press, San Diego.)

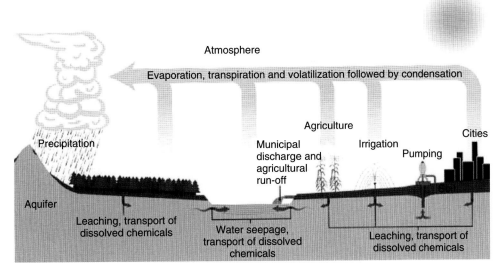

FIGURE 9.3   Processes that affect water quality. Leaching and transport of chemicals from land and the atmosphere continuously have an impact of surface and groundwater quality. (From Pepper I.L., Gerba C.P., Brusseau M.L. [1996] *Pollution Science,* Academic Press, San Diego.)

through ice formation, condensation, and precipitation are processes that control the distribution of water between these three states in the environment. In its liquid state, water can dissolve or contain other liquids, solids, and gases. Upon evaporation, water usually sheds all dissolved solids but can coexist in air with volatile compounds. As a gas, water vapor moves quickly through porous media such as soils, in response to gravitational and soil matrix energy potentials (see Chapter 12). Conversely, solid water (i.e., ice) does not move readily through the environment.

WATER QUALITY PARAMETERS

Chemical water quality is determined by the quantity and diversity of organic and inorganic chemicals residing within it. Likewise, microbial water quality is dictated by the presence or absence of beneficial and pathogenic microorganisms. Figure 9.4 shows the average concentration ranges of chemicals found in surface waters. Note that we accept that fresh surface water can and should contain some calcium (Ca), which is a necessary nutrient. The typical range of Ca is 1 to 250 mg L^{-1}. However, if mercury (Hg) is present at concentrations above 1 ng L^{-1}, we consider this water unfit for human consumption. Although hundreds of thousands of chemicals and living organisms can potentially be found in water, in reality only fewer than 300 chemicals and microbes are routinely monitored for in surface waters (see the next sections).

Figure 9.5 lists major water properties and related processes. Water quality and environmental processes are intimately related. The dynamic nature of most

water cycles, along with increasingly invasive anthopogenic activities, can quickly change important chemical, physical, and biological water quality properties.

Major changes in chemical and microbiological properties are directly linked to the suitability of water for human and animal consumption and for plant uptake. Conversely, changes in the physical states of water tend to affect natural water cycles at all scales. However, water temperature in a lake, for example, can also be used as an indirect indicator of total dissolved gases and of which species of microorganisms that may or may not live within it. Similarly, the electrical conductivity (EC) of water can be used to estimate the concentration of dissolved solids and even the ionic strength of water. All of these parameters are useful in chemical equilibrium and water quality modeling.

DRINKING WATER STANDARDS

Extensive water monitoring in the United States began with the Clean Water Act of 1948 and was followed by the Safe Drinking Water Act of 1974 (SDWA), with the amendments to the SDWA in 1989 and 1996. These acts resulted in the promulgation, with subsequent amendments, of Primary and Secondary Drinking Water Standards that are listed in Tables 9.1 to 9.3. Primary Drinking Water Standards are divided into six groups: inorganic and organic chemicals; disinfectants; disinfection by-products; radionuclides; and microorganisms. Note that Table 9.1 does not list all the organic chemicals, but groups them into pesticides, solvents, and a miscellaneous griyo. For a complete list of these chemicals, their

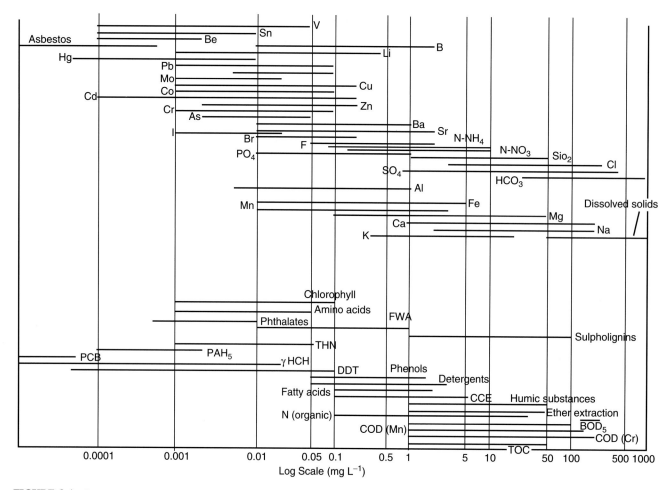

FIGURE 9.4 Average concentrations ranges of inorganic and organic substances in surface waters around the world. Except for anthropogenic chemicals like pesticides, it is difficult to determine what are the natural concentrations of many chemicals listed here because most of the water sources of the world have been affected by human activity. (Adapted Figs 1.5 and 1.6 from Dojlido & Best, 1993.)

health hazards, and contamination sources, see Table 16.2 in Chapter 16 or go to the Environmental Protection Agency (EPA) Website (www.epa.gov). Primary Drinking Water Standards are enforceable by EPA, as they are protective of human health. Conversely, Secondary Drinking Water Standards are nonenforceable guidelines designed to control contaminants that cause cosmetic effects such as tooth staining or produce undesirable aesthetic effects such as changes in taste, color, or smell. These are evolving standards that change in response to new scientific knowledge about the health effects of chemical and biological constitutents in drinking water. For example, in response to new epidemiological studies, standards for total trihalomethanes, which are disinfection by-products, and arsenic have been recently lowered from 100 to 80 μg L^{-1}, and from 50 to 10 μg L^{-1}, respectively. In addition, water disinfectants themselves have been added to the list because they are used to disinfect water and usually must have residual levels to protect drinking water supplies from the

source to the faucet. Recently, rapid developments in microbial detection techniques have added new waterborne pathogens to the list of drinking water standards (see Chapter 17).

NATIONAL RECOMMENDED WATER QUALITY CRITERIA

In response to section 304a of the Clean Water Act, water quality criteria for 157 priority toxic pollutants and for 45 additional nonpriority pollutants have also been established by the U.S. EPA. Water quality criteria that consider the concentration-based organoleptic or taste and odor effects of 23 chemicals on water are also available (U.S. EPA, 1999). These recommended standards are not directly enforceable by the EPA. Instead, they may be used as guidelines by tribes and states to develop environmental programs such as the National Discharge Elimination System Permits. These criteria can be used to

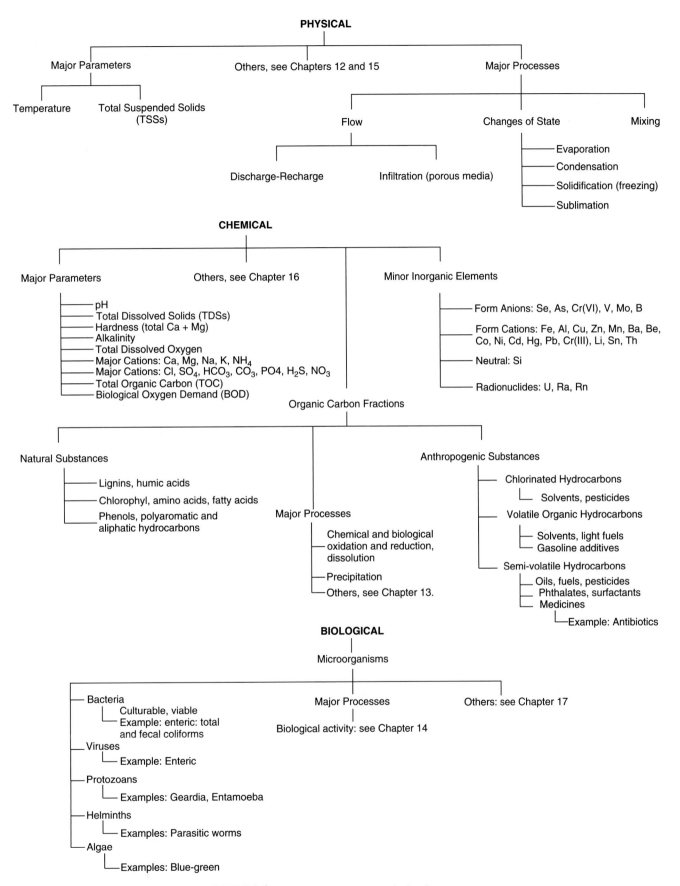

FIGURE 9.5 Major water properties and related processes.

TABLE 9.1
Primary Drinking Water Standards

Parameter	MCLs (mg L^{-1})	Parameter	MCLs (mg L^{-1})
Inorganic (17)		**Disinfection by-products (4)**	
Antimony	0.006	Total trihalomethanes	0.080
Arsenic	0.010	Total haloacetic acids	0.060
Asbestos	7H10^6 L^{-1} (fibers)	Bromate	0.010
Barium	2.0	Chlorite	0.8
Beryllium	0.004	**Radionuclides (5)**	
Cadmium	0.010	Radium 226 & 228 (total)	5 pCi L^{-1}
Chromium (total)	0.1	Gross alpha particles activity	15 pCi L^{-1}
Copper	1.3 TT	Gross beta particles activity	4 mrem year^{-1}
Cyanide (free)	0.2	Uranium (as of 12–2003)	0.030
Fluoride	4.0	**Disinfectants (4)**	
Lead	0.015	Chloramines	MRDL = 4.0
Mercury (inorg)	0.002	Chlorine gas	MRDL = 4.0
Selenium	0.05	Chlorine dioxide gas	MRDL = 0.8
Thallium	0.0005	**Microorganisms (7)**	See also Chapter 17
Nitrate-N	10.0	*Cryptosporidium*	99% removal TT
Nitrite-N	1	*Giardia*	99% removal TT
Fluoride	4.0	*Legionella*	No limit TT
Organic (53)		Viruses (enteric)	99% removal TT
Insecticides, herbicides	Total: 22 MCLs range from <0.001 to 0.7	Total coliforms	No more than 5% + TT
Solvents: Chlorinated, nonchlorinated	Total: 19 MCLs range from >0.001 to 0.07	Heterotrophic plate count	<500 BC mil^{-1} TT
Miscellaneous organics	Total: 12 MCLs range from 0.00003 μg L^{-1} to 0.1	Turbidity	1 NTU TT

Source: U.S. EPA: www.epa.gov/safewater/mcl.html.
MCL, Maximum concentrations level; *MRDL*, maximum residual disinfectant level; *TT*, treatment technique dependent.

TABLE 9.2
Primary Drinking Water Standards: Volatile and Synthetic Organic Chemicals

Volatile Organic Chemicals			
Parameter	MCLs (mg L^{-1})	Parameter	MCLs (mg L^{-1})
Benzene	0.005	Ethylbenzene	0.7
Carbon tetrachloride	0.005	Monochlorobenzene	0.1
p-Dichlorobenzene	0.075	Tetrachloroethylene	0.005
o-Dichlorobenzene	0.6	1,2,4-Trichlorobenzene	0.07
1,2-Dichloroethane	0.005	1,1,1-Trichloroethane	0.2
1,1-Dichloroethylene	0.007	1,1,2-Trichloroethane	0.005
cis-1,2-Dichlorethylene	0.07	Trichloroethylene	0.005
trans-1,2-Dichloroethylene	0.1	Total trihalomethanes	0.002
1,2-Dichloropropane	0.005	Vinyl chloride	0.10

Synthetic Organic Chemicals			
Parameter	MCLs (mg L^{-1})	Parameter	MCLs (mg L^{-1})
Acrylamide	TT	Glyphosate	0.7
Adipate (diethylhexyl)	0.4	Heptachlor	0.0004
Alachlor	0.002	Heptachlor epoxide	0.0002
Atrazine	0.003	Hexachlorobenzene	0.001
Benzo-a-pyrene	0.0002	Hexachlorocyclopentadiene	0.05
Carbofuran	0.04	Lindane	0.0002
Chlordane	0.002	Methoxychlor	0.04
Dalapon	0.2	Oxamyl (Vydate)	0.2
Dibromochloropropane	0.0002	Pentachlorophenol	0.001
Di(ethylhexyl)adipate	0.4	Picloram	0.5

(Continued)

TABLE 9.2 (Continued)

Synthetic Organic Chemicals (Continued)			
Parameter	MCLs (mg L^{-1})	Parameter	MCLs (mg L^{-1})
Di(ethylhexyl)phthlate	0.006	Polychlorinated byphenyls	0.0005
Dichloro-methane	0.005	Simazine	0.004
Dinoseb	0.007	Styrene	0.1
Diquat	0.02	Toluene	1.0
Endothall	0.1	Toxaphene	0.003
Endrin	0.002	Xylenes (total)	10.0
Epichlorohydrin	TT	2,4-D	0.07
Ethylene dibromide	0.00005	2,4,5-TP (Silvex)	0.05
2,3,7,8-TCDD (Dioxin)	0.00000003		

Source: US.EPA: www.epa.gov/safewater/mcl.html.
MCL, Maximum concentrations level; *TT*, treatment technique dependent.

TABLE 9.3
Secondary Drinking Water Standards

Contaminant	Level (mg L^{-1})	Contaminant effects
Aluminum	0.05–0.2	Water discoloration
Chloride	250	Taste, pipe corrosion
Color	15 color units	Aesthetic
Copper	1	Taste, porcelain staining
Corrosivity	Noncorrosive	Pipe leaching of lead
Fluoride	2.0	Dental fluorosis
Foaming agents	0.5	Aesthetic
Iron	0.3	Taste, laundry staining
Manganese	0.05	Taste, laundry staining
Odor	3 threshold odor number	Aesthetic
pH	6.5–8.5	Corrosive
Silver	0.1	Skin discoloration
Sulfate	250	Taste, laxative effects
Total dissolved solids	500	Taste, corrosivity, detergents
Zinc	5	Taste

Source: U.S. EPA: www.epa.gov/safewater/ncl.html.
Secondary standards are nonenforceable, maximum contaminant levels intended to protect "public welfare." Public welfare criteria include factors such as taste, color, corrosivity, and odor, rather than health effect.

set pollutant discharge limits for industries and utilities that discharge wastewaters into waterways. These criteria are protective of the water environment above all, but also consider human health effects (U.S. EPA, 1999). These two lists of pollutants include all of the chemicals listed in the Drinking Water Standards and many other chemicals of industrial and natural origins.

ECOREGIONAL NUTRIENT CRITERIA

These nutrient water quality criteria are an extension of water quality criteria discussed in the "National Recom-

mended Water Quality Criteria" section targeted to the protection of lakes, reservoirs, rivers, and streams within specific geographical locations or ecoregions of the United States. Specifically, total nitrogen (N), total phosphorus (P), chlorophyll, and turbidity may be regulated to prevent eutrophication and algal blooms. Numerous water sources in the United States are afflicted with long-term nutrient enrichment that results in low dissolved oxygen, cloudy or murky waters, and reduced beneficial fauna such as fish or excessive flora such as algae or aquatic plants. These nutrient criteria have been adapted to 13 distinct ecoregions that are defined based on climate, geology, and soil type (U.S. EPA, 2001). These criteria are evolving and may be modified for application within each ecoregion. The maximum allowable nutrient concentrations vary across ecoregions, ranging from 10 to 80 μg L^{-1} for total P, 0.1 to 2 mg L^{-1} for total N, 1 to 4 μg L^{-1} for chlorophyll, and 1 to 18 formazine turbidity unit (FTU) or nephelometric turbidity unit (NTU) for turbidity.

WATER QUALITY FOR AGRICULTURE

Agriculture depends on ample supplies of water that can sustain plant growth without affecting the soil environment. In fact, more than 60% of fresh water consumed in the United States is for food production. Nonetheless, there are no water quality standards for irrigated agriculture. There are only water quality criteria that are protective of crops and the soil environment. The total amount of soluble solids dissolved in water, which are often referred to as salts, is the first criterion used to determine the suitability of its use for crop production. With few exceptions most crops, including vegetables and fruits, are moderately tolerant to sensitive with respect to the total amount of dissolved solids present in irrigation water. Although grain and fiber crops like barley and cotton can grow

TABLE 9.4
Water Quality Parameters Measured and Restrictive Use Criteria[a]

Water Parameter	Typical Range in Irrigation Water (mg L^{-1})	Severe Restriction	Effects
Salinity:			
Total dissolved solids (TDS) −	0–2000	>2000	Plant water availability and yield
Electrical conductivity (EC$_w$)	0–3 dS m^{-1}	>3	
Sodium absorption ratio (SAR)	0–15	Variable[b]	Soil infiltration
pH	0–8.5	None[c]	Nutrient available at pH > 8
Sodium (Na$^+$)	0–920	200	Na$^+$ sensitive crops, SAR
Potassium (K$^+$)	0–78	None	Nutrient
Calcium (Ca^{++})	0–400	None	Nutrient, soil alkalinity, SAR
Magnesium (Mg^{++})	0–60	None	Nutrient, SAR
Carbonate (CO$_3^-$)	0–3	None	Soil alkalinity, SAR$_{adj}$[a]
Bicarbonate (HCO$_3^-$)	0–600	>500	Soil alkalinity, SAR$_{adj}$[d]
Chloride (Cl$^-$)	0–1070	>350	Cl$^-$ sensitive crops
Sulfate (SO$_4^-$)	0–920	None	Nutrient
Nitrate (NO$_3^-$)	0–44	>75	NO$_3^-$ sensitive crops, ground and surface water
Ammonium (NH$_4^+$)	0–6	None	Surface water
Phosphate (PO$_4^5$)	0–7	None	Surface water
Boron (B)	0–2	>3	B sensitive crops
Trace metals			Micronutrients essential for plant and animal growth, but can also be toxic at high concentrations
Micronutrients:			
Co, Cr, Cu, F, Fe, Li, Mn, Mo, Se, Zn	0−high μg L^{-1}	Variable[e]	Nutrient uptake
Toxic metals:			
Al, As, Hg, Be, Ni, Pb, Sn, Ti, W, V	0−low μg L^{-1}	Variable[e]	Toxic metals: nonessential, usually toxic to plants and animals

[a]Adapted from Food and Agriculture Organization of the United Nations (1985).
[b]Severe restriction varies with total salinity.
[c]None within the typical range of irrigation waters.
[d]Adjusted SAR.
[e]Recommended maximum concentrations vary with each element.

well when irrigated with slightly saline water with an EC of water (EC$_w$) of 5 deciSiemens (dS) m^{-1}, most other crops do not tolerate irrigation water with an EC above 4 dS m^{-1} (FAO, 1985). However, many other specific water quality parameters can affect crop yields and change the physical and chemical characteristics of soils. Table 9.4 lists the irrigation water quality parameters that should be monitored and their typical ranges in irrigation water.

Note that many of these parameters are also listed in the Drinking Water Standards tables with exceptions like Ca^{++}, Mg^{++}, Na$^+$, K$^+$, Cl$^-$, phosphate, boron, and trace elements. The sodium absorption ratio (SAR) together with EC$_w$ affect irrigated soil infiltration properties.

$$SAR\frac{Na}{[(Ca + Mg)/2]^{1/2}} \qquad (Eq. 9.1)$$

Note: Na, Ca, and Mg values should be in mole-equivalent units.

SAMPLING THE WATER ENVIRONMENT

The aquatic environment is especially vulnerable to pollutants. Two mechanisms are primarily responsible for the transport and subsequent contamination of water bodies: diffusion and advection (see Chapter 12). Diffusion is the result of the tendency of molecules to move from high to low concentration zones. Advection is the result of water moving in response to gravity or pressure forces. For example, if a barrel of pesticide is dumped into a lake, diffusive processes will cause the chemical to move away from its initial location, and advection forces such as waves and currents will move it in the direction of the water flow. Eventually, the concentrations of the pesticide may become relatively uniform throughout the lake. Thus, in theory, all of the water will eventually be equally contaminated. The same mechanisms act on other common forms of water pollution, such as runoff and pipe discharges from sewage treatment plants, ships, boats, and recreational vehicles.

Surface water environments, which are generally classified as oceans, streams, lakes, and reservoirs, are generally more homogeneous in composition than soils, yet this apparent homogeneity can be misleading. In fact, surface water bodies are often stratified in that they have layers composed of different temperatures and densities and with different chemical compositions. This means that a pollutant release into a body of water may or may not uniformly contaminate all the water within the lake as in the example given. Because pollutant concentrations and other chemical properties may change throughout a body of water, surface water monitoring must incorporate consideration of these heterogeneities.

Spatial Water Composition

When sampling bodies of water, we must consider the surface, volume, and dynamics of these systems. For example, flowing bodies of water should be sampled across the entire width of the channel because water generally moves faster in the center than at the edges. In cross-sectional sampling, flow velocities should be measured and recorded at each sampling location (see the

"Stream Flow Measurement" section). In enclosed bodies of water, such as lakes and lagoons, it is also very important to sample by depth, particularly where abrupt temperature changes or thermoclines occur.

Figure 9.6 shows that temperature changes occur not only with depth but also seasonally in lakes and reservoirs. In addition, the chemical composition of these bodies of water will also change significantly with depth. In the example shown in Figure 9.7, the distribution of arsenic species varies considerably with depth. Here a strong association could have been observed by measuring total dissolved oxygen levels and redox potential as a function of water depth.

Temporal Water Composition

Temporal sampling in dynamic water systems is common because the natural chemical composition and pollutant concentrations of these bodies can change quickly over hours or days. Frequent sampling intervals are warranted when the concentrations of pollutants are near critical regulatory levels. Frequent sampling at hourly, daily, or weekly intervals may be warranted at or near point-source

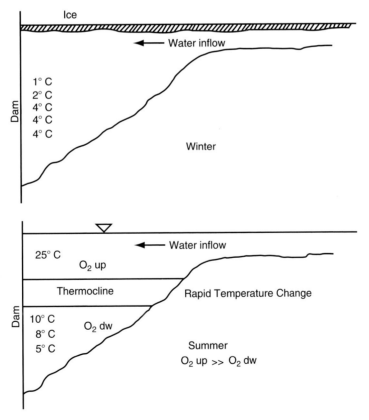

FIGURE 9.6 Typical distribution of spatial and seasonal changes in temperature in a deep impoundment. A thermocline layer (rapid temperature change) prevents mixing. Thus, the dissolved oxygen (O_2) content in the water below a thermocline is usually much lower than above. (Modified from Dojlido & Best, 1993, Fig. 1.25.)

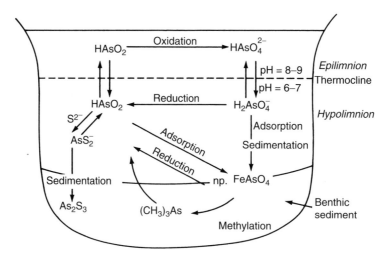

FIGURE 9.7 The arsenic cycle in a stratified lake. Arsenic changes oxidation states, and undergoes sorption/desorption and precipitation reactions (see Chapter 13), in water/sediment systems with changes in temperature, oxygen, pH, and organic carbon. (From Dojlido & Best, 1993, Fig. 2.21.)

discharges where violations through pollutant releases have occurred.

Point sources of pollution include natural drainage pathways and waterways such as rivers and creeks. Industrial and municipal discharge points or outfalls can affect water systems much more rapidly and unevenly. Here, monitoring at the point of discharge is the best way to limit and quantify pollutant impacts to water systems. In this instance, it is usually more important to prepare and initiate a monitoring plan with frequent sampling intervals. Less frequent sampling is usually done in

water environments that have no obvious point sources of pollutants. In these environments, seasonal variations are significant because the net volumes of water coming in and the surface evaporation change markedly during seasons (Figure 9.8).

Nonpoint sources of pollution may be in the form of spikes associated with precipitation events that produce inflow of contaminated runoff water. When the impacts of nonpoint source pollution are mitigated by natural mechanisms, their effects on water quality are difficult to detect. For example, acid rain on a lake is often quickly

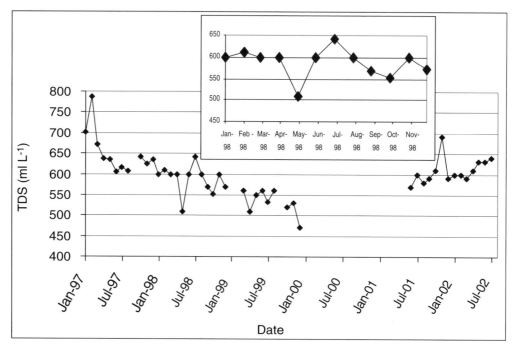

FIGURE 9.8 Historical (4.5 years) Colorado River water quality, total dissolved solids *(TDS)* measured at the Lake Havasu (AZ) inlet. Inset shows 1998 data with typical seasonal variations in TDS. Data from Central Arizona Project [CAP] Website www.cap-az.com.

neutralized by the buffering capacity of the natural water. In this case, pH is not a good predictor of what will happen with further acid inputs. Alkalinity measurements, on the other hand, may show a progressive decline in the buffering capacity of the water. These data could in turn be used to predict if future acid rain events may cause a sudden change in pH and thus provide the opportunity to avert a serious environmental impact to the lake. An unforeseen drop in pH may trigger a series of changes to the aquatic system that are difficult to reverse. These may include death of fish, reduction in photosynthesis by marine algae, reduction of natural biological cycles that control micronutrient and macronutrient concentrations, and increased concentrations of toxic metals in the water.

SAMPLING TECHNIQUES FOR SURFACE WATERS

We do not distinguish between destructive and nondestructive methods of sampling and analysis in water environments. Such a distinction is unnecessary because removing a small volume of water for a sample from a lake or a river, for example, does not disrupt or disturb these environments. Nonetheless, we do recognize two types of water sampling: real-time, *in situ* analysis, which is done on location, and sample removal, which is followed by analysis in the field or in a laboratory.

DIRECT *IN SITU* FIELD MEASUREMENTS

With the development of portable field analytical equipment, several water quality parameters or indicators can

TABLE 9.5
Water Quality Parameters Measured Directly in Water

Parameter	Associated Parameters	Method of Analysis/Probe
pH	Acid base [H$^+$]	Hydrogen-ion electrode
Salinity	Major soluble ions	Electrical conductivity
Dissolved oxygen	Carbon dioxide	Gas-sensitive membrane electrode
Redox potential	Free electrons	Platinum electrode
Turbidity	Suspended particles	Light scatter meter
Temperature	Thermal energy	Thermocouple electrode
Stream flow	Kinetic energy	Flow meter

now be quantified on location and in real time. Some general water quality parameters that can be measured with field equipment by inserting a probe directly into the water are listed in Table 9.5. This is possible because of the development of sensors that respond to changes in chemical and physical properties when in direct contact with water (Figure 9.9). Most sensors have some sort of a porous or semi-porous membrane that senses changes in the activities of the species of interest or undergo redox reactions. Electronic components convert these responses into electrical potentials (V), which can be measured. Water pH measurements require the use of a glass electrode coupled to a reference electrode (see Chapter 13 for a more detailed discussion of the principles of operation of these two electrodes). Briefly, Figure 9.10A shows the basic design equivalent for the electric circuit that these two electrodes form. When placed in solution, these two electrodes develop internal potentials (E″ and E′) that cancel each other. This leaves the solution potential E to be measured,

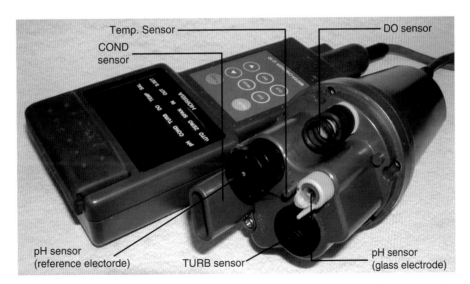

FIGURE 9.9 Field portable multiprobe water quality meter (manufacturer: Horiba Ltd.). The delicate multi-sensor array is covered with a protective sleeve. Each sensor requires a different level of care and periodic calibration. See diagrams of each sensor in Figure 9.10.

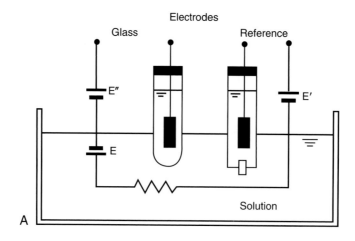

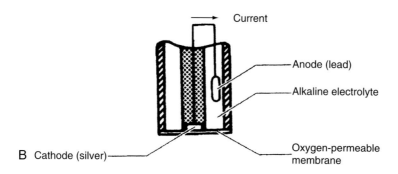

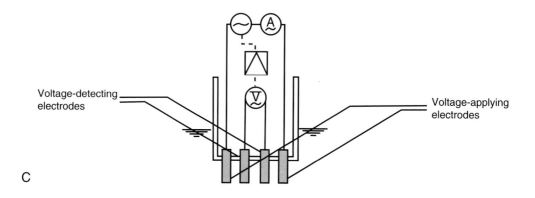

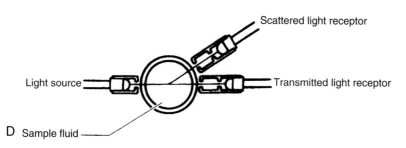

FIGURE 9.10   Diagrams of pH sensor (**A**), DO sensor (**B**), EC sensor (**C**), and turbidity sensor (**D**) pH and DO sensors have membranes sensitive to H^+ and O_2 in water that can be easily damaged and require maintenance. (Modified from Horiba U-10 water quality checker. *Instruction Manual.* 2nd Ed. 1991. Horiba Ltd., 1080 E. Duane Ave., Suite J., Sunnyvale, CA 94086.)

which is proportional to the H^+ activity of the solution. Dissolved oxygen (DO) is measured with electrodes that can undergo oxidation/reduction reactions in the presence of oxygen (see Chapter 13 for a discussion on the principles of redox reactions). Figure 9.10B shows a diagram of a DO electrode. Briefly, dissolved oxygen diffuses into the porous membrane, where it is reduced with water and electrons that are the product of lead oxidation at the anode. The overall reaction is written as follows:

$$2Pb + O_2 + 2H_2O + 4e^- \gg 2Pb^{++} \\ +4OH^- + 4e^- \quad \text{(Eq. 9.2)}$$

The resulting current flow is proportional to the molecular oxygen (O_2) concentration in the solution.

The principles of EC are also discussed in Chapter 13. Briefly, water conducts electrical current in direct response to the activities of electrolytes or ions present in solution. The four electrode configuration is commonly used to measure EC in water (Figure 9.10C). This electrode configuration minimizes electrode decomposition (degradation) that can result in polarized surfaces, which would produce erroneous EC readings.

Turbidity may be determined directly in water by measuring the amount of light the suspended particles scatter in water. Figure 10.9D shows a diagram of this principle in a portable probe that uses a low-power pulsating diode as an infrared light source, used to minimize particle color interferences. Water temperature can be measured easily with a thermistor temperature sensor (see Chapter 4 for the principle of operation of this sensor). Often this type of sensor is built into portable probes to provide temperature data and to allow for automatic temperature compensation for collection of pH, EC, and DO measurements, which are temperature dependent.

The parameters just discussed are routinely measured in most bodies of water, serving as indicators of water quality. Any deviations from "natural" ranges may suggest that a system is under stress. For example, low concentrations of dissolved oxygen usually indicate excessive biological activity due to unusually large inputs of nutrients, like nitrates and phosphates, and organic matter with a high biological oxygen demand. Thus, in lakes and streams, initial signs of eutrophication can be detected by measuring changes in oxygen concentrations.

STREAM FLOW MEASUREMENTS

Stream flow measurements are important to quantify the volumes of water that a flowing stream carries.

Together with water quality data, we can calculate the quantities of chemicals transported by the stream. Because water flow within streams changes temporally and spatially, defining the time and location of measurement is important. Selecting cross sections with low turbulence and smooth geometry is preferable. An average stream flow or discharge (Q) is computed by dissecting the channel into cross sections (W_1 to W_m), each no more than 5% of the total channel width, and measuring the flow (V) at two depths in each cross section. Depth ($D_{0.8}$) is at about 80% of each cross-sectional depth, and $D_{0.2}$ is at about 20% of the same cross-sectional depth (Figure 9.11A). Stream flow is computed with the following formula:

$$Q = \sum_m^1 \left[W \left(\frac{D_{0.8} + D_{0.2}}{2} \right) \left(\frac{V_{0.8} + V_{0.2}}{2} \right) \right] \quad \text{(Eq. 9.3)}$$

Stream flow meters like the one shown in Figure 9.11B can be used to measure stream flows. These meters have a free rotating propeller that can be inserted directly in shallow streams with velocities between 0.03 and $7.5 \, \text{m s}^{-1}$.

PARTICULATE-SEDIMENT MEASUREMENTS

In the previous section the direct measurement of turbidity with field probes was discussed. Turbidity, a secondary water quality standard, is important near stagnant water sources (see Table 9.2). However, fast-moving surface waters often carry large quantities of mineral sediments that degrade water sources directly by increasing water turbidity and indirectly by releasing sorbed nutrients and pollutants into the environment. Estimates of water-induced sediment losses are also important to determine the rate of erosion of upstream soil environment (see Chapter 15). The simplest way to measure water sediment loads is to collect a grab sample and to filter, dry, and weigh its entire sediment content. Because sediment size and transport are related to water velocity, the location of the water sample within a stream must also be considered, as explained in the previous section. The actual sediment measurement is best done under laboratory conditions. Briefly, preweighed 0.5-μm glass-fiber filters are used to filter all the water sample sediments with a suction-filtering device. Subsequently, the filtered sediment is dried at 105° C for several hours and weighed several times until the weight change is less than 0.5 mg. The data are reported as total suspended solids (TSS) in mg L^{-1}. Further physical and chemical characterization of dried sediment samples may be done with laboratory equipment.

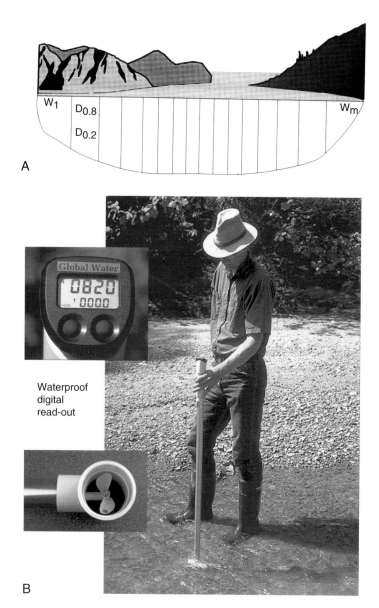

FIGURE 9.11 **(A)** Stream cross section where flow velocities are measured at two depths with a stream flowmeter, and the data are used to compute discharge with Equation 9.1. **(B)** Streamflow meter. Impeller revolutions are proportional to the water velocity. Flowmeters have a range of operation given in meters or feet per second. (**A,** From Meadows Company Catalog, 3589 Broad Street, Atlanta, GA 30341. **B,** From Ben Meadows Company, a division of Lab Safety Supply, Inc.)

In Situ Sample Analysis

Field portable water quality measurement kits now include the capability for on-site measurement of many specific parameters (see Table 9.4). Most modern field portable kits include all necessary sample handling, preparation, and storage components. Additionally, these kits also include battery-powered meters with probes, spectrophotometers, and portable data storage devices such as portable data-loggers and computers. In fact, the distinction between field portable and bench-top laboratory equipment is sometimes difficult to make (Figure 9.12).

In fact, some manufacturers market instruments that can be used in both the laboratory and the field. Numerous chemical parameters listed can be measured in the field with battery-powered equipment (see, for example, the Hatch Company catalog). Nevertheless, in many instances, the measurements are not as precise or accurate as those obtained with standard laboratory methods. Therefore the environmental scientist that chooses to use these portable field measurement kits must be aware of both their advantages and limitations.

Field portable kits can provide accurate concentration measurements of many general water quality parameters.

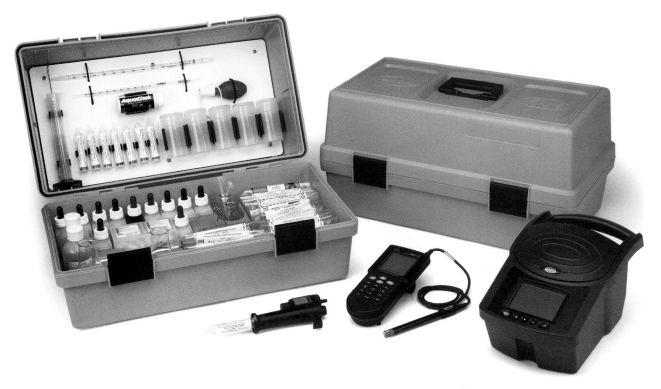

FIGURE 9.12 Field portable water quality kit. It includes one or more portable colorimeters, premeasured reagents, and step-by-step operating instructions for the analysis of multiple water quality parameters in the field. (Hach Company Catalog, P.O. Box 608, Loveland, CO 80539-0608.)

With few exceptions, these parameters are quantified by titration or by the addition of premeasured reagent mixtures to water samples and with a portable colorimeter or set of color wheels to quantify the intensity of the resulting color. Besides the obvious advantage of portability and near real-time measurements, the chemistry of these kits is usually based on scientifically sound wet chemistry color reactions, many of which have been incorporated into modern laboratory analytical chemistry. Another advantage of these kits is that all operating procedures are simple and clear, and all chemicals have been premixed when possible, leaving little room for operator error. However, these field kits have two major disadvantages. The first is that their precision and accuracy usually cannot match standard laboratory methods. Similarly, the detection limits of field portable kits are usually much higher than standard laboratory analytical procedures. This is particularly true in measuring trace inorganic and organic chemicals. In addition, if unusual matrix interferences are encountered, the operator may not be able to detect them. The reason is that the abbreviated operating procedures and premixed chemicals give the operator little latitude for modifications. The use of field portable kits should be accompanied by an awareness of unusual or extreme sample conditions such as color, temperature, turbidity, high salinity, and low pH. Split samples should

also be collected for laboratory analysis verification. The second major disadvantage of field portable kits is they lack automation, thus sample processing is slow compared with modern laboratory bench instruments.

WATER SAMPLING EQUIPMENT

Samples from flowing bodies of water can be collected with jars or bottles placed directly into the stream if depth-integrated sampling is not an issue. However, if the stream is more than 1 to 2 m deep, vertically discrete water samples may be needed. These can be collected with vertically nested water samplers. Water samples for sediment measurement require special samplers. Figure 9.13 shows a sediment water sampler designed to allow water to enter the bottle, while at the same time limiting turbulence inside the bottle. When the water level reaches point A, water collects in the bottle and stops when the water level reaches point B, when water pressures are equal at both points. A modified version of this sampler includes inlet and outlet tubes bent in direct alignment with the water flow. For changes in sediment loads to be measured over time, vertical nests of these sediment water bottles may be placed in rapidly rising water streams to collect water samples at discrete intervals.

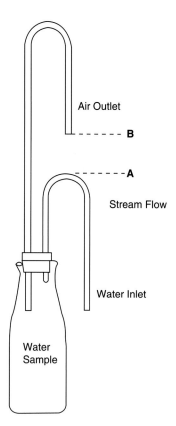

Air Outlet

- - - - - **B**

- - - - - **A**

Stream Flow

Water Inlet

Water
Sample

FIGURE 9.13   Water-sediment sample bottle. It may be placed at multiple depths and cross-sectional locations (nested), as needed, to measure total stream sediment discharge.

Offshore subsurface water samples can be collected with the use of a Van Dorn sampler in lakes and ponds. The sampler trap doors can be activated remotely to collect water samples at any depth. These and other similar devices allow for the collection of deep, discrete water samples directly into appropriate storage bottles. Weighted hoses attached to peristaltic pumps can also be used to collect water samples at discrete depth intervals.

Multiple composites obtained by bulking water subsamples can also be obtained with a churn splitter. However, sample handling requirements of each water quality parameter should be carefully considered when using water sampling and bulking equipment. For example, water samples collected for the analysis of pH, DO, metals, or volatile organic chemicals must not be bulked or exposed to air or plastic containers. It is also necessary to store water samples in appropriate containers that take into account the chemical compatibility with the material of the storage containers.

MISCELLANEOUS FIELD METHODS

Large volumes of water, at least 1 L, are needed for the analysis of trace organic chemicals in the laboratory.

Regulatory detection limits require the extraction and concentration of trace chemicals. For example, the determination of pesticides in water requires an extraction of these pesticides with an organic solvent, such as hexane, followed by concentration, which is done by evaporating most of the solvent. This necessary step usually results in a 1000-fold concentration of the pesticides if, for example, 1000 ml of water is extracted into 1 ml of hexane. This is typically followed by chromatography and spectroscopy analysis to separate and identify each pesticide found within the water sample. The obvious disadvantage of this procedure is the need to carry large volumes of samples from the field into the laboratory. This can result in increased costs due to sample containers, transportation, storage, and preservation considerations.

SOLID PHASE EXTRACTION

Recent advances in solid phase chemical separations have led to the development of solid phase cartridges and disks that can remove and concentrate many organic chemicals from water. Thus, a water sample can be readily extracted in the field following collection, and its chem-

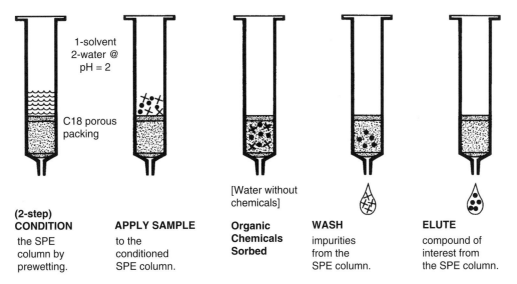

FIGURE 9.14   Solid phase extraction cartridge-sequence. Water sample is passed through a porous medium (C-18) with hydrophobic properties where organic chemicals are sorbed; later these can be flushed out (desorbed) with a small volume of a solvent prior to analysis. (Adapted from Whatman Laboratory Division Catalog, Whatman Inc., 9 Bridewell Place, Clifton, NJ 07014.)

ical contents are stored in a small cartridge or disk that weighs just a few grams. The extraction process is depicted in Figure 9.14. However, only the first two steps are carried out in the field, leaving the elution of the organic chemicals to be completed in the laboratory.

QUALITY CONTROL IN FIELD MEASUREMENTS AND SAMPLING

GENERAL CONSIDERATIONS

All instruments require routine maintenance and calibration. After purchasing a new field instrument, an operating protocol with several key components should be prepared. Operating protocols should include detailed calibration and check procedures, a detailed maintenance description, and short- and long-term storage procedures. In addition, a form to record instrument calibrations, sample collection, and a method for sample cross-referencing should be provided. This information should come from the instrument's manufacturer manual and, when applicable, should use information and steps provided in regulatory methods. Each instrument has its own specific maintenance, calibration, and storage requirements. Additionally, it is important to keep a maintenance logbook for each field instrument. At a minimum, this book should contain dated entries of instrument checks and replaced or repaired parts.

In general, instruments must not be subjected to extreme heat or cold, high humidity, or physical impacts. Because all field instruments are battery operated, remov-

ing the batteries during long periods of storage is desirable to prevent slow drainage or acid leakage.

IN SITU DIRECT MEASUREMENTS

Several important general water quality parameters should be measured directly to obtain unbiased results. Equipment for *in situ* measurements such as probes requires special attention in the form of periodic cleaning or reconditioning. Replacement is necessary when equipment develops slow response times or no longer has an adequate response range. To perform these maintenance tasks, follow an operating procedure written as described previously. However, this should be done before going to the field. Probes such as glass electrodes for pH and gas-permeable membranes for DO should be stored in suitable solutions when not in use. Long-term storage of delicate probe sensors must be done according to manufacturer specifications. Inadequate storage procedures and long periods of inactive use will affect the performance of probes.

In preparation for a field trip, all equipment should be checked and calibrated. This should include, at a minimum, a two-point calibration. Records of all maintenance and calibration checks should be kept including probe replacement, working range of the probe, and stability of the probe.

In the field, it is recommended that a one- or two-point calibration be performed daily. If probes are in continuous use, calibrations at regular intervals during the day are recommended. Detailed records should be kept of method, instrument model, time, location, water

Field Sample Data Collection Form

Method: _____   Location:_____

Reference:_____   Date:_____

Instrument Model:_____   Operator:_____

Sample D Number	Sample Description	Water Temp. (°C)	pH Reading (s.u.)	Date/ Time	Remarks
Reference	pH 4.0 @ 25°C				
Reference	pH 7.02 @ 25°C				
Check 1	6.5@20°C				
Replicate					

FIGURE 9.15 Field sample data collection form. Field sampling requires careful records of sample locations and conditions, equipment calibration, operator, and sample identification numbers.

temperature and ambient temperature, calibration checks, and replicate measurements with a form similar to that in Figure 9.15. Modern dedicated (turnkey) field probes often have multipoint and data storage capacity. It is not recommended that data be stored in dataloggers with volatile memory because power failure will result in data loss. In the field, probes should be carefully cleaned and dried or kept in storage solutions when required, after each use and before short-term storage. After a field trip, probes should again be thoroughly cleaned and dried and placed in long-term storage solutions when needed.

FIELD ANALYSIS OF WATER SAMPLES

In the "Particulate-Sediment Measurements" section, we introduced the use of portable water quality testing kits. The sophistication and portability of some pieces of modern equipment are such that they can be used in either the field or laboratory settings. Nonetheless, portable field monitoring equipment must be completely self-contained with battery-powered instruments, premixed reagents, and dedicated glassware. The operating procedures provided by the manufacturer should be followed carefully. Unlike field portable probes, these kits usually have complete instructions for each water quality analysis. At the very least, these instructions should include a method number and reference; a detailed stepwise operating procedure for the mixing of chemicals; and a calibration procedure when a spectrophotometer is used. Before a field trip, all portable kits must be checked thoroughly to ensure that instruments have new batteries, that they are operating properly, that all glassware is present and clean, and that sufficient reagent supplies are on hand. Usually these kits come with a fixed number of reagent packets, and once these are exhausted, no more samples can be analyzed. An adequate supply of glassware, sample collection bottles, and a good supply of deionized water to rinse all glassware after each use should be on hand. Kits should be opened in a shaded location with a flat surface preferably 1 m above ground in the field. Glassware, chemicals, and instruments should not be exposed to moisture, dust, or direct sunlight.

The prescribed instrument calibration(s) for each water quality parameter and accuracy with a quality assurance

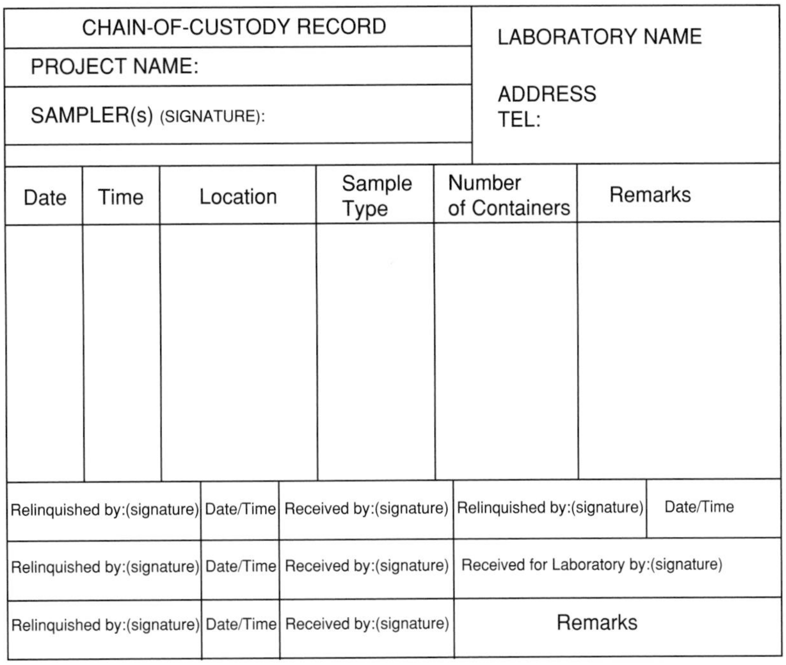

FIGURE 9.16 Chain-of-custody form. This form and copies of it must be kept throughout the sample transfers, including storage, shipping, and analyses. It originates during sample collection and terminates after final sample disposal.

(QA) control sample should be performed daily. Detailed, records for each water quality parameter method and replicate measurements should be kept with a form similar to that shown in Figure 9.16.

Efficiency and quality control dictates that whenever possible, one quality control (QC) check should be done every 10 to 20 water samples. Container requirements should be followed as listed in Table 9.5. However, sample preservation techniques may not be needed if samples are processed soon after collection. Therefore no more samples should be collected than can be processed in about 1 to 2 hours.

COLLECTION OF WATER SAMPLES FOR LABORATORY ANALYSIS

Trace levels of pollutants and other important minor water quality parameters are usually best quantified in the laboratory. Water samples should be stored, preserved, and transported with approved techniques. Careful protocols are necessary to maintain the chemical and biological properties of water samples. Table 9.6 presents a summary of sample containers and preservation techniques. For a complete list of parameter-specific protocols, the reader should refer to the *Code of Federal Regulations* (40 CFR 136 and 141), U.S. EPA publications such as SW-486, and *Standard Methods for Exam-*

ination of Water and Wastewater reference published by APHA, AWWA, WEF.

For sample contamination to be reduced, all samples should be collected in new bottles, vials, or jars provided by vendors that specialize in these types of containers. Often these vendors provide specific types of containers for each water quality parameter, including preserving agents, waterproof sample labels, and seal tags.

All water samples in the field must be transferred quickly from sampling containers to storage bottles, vials, or jars and sealed with zero headspace. To reduce sample degassing, it is very important that no air be trapped inside storage bottles. For example, subsurface water samples usually have higher carbon dioxide (CO_2) concentrations than those equilibrated with atmospheric air. After sampling, CO_2 will be lost from these samples, and this may result in lower than expected CO_2 concentrations and higher-than-expected pH values. Similarly, water samples contaminated with volatile organic compounds (VOCs) may experience significant chemical losses due to volatilization while exposed to air or inside a closed container with headspace. Therefore preserving the integrity of the field sample by reducing any gaseous exchanges and losses is critical in environmental water sampling.

Each sample container must be tagged with an appropriate label and a custody seal. Also, at this time, each sample must have a chain-of-custody form (Figure 9.16). After log-in, the sample should be placed inside a cooler with sufficient amounts of crushed ice or dry ice packs to cover the container. For the prevention of container ruptures, water samples must not be frozen. Adequate provisions must also be made to reduce transport time because the maximum sample holding time starts at the time of field collection.

SAMPLE STORAGE AND PRESERVATION

Currently, accurate and precise analysis of trace components or contaminants of water are done in the laboratory. Modern extraction, separation, and quantification techniques must be used following government-approved analytical protocols. For this reason, water samples collected in the field are usually stored and preserved for variable periods of time before the actual analyses are carried out. However, because of the dynamic chemical and biological nature of water, significant changes can take place in a water sample stored only for a few hours. For the reduction of these changes and the preservation of the integrity of the sample, strict sample storage and preservation guidelines have been developed. The Code of Federal Regulations 40, Part 136, lists specific types of containers, volumes, preservatives, and maximum holding times for more than 70 different water quality–related parameters. Table 9.6 lists a summary of the major types of sample containers and methods of preservation needed to store field samples. Exact provisions must be made to collect the correct number of samples in one or more bottles with the prescribed sample container

types and preservation methods. The methods of preservation include the additions of chemical antioxidants and strong acids for pH adjustment. Additionally, all samples are typically kept at 4° C in coolers or refrigerators after collection and during transport and storage. Freezing water samples is not recommended, but may be done in emergency situations for inorganic parameters only, if containers have 20% or more headspace. Not included in this table are the holding times for each of the samples collected. Most holding times vary from 24 hours to 40 days, depending on the parameters to be analyzed. Acidified water samples for metal analyses can be kept for up to six months. Nevertheless, some parameters like pH, dissolved oxygen, and chlorine can change quickly; thus these analyses should be done immediately. These parameters are best measured in the field with portable probes and chemical testing kits.

QUESTIONS

1. According to the principle of electrical neutrality, a water sample must have the same number of anions and cations. Assuming that a water sample contains $40\,mg\,L^{-1}$ of chloride, $100\,mg\,L^{-1}$ sulfate, and $10\,mg\,L^{-1}$ nitrate, estimate the maximum amount (in $ml\,L^{-1}$) of sodium that could be present in this water sample.

2. Is water below a thermocline more or less oxygen saturated and more or less dense than water above a thermocline? Explain your answer.

TABLE 9.6

Major Types of Water Sample Containers, Methods of Preservation, and Minimum Sample Volumes

Water Parameters	Sample Container	Preservation and Storage	Minimum Volume (ml)
pH, E.C. + alkalinity + major anions	Plastic[a] bottle	Keep cool at 4° C	100 + 100 + 200
Metal cations (except Hg and CrVI)	Plastic[a] bottle	Add nitric acid (pH < 2)	200
Pesticides + polynuclear aromatic hydrocarbons (PNAs) + surfactants, dioxins, phenols,	Glass bottle (clear or amber)	Add reducing agent[b] + keep cool at 4° C + adjust pH(variable)[c]	1000 or more for each group
Volatile organics (halogenated + nonhalogenated)	Glass vial (clear)	Add reducing agent[b] + keep cool at 4° C + adjust pH(variable)[c]	5 to 25 for each group
Coliforms (fecal and total) + fecal streptococci	Plastic bottle	Add reducing agent[b] + keep cool at 4° C	100 + 100

Note: Sample holding timers are variable (immediately up to 6 months).
[a]Use high density polyethylene (HDPE) plastic bottles.
[b]Sodium thiosulfate in low concentrations is added as an antioxidant.
[c]Hydrochloric acid or sulfuric acid is added to adjust pH or lower it below 2.

3. List five common water quality parameters that can be measured *in situ* with portable field probes. Rank the ease of measurement of each parameter (hint: consider the type of probe used to measure each parameter and problems associated with calibration and maintenance).

4. Water samples collected for metals analyses should be acidified. Why? Dark glass bottles should be used to store water samples for pesticides analyses. Why?

5. Which water quality parameters can be affected by these sample changes:
   a) gassing or degassing
   b) pH increase or decrease

6. Explain which water quality parameters are associated with:
   a) alkalinity
   b) eutrophication
   c) color
   d) turbidity

7. Explain the best use of the major advantages and disadvantages associated with:
   a) portable probes
   b) portable chemical analysis kits
   c) laboratory instruments
   d) solid phase extraction

REFERENCES AND ADDITIONAL READING

Csuros, M. (1994) *Environmental Sampling and Analysis for Technicians.* Lewis Publishers, Boca Raton, FL.

Dojlido, J.R., and Best, G.A. (1994) *Chemistry of Water and Water Pollution.* Ellis Horwood Publishers, NY.

FAO. (1985) *Water Quality for Agriculture.* Irrigation and Drainage Paper 29 Rev. 1. Edited by R.S. Ayers, and D.W. Westcot. Food and Agriculture Organization of The United Nations, Rome, Italy.

Greensberg, A.E., Trussell, R.R., and Clesceri, L.S. (1985–1995) *Standard Methods for the Examination of Water and Wastewater.* 16th, 17th, 18th, and 19th Editions. APHA/AWWA/WPCF. American Public Health Association. 1015 1st St. NW, Washington, DC.

Keith, L.H. (1996) *Compilation of E.P.A.'s Sampling and Analysis Methods.* 2nd Edition. Lewis Press Inc., Boca Raton, FL.

Keith, L.H. (1991) *Environmental Sampling and Analysis: A Practical Guide.* ACS publication, Columbus, OH.

Schlesinger, W.H. (1997) *Biogeochemistry: an analysis of global change.* 2nd edition. Academic Press, San Diego, CA.

van der Leeden, F., Troise, F.L., and Todd, D.K. (1990) *The Water Encyclopedia.* Geraghty & Miller Ground-Water Series. Lewis Publishers, Creisea, MI.

WRRC. (1995) *Field Manual for Water Quality Sampling.* University of Arizona, College of Agriculture. Water Resources Research Center. Tucson, AZ.

U.S. EPA. (1999) National Recommended Water Quality Criteria-Correction. U.S. Environmental Protection Agency. Office of Water 4304. EPA 822-Z-99-001. April 1999.

U.S. EPA. (2000) Drinking Water Standards and Health Advisories. U.S. Environmental Protection Agency. Office of Water 4304 EPA 822-B-00-001. Summer 2000.

U.S. EPA. (2001) Ecoregional Nutrient Criteria-Fact Sheet. U.S. Environmental Protection Agency. Office of Water 4304. EPA-822-F-01-001. December 2001.

10

MONITORING NEAR-SURFACE AIR QUALITY

A.D. MATTHIAS

Control and prevention of air pollution is a major environmental challenge facing modern society. Meeting this challenge is especially important to the millions of people who live and work in urban areas, where the burning of fossil fuels releases many pollutants that form smog-filled air. Unbeknownst to them, humans regularly breathe these pollutants along with other toxic chemicals, such as formaldehyde and benzene. However, the senses of sight and smell often betray the presence of these pollutants, especially when driving in rush hour traffic.

Although the problem of air pollution has beset most societies throughout recorded history, including the ancient Greek and Roman societies, it was not until the latter half of the 20th century that many countries took action to address the problem. In the United States, for example, through the Clean Air Act (CAA) Congress authorized the Environmental Protection Agency (EPA) to establish and enforce national emission standards and national ambient air quality standards (NAAQSs). Similar standards have been promulgated by Canada and many

European nations. Emissions and ambient air quality monitoring systems were developed and deployed in metropolitan, industrial, and rural areas, where air quality problems were historically most prevalent, to help enforce compliance with these standards. This chapter describes these monitoring systems, especially those in the United States. The chapter also addresses some of the important benefits these systems have provided in improving air quality.

This chapter includes a brief explanation of how atmospheric properties affect pollutant transport and fate to provide context for describing monitoring systems. In particular, the role of wind and turbulence in influencing horizontal and vertical transport of pollutants near the ground is emphasized. Sources, types, and trends for traditional air pollutants and hazardous chemicals are also described. In addition, a brief historical overview of legislative programs to control air pollution is presented. This is followed by examples of sampling methods and instruments commonly used for monitoring ground-level air quality.

STRUCTURE AND COMPOSITION OF THE ATMOSPHERE

The main layers of the atmosphere are the troposphere, the stratosphere, the mesosphere, and the thermosphere. Each layer has distinct compositional, flow, and thermal properties. The troposphere is the lowest layer and contains about 80% of the total mass of the atmosphere. The three layers above the troposphere are often collectively referred to as the earth's upper atmosphere. Because the troposphere is in contact with the ground surface, it is the layer that receives most human-caused pollution. Some pollution from high-flying aircraft and rocket engines are emitted directly into the upper atmosphere. Some pollutants emitted at the ground are carried upward by tropospheric winds and some eventually reach the stratosphere. Chlorofluorocarbons, which release chlorine that catalytically destroy ozone, are an example of ground-level pollutants that eventually reach the stratosphere. Tropospheric pollution, particularly pollution present within the atmospheric boundary layer (a sublayer of the troposphere extending a few hundred meters above the ground), is the primary focus of this chapter.

The troposphere extends upwards to about 15 km over the warm tropics. Over the poles it is considerably shallower, extending to only about 5 km above the ice caps. The troposphere is generally characterized by strong vertical mixing by means of winds and thermal updrafts throughout, which accords a relatively uniform or homogeneous chemical composition. Dry, "nonpolluted" troposphere air contains about 78% molecular nitrogen (N_2) and 21% molecular oxygen (O_2) by volume with the remaining 1% composed mainly of argon (Ar), carbon dioxide (CO_2), and other trace chemicals. Water vapor may constitute up to about 4% of the total density of the troposphere at discrete locations. The relatively low molecular mass of water vapor (about 18 g per mole) decreases the density of air (average molecular mass of about 29 g per mole) and enhances vertical mixing by buoyancy.

Air temperature typically decreases with increasing height in the troposphere. This decrease of temperature with increase of height is referred to as the environmental lapse rate (ELR). Typically, at mid latitude, the daytime air temperature may decrease by about 7 °C per kilometer. The nighttime ELR can differ considerably from the daytime ELR, when the temperature may increase with height at night.

The stratosphere extends above the top of the troposphere (i.e., the tropopause) to a height of about 50 km above ground level. The composition of the stratosphere differs somewhat from that of the well-mixed troposphere, with photodissociation or the splitting apart of O_2 yielding atomic oxygen (O) and ozone (O_3) in the upper stratosphere. The stratosphere is relatively free of human-caused pollution, with the exception of nitrogen oxides (NO_x; nitric oxide [NO] and nitrogen dioxide [NO_2]) exhausted from high-flying aircraft engines and other synthetic compounds including the chlorofluorocarbons (CFCs) that can leak from automobile and home climate control systems. The stratosphere is dynamically stable and stratified; air temperature increases by about 2° to 3°C per kilometer increase of height, with little or no turbulence.

Extending above the stratosphere to a height above ground of about 150 km are the mesosphere and thermosphere. These layers are distinguished by extremely low air densities with many components of the air photodissociated or ionized by the intense ultraviolet light from the sun. There is little or no pollution of human origin typically present within these layers.

Returning to the troposphere, the very bottom of the troposphere is termed the *atmospheric boundary layer*. This relatively thin sublayer extends to about a 1 km depth during the daytime and about 200 m during the nighttime. It occurs because of frictional interaction and heat exchange between the earth's surface and the troposphere. Within the boundary layer, convective transport of heat and mass including pollutants and water vapor, as well as momentum due to the result of wind, are strongly linked to the temperature and roughness of the earth's surface. Both forced and free convective transport occur simultaneously, especially near the surface.

Forced convection is turbulent motion generated by the bulk movement of the atmosphere over the earth's surface. Air flowing rapidly over a rough surface generates intense, nearly random turbulent movement of air parcels, which readily disperses pollutants. Free convection is most important over warm, aerodynamically rough surfaces, such as tall buildings and trees. Free convection is mainly associated with the warming of the earth's surface by sunlight. Buoyant plumes generated by this warming often extend the planetary boundary layer to a height of at least 1 km above the surface. This heating and resultant mixing of the air tends to rapidly disperse pollutants upward from surface sources. During daylight, the surface is relatively warm, and air temperature tends to decrease with increasing height, which makes the air relatively unstable and conducive for mixing pollutants.

At night under clear skies, the ground surface often quickly cools by radiative energy loss to the atmosphere, and air adjacent to the surface becomes cooler than the air higher up in the boundary layer. This energy exchange results in the formation of an air temperature inversion in which relatively warm air overlies a layer of relatively cool air in contact with the ground. Temperature inversions are associated with stability, which suppresses turbulent motion in the vertical direction (such as also occurs in the stratosphere). The nighttime temperature inversion conditions tend to trap pollutants near the ground. Air temperature inversions occur most often at night under clear sky conditions, when long wave radiant energy loss from the ground to the atmosphere is most efficient. Inversions can also occur because of cold air advection related to frontal activity, subsidence associated with high atmospheric pressure, and airflow on the lee side of mountainous terrain. Inversions are especially common in many western U.S. cities situated in valleys, such as Los Angeles, Phoenix, and Denver, that collect cool air near the ground. When flue gas for example, is released from a chimney the atmosphere near the chimney will either trap or disperse the gas, depending on several factors including the height of the chimney, the temperature of the flue gas, and the temperature and wind patterns above and downwind from the chimney. When air is flowing slowly past the chimney and the air temperature increases with height above the ground (i.e., an air temperature inversion is present) making the air stable, the pollution will tend to remain relatively near the level of the chimney (Figure 10.1). If, on the other hand, wind speeds are high and the air temperature is decreasing with height (making the air unstable), the pollution will be quickly dispersed vertically by turbulence and advected away horizontally from the chimney by the mean wind.

SOURCES AND IMPACTS OF AIR POLLUTION

AIR POLLUTANTS

Air pollutants are categorized by the U.S. EPA according to whether they are emitted directly to the atmosphere (primary pollutants) or formed by reactions involving primary pollutants and other constituents within the atmosphere (secondary pollutants).

Stationary sources include point and plane sources such as chimneys, landfills, oil refineries, and chemical plants. Nonstationary sources include most motor vehicles, including automobiles, boats, and aircraft. In addition to the six primary pollutant categories (Table 10.1) identified in earlier versions of the CAA, 189 hazardous pollutants were included in the 1990 CAA. These include highly toxic inorganic materials, such as mercury, and volatile organic compounds (VOCs), such as benzene, toluene, and xylenes (BTXs) (defined in Table 10.1).

Secondary air pollutants are formed from sunlight that results in photochemical reactions of O_2 with one or more of the primary air pollutants. O_3 and other oxidants

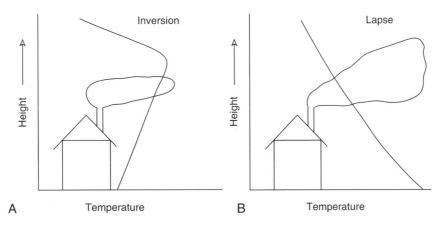

FIGURE 10.1 Air temperature variation with height above ground during nighttime inversion conditions (**A**) and daytime lapse conditions (**B**). Note the relative trapping of smoke during inversion conditions. (From A. Matthias.)

TABLE 10.1
Six Categories of Primary Air Pollution[a]

Pollutant	Designation	Health effects	Environmental effects
Carbon monoxide	CO	Irritant. Decreased exercise tolerance, possible effects from viruses.	
Lead	Pb	Chronic toxicity. Impaired blood formation, infant development.	
Nitrogen dioxide	NO_2	Irritant. Aggravation of respiratory disease.	Forms nitric acid. Precursor to ozone formation. Atmospheric discoloration.
Particulates (designated by size of particulate matter; $< 10\,\mu m$ and $< 2.5\,\mu m$)	PM_{10} and $PM_{2.5}$	Respiratory tract irritants. May damage lungs. Respiratory disease symptoms worsen; excess deaths.	Visibility.
Sulfur dioxide	SO_2	Irritant. Wheezing, shortness of breath, chest tightness.	Forms sulfuric acid and acid rain. Plant damage and odor.
Volatile organic compounds	VOCs	Toxic, carcinogenic.	Examples: benzene, xylene, toluene (referred to as BTX). Precursors to ozone formation.

[a]The EPA designates six categories of primary air pollution emitted directly into the atmosphere from stationary and nonstationary sources. Ozone is not considered to be a primary pollutant but is listed as a criteria pollutant (see Table 10.2).

in photochemical smog are examples of secondary air pollutants formed photochemically in the presence of carbon monoxide (CO), NO_x, and volatile hydrocarbons.

NATURAL SOURCES

Air pollution can originate from many natural sources, including wildfires that often release large amounts of smoke particles into the atmosphere. Volcanoes also release sulfate and other particles that may eventually become suspended throughout much of the atmosphere, including the stratosphere. Dust storms caused by strong winds blowing over nonvegetated, cultivated soil are another natural source of suspended atmospheric particulates. Dust generated by vehicle traffic over unpaved roads is another example (see Chapter 15).

ANTHROPOGENIC SOURCES

Most air pollution comes from the burning of fossil fuels for electrical energy generation, automobile transportation, industry, farming, home and business, and other activities. CO, for example, has long been known to be a harmful product of incomplete combustion of fossil fuel and is the single largest pollutant in the United States. It is has a relatively short atmospheric lifetime with most CO being oxidized to CO_2. Burning coal releases sulfur from the coal in the form of sulfur dioxide. Nitrogen in fossil fuels is released as NO_x (where $NO_x = NO_2 + NO$) when the fuels are burned; NO_x is also formed when O_2 and N_2 are combined at very high temperatures within internal combustion engines. Volatile hydrocarbons of high vapor pressure can volatilize directly from fuels, such as gasoline, to the atmosphere. Processed petroleum products used in backyard

barbecuing can also be a significant source of hydrocarbons. Anthropogenic sources of particles include effluent from smoke stacks, mining operations, and automobile exhaust. Particles are also emitted from vehicle tires as the vehicles are driven at high speeds.

URBAN

Many sources and types of air pollution exist within urban areas. Because motor vehicles are a primary source, pollutant concentrations tend to be highest along busy transportation corridors during morning and afternoon rush hour traffic. The emission of ozone precursors, NO_x and volatile hydrocarbons often leads to the production of ozone in the photochemical smog that is ubiquitous to most urban environments.

Urban air pollution concentrations increase when air temperature inversions are present, preventing dispersal of the pollutants away from the sources. Figure 10.2 illustrates the diurnal variation of CO measured at a busy traffic intersection in Tucson, Arizona. The peak concentrations during the early morning and early evening hours are associated with high traffic volume and air temperature inversion conditions over the city.

Spatial variation of air pollution within an urban area is illustrated in Figure 10.3, which shows the peak 1-hour average ozone concentrations for Los Angeles on a summer day. These variations are due to many factors, including automobile and industrial sources of precursor pollutants, air temperature, wind patterns, and topography.

How has air quality improved in the United States? Generally, air quality has improved since strict enforce-

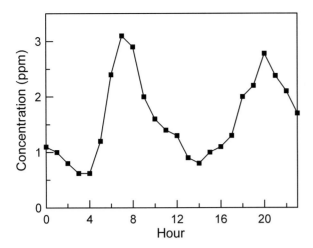

FIGURE 10.2 Mean diurnal variation of carbon monoxide during December 1998 at a busy traffic intersection in Tucson, Arizona. (From 1998 Annual Data Summary of Air Quality in Tucson, Arizona. Pima County Dept. of Environmental Quality report AQ-309)

ment of NAAQS began in the 1970s. Figure 10.4 shows for example, the general decreasing trend of the number days per year that the 1-hour standard for ozone was exceeded in the South Coast Air Basin within the Los Angeles metropolitan area. The figure shows how air quality in Los Angeles, though still among the worst in the United States, has improved considerably since the 1970s.

On a national level, many large urban areas do not comply with NAAQS. Figure 10.5 shows the counties (or parts of counties) that are designated by the U.S. EPA as nonattainment areas for ozone. These nonattainment areas are concentrated mainly in California, the Northeast, Midwest, and Texas. Also shown in Figure 10.6 are the nonattainment areas for particulate matter (PM_{10}), which are mainly in the western United States.

INDUSTRIAL

Air quality in urban and rural areas immediately downwind from stacks and other industrial sources of pollution can be impacted greatly by those sources. The pollutants most often associated with industrial sources are NO_x and sulfur dioxide (SO_2), which are produced from burning coal. These pollutants can be

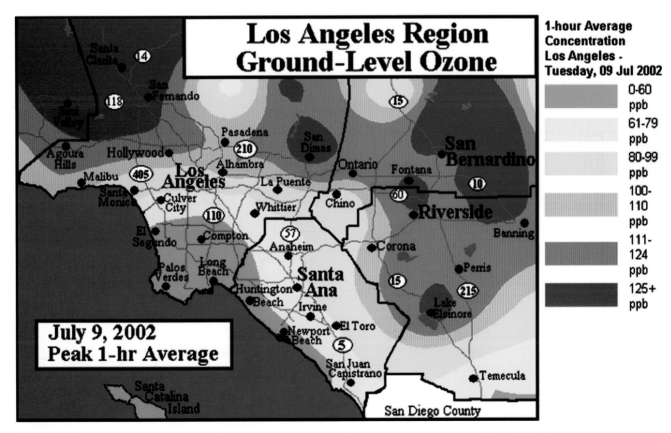

FIGURE 10.3 Peak 1-hour average ground-level ozone concentrations within the Los Angeles area on July 9, 2002. Highest concentrations are inland because ozone-forming conditions (sunlight, temperature, wind speeds, and ozone precursor chemical concentrations) are optimal there. Note concentrations greater than 120 ppb exceed the 1-hour NAAQS for ground-level ozone. (From the U.S. Environmental Protection Agency AirData Web site at www.epa.gov.)

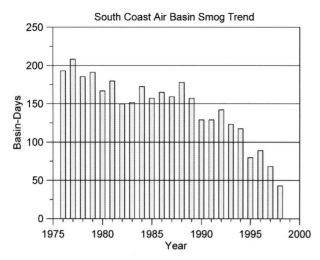

FIGURE 10.4 Number of days per year in which 1-hour ozone NAAQS was exceeded in the South Coast Air Quality Management District in the Los Angeles area between 1976 and 1998. (Courtesy South Coast Air Quality Management District.)

carried by the wind over distances of hundreds of kilometers, often being washed out of the atmosphere by rain or snow leading to acid precipitation. All types of pollutants, including toxic materials, such as mercury, are emitted from industrial sources. In the United States, about one third of all air pollution is produced from industrial sources.

RURAL

People living in rural areas often experience the effects of air pollution, which may come from sources locally or from distant urban or industrial centers. An important air pollutant often detected in rural areas is dust produced by agricultural tillage and harvest activities. In Arizona, for example, dust from fallow or cultivated fields can significantly deteriorate visibility for motorists (Figure 10.7). Open pit mines may also produce dust and other pollutants. Ozone and other pollutant concentrations can be relatively high in rural areas downwind from large urban centers. This is particularly a problem in parts of the southeastern United States. Relatively high ozone concentrations in these areas may be related in part to emissions of the ozone precursor NO_x formed in agricultural soils by denitrification and nitrification.

There are many additional localized sources including volatilization of insecticides, herbicides, and fungicides applied to agricultural fields. Emissions of compounds, such as hydrogen sulfide (H_2S), ammonia (NH_3), and methane (CH_4) from animal waste ponds and biosolid amended soils, may also be a significant local source of these gases in rural areas.

In some rural and wildland areas, air quality has declined in recent years. An example of this decline is within the Grand Canyon National Park. Figure 10.8 shows

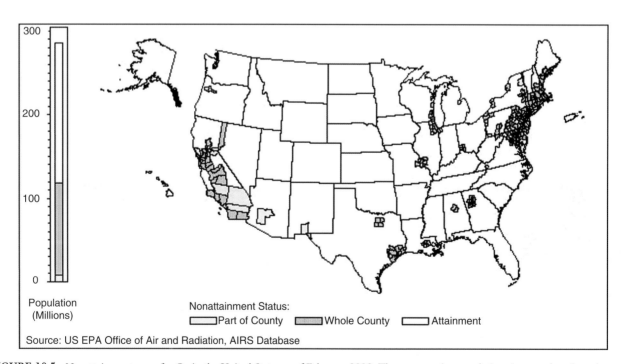

FIGURE 10.5 Nonattainment areas for O_3 in the United States as of February 2003. These nonattainment designations are based on the ozone 1-hour air quality standard. Designations will be made at a later time based on the ozone 8-hour standard that EPA defined in July 1997. (From the U.S. Environmental Protection Agency AirData Web site at www.epa.gov.)

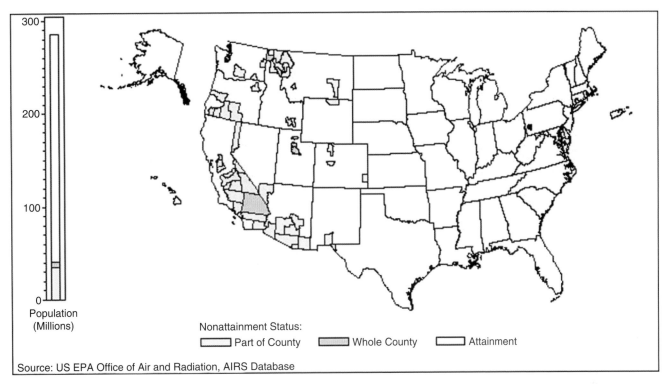

FIGURE 10.6 Nonattainment areas for PM_{10} in the United States as of February 2003. These nonattainment designations are based on the air quality standard for particulate matter smaller than 10 μm (PM_{10}). Designations will be made at a later time based on the standard EPA defined in July 1997 for particulate matter smaller than 2.5 μm ($PM_{2.5}$). (From the U.S. Environmental Protection Agency AirData Web site at www.epa.gov.)

FIGURE 10.7 Dust storm rises over Phoenix, Arizona, on Labor Day, 1972. No rain had fallen in the area for 153 days. (From the U.S. Environmental Protection Agency National Archives.)

FIGURE 10.8 Grand Canyon National Park viewed on clear *(top)* and polluted *(bottom)* days. (Courtesy the U.S. National Park Service through Air Resource Specialists, Inc.)

strikingly different air quality conditions observed at the Grand Canyon. The air pollution in the Grand Canyon is thought to be mainly a result of long distance transport of pollutants from Los Angeles and Las Vegas, as well as from nearby coal-fired electrical power generators.

HISTORY OF AIR POLLUTION CONTROL

EARLY RECOGNITION OF AIR POLLUTION AND ATTEMPTS AT CONTROL

Legislative efforts to control air pollution have evolved over a long period. Certainly, before any pollution-related laws were written, there first had to be a recognition of the pollution problem. Recognition slowly developed as air pollution problems and use of fossil fuels increased. Pollutant emissions from industry and the automobile ultimately became so acute by the middle of the 20th century that governments in North America and Europe took action to enact laws and regulations to improve air quality.

The first known governmental attempt to limit emissions of smoke occurred during the 14th century in London, where it was illegal to burn coal when Parliament was in session. That unfortunately did not prevent the use of coal, and London had such a serious pollution problem by the 17th century that ideas for preventing pollution were proposed to the King of England and to Parliament. The Industrial Revolution brought a whole new era of air pollution problems caused by the invention and use of the coal-fired steam engine. The introduction of the gasoline-powered automobile at the beginning of the 20th Century ushered in another era of pollution problems, which continues to this day. It was not, however, until major, deadly air pollution episodes in Donora, Pennsylvania, and in London in the mid-20th century that the U.S. federal government began concerted efforts to control air pollution.

In 1948 the residents of Donora experienced unusually high pollutant concentrations during unfavorable weather conditions, in which a strong air temperature inversion trapped sulfur dioxide and other pollutants from local steel factories and a zinc smelter for nearly 1 week. Nearly 600 of the 14,000 residents of Donora became ill, and 20 died because of the pollution. The people of London, 4 years later, experienced similar air temperature inversion conditions, which lasted a week. The conditions intensified the city smog to dire levels and caused the deaths of nearly 4,000 Londoners. The Donora and London episodes brought the growing problem of air pollution to the public's and policy makers' attention.

Air quality in the United States remains a concern even though it has mostly improved since the 1950s. In fact, nearly 50% of the U.S. population lives within counties that are not fully compliant with the NAAQS, although millions of tax dollars are spent annually for monitoring air quality in those counties. Regrettably, thousands of people may die prematurely each year because of long-term exposure to air pollution, and millions of dollars are spent for treatment of respiratory diseases, such as asthma, chronic bronchitis, and emphysema; that is, diseases induced, aggravated, or both, by air pollution.

In addition to the burdensome human costs, air pollution harms plants and animals, and this diminishes the earth's capacity to sustain life. Air pollution, especially in the form of suspended particles, also reduces visibility and diminishes our appreciation of the natural environment. Pollution also has an impact on the global climate by altering the radiative energy balance of the earth's atmosphere system.

Regulatory attention concerning air pollution has focused mainly on controlling ozone and the primary pollutants (see Table 10.1) that are common in urban environments because of the result of fossil fuel burning. Although relatively cleaner fuels are currently used, total pollutant emissions and therefore atmospheric pollutant concentrations remain high because of high overall consumption of these fuels. In recent decades, the challenges associated with controlling emissions of less common, but severely toxic materials, such as mercury, have begun to be addressed.

AIR QUALITY AND EMISSIONS STANDARDS

FEDERAL

Meaningful federal efforts in the United States to control and prevent air pollution did not begin until the mid-20th century. Californians were among the first to recognize the problem of air pollution and took steps to improve air quality before the 1950s. The federal government actually followed the lead of California in writing legislation aimed at reducing air pollution throughout the nation, motivated by concerns about the impacts of air pollution on natural and agricultural ecosystems and global climate. Efforts by the federal government to control air pollution nationally began in 1955 with enactment and promulgation of the Air Pollution Control Act. This act permitted federal agencies to aid those state and local governments who requested assistance to carry out research of air pollution problems within their jurisdictions. It also established the ongoing principle that state and local governments are ultimately responsible for air quality within their jurisdictions.

The principle that the federal government fund state and local governments for air quality monitoring and research began with the CAA of 1963. This act forced the federal government to take responsibility for establishing advisory air quality criteria based on consideration of possible health effects that could result from exposure to air pollution. The 1967 CAA defined four main approaches that are still a part of current air quality legislation in the United States. These are: (1) the establishment of air quality control regions (AQCRs) in the United States; (2) the development of specific mandated air quality criteria and control technology for industry; (3) adoption of NAAQ to protect public health and welfare; and (4) the establishment of State Implementation Plans (SIPs) to help ensure NAAQS compliance in each state.

The 1970, 1973, and 1975 CAA amendments established national emission limits for new or modified stationary sources, new motor vehicles, and hazardous pollutants (e.g., National Environmental Standards for Hazardous Air Pollutants [NESHAPs]). The 1970 CAA required the EPA to establish New Source Performance Standards (NSPSs) for industrial sources emitting pollutants for which NAAQS had been adopted. The NSPS chemicals include, for example, asbestos, beryllium, mercury, benzene, and radionuclides. The 1977 amendment to the CAA put forth two new regulatory provisions: the Nonattainment Program and the Prevention of Significant Deterioration (PSD) Program to improve air quality in urban areas.

The 1977 amendment also established the "bubble" strategy in which the total emissions of a pollutant (e.g., sulfur dioxide) from an industry or category of source should not exceed the limit set by the EPA. This strategy allows, for example, an electrical utility con-

siderable flexibility to balance emission increases at one power plant with emission decreases at another of its plants.

The 1990 CAA amendments strengthened air quality legislation in response to concerns about acid rain and nonattainment areas failing NAAQS by deadlines set forth in the 1977 CAA and included new civil and criminal enforcement penalties. New programs to be phased in over a 15-year period included efforts to limit regional and global air pollution such as acid rain and stratospheric ozone depletion; efforts to cut SO_2 emissions by 50% through a nationwide system of emission trading; and the reduction of emissions of air toxins (such as mercury) by at least 75% from 1987 levels.

Currently, the U.S. EPA is considering strengthening several provisions of the 1990 CAA. For example, recent progress on urban air quality may be reversed by the greater use of larger and less efficient motor vehicles, including popular sport utility vehicles (SUVs), pickup trucks, and minivans. Also, the EPA is considering requiring use of low-sulfur gasoline, because sulfur interferes with the operation of pollution control devices on vehicles. This could help achieve a possible 80% reduction in smog-producing chemicals emitted from vehicles in the United States. The EPA may also require automakers to include the larger vehicles in determining fleetwide average emissions. Efforts by the EPA to strengthen the 1990 CAA are being challenged within the U.S. judicial system, and the outcome is uncertain.

Current federal NAAQS, given in Table 10.2, for the six so-called criteria pollutants set limits to protect public health, including people especially sensitive to air pollution, such as the elderly and children. Secondary standards set limits to protect public welfare, including protection against decreased visibility, and damage to animals, plants, and buildings.

TABLE 10.2

National Ambient Air Quality Standards for the Six Criteria Pollutants

Pollutant	Average Time	Primary Standard	Secondary Standard
Carbon monoxide	8 hr	9 ppm (10 mg m^{-3})	None
	1 hr	35 ppm (40 mg m^{-3})	None
Ground level ozone	8 hr	0.08 ppm (155 μg m^{-3})	Same as primary
	1 hr[a]	0.12 ppm (232 μg m^{-3})	Same as primary
Lead	3 months	1.5 μg m^{-3}	Same as primary
Nitrogen dioxide	Annual	0.053 ppm (100 μg m^{-3})	Same as primary
PM$_{10}$	Annual	50 μg m^{-3}	Same as primary
	24 hr	150 μg m^{-3}	Same as primary
PM$_{2.5}$	Annual	15 μg m^{-3}	Same as primary
	24 hr	65 μg m^{-3}	Same as primary
Sulfur dioxide	Annual	0.03 ppm (80 μg m^{-3})	None
	8 hr	0.14 ppm (365 μg m^{-3})	None
	3 hr	None	0.5 ppm

[a]Ozone 1-hour standard applies only to areas that were designated nonattainment when the ozone 8-hour standard was adopted in July 1997.

For an area to be designated a nonattainment area, a standard for any of the criteria pollutants in Table 10.2 must be exceeded for several consecutive years. For the NAAQS for ozone to be exceeded, for example, the 1-hour average concentration may not exceed 0.12 parts per million (ppm) (volume/volume [v/v]) on more than 1 day per year, or the fourth highest 8-hour average concentration value of the year may not exceed the standard of 0.08 ppm. These standards are subject to review and revision, as has occurred most recently with standards for ozone and particulate matter.

The pollutant standard index (PSI) is the index used by the EPA to report air pollution levels to the public. The PSI is based on the 1-hour average daily maximum concentration for ozone, the 8-consecutive-hour average maximum for carbon monoxide, and the 24-hour mean for PM_{10}. PSI values can range from 0 to over 100. These values are converted to a numerical scale with easily interpreted descriptor terms: good, moderate, unhealthful, very unhealthful, and hazardous.

In summary, the CAA requires that the EPA do the following four things with regard to improving air quality in the United States:

1. Set NAAQS for pollutants harmful to public health and environment
2. Cooperate with states to ensure standards are met or attained
3. Ensure sources of toxic pollutants are well controlled
4. Monitor effectiveness of the air quality monitoring program

STATE AND LOCAL

Air quality standards of state and local governments must be at least as stringent as the federal NAAQS. California, for example, has the most stringent air quality standards in the world. There is even a trend toward strengthening the federal standard to match California's tough standards.

SAMPLING AND MONITORING CONSIDERATIONS

SPATIAL ASPECTS

In this section we focus mainly on the U.S. EPA program to monitor air pollution in the United States. The overall objectives of the EPA air quality monitoring program and air quality monitoring stations are listed in Boxes 10.1 and 10.2, respectively. All but four states have SLAMS (state and local air monitoring stations) and NAMS

BOX 10.1 *Air Quality Monitoring Program Objectives of the Environmental Protection Agency*

- To judge compliance with and/or progress made towards meeting ambient air quality standards
- To activate emergency control procedures that prevent or alleviate air pollution episodes
- To observe pollution trends
- To provide database for research evaluation of effects (urban, land use, transportation planning)
- To develop and evaluate abatement strategies
- To develop and validate air quality models

BOX 10.2 *The Four Air Quality Monitoring Stations Currently Used by the Environmental Protection Agency*

These are four basic types:

1. The state and local air monitoring stations (SLAMS) (about 4000 stations nationwide)
2. The national air monitoring stations (NAMS) (about 1080 nationwide)
3. The special purpose monitoring stations (SPMS)
4. The photochemical assessment monitoring stations (PAMS) (about 90 nationwide)

The SLAMS system is a large network of quasipermanent monitoring stations operated by state and local air pollution control agencies. The purpose of SLAMS is to help these agencies meet the requirements of their state implementation plan (SIP). NAMS is a subset of the SLAMS network with emphasis given to urban and multisource areas. These are the very important sites that monitor pollution in areas of maximum concentrations and high population density. A PAMS network is required by the EPA in each ozone nonattainment area that is rated serious, severe, or extreme. It is required to have two to five sites, depending on the population of the nonattainment area. SPMS provide for special studies by the state and local agencies to support SIP and other activities. SPMS are not permanently sited and can be easily adjusted to meet changing needs and priorities. They must meet all quality control and methodology requirements established for SLAMS monitoring.

(national air monitoring stations). The locations of the current monitoring stations are illustrated in Figure 10.9.

How are networks established to meet SIP requirements? Initially, the objectives of the network need to be identified. Objectives for SLAMS and NAMS networks generally include determining the highest concentrations that are expected to occur in the area covered by the network; determining the representative concentrations in areas of high population density; determining the impact on ambient pollution levels of significant sources or source categories; and determining the general background concentration levels. Site selection is obviously an important consideration when attempting to meet these objectives.

General EPA guidelines for selecting SLAMS and NAMS sites include the following:

- Locating sites in areas where pollutant concentrations are expected to be highest
- Locating sites in areas representative of high population density
- Locating sites near significant sources
- Locating sites upwind of region in order to determine amount of pollution incident on region

- Considering topographic features and meteorological conditions (e.g., valleys, inversion frequency)
- Considering the effects of nearby obstacles, such as buildings and trees, on airflow patterns

The number of stations recommended by the EPA for monitoring CO and particulates, for example, are given in Table 10.3. Once the sites are selected, consideration must be given to the control of the sampling environment. An air-conditioned trailer (Figure 10.10) is often necessary at each site to properly maintain analytic instruments, data recorders, and calibration equipment. This trailer requires access to electrical power, surge protectors, and backup power.

The sampling of air from outside the trailer for analysis by instruments inside the trailer requires the use of a pump to draw air through probes and manifolds. Sampling probes (Figure 10.10) and manifolds are generally made of nonreactive materials such as glass or Teflon. The siting of the probe must avoid interference or contamination from nearby objects, such as trees. For example, PM_{10} high-volume air samplers, which are operated outside the trailer, are generally set 2 to 7 m above ground and 5 to 10 m from roads. Monitoring equipment must be either an EPA reference method or equivalent certified

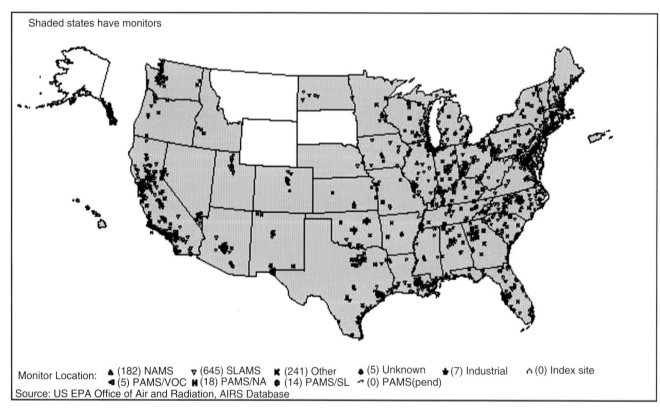

FIGURE 10.9 Locations of national air monitoring stations (NAMS), state and local air monitoring stations (SLAMS), and other stations for monitoring ground-level ozone within the United States for 2002. (From the U.S. Environmental Protection Agency AirData Web site at www.epa.gov.)

TABLE 10.3

Number of Stations Needed for Monitoring Carbon Monoxide and Particulates in an Urban Area

Population	Number of Stations
CO monitoring	
100,000	1
100,000–5 million	1 per 100,000
> 5 million	6 per 100,000
PM$_{2.5}$ monitoring	1 per 250,000

(From 40 CFR part 56.17, EPA July 1975.)

FIGURE 10.10 An air quality monitoring station near a residential area in Tucson, Arizona. Meteorological sensors and particulate matter samplers are visible in photograph. Note the air sampling probe *(insert)* through which air is drawn to analytical instruments inside the trailer. (From A. Matthias.)

by the EPA. Other considerations in selecting the equipment include reliability, vendor support, ease of use, cost, and type of recorder (e.g., analog strip chart, automated data logger; see Chapter 9).

The frequency at which the pollutants are monitored and the duration of the monitoring period must also be considered. Generally, at least 1 year of monitoring is required for studies focused on PSD of the environment. The SPMS (special purpose monitoring stations) may be short term. Longer-term studies are generally required for urban NAMS and SLAMS. Most pollutants are monitored continuously to obtain 1-hour, 8-hour, or 24-hour average or total concentration values (see Table 10.2). The data are reported to the EPA on a monthly

basis. The reports and queries are available to the public on the World Wide Web from the EPA Air Quality Subsystem (http://www.epa.gov/airs/ags.html).

Quality assurance involves several procedures. The analytic instrument(s) must be calibrated at regular intervals specified by the EPA using EPA traceable protocols and standards. Zero and span checks of the analyzer must also be done according to a predetermined schedule. Determination of the precision of the instrument must be done approximately every 2 weeks, with a known concentration of the pollutant. Finally, an independent performance audit must be done once per year. During this audit, the regulatory agency will check the instrument at four known concentrations. The instrument satisfactorily passes the audit if the accuracy is within 15% for all four concentrations. Most NAMS require a minimum of meteorological instruments, such as a radiation-shielded thermometer to measure air temperature, an anemometer and wind vane to record wind speed and direction, and a pyranometer to measure solar radiation (needed for ozone monitoring; see Figure 10.10).

In addition to air quality monitoring networks operated by the EPA in cooperation with state and local governments, there are other specialized national and international networks involved in pollution-related climate change research. Examples include the Euroflux and Ameriflux networks established to monitor surface fluxes and concentrations of CO_2 throughout a wide range of ecosystems in Europe and North America.

TEMPORAL ASPECTS

As stated in the previous section, the monitoring of air pollutant concentrations by SLAMS and NAMS is generally done on a continuous basis. Air pollutant means are computed on an hourly and daily (24-hour) basis for comparison with NAAQS.

AIR SAMPLING

SAMPLE COLLECTION

BAGS

Most SLAMS and NAMS networks in the United States sample air automatically. In other words, samples of air are diverted automatically into instruments for continuous measurement of pollutant concentrations. In some applications, however, air samples are collected in sample bags for later analysis in a laboratory. An example of the use of this type of sampling would be the collection of samples within a building to check for the presence of asbestos. Bags generally used are made of inert material such as Teflon.

CHAMBERS

Chambers are often used to measure fluxes of gaseous pollutants, such as NO_2, emitted from soils to the atmosphere. Generally the mass of gas emitted per unit area per unit time is very small, which necessitates the use of chambers. Chamber enclosures are deployed over the soil surface. During the time that the enclosure is present over the soil, gases emitted from the soil accumulate within the head space of the chamber. The rate of change of concentration of the gas of interest within the chamber is measured and used to calculate the flux. Chambers are also often used to measure emissions of toxic gases, such as hydrocarbons, from hazardous waste disposal sites.

SAMPLING FLUE GAS FROM STACKS

Flue gases are sampled directly from stacks and chimneys with special heat-resistant probes and manifolds. Sampling flow rates are measured accurately with mass flow meters, and samples are generally analyzed *in situ* for sulfur dioxide and other common flue gases with EPA-approved analytical techniques.

LABORATORY METHODS

The following are brief descriptions of analytical instruments and methods used to measure various air pollutants at EPA air quality monitoring stations, such as those that are part of NAMS. These instruments and methods are also used to measure analytes in flue gas from stacks and other industrial sources.

This section begins with a relatively detailed description of chemiluminescent instruments used to measure NO_x in air. This description is given as an example of the complexity of instruments and methods used for air monitoring. Brief descriptions of the instruments used to measure concentrations of criteria pollutants other than NO_x are also given.

NO_2: EPA REFERENCE METHOD

The chemiluminscent analyzer (e.g., Thermo Environmental Instrument Model 42 analyzer) uses the chemiluminescent reaction between NO and ozone to determine levels of total NO_x with NO_2 determined by subtraction (Figure 10.11). Air is drawn into the analyzer and divided into two streams. The NO_2 in one stream is reduced completely to NO by a molybdenum

(Mo) catalyst (converter) heated to about 325° C. The NO is then reacted in a low-pressure reaction chamber with ozone generated artificially by an ozonizer-forming NO_2 molecules. Some of the newly formed NO_2 molecules are in an excited state and emit light (photons) on returning to their ground state. This light is detected and measured with a photomultiplier tube, which produces a voltage signal proportional to the light intensity. The second stream of air bypasses the Mo catalyst, and the NO present is reacted with ozone producing light, as previously described. The actual air concentration of NO_2 is then computed as the difference between the converted and nonconverted NO_2 readings. Artificial ozone is produced by an ozonizer (corona discharge or ultraviolet [UV] lamp) and introduced continuously into the reaction chamber of the analyzer. A major disadvantage of the analyzer is that in addition to NO_2, other nitrogen species, such as peroxyacetyl nitrate (PAN) and nitrous acid (HNO_2), can also be reduced to NO by the Mo catalyst. The effect of this interference, however, on the measured NO_2 concentration is generally small in all but the most polluted urban atmospheres. The chemiluminescent analyzer is specific to NO_x and is most sensitive when operated at low air pressures of 5 to 25 millibars (mb) within the reaction vessel. The minimum detection limit of the Model 42 analyzer is 500 parts per trillion (ppt) NO_2 as specified by the manufacturer.

EQUIVALENT METHOD

A second type of chemiluminescent instrument measures NO_2 directly without the need to convert the NO_2 to NO. The liquid phase luminol NO_2 analyzer (e.g., Model LMA-3 Luminox manufactured by Scintrex/Unisearch Limited) is based on the chemiluminescent reaction between NO_2 and luminol (5-amino-2,3-dihydro-1,4 phthalazinedione) in solution. The light intensity of the chemiluminescence is measured by a photomultiplier tube and converted to a voltage signal. The instrument is small and thus portable. It has a high sensitivity to NO_2 and a fast response time. The manufacturer-specified minimum detection limit is 10 ppt. The instrument, however, has a nonlinear response for NO_2 concentrations below 3 parts per billion (ppb).

In addition to NO_2, other species including ozone and PAN oxidize the luminol solution, which interfere with the measurement of NO_2.

VOLATILE ORGANIC HYDROCARBONS

Gas-liquid column chromatography with flame ionization detector is generally used to measure concentrations of volatile organic carbon compounds. Methods used are specific to the pollutant under study.

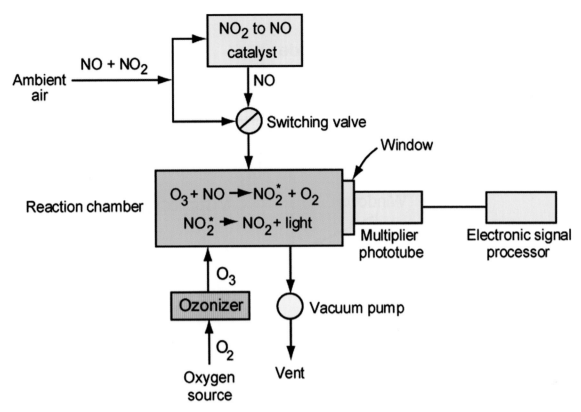

FIGURE 10.11 Schematic of chemiluminescent analyzer used to measure NO and NO$_2$ concentrations in air. Within the reaction chamber NO reacts with O$_3$ (produced by ozonizer) to form NO$_2$, which emits light detected by the multiplier phototube. The intensity of the light emitted is proportional to the concentration of the NO in the air drawn through the reaction chamber. NO$_2$ concentration in air is determined as the difference between measurements made with and without catalytic conversion of NO$_2$ to NO. (From A. Matthias.)

SO$_2$: EPA REFERENCE METHOD (WET CHEMICAL METHOD)

Sulfur dioxide in air is drawn through a solution of potassium tetrachloromercurate—the two chemicals react to form a complex new molecule. The complex is subsequently reacted with pararosoaniline dye and formaldehyde, and the transmission of light through the resulting solution is directly related to the sulfur dioxide concentration.

EQUIVALENT METHOD (UV FLOURESCENCE SPECTROSCOPY)

UV light is pulsed into a reaction chamber containing air and sulfur dioxide (Figure 10.12). Some electrons in the sulfur dioxide are raised to an excited state by the UV light and then quickly decay to the ground state. During the transition back to the ground state, characteristic photons in the infrared (IR) light range are emitted and measured by a photomultiplier tube, which converts the photon energy to a voltage signal (see Chapter 9). The signal is proportional to the concentration of SO$_2$ in air.

PM$_{10}$ AND PM$_{2.5}$: EPA REFERENCE METHOD

Inertial separation with multiple chambers are used to divide particulates into one or more size class with the 10-μm–diameter limit (Figure 10.13). Each particle size is captured on separate filter paper, which is subsequently weighed to estimate the amount of particulates retained per unit volume of air.

EQUIVALENT METHODS

These methods use high-volume samplers, dichotomous samplers, beta gauge, Rupprecht & Patashnick sequential air sampler. These methods are based on principles similar to the reference method.

CO: EPA REFERENCE AND EQUIVALENT METHODS

These methods use nondispersive (single wavelength) IR photometry and are based on Beers-Lambert law and property of CO to readily absorb photons of 46 μm wavelength (Figure 10.14).

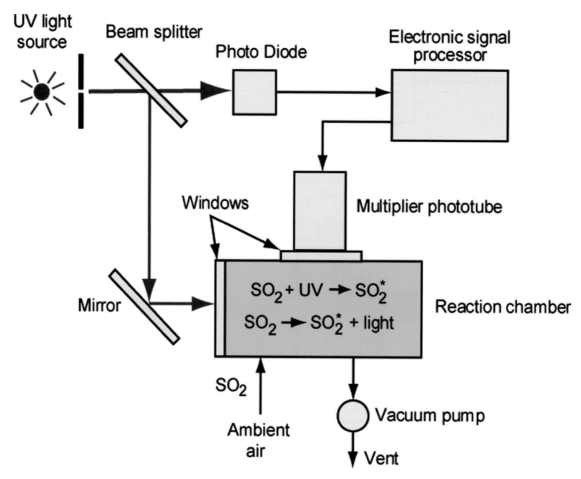

FIGURE 10.12 Ultraviolet (UV) fluorescence spectrometer for SO_2 measurement. The UV light directed into the reaction chamber causes the SO_2 to emit light detected by the multiplier phototube. The intensity of the light emitted is proportional to the concentration of the SO_2 in the air drawn through the reaction chamber. (From A. Matthias.)

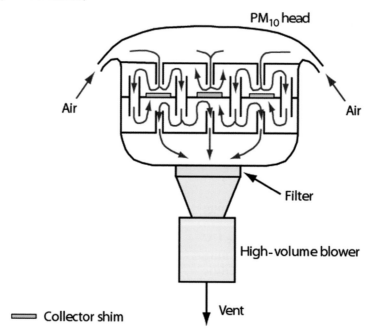

FIGURE 10.13 Schematic of particulate matter (PM_{10}) sampler showing flow patterns of air drawn through the instrument. Large particulate matter that is greater than $10\,\mu m$ impact upon the collector shims. Particulate matter of size less than $10\,\mu m$ collects on the filter. (From A. Matthias.)

O₃: EPA REFERENCE METHOD

This method is based on reaction of O_3 with ethylene, which produces photons (light) that are detected and converted to a voltage signal with photomultiplier tube. The resulting electron volt (eV) is used to estimate the concentration of O_3 in air.

EQUIVALENT METHOD

UV photometry. It is on Beers-Lambert law and property of O_3 to readily absorb UV photons.

LEAD (P)—EPA REFERENCE AND EQUIVALENT METHOD

Pb-containing aerosols are collected on filter paper with a high-volume air sampler. Atomic absorption spectrophotometry is then used to analyze for Pb.

DIRECT AIR MEASUREMENTS

CONTINUOUS OPEN PATH SPECTROPHOTOMETRIC METHODS ULTRAVIOLET, INFRARED

DIFFERENTIAL OPTICAL ABSORPTION SPECTROMETER

The differential optical absorption spectrometer (DOAS) systems are based on the Beers-Lambert relationship that relates pollutant concentrations to absorption of light by the pollutant. Different pollutants have different abilities to absorb light at different wavelengths. DOAS systems make use of this principle to monitor stack emissions (concentration times flow rate) and integrated pollutant concentrations in the open atmosphere. Some

pollutants readily absorb UV, visible, or IR light. The Swedish-made Opsis system is an example of a U.S. EPA–approved DOAS system for remote monitoring of air pollution. The Opsis system is used for cross-stack monitoring of pollutants (Figure 10.15) and for monitoring pollutants in urban areas and airports. DOAS systems have also been developed for *in situ* monitoring of pollutants in exhaust from moving motor vehicles and aircraft engines.

A typical DOAS system has a light source, spectrometer, and computer. The light source, such as a xenon lamp, is used to simultaneously emit in UV, visible, and IR wavelengths. Light from the lamp is directed through the atmosphere toward a grating spectrometer, which splits the light into discrete wavelength intervals. According to the Beer-Lambert law, the light intensity is diminished by absorption by pollutants and is converted by portable computer to an absorption spectra. The spectra at each wavelength interval of interest are divided by a reference absorption spectra produced when no pollutant is present to absorb radiation. The object is to compare the measured absorption spectra with library spectra obtained from known concentrations across an optical path. In theory, a DOAS measurement should produce an accurate integrated pollution concentration; however, dust and other gaseous radiation absorbers may interfere.

OPEN PATH FOURIER TRANSFORM INFRARED SPECTROSPCOPY

Open path Fourier transform infrared (op FTIR) spectroscopy is similar to DOAS and is based on the Beer-Lambert relationship that relates pollutant concentration to absorption of IR radiation by pollutants. The op FTIR spectroscopy makes possible simultaneous measurement of multiple pollutant components across an optical path. This multicomponent feature of op FTIR makes it a useful tool for monitoring various pollutants

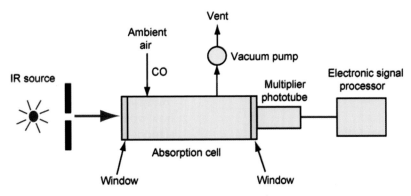

FIGURE 10.14 Infrared photometer based on Beers-Lambert law for measuring CO concentrations in air. CO absorbs selected wavelengths of infrared light directed into the absorption cell. The CO concentration is proportional to the amount of light absorbed. (From A. Matthias.)

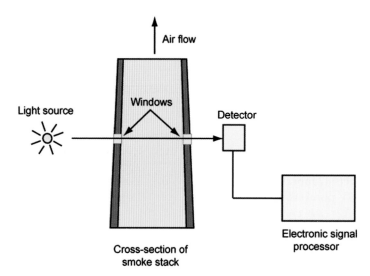

FIGURE 10.15 Cross-stack monitoring of pollutants by a differential optical absorption spectrometer (DOAS) system. Concentrations of flue gas pollutants are determined by measuring the attenuation of light at selected discrete wavelength intervals directed through the windows in the stack. (From A. Matthias.)

emitted from sources including hazardous waste sites and landfills.

The heart of the op FTIR system is a movable mirror spectrometer, which converts the incoming light into an interferrogram composed of interference patterns. The patterns are then converted with Fourier transform computer software into an absorption spectrum. The op FTIR system is operated in one of two modes: monostatic and bistatic (Figure 10.16). In the monostatic mode, the IR source is next to the spectrometer. Radiation from the IR source passes through the optical path and is reflected back to the spectrometer by a mirror. The mirror can be as close as a few meters or as far away as about 500 m. Thus, in the monostatic mode, radiation passes twice through the open path. In the bistatic mode, the IR source is positioned on one end of the open path opposite the spectrometer. There is no reflective mirror, and thus IR passes through the air only once. The advantage of the bistatic mode is that the intensity of IR received at the spectrometer is generally higher than in the monostatic mode. The monostatic mode, however, does not require electrical power for an IR source far removed from the spectrometer.

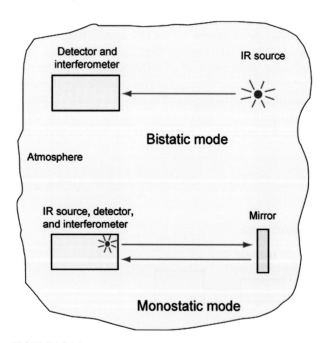

FIGURE 10.16 Open-path Fourier Transform Infrared (FTIR) spectrometer system in configured bistatic *(top)* and monostatic *(bottom)* operating modes. FTIR simultaneously measures concentrations of several pollutants that absorb infrared radiation. (From A. Matthias.)

QUESTIONS

1. What are the six "criteria" air pollutants? Which one occurs most commonly during summertime? Why?

2. What is SLAMS? What is its purpose?

3. If you were responsible for locating air quality monitoring stations in a city with about 500,000 residents, how would you go about deciding where to place the stations? How many stations would you need?

4. Thermo Environmental Instruments Inc. is an example of a manufacturer of several important instruments used in air quality monitoring. The URL for Thermo Environmental Instruments Inc. is http://

www.thermoei.com. Go to that web site and read about the DOAS 2000 and Model 42C instruments manufactured by that company.

a. What atmospheric pollutants are measured by the DOAS 2000 instrument? How does the instrument measure these gases? How does dust and rain interfere with these measurements?

b. Why do you think the DOAS 2000 instrument is not commonly used as part of SLAMS or NAMS?

c. Briefly describe how the Model 42C measures NO_2, NO, and $NO_x (= NO + NO_2)$ concentrations within air samples.

5. To gain experience in obtaining air quality information from the EPA Air Quality Subsystem, we would like for you to determine if any of the national ambient air quality standards (NAAQS) have been exceeded in your county during the past year. Go to the subsystem at http://www.epa.gov/airs/aqs.html. Use the menu to select your state, county, and the year for each criteria pollutant. Use the tabulated results to determine if any NAAQS have been exceeded. If so, which one(s) and how often? Also, from the Air Quality Subsystem, determine what locations, if any, in your state have been designated as nonattainment areas for PM_{10}.

REFERENCES AND ADDITIONAL READING

Clarke, A.G. (1998) *Industrial Air Pollution Monitoring.* Chapman & Hall. London.

Couling, S., ed. (1993) *Measurement of Airborne Pollutants.* Butterworth, Heinemann, Oxford, England.

E. Roberts Alley & Assoc. (1998) *Air Quality Control Handbook.* McGraw-Hill. New York.

Erbes, R.E. (1996) *Air Quality Compliance.* John Wiley & Sons, Inc. New York.

Greyson, J. (1990) *Carbon, Nitrogen, and Sulfur Pollutants and Their Determination in Air* and *Water. Marcel Dekker*, Inc. New York. See EPA Web site at http://www.epa.gov/oar/oaqps/qa/monprog.html.

11

REMOTE SENSING FOR ENVIRONMENTAL MONITORING

A.R. HUETE

From its origins, the Earth has experienced change as a natural process over time scales ranging from years to thousands of years and more. There are strong scientific indications that natural change is being accelerated by human intervention, resulting in rapid land cover conver-

sions; reductions in biodiversity and habitat; degraded water quality; and changes in the composition of the atmosphere, including climate warming. Neither the short-term effects of human activities nor their long-term implications are fully understood. However, if climate and human interventions with the environment are to be understood, more effective and advanced environmental research and monitoring tools need to be developed and implemented. Remote sensing from space-based platforms is one such powerful tool for environmental monitoring of the Earth's surface and atmosphere. Satellite sensor systems, with their synoptic and repetitive coverage of the land surface, are increasingly being relied on to characterize and map the spatial variation of relevant surface properties for environmental applications. Remote sensing may be the only feasible means of providing such spatially distributed data at multiple scales, and on a consistent and timely basis.

In December 1999 and April 2002, NASA launched the Earth Observing System (EOS) "Terra" and "Aqua" platforms, with 10 instruments onboard to observe and measure the state of the Earth system. The objective of the EOS program is to develop the ability to *monitor and predict* environmental changes that occur both naturally and as a result of human activities. For this to be accomplished, satellite sensors provide measurements of the global and seasonal distribution of key Earth surface and atmosphere parameters, such as land use, land cover, surface wetness, snow and ice cover, surface temperature, clouds, aerosols, fire occurrence, volcanic effects, and trace gases. These measurements brought forward a new era for the detection of environmental change enabling one to monitor droughts, deforestation, overgrazing, and disruptions due to volcanic activity, flooding, large-scale fires, and disease outbreaks. In this chapter, we examine how remotely sensed data and satellite-based systems are used to characterize and monitor the Earth's surface and provide information for environmental studies related to ecosystem sustainability, drought mitigation, human health, biogeochemical and carbon cycling, erosion and sediment yield, and water quality.

PHYSICAL PRINCIPLES OF REMOTE SENSING

Remote sensing refers to the acquisition of biophysical and geochemical information about the condition and state of the land surface by sensors that are not in direct physical contact with it. This information is transmitted in the form of electromagnetic radiation (EMR), and the source of this energy may be "passive" (originating from the sun and/or the Earth) or "active" (energy is artificially generated as in radar). The relationship between the "source" signal or irradiance interacting with the surface and the reflected, "received" signal at the sensor (Box 11.1) provides the information that is used in remote sensing to characterize the Earth's surface (Figure 11.1). The remotely sensed signal is composed of energy representing different wavelengths over the electromagnetic spectrum (Figure 11.2). Remote sensing is an extension of our vision in two respects: (1) it extends our vision to higher altitudes above the surface for synoptic viewing of large areas, and (2) it extends our vision beyond the visible range to different parts of the electromagnetic spectrum.

The most important regions of the electromagnetic spectrum for environmental remote sensing are listed in Table 11.1, in order of increasing wavelengths. The ultraviolet portion of the spectrum is useful in atmosphere pollution studies including ozone depletion monitoring. The visible portion of the spectrum is how human beings "see" reflected energy patterns of the Earth and is useful in vegetation, soil, ice, and ocean monitoring, with the use of pigment and mineralogical properties of surface materials. The near-infrared region is most useful in assessing the quantity of biomass at the surface of the earth, whereas the middle-infrared region is most sensitive to leaf moisture contents in vegetation. Combined, the

---

BOX 11.1 *Energy Interactions at the Earth's Surface*

The energy that interacts with the surface is known as *irradiance (E)* and is measured in watts per square meter (w/m^2), whereas the energy that is reflected and scattered away from the surface in all directions is called *exitance (M)*, also measured in w/m^2. The ratio of exitance and irradiance is termed the *albedo* or *hemispherical reflectance* of the surface,

$$\rho_{\text{hemispherical}} = M/E,$$

a dimensionless quantity that varies between 0 and 1. The incoming irradiance may also be absorbed and transmitted by the surface materials, which upon ratioing, yields absorptance (α) and transmittance (τ) such that:

$$\alpha = \text{energy absorbed}/E,$$

$$\tau = \text{energy transmitted}/E, \text{ and}$$

$$\rho + \alpha + \tau = 1.$$

Typically, only a portion of the surface-reflected energy that is within the field of view of the sensor is measured. This is known as *directional reflectance* and has the formula:

$$\rho_{\text{directional}} = \pi L/E$$

where L is the surface-leaving radiance within the field-of-view of a sensor, with units of watts per square meter per steradian ($w/m^2 sr$).

---

ultraviolet, visible, near-infrared and middle-infrared regions are also referred to as *shortwave radiation*. The thermal infrared region, also known as *longwave radiation*, is useful in assessing the temperature of the Earth's surface and has been used in vegetation stress detection and thermal pollution studies. The microwave region is cloud transparent and has been used to map vegetation and measure soil moisture.

There are also important atmospheric influences that alter the solar irradiance reaching the Earth's surface, as well as the reflected energy reaching a spaceborne or airborne sensor. Molecules, particulates, and aerosols scatter incoming and surface reflected solar radiation,

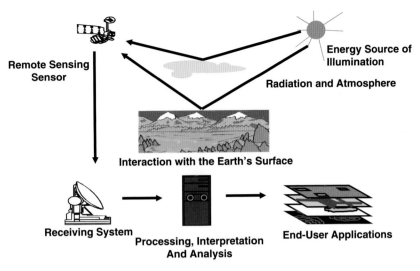

FIGURE 11.1   Basic components of remote sensing detection and interpretation.

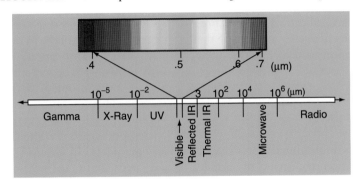

FIGURE 11.2   Diagram of the electromagnetic spectrum. (Courtesy NASA.)

with the shortest wavelengths (blue) being most affected, and this process results in skies appearing blue. There are also gases in the atmosphere that trap and absorb the solar radiation, including water vapor (H_2O), carbon dioxide (CO_2), ozone (O_3), nitrous oxide (N_2O), carbon monoxide (CO), methane (CH_4), and oxygen (O_2) (Figure 11.3). These gases reduce the overall trans-

missive properties of the atmosphere and interfere with the measurement and monitoring of the land surface. Thus, for the potential of satellite data for land-monitoring applications to be fully realized, data are typically "corrected" for atmospheric interference of background.

OPTICAL PROPERTIES OF EARTH SURFACE MATERIALS

Much has been learned about the optical properties of Earth surface materials, such as soil, vegetation, litter, snow, and water, through extensive laboratory analyses and field-based radiometric studies. The spectral composition of surface reflected energy contains information related to (1) the object's biogeochemical composition (mineral and organic) from the way light is reflected and absorbed; (2) the object's morphology (structure, size, roughness) from the way it is illuminated and shadowed; and (3) moisture content. When the

TABLE 11.1

Regions of the Electromagnetic Spectrum Used in Environmental Monitoring

Spectral Region	Wavelengths	Application
Ultraviolet (UV)	0.003 to 0.4 μm	Air pollutants
Visible (VIS)	0.4 to 0.7 μm	Pigments, chlorophyll, iron
Near infrared (NIR)	0.7 to 1.3 μm	Canopy structure, biomass
Middle infrared (MIR)	1.3 to 3.0 μm	Leaf moisture, wood, litter
Thermal infrared (TIR)	3 to 14 μm	Drought, plant stress
Microwave	0.3 to 300 cm	Soil moisture, roughness

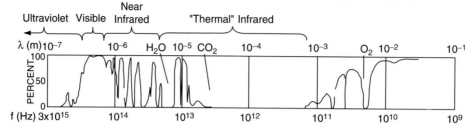

TRANSMISSIVITY OF A STANDARD ATMOSPHERE

FIGURE 11.3 Atmosphere influences on solar radiation reaching the Earth's surface. (Courtesy NOAA.)

reflected energy is plotted over various wavelengths, a "spectral signature" is traced out with a unique characteristic shape for the material being remotely sensed. The aim of remote sensing is to exploit and model these signatures and patterns of surface energy interactions. This in turn allows scientists to extract information about the biophysical and biochemical character of the surface, to map this information, and to monitor their changes over time.

BIOGEOCHEMICAL PROPERTIES

Spectral reflectance signatures result from the presence or absence, as well as the position and shape of specific absorption features, of the surface. Examples of spectral signatures for soils, litter, and vegetation are shown in Figure 11.4. In the case of vegetation, light absorption

by leaf pigments dominates the reflectance spectrum in the visible region (400–700 nm). Chlorophyll pigments a and b selectively absorb blue (400–500 nm) and red (600–700 nm) wavelengths for photosynthesis. There is less absorption over the "green" wavelengths (500–600 nm) and thus the green appearance of healthy vegetation. The yellow to orange-red pigment, carotene, has strong absorption in the blue wavelengths (400–500 nm). The red and blue pigment, xanthophyll, also absorbs strongly in the 400–500–nm range and is responsible for various deciduous leaf colors. Stressed vegetation will give off a different spectral signature corresponding to the effect of the stress on the various leaf pigments. In contrast, leaf pigments and cellulose are transparent to near-infrared wavelengths (700–1300 nm), and leaf absorption is small. Most of the energy is transmitted and reflected, dependent on leaf structural characteristics, which results in a high near-infrared

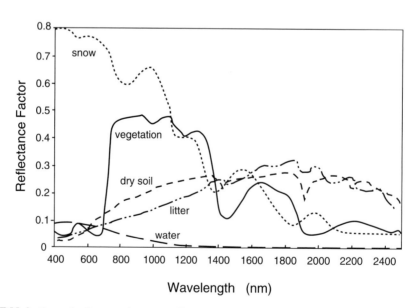

FIGURE 11.4 Spectral reflectance signatures of healthy vegetation, dry soil, gray grass litter, water, and snow.

(NIR) plateau. The sharp rise in reflectance between the red and NIR regions is known as the *red edge* and is used in plant stress detection. The middle-infrared (MIR) region (1300–2500 nm) is dominated by soil and leaf water absorption, particularly at 1400 and 1900 nm with reflectance increasing when leaf liquid water content decreases (see Figure 11.4).

Soils are complex mixtures of a number of mineral and organic constituents and have spectral signatures unlike those of vegetation. The most important constituents controlling the shape of soil spectral signatures are iron and organic matter content, as well as moisture and salt contents. Stoner and Baumgardner (1981) analyzed a large geographic range of soils (485 soils), using the spectral range of 0.50 to 2.5 μm, and documented five unique soil spectral signatures primarily related to their organic matter and iron oxide contents and modulated by their textures. Soil reflectances tend to increase with increasing wavelength from 0.4 to 1 μm because of iron oxide absorptions at the shorter wavelengths. Many studies have also shown the relationships between soil color and remotely sensed optical measurements, which enable the coupling of extensively published soil color information with remote sensing data. Spatial and temporal variations in soil surface color yield important clues to land degradation processes such as salinization, erosion, and drainage status of a soil. The presence of grayish soil colors, for example, is indicative of poor drainage and waterlogging.

Nonphotosynthetic vegetation (NPV), such as grass litter, also has unique spectral signatures that vary with their stage of decomposition. As the chlorophyll pigments degrade, the red reflectance increases, producing a yellow senescent appearance to vegetation. With time, the more persistent carotenes degrade, and the blue reflectances increase causing a "gray" appearance, as in gray grass litter (see Figure 11.4). The signal in the MIR region also rises and produces unique absorption features related to lignins. Eventually, litter will fully decompose and become part of the soil organic signature.

OPTICAL–GEOMETRIC INTERACTIONS

Spectral signatures are modified at the landscape level by optical-geometric interactions, whereby incident radiation is scattered in accordance with the three-dimensional structure of the surface. Vegetation canopy structure describes the three-dimensional arrangement of leaves and includes such canopy attributes as canopy architecture, leaf angle distribution, ground cover fraction, leaf morphology, vegetation spatial heterogeneity, and shadows. Likewise, particle size distribution and surface-height variation (roughness) are the most important factors influencing the scattering properties of soils. They cause a decrease in reflectance with increasing size of "roughness elements," with coarse aggregates containing many interaggregate spaces and shaded surfaces. Smooth, crusted, and structureless soils generally reflect more energy and are brighter. Clay-like soils, despite having a finer particle size distribution, tend to be darker than sandy soils because clay aggregates behave as larger, "rougher" surfaces. Optical-geometric scattering from the surface also varies with the positions of the sensor (viewing angle) and the sun (solar zenith angle) relative to the surface. As the sensor view or solar zenith angle changes, the amount of shadow on the land surface and viewed by the sensor will change causing the reflected signal to vary. This yields useful information in characterizing the structural properties of soil and vegetation canopies.

SOIL MOISTURE

Soil moisture is an important component of the global energy and water balance and is needed as input into various hydrological, meteorological, plant growth, biogeochemical, and atmosphere circulation models. Although there are significant limitations in the use of remote sensing techniques for soil moisture assessment, remote sensing is increasingly being used for its spatial coverage and ability to integrate the spatial variability within an area. In the optical portion of the spectrum, the major effect of adsorbed water on soil reflectance is a decrease in reflected energy, which makes soils appear darker when moistened, particularly in the water absorption bands centered at 1.45 μm and 1.9 μm. Passive and active microwave approaches can accurately measure surface soil moisture contents in the top 5 cm and appear promising for measuring soil moisture. The theoretical basis for microwave remote sensing measurements of soil moisture result from the large difference between the dielectric properties of water and dry soil. Thus, an increase in soil moisture content results in higher dielectric constants. For active microwave remote sensing of soils, the measured radar backscatter signal, σ_s^o, is related directly to soil moisture but is also sensitive to surface roughness. Currently, global near-real-time soil moisture indices are compiled with microwave remote sensing information, although such indices can only provide a rough indication of soil wetness ($\pm 10\%$) and are best suited to identifying areas of extreme moisture conditions.

REMOTE SENSING AT THE LANDSCAPE SCALE

There is a wide array of airborne and satellite sensor systems currently available for environmental studies of the land surface. These are able to measure the reflected and emitted (thermal) energy from the land surface over a wide range of spatial and temporal scales and over various wavebands and sun-view geometries. At the landscape level, it is much more difficult to interpret the remote sensing signal because of land surface heterogeneity and the presence of "mixed" pixels, in which soil, plant, litter, and water may be present within the same pixel. Resolving the extent and spatial heterogeneity of these surface components are important for assessments of landscape, hydrological, and biogeochemical processes. For example, soil erosion processes are dependent on the amount and distribution of "protective" vegetation and litter cover, whereas carbon and nitrogen uptake and evapotranspiration are dependent on the green vegetation fraction.

SENSOR CHARACTERISTICS

The discrimination of surface features is not only a function of the biogeochemical and optical-geometric properties of the surface, but also sensor characteristics such as the number of and location of wavebands, spatial resolution, and instrument calibration. Spaceborne sensors may be characterized in terms of their spatial, spectral, temporal, and radiometric resolutions, with finer resolutions generally enhancing the detection and monitoring of Earth surface features (Box 11.2).

Most of the satellite sensor systems used in environmental remote sensing are "polar orbiting" (Figure 11.5). These are generally at low orbits (600–950 km above the surface) and have pixel sizes in the range of 1 m to ~1 km, with successive orbits close to the poles and constant equatorial crossing times. For environmental applications two types of polar orbiting sensor systems can be differentiated: the fine resolution sensors with small pixel sizes (<100 m) but low repeat frequencies (>16 days) and moderate resolution sensors with coarser pixel sizes (250 m to 1 km) and higher repeat frequencies (~1–2 days). The moderate resolution sensors are useful in seasonal and time series applications at regional to global scales, whereas the fine resolution studies are more useful in local environmental applications.

FINE RESOLUTION SENSORS

The Landsat series of satellites has been the most widely used in remote sensing of the environment. Land-

sat orbits the Earth in a near-polar, sun-synchronous pattern at a nominal altitude of 705 km and 16-day repeat cycle. The Landsat program began in 1972 with the Multispectral Scanner (MSS) sensor consisting of four broad bands in the visible and NIR regions, with a pixel size of ~80 m and a repeat cycle of 18 days (Table 11.2). The fourth and fifth Landsat satellites, launched in 1982 and 1984, respectively, included the Thematic Mapper (TM) sensor, in addition to the MSS sensor. The TM sensor has seven spectral bands, including two in the MIR region and one in the thermal region, and a pixel size of 30 m, which allowed for greatly increased multispectral characterization of the land surface. Landsat TM bands 5 (1.55–1.75 μm) and 7 (2.08–2.35 μm) are within the water-sensitive region and offer potential soil water indicators. The Enhanced Thematic Mapper (ETM+), launched in April 1999, further improved Landsat

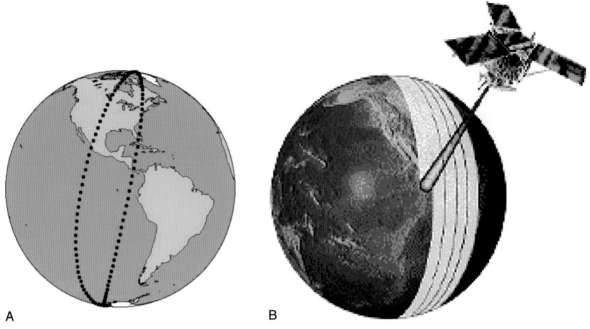

FIGURE 11.5 **(A)** Examples of sun-synchronous polar orbit. **(B)** As the satellite orbits pole to pole, the earth rotates toward the east, enabling global coverage. (Courtesy NASA.)

TABLE 11.2
High-Resolution Sensors in Use for Environmental Studies

Sensor	Pixel Size (m)	Blue (nm)	Green (nm)	Red (nm)	NIR (nm)	MIR (μm)	Thermal (μm)
Landsat MSS	80		500–600	600–700	700–800, 800–1100		
Swath 185 km							
Landsat TM	30	450–520	520–600	630–690	760–900	1.55–1.75	10.4–12.5
Swath 185 km	120 (TIR)					2.08–2.35	
Landsat ETM+	30	450–515	525–605	630–690	775–900	1.55–1.75	10.4–12.5
Swath 185 km	60 (TIR)					2.09–2.35	
SPOT 1, 2 (XS)	20		500–590	610–680	790–890		
SPOT-4 (HR VIR)	10 (pan)					1.58–1.75 (SPOT-4)	
ASTER	15 (VIS/ NIR)		520–600	630–690	760–860	1.6–1.7	8.125–8.475
Swath 60 km	30 (MIR)					2.145–2.185	8.475–8.825
	90 (TIR)					2.185–2.225	8.925–9.275
						2.235–2.285	10.25–10.95
						2.295–2.365	10.95–11.65
						2.36–2.43	
Ikonos	4	450–520	520–600	630–690	760–900		
Swath 13 km	1 (pan)						
QuickBird	2.44	450–520	520–600	630–690	760–900		
	0.61 (pan)						

capabilities with an improved 15-m panchromatic band and a 60-m thermal band (see Table 11.2).

The Landsat satellites have imaged the Earth for nearly three decades, enabling environmental and global change researchers the opportunity to quantitatively assess land use patterns and land cover changes. Land cover is subject to change through natural cycles (droughts, fires, succession, floods, volcanic activity) and human land use activities such as agriculture, forestry, grazing, and urbanization. With a doubling of the human population over the past half-century, landscapes are subject to increasing levels of stress, which often lead to land degradation with accelerated soil erosion, loss of biodiversity and water quality, and increasing human health risks. Land cover conversion is often directly observable with remote sensing, such as when forest is converted to pasture in the case of deforestation. As much as 60% of the global terrestrial surface demonstrates some degree of large-scale conversion, and an important goal of satellite sensor systems is to determine the rate at which anthropogenic land surface alteration is occurring within the global environment. Examples of Landsat image applications in land surface change detection assessments include the following:

- Deforestation in the state of Rondonia, Brazil (Figure 11.6). Landsat images between 1975 and 1992 show the extent of deforestation with systematic cutting of the forest vegetation along major roads, followed by forest cutting along secondary roads eventually creating the "fishbone" pattern shown in the 1992 image.

- The shrinking of the Aral Sea (Figure 11.7). The Aral Sea in southern Uzbekistan is a large freshwater lake which has lost 50% of its area and 75% of its water volume since 1960. Some of this loss may be due to decreased annual rainfall in that part of Asia, but the diminution is also caused by diversion of water from rivers feeding this inland lake for use in agriculture elsewhere. This has had an impact on the livelihood of coastal residents through the loss of suitable conditions for growth of cotton and a great drop-off in fish supply. The three Landsat images, taken in 1973, 1987, and 2000, clearly reveal the reduction in water-covered area and the increase in land area along a part of the Aral Sea.

STEREOSCOPIC IMAGERY

The 1986 launch of the French CNES satellite, Systeme Probatoire pour l'Observation de la Terre (SPOT) with the high-resolution visible (HRV) sensor, greatly improved surface terrain characterization by providing

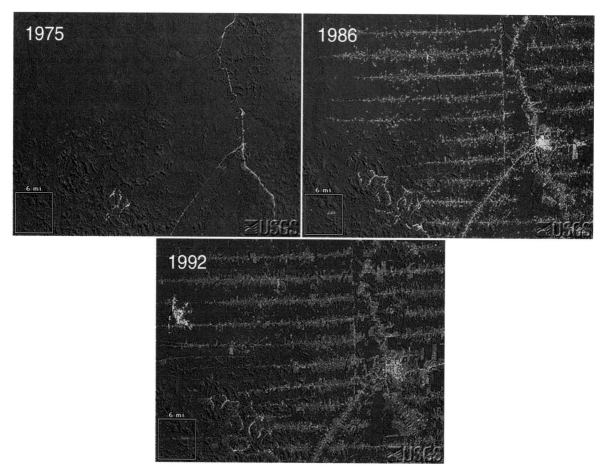

FIGURE 11.6   Deforestation analysis with Landsat imagery.

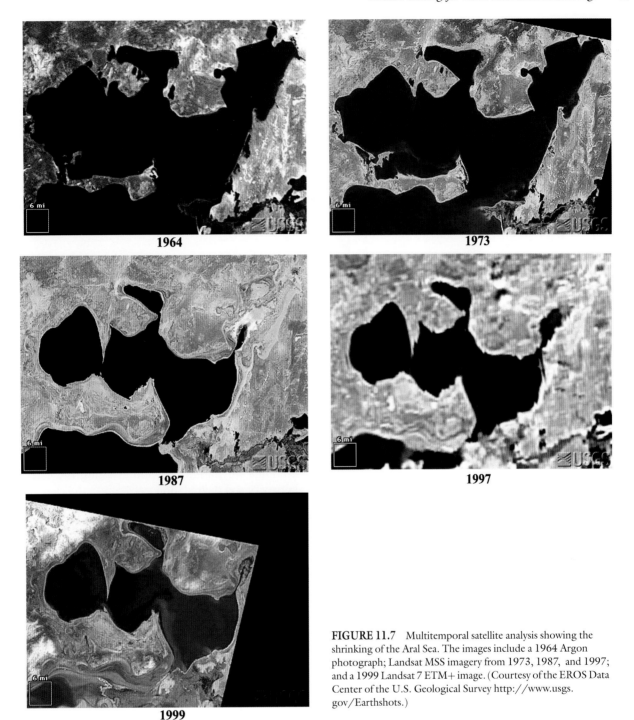

1964

1973

1987

1997

1999

FIGURE 11.7 Multitemporal satellite analysis showing the shrinking of the Aral Sea. The images include a 1964 Argon photograph; Landsat MSS imagery from 1973, 1987, and 1997; and a 1999 Landsat 7 ETM+ image. (Courtesy of the EROS Data Center of the U.S. Geological Survey http://www.usgs. gov/Earthshots.)

both fine spatial resolution (10-m panchromatic band and 20-m multispectral bands) and stereo-capabilities with a multidirectional (±27°) sensor (see Table 11.2). The SPOT satellites have a repeat cycle of 26 days. However, the multidirectional capabilities of the HRV sensor allow for more frequent observation opportunities. The

multiangular SPOT sensor system greatly improves the mapping of geomorphic features, such as drainage patterns, floodplains, terraces, relief, soil erosion potential, landscape stability, surface roughness, and other structural features (Figure 11.8). The Advanced Spaceborne Thermal Emission and Reflection Radiometer

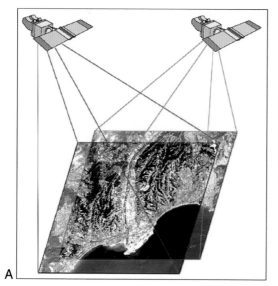

FIGURE 11.8   **(A)** SPOT stereoscopic data acquisition. **(B)** Example in Seoul, South Korea.

(ASTER), onboard the Terra platform, also offers fine spatial resolution and stereoscopic imagery. The instrument has three bands in the visible and NIR regions at 15-m pixel resolution. In addition, there are six bands in the MIR region with 30-m pixel resolution and five bands in the thermal infrared region with 90-m resolution. A backward-viewing telescope provides same-orbit stereo data in a single, NIR channel at 15-m resolution (see Table 11.2; Figure 11.9).

As we proceed to finer spatial resolutions, one can discern vegetation structural features, urban features (e.g., power plants), flooding, and finer-scale drainage patterns. The Space Imaging Ikonos and DigitalGlobe QuickBird instruments were launched in September 1999 and October 2001, respectively, as commercial satellites. Ikonos is multidirectional ($\pm 26°$) and can image a given area of land at 1-m (panchromatic) and 4-m (multispectral) resolutions. QuickBird images the Earth at 61-cm resolution with a panchromatic band and 2.44-m resolution in four spectral bands (blue, green, red, and NIR). At such fine, "hyperspatial" resolutions, Ikonos and QuickBird data offer the opportunity to provide a more complete representation of the landscape surface processes and disaster events.

LAND DEGRADATION APPLICATIONS

Spaceborne and airborne remote sensing observations have been successfully used as a starting point in the monitoring and control of land degradation and desertification. Land degradation is a major socioecological problem worldwide that influences critical environmental issues such as food security, reduced productivity, diminished quality of fresh water resources, loss of biodiversity,

and global climate change. The International Convention to Combat Desertification (CCD) defines desertification as land degradation in arid, semi-arid, and dry sub-humid areas because of human activity and/or climate variations. General information and data regarding the degree and extent of land degradation and the resulting impacts remain poorly understood. Remote sensing can play a major role by providing a quantifiable and replicable technique for monitoring and assessing the extent and severity of soil degradation.

There are many indicators and early warning signals of land degradation and desertification, which lend themselves to remote sensing-based monitoring (Box 11.3). These include loss of vegetative cover, changes in land-cover type, increases in albedo, wind and water erosion (Box 11.3), salinization (Figure 11.10), and increased spatial variability of surface properties (landscape instability).

HYPERSPECTRAL SENSORS

Hyperspectral imaging sensors, with a large number of contiguous bands, carry valuable diagnostic information about the Earth's surface and have improved the feasibility of unambiguously identifying numerous soil and plant absorption features, related to mineralogy, liquid water, chlorophyll, cellulose, and lignin content. The Airborne Visible-Infrared Imaging Spectrometer (AVIRIS) sensor is of particular interest for surface characterization. AVIRIS operates in the 400- to 2450-nm region collecting 224 spectral bands with a nominal 10-nm spectral response function. This enables one to observe high-resolution spectra associated with unique absorption features, which are typically lost in coarser waveband

FIGURE 11.9 ASTER stereoscopic view of Grand Canyon, Arizona. (Courtesy the Terra ASTER Science Team, NASA/GSFC.)

BOX 11.3 *Soil Erosion*

Spaceborne multispectral data have been used extensively in conjunction with aerial photos and ground data for mapping and deriving information on eroded lands. Remote sensing provides temporal and spatial information that can be coupled with soil erosion models, such as measures of the vegetation "protective" cover, soil moisture, land use, digital elevation data, and sediment transport. At the finest spatial resolutions, remote sensing can provide detailed information on linear erosion features such as gullies and sand dune formations (Figure 11.11). Remote sensing data thus have good potential for providing an objective and rapid surveying technique to aid in the assessment of spatial variations in soil loss in ecosystems.

---

data such as from Landsat TM, SPOT, and Advanced Very High Resolution Radiometer (AVHRR). Numerous studies have applied hyperspectral imagery in assessing the foliar chemistry of vegetation canopies such as measuring the levels of potassium in the needles of black spruce, which reflected the potassium distributions in the soil rooting zone (Cwick *et al.*, 1998). In November 2000, the Hyperion Hyperspectral Imager onboard the Earth Observing-1 (EO-1) was launched as part of NASA's New Millennium Program Earth Observing series designed to test new technologies for Earth System Science studies. This is the first spaceborne hyperspectral sensor and provides 220 contiguous bands covering the spectrum from 400 to 2500 nm.

HYPERSPECTRAL MIXTURE ANALYSIS AND MODELS

Approximately 70% of the Earth's terrestrial surface consists of open vegetation canopies overlying various proportions of exposed soil, litter, snow, and understory

FIGURE 11.10   True color composite ASTER image of the Colorado River Delta Region, Mexico, showing salinity accumulation resulting from the controlled flow of the Colorado River by upstream dams (September 8, 2000). (Image courtesy NASA/GSFC/MITI/ERSDAC/JAROS, and U.S./Japan ASTER Science Team.)

backgrounds. Spectral mixture models are commonly used in hyperspectral remote sensing analysis to separate measurements into their soil, nonphotosynthetically active vegetation (NPV), and photosynthetically active vegetation (PV) components. Resolving the extent of these components has utility in a variety of applications, including biogeochemical studies, leaf water content, land degradation, land cover conversions, fuel wood assessment, and soil and vegetation mapping. The optical properties and mixture proportions also vary seasonally, and with land cover conversions (Adams *et al.,* 1995).

Spectral mixture analysis (SMA) generally involves the following steps: (1) determine how many spectral constituents are present in an image or data set; (2) identify the physical nature of each of the constituents or "endmembers" within a pixel; and (3) derive the fractional amounts of each component in each pixel. The first step is generally accomplished with principal components analysis (PCA). The second step is achieved with an endmember analysis whereby various reference spectra, cataloged in spectral libraries, are used to model the image. Typical endmembers used in remote sensing include green vegetation, soil, NPV, and shade. The measured spectral

FIGURE 11.11   Fine resolution color composite Ikonos image (red, NIR, green at 4-m resolution) of sand dune "medanal" vegetation community at the Ñacuñán Reserve, Mendoza, Argentina (June 2001).

response of a pixel is equal to the weighted sum of the unique reflecting surface features:

$$d_{ik} = \sum_{j=1}^{n} r_{ij} c_{jk} + \varepsilon_{-} \qquad \text{(Eq. 11.1)}$$

Where d_{ik} is the measured spectral response of the spectral mixture k in waveband i,

n is the number of independent reflecting features in the mixture,

r_{ij} is the response of feature j in waveband i,

c_{jk} is the relative amount of feature j in spectral mixture k, and

ε_{-} is the residual error.

In matrix notation, Equation 11.1 is expressed as

$$[D] = [R][C] + [\varepsilon] \qquad \text{(Eq. 11.2)}$$

Where $[D]$ is the spectral data matrix,

$[R]$ is the response or reflectance matrix of the spectral constituents (endmembers),

$[C]$ is the component contributions or loadings matrix, and

$[\varepsilon]$ are the residual errors.

Residual images are a quick way to find "outlying" pixels not adequately explained by the mixture model and may indicate the presence of unknown or unusual reflecting components. Residuals in the middle infrared (MIR) region, attributable to cellulose and lignin in the vegetation, have been used to resolve and separate the NPV signal from soil. Residual images also contain the bulk of the noise and error in images, which generally lack spatial structure. The products of mixture modeling include a set of component fraction images and a set of residual images containing the differences between modeled and measured data.

POLLUTION APPLICATIONS

Remote sensing offers the opportunity to assess the sustainability of ecosystems subject to pollution stress and investigate the reversibility of damage. The increased use of hazardous substances and disposal on land has created a growing need for remote sensors that work in hazardous areas. Hazardous waste cleanup at many locations, especially large sites, throughout the United States are increasingly utilizing AVIRIS imagery to produce relatively inexpensive site thematic maps to aid in remediation. In one example, AVIRIS imagery was used to map sources of acid mine drainage and heavy-metal contamination at the California Gulch Superfund Site and at a Superfund site in Leadville, Colorado, in the Central Rockies (Figure 11.12). With the aid of spectral mixture models to recognize the spectral signatures of minerals, contaminants, and sources of acid mine drainage, waste rock, and mine tailings rich in pyrite and other sulfides were found to be dispersed over a 30 km² area including the city of Leadville. Oxidation of these sulfides releases lead, arsenic, cadmium, silver, and zinc into snowmelt and thunderstorm runoff, which drains into the Arkansas River. AVIRIS imagery was also used to identify waste rock and tailings from the Rodalquilar gold-mining areas in southern Spain and Coeur d'Alene River Valley, Idaho, by mapping a variety of ferruginous materials that contained trace elements that were environmentally toxic.

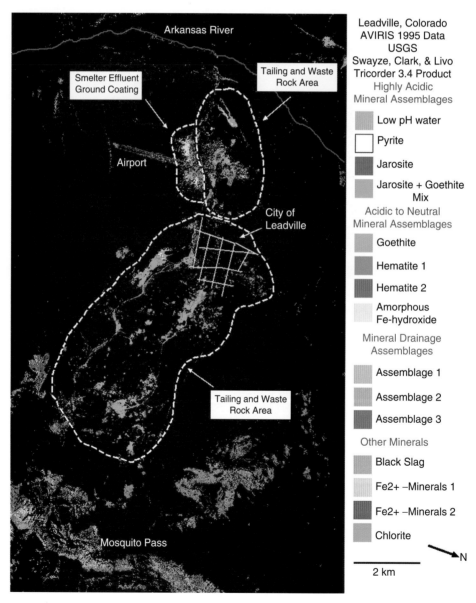

FIGURE 11.12   Mapping of waste rock and mine tailings at the Leadville Superfund site in Colorado. (Courtesy U.S. Geological Survey.)

Although remote sensing could not directly detect the trace metals, it was used to map the minerals that hosted these metals. The AVIRIS instrument was flown over the World Trade Center disaster to detect and assess asbestos levels in the burning aerosols.

Thermal imaging sensors onboard spacecraft (Landsat, ASTER) and aircraft are able to map and monitor thermal discharges from power plants into surface water bodies and ensure temperature ranges acceptable to aquatic habitats (Figure 11.13). They can also detect vegetation stress symptoms associated with soil and groundwater contamination. In the field of pollution sensing and monitoring, lightweight, remote sensor systems that can be hand-carried, deployed on vehicles, or flown on aircraft, are being developed to detect chemicals and buried

containers beneath the surface as well as surface waste materials, and groundwater contamination plumes. These sensors include ground-penetrating radar, magnetometry, and nuclear spectrometry systems such as gamma spectrometry that measure radionuclide signatures emitted from radioactive elements in rocks.

MODERATE RESOLUTION SENSORS

Moderate resolution satellites are useful in monitoring the dynamics of the Earth's surface because of their frequent temporal coverage. Through repetitive "global" measurements from satellites, mapping can include the dynamical aspects of the Earth's surface. Synoptic and repetitious satellite monitoring of the Earth

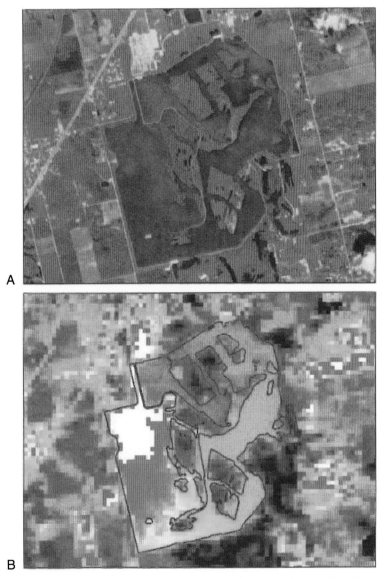

FIGURE 11.13   **(A)** Color composite and **(B)** thermal image of power plant and effluent into the water from ASTER (the red colors are the highest temperatures). (Courtesy the Terra ASTER Science Team.)

provides time series data sets made in a highly consistent manner facilitating surface analysis for change detection. Terrestrial biogeochemical, net primary production, fire, and hydrological models often require high temporal frequency data sets depicting seasonal patterns of greening and drying of vegetation. Seasonal data sets are very difficult to acquire with fine resolution sensors that typically have 16-day repeat cycles and are plagued with cloud contamination problems. Moderate resolution sensors have 1- to 2-day viewing capabilities that permit the "compositing" of multiple-day images into cloud-free image products at weekly to biweekly temporal resolutions. In the following sections I examine the integration of systematic and temporal satellite coverage into various environmental studies involving land cover change, land degradation, water quality and pollution, salinization, drought detection and monitoring, landscape epidemiology, carbon balance, and climate change.

Moderate resolution satellite systems useful in multitemporal land monitoring studies include the NOAA-AVHRR, SeaWiFS, MODIS, SPOT-4 VEGETATION, the Multispectral Medium Resolution Scanner (MMRS) onboard SAC-C, and the ADEOS-II Global Imager (GLI) (Table 11.3). The NOAA satellites contain the Advanced Very High Resolution Radiometer (AVHRR) sensor with a visible and near infrared channel as well as three additional thermal channels in the range from 3.55 to 12.5 μm. The AVHRR orbits at an altitude of 833 km with a wide field of view (±55.4°) providing a repeat cycle of 12 hours, or twice a day. NOAA satellite imagery has been used in Saharan desertification studies, famine early warning detection, and coral reef health and monitoring.

The Sea-viewing Wide Field-of-view Sensor (SeaWiFS) was launched in August 1997 and also provides global monitoring image data sets in eight narrow color bands, between 400 and 900 nm, at 1.1 km resolution, yielding useful information on coastal zone productivity, coastal sediments, ocean color, and vegetation production. The Moderate Resolution Imaging Spectroradiometer (MODIS) is a key instrument onboard the Earth Observing System (EOS) Terra and Aqua platforms, designed to monitor the Earth's atmosphere, ocean, and land surface with a set of visible, NIR, MIR, and thermal channels. MODIS images the Earth in 36 bands with spatial resolutions of 250 m, 500 m, and 1 km with a 2-day repeat cycle and over a wide (±55°) field-of-view, scanning 2300 km. MODIS includes a set of thermal bands applicable to fire detection and surface temperature monitoring.

SOIL EROSION STUDIES

One of the earliest space observations made from the Space Shuttle was the "bleeding island" effect, whereby eroded, red soil sediments from dissected landscapes in deforested portions of Madagascar were being transported by rivers into the Mozambique Channel and Indian Ocean (Wells, 1989). The alarming rate of soil loss, exceeding $100\,\mathrm{Mg\,ha^{-1}\,yr^{-1}}$, from the Loess Plateau of China, drained by the Yellow River, is observed quite well with MODIS satellite imagery (Figure 11.14). Generally, the fertile "topsoil" is eroded and deposited into waterways, resulting in loss of arable land and the silting up of reservoirs (see Chapter 15). Remote sensing and geographic information systems (GISs) (see Chapter 6) offer the opportunity for the development of dynamic, distributed modeling and simulation techniques to study and assess soil erosion rates. GIS further provides a method to incorporate land use and socioeconomic

TABLE 11.3
Medium and Coarse Resolution Sensors Used in Environmental Monitoring

Sensor	Pixel Size, (m)	Blue (nm)	Green (nm)	Red (nm)	NIR (nm)	MIR (μm)	Thermal (μm)
SPOT-4 VEGETATION Swath 2250 km	1150	430–470		610–680	780–890	1.58–1.75	
NOAA-AVHRR Swath 2700 km	1100			570–700	710–980		3.5–3.93 10.3–11.3 11.5–12.5
SeaWiFS Swath 2801 km	1100	402–422 433–453 480–500	500–520 545–565	660–680	745–785 845–885		
MODIS Swath 2330 km	250, 500, 1000	459–479	545–565	620–670	841–876	1.23–1.25 1.628–1.652 2.105–2.155	3.66–3.84 3.929–3.989 4.02–4.08 10.78–11.28 11.77–12.27
SAC-C/MMRS Swath 360 km	175	480–500	540–560	630–690	795–835	1.55–1.70	

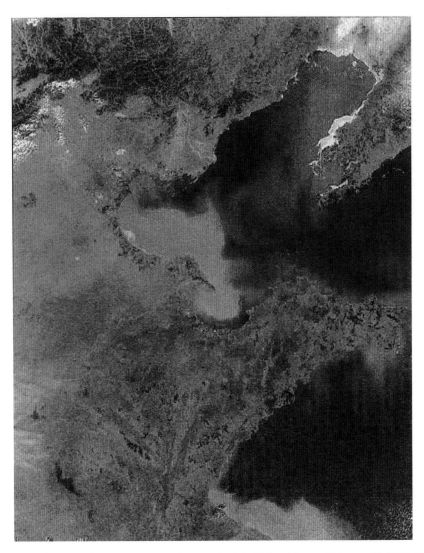

FIGURE 11.14 MODIS image showing sediment transport at the mouth of the Yellow River on February 28, 2000. Soil erosion from the Loess Plateau is proceeding at a very high rate. (Courtesy Jacques Descloitres, MODIS Land Rapid Response Team, NASA/GSFC.)

factors, such as demographic pressures, into erosion assessments.

Woodland depletion due to harvesting of fuel wood around villages in semiarid Africa have also resulted in increased water- and wind-induced erosion, which combined with persistent drought, are responsible for the aggravated African dust transport known as *Harmattan dust storms* (Figure 11.15). Increased sediment loads also alter environmental conditions of coastal zones and have been proposed as one potential contributor to coral bleaching and degradation. Dust can rise to an indefinite height in the atmosphere and exert strong environmental impacts on the Earth's climate, as well as have strong effects on ocean sedimentation, soil formation, groundwater quality, and the transport of airborne pathogens. Most of this dust is produced in arid areas with the Saharan desert as the world's largest supplier of dust,

releasing an estimated 25 to 50 million metric tons of dust over the Atlantic Ocean per year. SeaWiFS imagery has been used to analyze the formation and transport patterns of "Asian dust" from the Gobi desert, which can affect air quality as far away as North America.

VEGETATION SPECTRAL INDICES

Vegetation is a sensitive indicator of many ecosystem properties influencing the energy balance, climate, hydrology, and biogeochemical cycling. Remote sensing observations offer the opportunity to monitor, quantify, and investigate large-scale changes in the response of vegetation to human actions and climate. Spectral vegetation indices are radiometric measures of vegetation amount and condition that use the spectral contrast in vegetation canopy–reflected energy in the red and NIR

FIGURE 11.15   True color composite SeaWiFS image at 1-km resolution showing the Saharan dust storms (March 6, 1998). (Courtesy of SeaWiFS Project, NASA/Goddard Space Flight Center and ORBIMAGE.)

portions of the electromagnetic spectrum (see Figure 11.4). Spectral vegetation indices are designed for monitoring and change detection of the spatial and temporal variations of vegetation composition and photosynthetic activity. They also estimate various vegetation parameters such as biomass, leaf area, photosynthetic activity, and percent vegetation cover.

The normalized difference vegetation index (NDVI) has been extensively used to map spatial and temporal variations in vegetation. The NDVI is a normalized ratio of the NIR and red bands,

$$NDVI = \frac{\rho_{NIR} - \rho_{RED}}{\rho_{NIR} + \rho_{RED}} \qquad \text{(Eq. 11.3)}$$

where ρ_{NIR} and ρ_{RED} are the surface reflectances. Typically, values range from zero to one, with zero depicting hyperarid deserts with no vegetation, and the highest values depicting high biomass areas such as rainforests or dense croplands. Figure 11.16 shows an NDVI image of the lower Colorado River Delta along the United States–Mexico border.

There are also many improved variants to the NDVI equation that reduce noise and enhance sensitivity to vegetation detection. The enhanced vegetation index (EVI) was developed to optimize the vegetation signal with improved sensitivity in high biomass regions and improved vegetation monitoring through a decoupling of the canopy background signal and a reduction in atmosphere influences. The equation takes the form,

$$EVI = G * \frac{\rho_{NIR} - \rho_{red}}{\rho_{NIR} + C_1 * \rho_{red} - C_2 * \rho_{blue} + L} \qquad \text{(Eq. 11.4)}$$

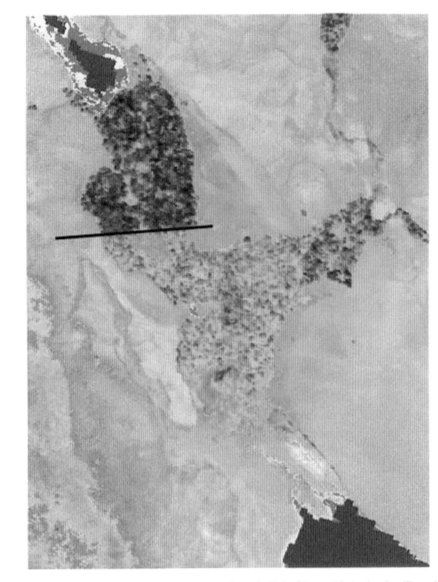

US

Mex

FIGURE 11.16   MODIS NDVI image of Colorado River Delta region along the United States–Mexico border. (From A. Huete.)

where ρ_λ are atmospherically corrected or partially atmosphere corrected surface reflectances, L is the canopy background adjustment, and C_1 and C_2 are the coefficients of the aerosol resistance term, which uses the blue band to correct for aerosol influences in the red band.

The coefficients adopted in the EVI algorithm are
$L = 1$,
$C_1 = 6$, $C_2 = 7.5$, and
G (gain factor) $= 2.5$

Vegetation index (VI) products are currently used in operational monitoring of vegetation by various agencies to address food security, crop production, fire probability, and climate change (Justice *et al.*, 1998). The NDVI and the EVI are standard satellite products of the MODIS sensor. An example of a global EVI image is shown in Figure 11.17 depicting photosynthetic vegetation activity in early March 2003.

URBANIZATION

Urbanization patterns and expansion can be seen from merged nighttime "city lights" imagery from the Defense Meteorological Satellite Program Operational Line-scan System (DMSP OLS) (Imhoff *et al.*, 1997). This type of imagery is used to estimate human population changes and to quantify the impact of urban sprawl on the natural environment (Figure 11.18).

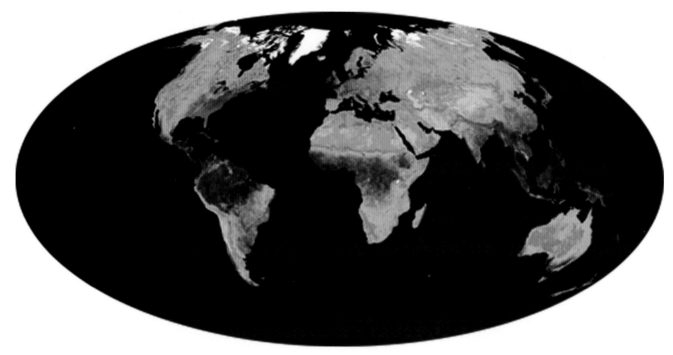

FIGURE 11.17   MODIS EVI image of global seasonality patterns. (From A. Huete.)

DROUGHT

Combining land surface temperatures (T_s) with VIs is of great interest in drought monitoring and plant stress detection. When soil moisture is abundant, plants are able to transpire water at maximal rates for the given meteorological conditions as their leaf temperatures are generally cooler than the surrounding air temperature. As the soil dries, it becomes more difficult for plants to extract the necessary water to meet atmosphere evaporative demands, and transpiration is reduced, resulting in higher leaf temperatures than for those plants with ample water supply. A deficit of soil moisture during critical periods of the growth cycle can result in significant reductions in forage yields and net primary production and an increase in fire risk. Improved remote sensing from satellites, as well as the use of thousands of daily *in situ* precipitation measurements, has dramatically improved drought-monitoring capabilities.

Future global soil moisture and drought monitoring systems will consist of some combination of *in situ*, model estimates, and derived estimates from satellite remote sensing. Satellite thermal imagery combined with NDVI has demonstrated direct relationships between sea surface temperature variations in the Atlantic and Pacific Oceans and large-scale atmospheric circulation patterns that bring moisture or drought conditions.

APPLICATIONS OF REMOTE SENSING IN ENVIRONMENTAL HEALTH AND TOXICOLOGY

Wind-blown dust from desert regions, emissions from industrial activity, sea salt spray from the ocean, and biomass burning are important sources of atmospheric aerosol particles. Remote sensing of aerosol particles from space can reveal the sources, transport pathways, and sinks of these particulates. In Figure 11.19, smoke fire plumes from a fire in Mexico are readily observed from space, spewing out large quantities of gases and organic particulates. The heavy white areas are clouds and dense aerosols, and the lighter white areas represent smoke and haze. Satellites provide observation capability for monitoring different fire characteristics such as fire susceptibility, active fires, burned area, smoke, and trace gases. The smoke and aerosol particles from the 1998 fires in Mexico and Central America were carried long distances by wind currents, reaching as far north as Wisconsin, and over the Gulf Stream causing health alerts in Texas. The Terra MOPITT (Measurements of Pollution in the Troposphere) sensor is one of many sensors designed to monitor and map ozone, carbon monoxide, and aerosols in the earth's atmosphere. Figure 11.20 is an example of an image depicting carbon monoxide levels over western North America.

FIGURE 11.18 Nighttime city lights satellite image with DMSP sensor depicting urban growth patterns on a global basis. (Courtesy NOAA-DMSP.)

FIGURE 11.19 Fire and smoke plumes in Mexico imaged by SeaWiFS, June 5, 1998. (Courtesy Sea WiFS Project, NASA/Goddard Space Flight Center and ORBIMAGE.)

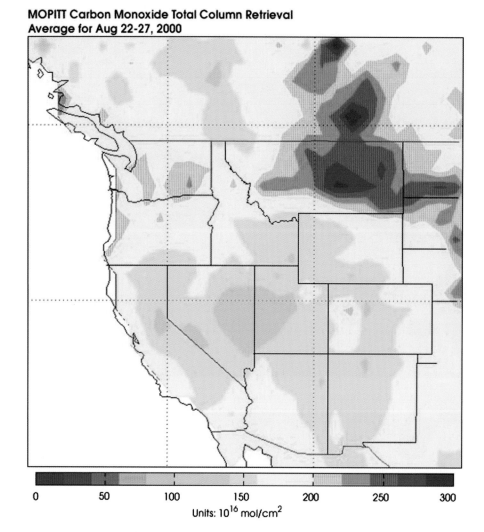

MOPITT Carbon Monoxide Total Column Retrieval
Average for Aug 22-27, 2000

0   50   100   150   200   250   300

Units: 10^{16} mol/cm^2

FIGURE 11.20   Averaged carbon monoxide levels measured by the Terra MOPITT sensor from August 22 to 27, 2000 for western North America. (Courtesy the MOPITT Science Team, NASA/Goddard Space Flight Center.)

An increasing number of medical and epidemiological scientists have used remote sensing data from Landsat, SPOT, and AVHRR for monitoring, surveillance, or risk mapping, particularly of vectorborne diseases. The introduction of new sensor systems have allowed assessment of a greater range of environmental factors such as moisture and temperature that promote disease transmission, vector production, and the risk for human-vector contact. Example applications of remote sensing data used in monitoring spatial and temporal patterns of infectious diseases are listed in Table 11.4. When combined with spatial modeling and GIS, the potential for operational disease surveillance and control appears promising (Beck *et al.*, 2000). Satellite and weather imagery already create an early warning system capable to detect potential outbreaks of water and soilborne diseases and human settlement risks.

Remotely sensed data are primarily used to assess the environmental conditions for various animal- or insect-spread outbreaks such as Rift Valley Fever in sub-Saharan Africa. Satellite images show how El Niño drove flooding rains over the Horn of Africa between December 1996 and December 1997. Similar studies have been made in the American Southwest where a wet winter spurred populations of mice capable of spreading the hantavirus (Figure 11.21). Nearly all vectorborne diseases are linked to the vegetated environment during some aspect of their transmission cycle, and in most cases, this vegetation habitat can be sensed remotely from space. Soil moisture is indicative of suitable habitats for species of snails, mosquito larvae, ticks, and worms. Several types of sensors can detect soil moisture, including SAR (synthetic aperture radars), MIR, and thermal-IR.

TABLE 11.4

Example Applications Using Remote Sensing Data to Map Disease Vectors[a]

Disease	Vector	Location	Sensor	Reference
Rift Valley Fever	*Aedes* and *Culex* spp.	Kenya	AVHRR	Linthicum *et al.*, 1990
Malaria	*Anopheles albimanus*	Mexico	Landsat TM	Pope *et al.*, 1993
		Belize	SPOT	Roberrts *et al.*, 1996
		Mexico	Landsat TM	Beck *et al.*, 1994
Schistosomiasis	*Biomphalaria* spp.	Egypt	AVHRR	Malone *et al.*, 1994
Lyme disease	*Ixodes scapularis*	New York, USA	Landsat TM	Dister *et al.*, 1994
Eastern equine encephalomyelitis	*Culiseta melanura*	Florida, USA	Landsat TM	Freier, 1993
Cholera	*Vibrio cholerae*	Bangladesh	SeaWiFS	Huq and Colwell, 1995

[a]Beck *et al.* (2000).

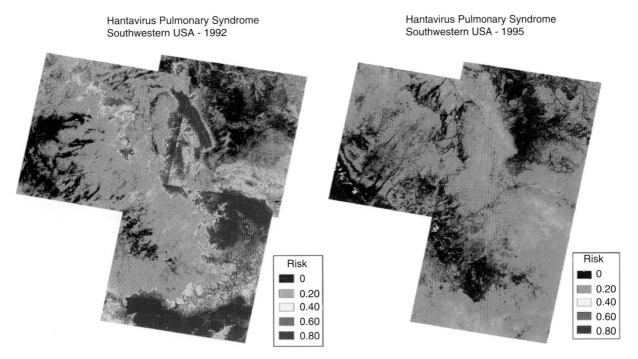

FIGURE 11.21   Hantavirus risk assessment by satellite imagery in southwestern United States for 1992 *(left)* and 1995 *(right)*. (Courtesy National Center for Infectious Diseases, Centers for Disease Control and Prevention [Glass *et al.*, 2000].)

Landsat TM imagery has been used to characterize the spatial patterns of key components of the Lyme disease transmission cycle in New York such as mapping relative tick abundance on residential properties by using the spectral indices of greenness and wetness. One can also relate forest patch size and deer distribution, with the white-tailed deer serving as a major host of the adult tick as well as its primary mode of transportation.

Remote sensing has also been used to monitor coastal environments. Different satellite sensors have been combined to characterize and map environmental variables in the Bay of Bengal that are associated with temporal patterns of cholera cases in Bangladesh. Sediment loads transported to the Bay of Bengal by the Ganges and Brahmaputra Rivers include nutrients that support plankton blooms, an important marine reservoir of *Vibrio cholerae.*

QUESTIONS

1. **(a)** Explain how shortwave, longwave, and microwave radiation are useful in environmental remote sensing.

   (b) Name the four spectral regions included in shortwave radiation and explain how each region is useful in environmental studies.

2. What are the red reflectance and red absorptance of a clay loam soil if $100 \, w/m^2$ of red solar energy reaches the soil surface and $75 \, w/m^2$ are absorbed? (Note that soils are considered opaque and thus have no transmittance.) What is the reflected radiance of this soil in the red spectrum?

3. What are spectral reflectance signatures, and why do they differ for soils and vegetation? What is the difference between dry and wet soil spectral signatures? Which materials absorb the most in the middle-infrared in Figure 11.4, and why is absorption so high?

4. How do optical-geometric energy interactions influence a spectral signature? For vegetation, give an example of a structural and a biogeochemical property.

5. What is the difference between a fine-resolution sensor and a coarse-resolution sensor in terms of spatial and temporal resolutions? Name the sensor that has the finest spatial resolution and the sensor with the coarsest spatial resolution.

6. How can satellite images be used for land surface change detection? Name one satellite sensor that you might use to observe changes from the 1970s to the present.

7. What are spectral mixture models, and why are they needed for some remote sensing-based environmental applications? Give an example of the use of spectral mixtures for environmental studies.

8. What are spectral vegetation indices, and how are they used for environmental remote sensing applications?

9. Give an example of the use of thermal remote sensing for pollution studies. Also, give an example of combining thermal and vegetation indices for environmental studies.

10. How is remote sensing used in early warning systems for detection of outbreaks of waterborne and soil-borne diseases?

REFERENCES AND ADDITIONAL READING

Adams, J.B., Sabol, D.E., Kapos, V., Almeida Filho, R., Roberts, D.A., Smith, M.O., and Gillespie, A.R. (1995) Classification of multispectral images based on fractions of endmembers: Applications to land-cover change in the Brazilian Amazon, *Remote Sensing of Environment,* **52,** 137–154.

Beck, L.R., Lobitz, B.M., and Wood, B.L. (2000) Remote sensing and human health, *Emerging Infectious Diseases,* **6,** 217–226.

Cwick, G.J., Aide, M.T., and Bishop, M.P. (1998) Use of hyperspectral and biochemical data from black spruce needles to map soils at a forest site in Manitoba, *Canadian Journal of Remote Sensing,* **24,** 187–219.

Glass, G.E., Cheek, J.E., Patz, J.A., Shields, T.M., Doyle, T.J., Thoroughman, D.A., Hunt, D.K., Enscore, R.E., Gage, K.L., Irland C., Peters, C.J., Bryan, R. (2000) Using remotely sensed data to identify areas at risk for hantavirus pulmonary syndrome, *Emerging Infectious Diseases,* **6,** 238–247.

Imhoff, M.L., Lawrence, W.T., Elvidge, C.D., Paul, T., Levine, E., Privalsky, M.V., and Brown, V. (1997) Using nighttime DMSP/OLS images of city lights to estimate the impact of urban land use on soil resources in the United States, *Remote Sensing of the Environment,* **59,** 105–117.

Justice, C., Hall, D., Salomonson, V., Privette, J., Riggs, G., Strahler, A., Lucht, W., Myneni, R., Knjazihhin, Y., Running, S., Nemani, R., Vermote, E., Townshend, J., Defries, R., Roy, D., Wan, Z., Huete, A., van Leeuwen, W., Wolfe, R., Giglio, L., Muller, J-P., Lewis, P., & Barnsley, M. (1998) The Moderate Resolution Imaging Spectroradiometer (MODIS): land remote sensing for global change research. *IEEE Transactions on Geoscience and Remote Sensing,* **36,** 1228–1249.

Stoner, E.R., and M.F. Baumgardner (1981) Characteristic variations in reflectance of surface soils. *Soil Science Society of America Journal,* **45,** 1161–1165.

Ustin, S.L., Smith, M.O., Jacquemoud, S. (1999) Geobotany: Vegetation mapping for earth sciences. In Rencz, A.N. (ed.). *Remote Sensing for the Earth Sciences,* John Wiley and Sons, New York.

Wells, G. (1989) Observing earth's environment from space, In Friday, L., and Laskey, R. (eds.), *The Fragile Environment,* Cambridge University Press, New York, Ch. 8: 148–192.

12

ENVIRONMENTAL PHYSICAL PROPERTIES AND PROCESSES

I. YOLCUBAL, M.L. BRUSSEAU, J.F. ARTIOLA, P. WIERENGA, AND L.G. WILSON

The physical properties of soils play a major role in controlling fluid flow and the transport and fate of contaminants in the subsurface environment. This chapter describes the primary soil properties, such as texture, bulk density, porosity, water content, and hydraulic conductivity, and their measurements in the laboratory and field. It also briefly presents the principles of fluid flow in porous media. Basic information on surface water flow and airflow are covered in Chapters 9 and 10, respectively.

MEASURING BASIC SOIL PROPERTIES

SOIL TEXTURE

Soil texture is one of the most fundamental properties of a soil. Qualitatively, the term *soil texture* describes the "feel" of the soil material, whether coarse and gritty or fine and smooth. Quantitatively, however, this term represents the measured distribution of particle sizes, and the relative proportions of the various size ranges of particles in a given soil. Several soil particle size classifications exist. The most widely used one, developed by the U.S. Department of Agriculture (USDA), is presented in Table 12.1 and graphically in Figure 12.1. Based on the USDA system, soil particle sizes are separated into four groups: gravel, sand, silt, and clay. Soil textural class names are determined by the relative mass percentages of sand, silt, and clay-sized particles in the soil.

TABLE 12.1
USDA Classification of Soil Particle Size

Type	Diameter (mm)
gravel	> 2
sand	0.05–2
very coarse sand	1–2
coarse sand	0.5–1
medium sand	0.25–0.5
fine sand	0.10–0.25
very fine sand	0.05–0.10
silt	0.002–0.05
clay	< 0.002

(From Soil Survey Staff, 1975.)

Particle-size analysis is often used to measure the relative percentages of grain sizes comprising the soil. For coarser materials, particle-size distribution is determined by conducting a sieve analysis, in which a soil sample is passed through a series of sequentially smaller screens down to a particle diameter of 0.05 mm. The mass retained on each screen is calculated and divided by the total soil mass to determine the relative contribution of that fraction. Particle-size distributions of finer materials (<0.05 mm) are determined by using the method of sedimentation based on the relative settling velocity of different particle sizes in aqueous suspension (Hillel, 1998). Gee and Bauder (1986) present a more comprehensive description of methods for particle-size analysis.

The texture of a soil sample can be estimated in the field by touch, which is done by rubbing a moist sample between the fingers (Figure 12.2). For instance, sand has a loose, grainy feel and does not stick together. Individual sand grains can be seen with the naked eye. Loams are generally soft and break into small pieces, but will tend to stick together. When moist, sand grains in loam cannot be felt. Clays exhibit a plastic behavior and can be molded when moist. When dessicated, clays typically form cracks.

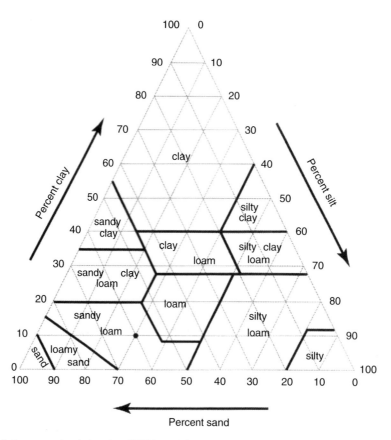

FIGURE 12.1   Soil texture triangle based on USDA particle-size classification. (Courtesy U.S. Soil Conservation Service.)

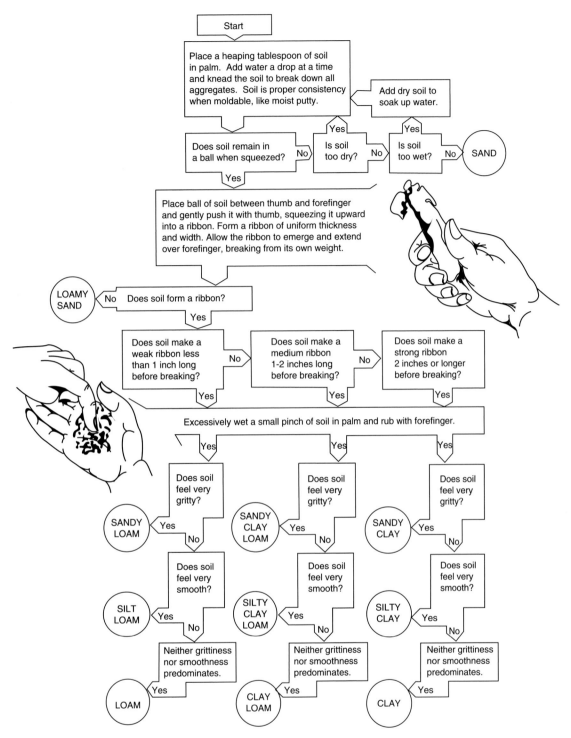

FIGURE 12.2 Flowchart for determining soil textures by feel. (Modified from Thien, S. [1979] J Agron Educ 8:54-55.)

BULK DENSITY

Mathematically, soil bulk density is defined as the ratio of dry mass of solids to bulk volume of the soil sample:

$$\rho_b = \frac{M_s}{V_T} = \frac{M_s}{V_s + V_w + V_a} \qquad \text{(Eq. 12.1)}$$

where

ρ_b = Soil bulk density $[\text{M L}^{-3}]$

M_s = Dry mass of solid $[\text{M}]$

V_s = Volume of solids $[\text{L}^3]$

V_w = Volume of water $[\text{L}]$

V_a = Volume of air $[\text{L}^3]$

V_T = Bulk volume of soil $[\text{L}^3]$

The bulk volume of soil represents the combined volume of solids and pore space. In Système Internationale (SI) units, bulk density is usually expressed in g cm^{-3} or kg m^{-3}. Bulk density is used as a measure of soil structure. It varies with a change in soil structure, particularly due to differences in packing. In addition, in swelling soils, bulk density varies with the water content. Therefore it is not a fixed quantity for such soils.

Determining bulk density of a soil sample involves several steps. First, the volume of sample as taken in the field is measured. Depending on which method is used for measuring bulk density (either core, clog, excavation, or radiation method), the bulk volume is determined differently. Then, the soil sample is dried at 105° C and weighed to determine its dry mass. For more detailed descriptions of measuring bulk densities, refer to Blake and Hartge (1986).

POROSITY

Porosity (n) is defined as the ratio of void volume (pore space) to bulk volume of a soil sample:

$$n = \frac{V_v}{V_T} = \frac{V_v}{V_w + V_s + V_a} \qquad \text{(Eq. 12.2)}$$

where n is the total porosity [-]; V_v is the volume of voids [L^3]; V_T is the bulk volume of sample [L^3]. It is dimensionless and described either in percentages (%) or in fraction where values range from 0 to 1. The general range of porosity that can be expected for some typical materials is listed in Table 12.2.

Porosity of a soil sample is determined largely by the packing arrangement of grains and the grain-size distribution (Figure 12.3). Cubic arrangements of uniform spherical grains provide the ideal porosity with a value of 47.65%. Rhombohedral packing of

similar grains presents the least porosity with a value of 25.95%. Because both packings have uniformly sized grains, porosity is independent of grain size. If grain size varies, porosity is dependent on grain size, as well as distribution.

Total porosity can be separated into two types: primary and secondary. Primary porosity describes the pore spaces between grains that are formed during depositional processes, such as sedimentation and diagenesis. Secondary porosity is formed from postdepositional processes, such as dissolution, reprecipitation, and fracturing. Fractures, root channels, and animal burrows are some examples of secondary porosity. Total porosity is the sum of the primary and secondary porosities.

The porosity of a soil sample or unconsolidated sediment is determined as follows. First, the bulk volume of the soil sample is calculated from the size of the sample container. Next, the soil sample is placed into a beaker containing a known volume of water. After becoming saturated, the volume of water displaced by the soil sample is equal to the volume of solids in the soil sample. The volume of voids is calculated by subtracting the volume of water displaced from the bulk volume of the bulk soil sample.

In a saturated soil, porosity is equal to water content since all pore spaces are filled with water. In such cases, total porosity can also be calculated by weighing the saturated sample, drying it, and then weighing it again. The difference in mass is equal to the mass of water, which, using a water density of 1 g cm^{-3}, can be used to calculate the volume of void spaces. Porosity is then calculated as the ratio of void volume and total sample volume.

Porosity can also be estimated using the following equation:

$$n = 1 - \frac{\rho_b}{\rho_d} \qquad \text{(Eq. 12.3)}$$

TABLE 12.2
Porosity Values of Selected Porous Media

Type of Material	n (%)
Unconsolidated media	
Gravel	20–40
Sand	20–40
Silt	25–50
Clay	30–60
Rock	
Dense crystalline rock	0–10
Fractured crystalline rock	0–20
Karst limestone	5–50
Sandstone	5–30
Shale	0–10

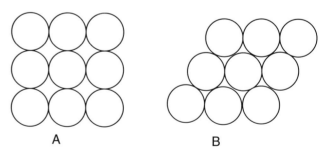

A   B

FIGURE 12.3 Cubic (**A**) and rhombohedral (**B**) packing of uniform-sized spheres. (Adapted from Leap, D.I. [1999] Geological occurrence of groundwater. In *The Handbook of Groundwater Engineering.* Jacques Delleur Editor, CRC Press, Boca Raton, FL.)

where ρ_b is the bulk density of soil $[M\ L^{-3}]$ and ρ_d is the particle density of soil $[M\ L^{-3}]$. A value of 2.65 g cm^{-3} is often used for the latter, based on silica sand as a primary soil component. More detailed descriptions of methods for measuring soil porosity are presented by Danielson and Sutherland (1986).

Void ratio (e), which is used in engineering, is the ratio of volume of voids to volume of solids:

$$e = \frac{V_v}{V_s} \qquad (Eq.\ 12.4)$$

The relationship between porosity and void ratio is described as:

$$e = \frac{n}{1-n} \qquad (Eq.\ 12.5)$$

It is dimensionless. Values of void ratios are in the range of 0 to 3.

TEMPERATURE

Soil temperature is often a significant factor, especially in agriculture and land treatment of organic wastes, because growth of biological systems is closely controlled by soil temperature. In addition, soil temperature influences the physical, chemical, and microbiological processes that take place in soil. These processes may control the transport and fate of contaminants in the subsurface environment. Soil temperature can be easily measured by using a thermometer. Some of the thermometers normally used in soil work include mercury or liquid in glass, bimetallic, bourdon, and electrical-resistance thermometers. The selection of the appropriate thermometer for an application is based on its size, availability, and accessibility to the measurement location, and the degree of precision required (Taylor and Jackson, 1986).

For precise temperature measurements, thermocouples are preferred because of their quick response to sudden changes in temperature and ease of automation (see Chapter 4). Soil temperature is influenced by solar radiation, daily and monthly fluctuations of air temperatures as well as vegetation, amount of precipitation, etc. For accurate measurements of soil temperature, measuring instruments should be protected from solar radiation, wind, and precipitation. Taylor and Jackson (1986) provide detailed descriptions of methods and instruments available for measuring soil temperature.

SOIL-WATER CONTENT

BASIC CONCEPTS

Soil-water content can be expressed in terms of mass (θ_g) or volume (θ_v). Gravimetric (mass) water content is the ratio of water mass to soil mass, usually expressed as a percentage. Typically, the mass of dry soil material is considered as the reference state; thus:

$$\theta_g\% = [(\text{mass wet soil} - \text{mass dry soil})/ \\ \text{mass dry soil}] \times 100 \qquad (Eq.\ 12.6)$$

Volumetric water content expresses the volume (or mass, assuming a water density, ρ_w, of 1 g cm^{-3}) of water per volume of soil, where the soil volume comprises the solid grains and the pore spaces between the grains. When the soil is completely saturated with water, θ_v should generally equal the porosity. The relationship between gravimetric and volumetric water contents is given by:

$$\theta_v = \theta_g \left[\frac{\rho_b}{\rho_w} \right] \qquad (Eq.\ 12.7)$$

A related term that is often used to quantify the amount of water associated with a sample of soil is "saturation" (S_w), which describes the fraction of the pore volume (void space) filled with water:

$$S_w = \frac{\theta_v}{n} \qquad (Eq.\ 12.8)$$

THERMOGRAVIMETRIC METHOD FOR MEASURING SOIL-WATER CONTENT

The simple, basic method for measuring soil-water content is based on thermogravimetric analysis of samples in the laboratory. To measure θ_g, a sample is collected from the field, weighed, dried in an oven (105° C), and then weighed again. The final weight represents the dry soil mass, and the difference between the initial (wet) and final weights represents the mass of water. A similar procedure is used to measure θ_v, with the exception that the volume of the soil sample must also be measured. Thus, for this case, the soil sample must be collected in an undisturbed state to accurately preserve the soil bulk density. This is often accomplished by collecting the sample with a brass cylinder. The methods used to collect soil samples for measurement of water content are described in Chapter 7. Although this method is accurate, the soil samples must be physically removed from the field, creating a significant disturbance. The destructive sampling required for this method precludes future measurements at the same location. In addition, obtaining soil cores from depths below the root zone (~3 m) can be expensive when mechanical coring devices (see Chapter 7) are required.

NONDESTRUCTIVE SOIL-WATER CONTENT MEASUREMENTS

Several methods are available for repeated *in situ* measurements of soil-water content that do not require the physical removal of soil samples, and thus minimally disturb the soil environment. Some of these methods are costly but offer the advantage of allowing repeated and automated soil-water content measurements at the same location, after a one-time installation disturbance. The following sections will discuss the theory and application of two widely used devices for monitoring soil-water content. The last section will briefly present other methods for measuring soil-water content.

NEUTRON THERMALIZATION PROBE

The neutron thermalization probe method takes advantage of elastic collisions between two bodies (particles) of similar mass that result in a significant loss of energy. Neutron particles have the same mass as hydrogen (H) atoms found in the nucleus of a water molecule. To measure soil-water content, high-energy (fast) neutrons, generated by a radioactive americium-beryllium source contained in the probe, are emitted into the soil as the probe is lowered into the access tube (Figure 12.4). The fast neutrons are slowed when they collide with H nuclei of the soil water surrounding the access tube. Some of the low energy neutrons that are not scattered away return to the access tube. The slow neutrons that reach the original source can then react with the boron trifluoride (BF_3) detector, located at the base of the neutron source. The detector converts the low-energy neutrons into an electrical impulse that is recorded as a count in the neutron probe box. By an appropriate calibration relationship, counts are related to soil-water content. Thus a discrete value of soil-water content is obtained for each sampling interval.

Neutrons emitted from the source can interact with soil organic matter, as well as with water, because it contains H atoms. Thus false water-content readings may be generated for soils with high organic matter content. Soil minerals such as clays and metal oxides can also scatter

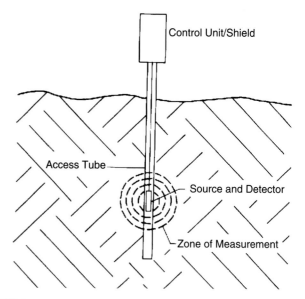

FIGURE 12.4 Schematics of soil-water content measurement with neutron probe. (From American Society of Testing Materials. ASTM D3404, Standard Practices for Measuring Matric Potential in the Vadose Zone Using Tensiometers. Annual Book of ASTM Standards, Section 4, Construction, Vol 04.08, Soil and Rock, Dimension Stone, Geosynthetics. ASTM International, W. Conshohocken, PA.)

neutrons with inelastic collisions. Because of these possible multiple interactions, neutron probes must be calibrated at each location before use. This requires the collection of soil cores for volumetric water content analysis, using soil coring devices at a location or locations near the neutron access tube. Neutron probe calibration is time consuming, as it must be done at multiple water contents. Typically, the θ_v is regressed against probe counts (fixed time interval). However, once the calibration process is completed, this method offers a repeatable and nondestructive method of measuring water content along the access tube depth. The projected soil sphere of influence varies from about 15 cm for saturated soils to about 50 cm for dry soils. Therefore the spatial (depth) resolution of this method is limited to no less than 15-cm intervals. Some of the advantages and disadvantages of the neutron thermalization method are presented in Table 12.3.

TIME DOMAIN REFLECTOMETRY (TDR)

The TDR method is another nondestructive approach to measuring soil-water content. This technique consists of a cable tester instrument, and a metal probe that is inserted directly into the soil (Figure 12.5). The TDR method is based on propagation of electromagnetic waves (generated by the cable tester), which are launched through the coaxial cable into the metal probe. The waves then travel down the steel waveguides of the probe and through the soil. The cable tester measures

TABLE 12.3
Advantages and Disadvantages of Nondestructive Methods Used for Measuring Soil-Water Content

Measurement Type	Advantages	Disadvantages
Tensiometer	• Inexpensive and easy to install • Simple operation, gauge readings can be automated • No calibration required • Point measurement	• Will not work in dry or nearly dry soils • Requires maintenance; water must be added periodically • Depth limitations, mechanical gauges (10 m?), transducers (same?)
Neutron thermalization probe	• Widely accepted • Rapid and repeatable • Nondestructive • Accuracies of $\pm 1\%$ can be obtained with long counting times (5–20 seconds) • Useful to measure water content changes and wetting fronts over time and by depth • May be set to trigger an alarm at specified water content. This is useful to monitor leakage below landfills and storage lagoons • Useful to determine soil profile water storage (plant water reserve) by collecting water content at fixed intervals (usually 12.5–25 cm) • Can be installed horizontally, vertically, or at an angle	• Moderate initial expense (capital equipment) • Calibration necessary for each soil • Operator requires license and training to handle a radioactive source • Radioactive source must be handled carefully • Temperature-dependent count ratio method requires standardized counts • Not suitable for near surface (< 30 cm) measurements
Time domain reflectometry	• Widely accepted • Minimum soil disturbance, no access tubes needed, probe can be inserted directly into the soil profile • Probes can be installed in any direction and even buried • Very rapid, reliable, and repeatable • Can be automated and multiplexed. One cable tester can be used to measure water content from multiple probes. Continuous monitoring • Calibration not necessary for rough estimates of soil water contents; see Topp's equation • Capable of very good accuracy ($\pm 2\%\ \theta_v$) • Excellent spatial resolution, probes can vary in length (5–100 cm)	• Calibration necessary if very accurate water content values are needed • High initial expense (cable tester is expensive but probes are not) • Temperature-dependent cable resistance affects precision • Cable length and depth is limited to 50 m (cable tester to probe distance) • Direct insertion into coarse-grained soils and sediments may damage probe or change electrode symmetry

the time for the waves to travel to the end of the probe and back. Longer travel times correspond to higher water content and vice versa. The travel time (propagation velocity) of the electromagnetic wave along the probe is a function of the soil dielectric constant (ε), which, in turn, is a function of the volume of water in the soil. Pure water has a ε of about 78, whereas other soil components have much lower ε values (soil minerals $\simeq$ 3–5, air $= 1$). Therefore soil ε is very sensitive to soil-water content, and relatively insensitive to soil mineral composition.

The governing equation that relates the soil bulk dielectric constant to the electromagnetic wave propagation velocity (v) (where $v = \frac{2L}{t}$) is defined as follows:

$$\varepsilon_b = \left(\frac{c}{v}\right)^2 = \left(\frac{ct}{2L}\right)^2 \qquad \text{(Eq. 12.9)}$$

Where

$\varepsilon_b =$ soil bulk dielectric constant
$c =$ speed of light
$t =$ is the travel time for the wave to traverse (down and back) the probe
$L =$ the length of the probe

The first application of this nondestructive technique for monitoring water content in soils was reported in 1980 by Topp *et al*. He developed a nearly universal regression equation relating θ_v to ε_b that is often used to estimate soil-water content in soils:

$$\theta_v = -5.3 \times 10^{-2} + 2.92 \times 10^{-2}\varepsilon_b \\ - 5.5 \times 10^{-4}\varepsilon_{b^2} + 4.3 \times 10^{-6}\varepsilon_{b^3} \qquad \text{(Eq. 12.10)}$$

This equation provides a good estimation of soil-water content ranging from dry to near saturation ($\theta_v \simeq 2\%$ to 50%). Some of the advantages and disadvantages of TDR are presented in Table 12.3.

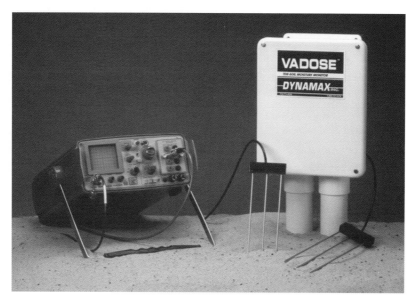

FIGURE 12.5 Time domain reflectometry unit. (Courtesy of Dynamax Inc.)

OTHER NONDESTRUCTIVE METHODS

Other methods, less widely used than those previously mentioned, are available for measuring soil-water content in the vadose zone. Some of these methods will be briefly described in this section.

Tensiometers are also used as an alternative method to determine soil-water content. Tensiometers actually measure soil-water tension in an unsaturated soil matrix, which can be related to soil-water content with suitable water retention curves. Water-retention curves describe the variation of soil-water content with soil-water tension and are highly nonlinear S-shaped curves. Using an appropriate calibration relationship representing the water retention curve obtained for the specific soil in question, one can estimate soil-water content at a given soil-water tension. A detailed description of a tensiometer is given in the "Measuring Hydraulic Gradients in the Unsaturated Zone" section.

Electrical resistance sensors measure the electrical resistance between two electrodes encased in a porous gypsum, nylon, or fiberglass block. The blocks are buried into the soil and allowed to equilibrate with the surrounding water. The electrical resistance between the two electrodes is a function of the amount of water that soaks into the blocks, which is similar to the soil-water content. As the soil dries, the electrical resistance increases and vice versa. These sensors are useful for intermediate soil-water contents. They are low cost and easy to use. However, gypsum blocks are not very accurate, and can clog with salts, increasing the time required for equilibration.

Heat dissipation and thermocouple psychrometer sensors are more sophisticated methods suited to measure soil-

water content in extreme environments. These methods utilize the heat-dissipating properties (rates) of metal probes or filaments inserted in the soil environment, as heat dissipation increases with water content. Both techniques require complex electronic data acquisition and processing. Thermocouple psychrometer sensors are useful in measuring soil-water content in dry to very dry environments, where other techniques fail. Heat-dissipation sensors are new, and offer a promising way to measure soil-water content over a wide range of operating conditions. Both methods require calibration and expensive equipment that can be automated for continuous monitoring. For more detailed information about the methods described in these sections, refer to Wilson *et al.* (1995).

Surface geophysical techniques, including *ground penetrating radar (GPR)* and *electromagnetic (EM) induction*, are also used for nondestructive measurement of subsurface moisture. GPR and EM measurements allow areal mapping of soil moisture with a high lateral resolution along a profile (Stephens, 1996). GPR is based on the transmission of high-frequency waves (10–1000 MHz) into the ground to monitor subsurface conditions. EM uses a coil of wire placed above ground to transmit a current at audio frequency into soil. For both techniques, reflection of the signal occurs where electrical properties of the soil change. Electrical properties of the soil (dielectric and electrical conductivity) are functions of the soil and rock matrix, soil-water content, and the type of pore fluid. The effective depth of measurement is 30 m in materials with low electrical conductivities, such as dry coarse sand. Penetration depth of GPR is limited significantly by the presence of clays. The effective measurement depth of EM is up to 60 m depending on coil spacing (Stephens, 1995). The effectiveness of electromagnetic measurement decreases at low soil electrical conductivities. GPR and EM can also be used for detecting and mapping contaminant plumes, buried wastes, and drums.

SUBSURFACE WATER FLOW

Characteristics of water flow in subsurface environments depend on the physical and hydraulic properties of the subsurface. Therefore, before discussing equations that govern water flow, we should be familiar with the major zones that exist in a subsurface profile. A schematic of a typical subsurface profile is presented in Figure 12.6.

The vadose zone, also known as the zone of aeration or unsaturated zone, represents a region extending from ground surface to a water table. The water table is defined as a water surface that is at atmospheric pressure. In the vadose zone, pores are usually partially filled with water, with the remainder of pore spaces taken up by air and, possibly, nonaqueous liquids. Water stored in this zone is held in the soils by surface and capillary forces acting against gravitational forces. Molecular forces hold water like a thin film around soil grains. Capillary forces hold water in the small pores between soil grains. This water is unable to percolate downward (it is held against the force of gravity). Water pressure in the vadose zone is less than atmospheric pressure (P<0). Water in excess of that held by surface and capillary forces is free to move under the force of gravity. In the vadose zone, since pores are partially filled with water, the soil-water content is less than porosity. The thickness of the vadose zone varies from essentially zero to a meter in tropical regions to a few hundred meters in arid regions, depending on the climate, soil texture, and vegetation.

The capillary fringe is the region above the water table where water is pulled from the water table by capillary forces. This zone is also called the tension-saturated zone. The thickness of this zone is a function of grain size distribution, and varies from a few centimeters in coarse-grained soils to a few meters in fine-grained soils. The water content in this zone ranges from saturated to partially saturated, but fluid pressure acting on the water is less than atmospheric pressure (P<0).

The region beneath the water table is called the saturated or phreatic zone. In this zone all pores are saturated with water, and the water is held under positive pressure. In this zone soil-water content is constant and does not vary with soil depth (unless the soil is heterogeneous). Since all pores are filled with water, soil-water content is equal to porosity, except when organic, immiscible liquids are also present in the pore spaces.

PRINCIPLES OF SUBSURFACE WATER FLOW

A fluid, at any point in the subsurface, possesses energy in mechanical, thermal, and chemical forms. For

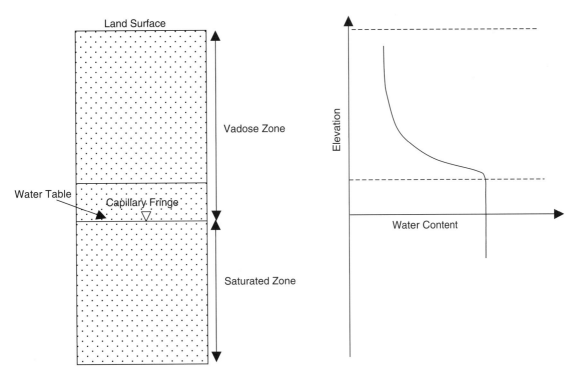

FIGURE 12.6 Schematic of subsurface and soil moisture profile.

groundwater flow, assuming single-phase flow and relatively constant temperatures, contributions of chemical and thermal energies to the total energy of a fluid can be considered minimal. Therefore we can simply state that water flow through porous media is a mechanical process. As Hubbert (1940) pointed out, fluid flow through porous media always occurs from regions where energy per unit mass of fluid (fluid potential) is higher to regions where it is lower. From fluid mechanics, the energy level of water at a point in the subsurface is characterized by the Bernoulli equation:

$$\frac{1}{2}v^2 + gz + \frac{P}{\rho} = \text{constant} \qquad (\text{Eq. } 12.11)$$

where P is the fluid pressure $[\text{M L}^{-1} \text{ T}^{-2}]$, ρ is the density of fluid $[\text{M L}^{-3}]$, z is the elevation of fluid level from a reference point [L], and v is the fluid velocity $[\text{L T}^{-1}]$. The first term in the Bernoulli equation represents kinetic energy of the fluid, the second term represents the potential energy, and the third term represents energy of fluid pressure. The Bernoulli equation simply states that total energy per unit mass of flowing fluid, at any point in the subsurface, is the sum of the kinetic, potential, and fluid-pressure energies and is equal to a constant value.

For water flow in porous media, kinetic energy is generally negligible since pore-water velocities are very small. Therefore the Bernoulli equation reduces to:

$$gz + \frac{P}{\rho} = \text{constant} \qquad (\text{Eq. } 12.12)$$

Dividing each side of the equation by g results in:

$$z + \frac{P}{\rho g} = \text{constant} \qquad (\text{Eq. } 12.13)$$

Under hydrostatic conditions, the pressure at a point in the fluid is equal to:

$$P = \rho g \psi \qquad (\text{Eq. } 12.14)$$

Substituting this into the previous equation produces:

$$z + \psi = h \qquad (\text{Eq. } 12.15)$$

Thus the total hydraulic head (h) is equal to the sum of the elevation head and pressure head, where z is elevation head and Ψ is the pressure head. Each head has a dimension of length [L] and is generally expressed in meters or feet. This basic hydraulic head relationship is essential to

an understanding of groundwater flow. Water flow in a porous medium always occurs from regions in which hydraulic head is higher to regions in which it is lower.

In the laboratory, hydraulic head is measured using manometers in which the elevation of water level can be determined. In the field, piezometers are used to measure hydraulic head at a distinct point in an aquifer, which will be discussed in a following section. Figure 12.7 shows the head relationship for the manometer in a saturated sand column. The value of z represents the distance between the measurement point in the manometer and a reference datum. The value of Ψ represents the distance between the point of measurement and the water level in the manometer. The value of h represents the elevation of water from a reference datum. Sea level is often taken as the reference point where z = 0, although some workers use the elevation of land surface as the reference datum and others use the base of the lowest piezometer as z = 0.

DARCY'S LAW

In 1856 a French hydraulic engineer, Henry Darcy, established a law that bears his name, based on studies of water flow through columns of sand, similar to the schematic shown in Figure 12.8. In Darcy's experiment, the column is packed with homogeneous sand and plugged on both ends with stoppers. Water is introduced into the column under pressure through an inlet in the stopper, and allowed to flow through until all the pores are fully saturated with water, and inflow and outflow rates are equal. Water pressures along the flow path are measured by the manometers installed at the ends of the column. In his series of experiments, Darcy studied the relation between flow rate and the head loss between the inlet and outlet of the column. He found that:

1. The flow rate is proportional to the head loss between the inlet and outlet of the column:
   $$Q \propto (h_a - h_b)$$

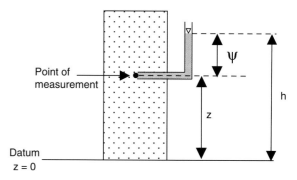

FIGURE 12.7 Hydraulic head *(h)*, elevation head *(z)*, and pressure head *(Ψ)* measurements in a laboratory column.

2. The flow rate is inversely proportional to the length of flow path: $Q \propto \frac{1}{dl}$
3. The flow rate is proportional to the cross-sectional area of the column: $Q \propto A$

Mathematically, these experimental results can be written as:

$$Q = KA\left(\frac{h_b - h_a}{dl}\right) = KA\left(\frac{dh}{dl}\right) \qquad \text{(Eq. 12.16)}$$

where
Q = flow rate or discharge $\left[L^3\,T^{-1}\right]$
A = cross-sectional area of cylindrical column $\left[L^2\right]$
$h_{a,\,b}$ = hydraulic head [L]
dh = head loss between two measurement points [L]
dl = the distance between the measurement locations [L]
dh/dl = hydraulic gradient [-]
K = proportionality constant or hydraulic conductivity $\left[L\,T^{-1}\right]$

This empirical equation is commonly known as Darcy's law. Rearranging the terms, we can rewrite Darcy's law as:

$$\frac{Q}{A} = q = K\frac{dh}{dl} \qquad \text{(Eq. 12.17)}$$

where q is called specific discharge, or Darcy velocity. It has units of velocity $\left[L\,T^{-1}\right]$. This is an apparent velocity because Darcy velocity represents the total discharge over a cross-sectional area of the porous medium. Cross-sectional area is a total area that includes both void and solid spaces. However, water flow occurs only in the connected pore spaces of the cross-sectional area. Therefore to determine the actual mean water velocity, specific discharge is divided by the porosity of the porous medium:

$$v = \frac{q}{n} \qquad \text{(Eq. 12.18)}$$

where v is the pore-water velocity or average linear velocity. Pore-water velocity is always greater than Darcy velocity.

In summary, Darcy's law is an empirical law, obtained experimentally without a rigorous mathematical proof. Darcy's law simply states that specific discharge in a porous medium is linearly proportional to hydraulic gradient and is in the direction of decreasing head. The validity of Darcy's law depends on whether flow in a porous media is laminar or turbulent. This is determined by the magnitude of the Reynolds number, which defines the ratio of inertial forces to viscous forces in a porous medium. Reynolds number is a dimensionless number and expressed as:

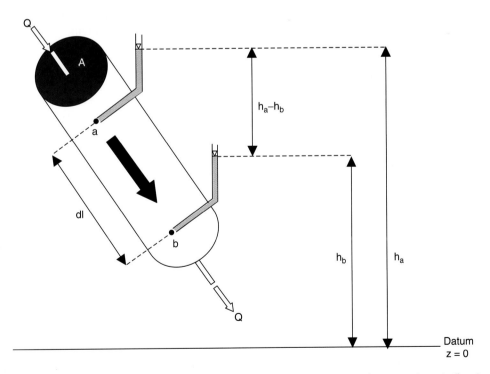

FIGURE 12.8 Schematic picture of Darcy's experimental apparatus, except that original apparatus is vertically oriented.

$$R_e = \frac{\rho q d}{\mu} \qquad \text{(Eq. 12.19)}$$

where

R_e is the Reynolds number [-]
ρ is the fluid density $[M\ L^{-3}]$
q is the discharge velocity $[L\ T^{-1}]$
d is the diameter of pores [L]
μ is the dynamic viscosity of fluid $[M\ T^{-1}\ L^{-1}]$

Since it is difficult to measure the diameter of pores for porous media, the average grain diameter is often used. Darcy's law is assumed valid if laminar flow conditions are present in the porous medium. Flow in porous media is laminar if Reynolds number is between 1 to 10. If the number is more than 10, flow shows a turbulent character, and Darcy's law becomes invalid. In natural groundwater conditions, flow through porous media is often considered as laminar because groundwater velocities are usually very small, except in the fractures and large conduits. For larger hydraulic gradients and velocities, observed in areas such as near pumping wells, flow does depart from the linear relationship of Darcy's law. Darcy's law is also invalid at extremely small hydraulic gradients, observed in clays. In clays, there is a threshold gradient that must be exceeded before flow occurs.

HYDRAULIC CONDUCTIVITY

The proportionality constant in Darcy's law, which is called hydraulic conductivity or coefficient of permeability, is a measure of the fluid transmitting capacity of a porous medium and is expressed as:

$$K = \frac{k \rho g}{\mu} \qquad \text{(Eq. 12.20)}$$

where k is the intrinsic or specific permeability $[L^2]$, ρ is the fluid density $[M\ L^{-3}]$, g is the acceleration due to gravity $[L\ T^{-2}]$, and μ is the dynamic viscosity of the fluid $[M\ T^{-1}\ L^{-1}]$. Hydraulic conductivity has a dimension of velocity $[L\ T^{-1}]$, and is usually expressed in m s^{-1}, cm s^{-1}, or m day^{-1} in SI units or ft s^{-1}, ft day^{-1}, or gal day^{-1} ft^{-2}.

As indicated by Equation 12.20, hydraulic conductivity depends on properties of both the fluid and porous medium. Fluid properties include fluid density and dynamic viscosity. Both fluid properties vary with temperature. Therefore K may also depend on the fluid temperature indirectly. However, since temperature changes in groundwater are minimal, the influence of fluid

temperature on K may be considered negligible. Intrinsic permeability is a property that solely depends on the physical properties of the porous media, and is represented by the expression called the Hazen approximation:

$$k = C(d_{10})^2 \qquad \text{(Eq. 12.21)}$$

where C is shape factor [-], and d_{10} is the effective grain diameter [L].

C is a constant value that represents the packing geometry, grain morphology, and grain-size distribution. The value of C ranges between 45 for clays and 140 for pure sand. Therefore quite often, a value of C = 100 is used as an average. d_{10} is the diameter for which 10% (by weight) of the sample has grain diameters smaller than that diameter, as determined by sieve analysis. The Hazen approximation is applicable to sand with an effective mean diameter between 0.1 and 3.0 mm. Intrinsic permeability (k) has dimensions of square feet (ft^2), square meters (m^2), or square centimeters (cm^2). In the petroleum industry, darcy is also used as a unit of intrinsic permeability:

$$1 \text{ darcy} = 9.87 \times 10^{-9} \text{ cm}^2 \qquad \text{(Eq. 12.22)}$$

As previously noted, porous-media properties that control K include pore size, grain size distribution, and geometry and packing of grains. Among those properties, the influence of grain size on K is dramatic since K is linearly proportional to the square of grain diameter. The larger the grain diameter, the greater is the hydraulic conductivity. For example, hydraulic conductivity of sands ranges from 10^{-1} to 10^{-4} cm s^{-1}, whereas the hydraulic conductivity of clays ranges from 10^{-7} to 10^{-9} cm s^{-1}. The values of saturated hydraulic conductivity in soils vary within a wide range of several orders of magnitude, depending on the soil material. The range of values of hydraulic conductivity and intrinsic permeability for different media is given in Figure 12.9.

WATER FLOW IN UNSATURATED CONDITIONS

Up to this point, the concepts of water flow and hydraulic conductivity have been developed with regard to a saturated porous medium. Now, let's discuss the concept for unsaturated flow conditions in porous media. In the vadose zone, hydraulic conductivity is not only a function of fluid and media properties, but also the soil-water content, and is described by the following equation:

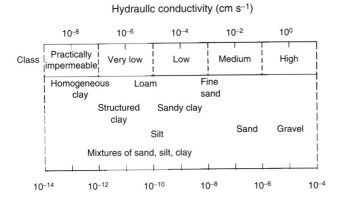

FIGURE 12.9 Hydraulic conductivity at saturation. (Adapted from Klute, A., and Dirksen, C. [1982] Methods of soil analysis. Part I—physical and mineralogical method. Soil Science Society of America, Madison, WI.)

$$K(\theta) = Kk_r(\theta) = \left(\frac{k\rho g}{\mu}\right)k_r(\theta) \qquad \text{(Eq. 12.23)}$$

where

$K(\theta)$ is the unsaturated hydraulic conductivity,

K is the saturated hydraulic conductivity, and

$k_r(\theta)$ is the relative permeability or relative hydraulic conductivity.

Relative permeability is a dimensionless number that ranges between 0 and 1. When $k_r(\theta)$ is equal to unity, it means all the voids are fully saturated with water. However, in the vadose zone, voids are partially filled with water because of entrapped air. Therefore unsaturated hydraulic conductivity is always lower than saturated hydraulic conductivity.

Unlike saturated hydraulic conductivity, unsaturated hydraulic conductivity is not constant and varies with the moisture content of soil. In the vadose zone, both unsaturated hydraulic conductivity and pressure head are functions of soil-water content, and thus the unsaturated hydraulic conductivity is a function of the pressure head. As the moisture content of soil decreases, so does Ψ and $K(\theta)$. In the vadose zone, a small drop in the water content of a soil, depending on its texture, may result in a dramatic decrease (e.g., 10^3, 10^6) in the unsaturated hydraulic conductivity of the soil. As we discussed earlier, the hydraulic conductivity of sands is always greater than that of clays for a saturated porous media. However, in the vadose zone, this relationship may not always hold true. For instance, during drainage of a soil, larger pores drain first and the residual water remains in the smaller pores. Since sand has a larger pore size than clay, it will drain sooner. Consequently, at relatively low soil-water contents or pressure heads, most of the voids of clays will remain saturated relative to sand. Therefore at lower soil-water contents or pressure heads, unsaturated

hydraulic conductivity of a clay unit may be greater than that of a sand unit.

MEASURING HYDRAULIC CONDUCTIVITY

Hydraulic conductivity of soil is a critical piece of information required for evaluating aquifer performance, characterizing contaminated sites for remediation, and determining the transport and fate of contaminant plumes in subsurface environments. For example, for water management issues, one needs to know the hydraulic conductivity to calculate the water transmitting and storage capacities of the aquifers. For remediation applications, knowledge of K distribution of contaminated soils is necessary for calculating plume velocity and travel time, to determine if the plume may reach a down gradient location of concern.

Hydraulic conductivity of soils can be measured in the laboratory as well as in the field. Laboratory measurements are performed on either disturbed or undisturbed samples that are collected in the field. Obtaining an undisturbed core sample is usually possible for consolidated materials, such as rocks, but is difficult for unconsolidated sediments, such as sand and gravel. Measurements made for a core sample represent that specific volume of media. That sample is then assumed representative of the field site. However, a single sample will rarely provide an accurate representation of the field because of the heterogeneity inherent to the subsurface. Thus a large number of samples could be required to characterize the hydraulic conductivity distribution present at the site. Laboratory measurements are inexpensive, quick, and easy to make compared to field measurements. They are often used to obtain an initial characterization of a site before on-site characterization is initiated.

LABORATORY METHODS

There are numerous laboratory methods that have been used to measure both saturated and unsaturated hydraulic conductivities of soils. Three methods are commonly used in the laboratory to measure or estimate the hydraulic conductivities of saturated soils. These include constant head and falling head permeameter methods, and estimation methods based on particle-size distribution, which will be briefly discussed in the next paragraph. For detailed information, refer to Klute and Dirksen (1986) and Fetter (2000).

Laboratory methods for determining unsaturated hydraulic conductivity of soils are categorized into two groups: steady-state flow methods and transient flow methods. Steady-state flow methods are, in fact, similar to permeameter tests in saturated soils with the exception that they are conducted under constant tension applied to the ends of the permeameter. Transient flow methods include: the instantaneous profile method; the Bruke-Klute method; the pressure-plate method; the one-step outflow method; and the ultracentrifuge method. Determining unsaturated hydraulic conductivity using these different methods is an advanced topic; therefore it is not discussed in this section. For information about methods for measuring unsaturated hydraulic conductivity, refer to Klute and Dirksen (1986) and Stephens (1996).

PERMEAMETER METHODS

The constant-head permeameter method is usually used for noncohesive soils, such as gravels and sands, whereas the falling-head permeameter method is used for cohesive soils, such as clays. The soil sample is placed in a chamber that is usually cylindrical and outfitted with a porous plate at both ends, which allows only water to pass through the chamber. The soil sample is fully saturated with water that is deaerated to prevent entrapment of air in the pores. The soil sample is tightly packed to prevent preferential flow along the walls of the permeameter.

For constant-head permeameters, heads at both ends of the permeameter are kept constant during the test by continuously supplying water into the permeameter (Figure 12.10A). It is important to use hydraulic gradients similar to those in the field. Hydraulic head loss in the permeameter should never be more than half of the sample length to prevent turbulence flow, which invalidates Darcy's law. Flow-rate measurements are recorded periodically throughout the test. The hydraulic conductivity of the soil sample can be obtained from Darcy's law:

$$K = \frac{VL}{Ath} \qquad \text{(Eq. 12.24)}$$

where

K is the saturated hydraulic conductivity $[L\ T^{-1}]$
L is the length of sample [L]
A is the cross-sectional area of the permeameter $[L^2]$
V is the volume of water collected in time t $[L^3]$
h is the constant hydraulic head [L].

For the falling-head permeameter, the water level in the falling-head tube is allowed to drop (Figure 12.10B), and then head measurements are recorded frequently until no flow occurs in the permeameter. Hydraulic conductivity of the saturated soil sample can be obtained from the relationship:

$$K = \frac{d_t^2 L}{d_c^2 t} \ln\left(\frac{h_0}{h}\right) \qquad \text{(Eq. 12.25)}$$

where

d_t is the inside diameter of falling-head tube [L]
d_c is the inside diameter of permeameter [L]
L is the sample length [L]
h_0 is the initial hydraulic head in the falling-head tube [L]
h is final hydraulic head in the falling-head tube at time t
t is the time that it takes for head to change from h_0 to h [T].

PARTICLE SIZE ESTIMATION

Hydraulic conductivity of saturated soils can also be estimated from their particle-size distribution. Several equations relate hydraulic conductivity of saturated soils to their mean particle size. One of the commonly known equations is the Hazen approximation, which was discussed earlier in this chapter. Shepherd (1989) also developed a general equation based on the analysis of 19 sets of published data relating hydraulic conductivity to grain size:

$$K = Cd_{50}^{j} \qquad \text{(Eq. 12.26)}$$

where

C is the shape factor
d_{50} is the mean particle diameter [L]
j is the exponent [-].

The value of j varies from 1.1 for texturally immature sediments, to 2.05 for texturally mature sediments (i.e., well-sorted sediments with uniformly sized grains having high roundness and sphericity). The Kozeny-Carman equation is another widely used relationship, and is given by:

$$K = \left(\frac{\rho g}{\mu}\right)\left[\frac{n^3}{(1-n)^2}\right]\left(\frac{d_m^2}{180}\right) \qquad \text{(Eq. 12.27)}$$

where

ρ is the density of water $[M\ L^{-3}]$
g is the acceleration due to gravity $[L\ T^{-2}]$
μ is the dynamic viscosity
n is the porosity $[M\ T^{-1}\ L^{-1}]$
d_m is the representative grain size [L].
For other equations, refer to Batu (1998).

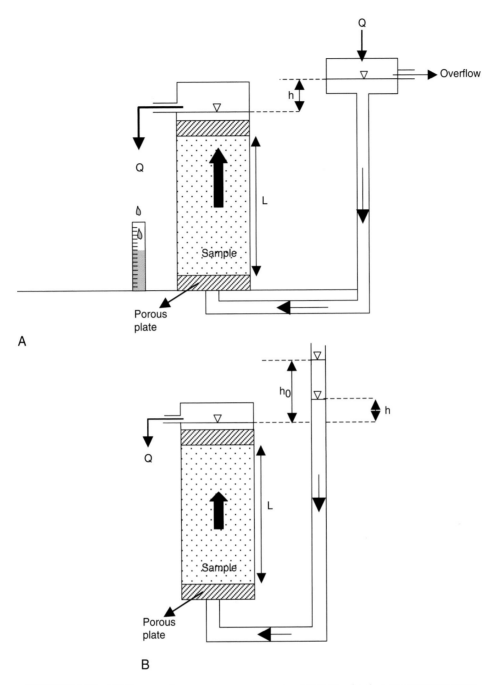

FIGURE 12.10 (**A**) Constant-head permeameter apparatus. (**B**) Falling-head permeameter apparatus.

VADOSE ZONE METHODS

There are numerous techniques for measuring hydraulic conductivity in the vadose zone. Tension-disc and double ring infiltrometers are two of the most widely used field methods and are discussed in the following section. For detailed information about the methods, refer to Green *et al.* (1986).

TENSION DISC INFILTROMETER

The tension disc infiltrometer is a device that measures both saturated and unsaturated hydraulic conductivity of topsoils over a range of negative pressures under field conditions. An illustration of a tension disc infiltrometer is shown in Figure 12.11. In this method, water supplied from a graduated reservoir is allowed to infiltrate into soil through a porous membrane at water pressures

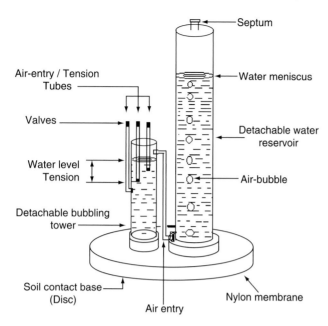

FIGURE 12.11   Tension disc infiltrometer apparatus. (From Wyseure, G.C.L., Sattar, M.G.S., Adey, M.A., and Rose, D.A. Determination of unsaturated hydraulic conductivity in the field by a robust tension infiltrometer. Dept. of Agriculture and Environmental Science, University of Newcastle [http://www.agr.kuleuven.ac.be/facdid/guidow/Tensio.htm].)

that are negative relative to atmospheric pressure. The required water pressure at the base of the membrane is established by adjusting the height of water in the second, smaller bubble tower with a movable air-entry tube.

The range of tensions that can be set is limited to a quite small range: from 0 to −30 cm of H_2O. A thin layer of fine sand is placed on the soil surface to establish a good hydraulic contact between the disc and soil. If the disc is not leveled, small pressure differences between the higher and lower ends of the membrane can cause leaks and inaccuracies in tension measurements. Infiltration is initiated at low tension. After steady state flux is achieved, a subsequent higher tension is applied. Water flux rate can be measured by simply recording the change in water level in the graduated reservoir over time. Unsaturated hydraulic conductivity of soil is equal to field saturated hydraulic conductivity under a small range of water pressures below atmospheric pressure. By setting the boundary pressure head to zero or close to zero, one can also measure the field saturated hydraulic conductivity of soil with a tension disc infiltrometer.

There are several methods to calculate the hydraulic conductivity from tension disc data. The standard analysis uses Wooding's (1968) solution for three-dimensional infiltration where the flux into a circular wetted area of radius r, at steady state, is:

$$q_\psi = \frac{Q}{\pi r^2} K_s \exp\left(\alpha\psi\right)\left(1 + \frac{4}{\pi \alpha r}\right) \qquad \text{(Eq. 12.28)}$$

where

$q_{(\Psi)}$ is the specific discharge at a given Ψ
Q is the water flux into soil
r is the disc radius
α is a parameter
Ψ is the pressure head set for the infiltration boundary condition
K_s is the field saturated hydraulic conductivity.

In this equation there are two unknown parameters, K_s and α. Consequently, to solve the equation for these parameters we need to carry out steady-state infiltration tests for at least two different tensions or disc radii. This solution is based on the assumption that the Gardner relationship (Gardner, 1958), where $K(\psi) = K_s \exp(\alpha\psi)$, is valid. At pressure heads relatively close to zero ($\psi \leq 0$), this relation reduces to $K(\psi) = K_s$, where unsaturated hydraulic conductivity is equal to field saturated hydraulic conductivity. One drawback to the use of Wooding's solution is the use of Gardner relation that is only valid at pressure heads relatively close to zero. However, this is a valid assumption for optimum tension range of the tension disc infiltrometer. Another limitation of Wooding's solution is the difficulty of establishing steady-state flow conditions in the field, and the time required to reach that condition. These limitations have been overcome by other solutions developed by Smettem *et al.* (1994) and Haverkamp *et al.* (1994) for transient three-dimensional infiltration.

Tension disc infiltrometers are easy to operate, transport, and maintain. As compared to ring infiltrometers, tension disc infiltrometers provide faster measurements, require less volumes of water, and cause minimal disturbance of the soil surface. Another advantage of this instrument is that it can provide a measure of hydraulic conductivity of the soil matrix, without the effects of macroporosity, such as root channels and wormholes. By providing water to soil surface at small tensions, the large pores can be selectively excluded from the hydraulic conductivity measurements.

DOUBLE RING INFILTROMETER

The double ring infiltrometer is another way of determining the field saturated hydraulic conductivity of the surface soil. It consists of open-ended inner and outer rings that are driven concentrically into the soil at a given depth, as shown in Figure 12.12. The diameters of the rings are typically 20 cm for the inner ring, and 30 cm for the outer ring. The penetration depth of the metallic rings is usually limited to ~5 cm to minimize soil disturbance. Each ring is supplied with a constant head of water from a Mariotte bottle. As water infiltrates into the soil from the inner ring, an equal amount of water flows into the inner ring from the Mariotte bottle to

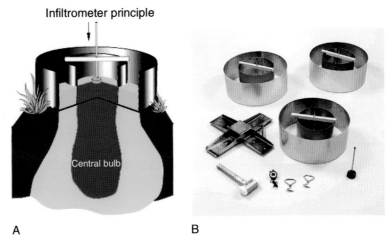

FIGURE 12.12 **(A)** Principles of the double ring infiltrometer. **(B)** Double ring infiltrometer components (**A**, From http://www.sdec-france.com. **B**, From www.eijkelkamp.com.)

keep the head constant. Infiltration rate is measured in the inner ring only. The rate of water flow is determined by measuring the change in the water level in the inner ring over an interval of time. An infiltration rate is then calculated by dividing flow rate by the cross-sectional area of the inner ring. This process is continued until the infiltration rate becomes steady. The steady infiltration rate measured with a double ring infiltrometer is often equal to the saturated hydraulic conductivity. The reason for using two rings is to reduce the lateral divergence of flow under the inner ring resulting from three-dimensional flow conditions, so that truly vertical flow will occur in the inner ring. This may be achieved by keeping the height of water at the same level in both rings, and by using large-diameter rings.

Compared with the tension disc infiltrometer, double ring infiltrometers are a more economical way of measuring saturated hydraulic conductivity of topsoils. However, a major drawback of the double ring infiltrometer is the overestimation of saturated hydraulic conductivity in soils having macropores such as wormholes and cracks. Because water is ponded over the soil surface at atmospheric pressure, a large portion of this water may infiltrate into these macropores, and thus result in a very large saturated hydraulic conductivity value, which is not representative of the soil matrix.

SATURATED ZONE METHODS

SLUG OR BAIL TEST

A slug or bail test measures the recovery of head in a well after a sudden change of water level has occurred in the well (Figure 12.13). This sudden change of head at the beginning of the test can be created by introducing a displacement object or equivalent volume of water into

the well (slug test), or by removing a volume of water from the well (bail test). Then, the rate at which the water level drops or rises in the well (recovery) is measured. From the recovery rate of head in the well, one can determine the hydraulic conductivity of the material in the immediate vicinity of the well using appropriate methods such as the Hvorslev's method, Cooper's method, or the Bouwer and Rice method for analyses of slug-test data. Further information about these methods can be found in Freeze and Cherry (1979); Buttler (1997); and Fetter (2001).

Slug/bail tests are easy to perform, relatively inexpensive, and require less time to complete as compared to alternative methods. Slug/bail tests are performed using a single monitor well. Pumping and observation wells are not required for the test. Analysis of slug/bail test data is also very straightforward.

On some occasions, slug/bail tests are an alternative testing method to the aquifer test described in the next section. For example, slug/bail tests are the best option for determining hydraulic properties of less permeable materials, for which aquifer tests are of limited use. One of the major advantages of the slug test is that it can be performed in a contaminated aquifer without removing any contaminated water.

The major limitations of the slug and bail tests are that the test estimates are highly affected by the properties of the well, such as well development and well intake. Well development is usually done after completion of a monitor well, and the aim is to remove the fine materials such as clays and silts making up borehole skin, by pumping water out until it is clear. Incomplete well development before a slug test may result in artificially low hydraulic conductivity estimates. If the well points and screens are clogged, measured values may be biased.

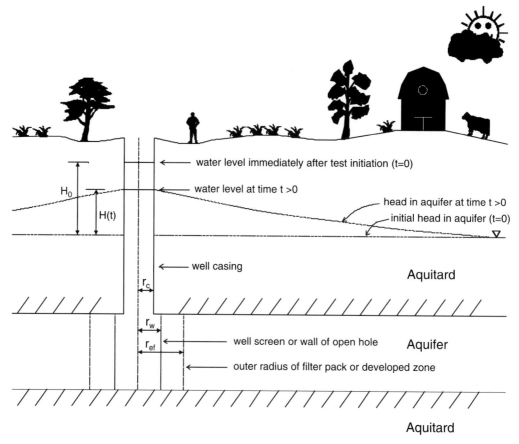

FIGURE 12.13   Schematics of slug test in a monitoring well that is fully screened across a confined aquifer. Slug test is initiated at t = 0; figure not to scale. (From Butler, J.J. [1997] *The design, performance, and analysis of slug tests.* Lewis Publishers, Boca Raton, Fla.)

PUMPING OR AQUIFER TESTS

The term *pumping test* is applied to tests for characterizing (1) the hydraulic properties of a well, and (2) the hydraulic properties of an aquifer. Hydraulic properties of a well include well losses and well efficiency. Aquifer hydraulic properties of interest include the ability of an aquifer to transmit, store, and release water. An aquifer is simply defined as a saturated-permeable geological formation, capable of yielding groundwater to wells at a rate sufficient to support a specified use, such as for potable water supply. The two major types of aquifers that exist in the subsurface are unconfined and confined aquifers. An unconfined aquifer, or water-table aquifer, is an aquifer whose upper surface is unconfined (water table) and is under atmospheric pressure. A confined aquifer is an aquifer bound between two aquitards (a formation of low-permeability materials), and whose groundwater is under pressure greater than atmospheric. During aquifer-pumping tests, the aquifer system is stressed by pumping water in a test well, producing a decrease in water (unconfined), or potentiometric surface (confined) levels in the aquifer. These changes in hydraulic head levels are monitored and recorded in the pumping well and associated observation wells or piezometers (Figure 12.14) over the test period. If the test has been properly designed and conducted, aquifer transmissive and storage properties such as hydraulic conductivity, transmisivity (T), and storativity (S) are determined using analyses appropriate to the physical properties of the system. Pumping test measurements are representative of a relatively large volume of subsurface, and thus data from aquifer pumping tests provide more representative estimates of the hydraulic properties of an aquifer than do slug/bail tests that sample only a limited volume of aquifer. Aquifer-pumping tests are important when planning water supply and aquifer remediation projects.

Aquifer-pumping tests for determining aquifer hydraulic properties involve pumping a well at a constant rate, measuring water level responses in observation wells or piezometers during pumping and recovery, and utilizing appropriate methods to calculate T and S (Table 12.4).

Detailed information on planning for an aquifer-pumping test is included in the references. The following is a summary of requirements (Osborne, 1993):

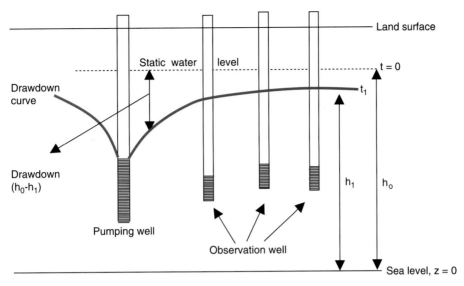

FIGURE 12.14 Drawdown variation with distance from a pumping well.

1. Hydrogeologic reports in the project area to provide background T and S values and to aid in determining discharge rates and test duration. Data from nearby wells are particularly relevant.
2. Driller's logs and geophysical data for the pumping well and observation wells or piezometers. Such data aid in determining the thickness of the aquifer being tested.
3. Location of recharge sources, such as streams and lakes, that may interfere with the test.
4. Long-term water levels in the aquifer to assist in analyzing test data.
5. Measured distances between the pumping well and the observation wells.
6. Surface elevations of all wells and their latitude and longitudes (e.g., using a global positioning system [GPS]).

There are two phases during an aquifer-pumping test: a pumping phase and a recovery phase. During the pumping phase, the discharge rate of the pumping well should remain essentially constant. Preferably, the discharge rate should not vary by more than plus or minus 5%. Flows should be adjusted using flow control valves, rather than throttle adjustments.

Within the pumping phase, there are two classes of tests: steady state and nonsteady state. During steady-state tests, pumping is continued until there is a negligible change in water levels in monitor wells. These levels may not represent long-term equilibrium conditions. Steady-state tests are used in determining values of transmisivity. Non–steady-state tests are generally long-term tests, providing estimates of T and S. During both tests, the difference between water levels in the measuring points at a particular time and the static water is termed *drawdown*

TABLE 12.4
Components of a Typical Aquifer-Pumping Test

Components	Description
Well	Preferably, the well should be perforated throughout the aquifer region of interest. It should be constructed and completed to ensure minimal well losses that may interfere with aquifer-pumping tests. Well hydraulic characteristics are determined by step-drawdown tests.
Pumps	Alternatives include deep-well turbine pumps, submersible pumps, and airlift pumps. Submersible pumps are preferred. Power sources include gasoline or diesel engines or electricity.
Water-level measuring devices	Alternatives include steel tapes, electric sounders, and pressure transducers.
Flow-measuring devices	Alternative methods include orifice plates, propeller in-line flowmeters, and flumes and weirs.
Observation wells	Existing wells may be used for this purpose provided they terminate in the aquifer formation of interest and their perforations or well screens are unclogged. At least one well is required in determining S, as well as T. Several wells are useful for locating anisotropic regions.
Water disposal	An adequate drainage system is required to ensure that discharged water does not recirculate into the aquifer being tested and interfere with test results. If there are quality concerns with the discharge water, treatment may be required.

(Adapted from Osborne, P.S. [1993] Suggested operating procedures for aquifer pumping tests. Environmental Protection Agency.)

(see Figure 12.14). During the recovery phase, water-level data obtained after pumping has stopped and water levels are rebounding are used to determine T. This provides an independent check on T values calculated from the pumping phase.

The following are the basic steps in conducting an aquifer-pumping test (Osborne, 1993):

1. Determine the initial water levels in all wells.
2. For the pumping phase, set t = 0 at the start of pumping.
3. Continuously monitor pump discharge rate and flow volume.
4. Monitor water levels in wells at frequencies depending on distance from the pumping well.
5. For the recovery phase, set t = t' as the time when pumping has ceased.
6. Monitor water levels (residual drawdown) during recovery.
7. Correct water-level data for long-term regional trends.
8. Plot data.
9. Analyze and interpret data.

During recovery tests, water levels are monitored as a function of the time after pumping was stopped.

MEASURING HYDRAULIC HEADS

Hydraulic head measurements are essential pieces of information that are required for characterizing groundwater flow systems (i.e., direction and hydraulic gradient of flow), determining hydraulic properties of aquifers (i.e., K, T, and S), and evaluating the influence of pumping on water levels in a region. Piezometers are used to measure the hydraulic head at distinct points in saturated regions of the subsurface. A piezometer is a hollow tube, or pipe, drilled or forced into a profile to a specific depth. Water rises inside the tube to a level corresponding to the pressure head at the terminus. The level to which water rises in the piezometer, with reference to a datum such as sea level, is the hydraulic head (Figure 12.15). In the field, a common way of measuring head in the piezometer requires depth-to-water measurement in a piezometer, and a knowledge of the elevation of the top of the piezometer casing. Hydraulic head in the piezometer is then calculated by subtracting depth to water from the elevation of the top of the piezometer casing.

Data from piezometers terminating in depth-wise increments provide clues on the vertical flow direction of water in saturated regions. Groupings of piezometers installed in different boreholes are referred to as clusters. Piezometers installed in the same borehole are referred to

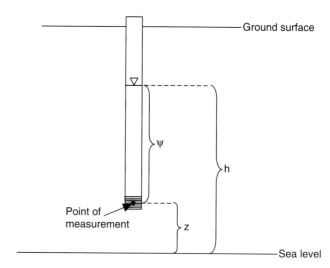

FIGURE 12.15 Concept of hydraulic head *(h)*, elevation head *(z)*, and pressure head *(Ψ)* in a piezometer.

as nests. In contrast to clustered piezometers, there is concern for tests that uncertain seal placement may cause side leakage between units along the borehole, affecting the accuracy of readings. Another alternative is modular-type, multilevel monitoring installations, with packers separating measurement ports.

Water-level measurements in nested piezometers are used to define vertical hydraulic gradients. Similarly, water-level measurements in a network of production or monitor wells are used to define the potentiometric surface of regional groundwater. Knowledge of this surface is required to define hydraulic gradients and flow directions. Water-level measurements are also an integral component of aquifer-pumping tests. This section summarizes methods for measuring well and piezometer water levels.

WATER LEVEL DEVICES

Common methods for measuring depth to water in wells include (1) steel tapes; (2) electric sounders; (3) bubblers; (4) pressure transducers; (5) float-operated water level recorders; and (6) acoustic methods.

STEEL TAPES

Using steel tapes is the simplest method for measuring water levels in wells. Tapes graduated in units of hundredths of feet/meters offer great accuracy. To measure water levels, blue carpenters' chalk is applied to the lower end of the dried tape, and the tape is lowered into the well to a predetermined depth and held at an even foot (or meter) mark at the top of the well. The difference between the surface reading and the reading on the wetted chalk line is the depth to water (Figure 12.16A). The elevation of the water level is readily calculated from the elevation at the top of the well. One problem with this method is that it requires a prior estimate of the water level. This can be approximately determined by attaching a bell-shaped fitting on the end of the tape and listening for a "plopping" sound when the fitting contacts the water surface. A second problem occurs when there is cascading water in the casing, causing smearing of the chalked section. This problem is avoided by installing a smaller-diameter pipe within the larger casing.

ELECTRIC SOUNDERS

Another difficulty with the tape method is that the tape must be withdrawn from the tube between readings. This time delay between readings is a problem when water levels are fluctuating rapidly, as for example, during aquifer-pumping tests. In contrast, electric sounders allow readings to be taken without withdrawing the probe from the hole. However, a dedicated sounder is required for each well, increasing costs. The basic components of an electric sounder are a probe, a reel-mounted cable with depth-wise markings, a 9- or 12-volt battery, and a water-level response indicator (e.g., milli-amp meter) (Figures 12.16B and 12.17). In some units the cable includes an insulated two-conductor wire. In other versions the cable includes a single wire, and a second wire is attached to the steel casing of the well. When the probe contacts the water, the electric circuit is completed, causing a response on the water level indicator (e.g., a deflection of a needle on a milli-amp meter). The depth to water is determined by subtracting the distance from the point of measurement (e.g., top of well) to the next marking on the cable. Readings may be affected by cascading water. This problem is avoided by inserting a smaller-diameter pipe inside the larger casing.

BUBBLERS

The bubbler method of sensing water levels in a well is based on the principle that the air pressure required to displace a column of water from a submerged tube is directly proportional to the height of displaced water column (Figure 12.16C). If L is the length of the air line below a land surface datum, d is the depth to water in the well below land surface datum, and h is the distance between the water surface and the tip of the air line, then $d + h = L$, or $d = L - h$. Table 12.5 lists some useful equivalency units to convert gage data to equivalent water depths.

For shallow systems, a simple arrangement, such as a hand pump with an air pressure gage, can be used to estimate water levels. More sophisticated methods are available on the market for greater control, including those with compressed air generators, microprocessor controlled pressure sensors, data loggers, and telemetry systems for transmission of data to a remote central location.

PRESSURE TRANSDUCERS

The advent of modern electronics has greatly improved the ease and efficiency of measuring water levels in wells. Pressure transducers are now routinely installed at the base of wells (Figure 12.16D) to monitor water levels in remote locations. A commercially available pressure transducer consists of a sensitive diaphragm connected to a vibrating wire strain gage (Figure 12.18). Vibrating wire transducers operate on the principle that the frequency of vibration of a wire under tension varies with the tension on the wire. One end of the wire is clamped to the transducer body, while the other end is attached to a movable diaphragm. The wire is vibrated by a magnetic coil, and the resultant frequency depends on the flexion of the diaphragm and therefore on the water pressure exerted against the diaphragm. The equivalent frequency signal is transmitted through a cable to a read-out module at land surface. Knowing the depth of the transducer sensor, the pressure data are converted to depth to water below a surface reference point. Alternative surface components include a data logger, cellular phone package, and a phone modem. Cellular-phone technology and satellite telemetry allow the data to be downloaded at a remote central location. Programming instructions and data corrections can also be sent by telephone. Components are housed in a suitable enclosure for protection. Transducers have been installed to depths of over 300 m (~1000 ft) below land surface, and changes in water levels of as little as 0.2 mm (0.008 in) can be measured. Water level data obtained from the described methods should be corrected for changes in barometric pressure.

FLOAT-OPERATED WATER-LEVEL RECORDERS

Float-operated water-level recorders have been a basic technique for monitoring and recording water level fluctuations in wells for many years (Stevens Water Monitoring Systems, www.stevenswater.com). Components include a cylindrical drum attached to a pulley, a float and counterweight, a beaded cable, and a clock-driven pen (Figure 12.19). The uniformly distributed beads match recessed areas in the float pulley to ensure positive response of the pulley to water level changes. Before taking a recording, a chart paper is wrapped around the drum. The drum rotates as the pulley moves up or down

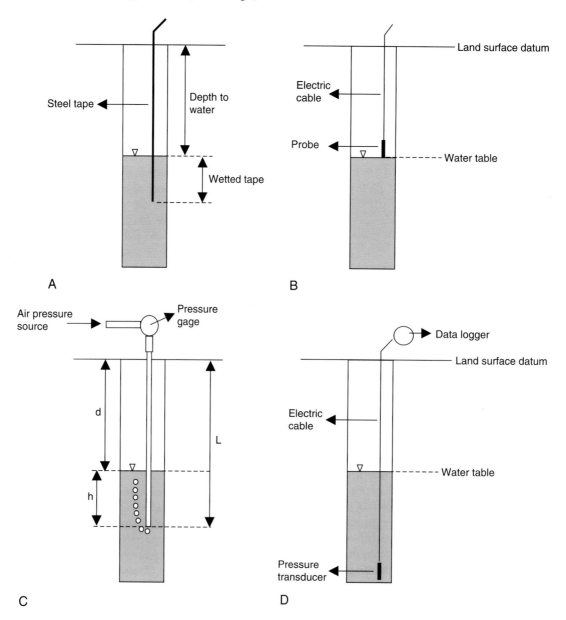

FIGURE 12.16   **(A)** Steel tape method of measuring depth to water in a well. Depth to water is the difference between the land surface datum and wet/dry interface on tape. **(B)** Electric sounder method. Depth to water is the reading on the tape when electric contact is made with the water surface. **(C)** Bubbler method. **(D)** Pressure transducer method. (**A**, **B**, and **D**, Adapted from Roscoe Moss Co. [1990] *Handbook of ground water development.* New York, Wiley. **C**, Adapted from Driscoll, F.G. [1986] *Groundwater and wells. St.* Paul, Minn, Johnson Screens.)

FIGURE 12.17   Electric sounder. (Courtesy Heron Instruments Inc., Burlington, Ontario, Canada.)

TABLE 12.5
Some Useful Equivalency Units to Convert Gage Data to Equivalent Water Depths

Unit	psi	Ft of Water	m of Water	Atmospheres
1 pound per square inch (psi)	1.0	2.31	0.704	0.0681
1 foot of water	0.433	1.0	0.305	0.02947
1 meter of water	1.421	3.28	1.0	0.0967
1 atmosphere at sea level	14.7	33.09	10.34	1.0

(Taken from Driscoll, F.G. [1986] *Groundwater and wells.* St. Paul, Minn, Johnson Screens.)

FIGURE 12.19   Float-operated water-level recorder. (Courtesy Rickly Hydrological Co., Columbus, Ohio.)

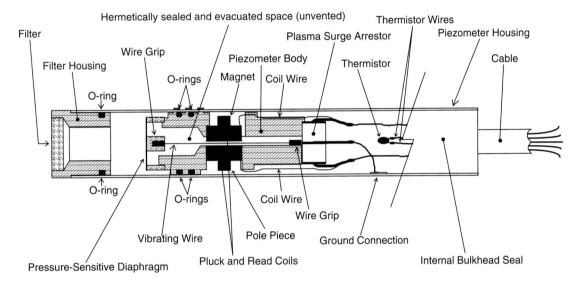

FIGURE 12.18   Commercially available vibrating wire pressure transducer. (Courtesy Geokon Inc., Lebanon, NJ.)

in the well in response to water level changes. Concurrently, the pen advances across the drum at a rate governed by gear selection. The resulting graph shows the water level changes as a function of time. A variety of clocks are available, including spring-driven clocks, AC or DC clocks, and weight-driven clocks. The recorder assembly is mounted on a well with the float, tape, and counterweight hanging inside the well. The metal base has four legs for support. A sheet metal cover is used for protection from the elements. These recorders are more suited to larger diameter wells than the size normally used as piezometers.

In an updated version of the basic design, the shaft pulley is attached to a precision potentiometer, or shaft encoder, in lieu of the cylindrical drum. The potentiometer converts mechanical changes produced by pulley rotation into voltage changes. A data logger converts the potential voltage changes to water level change. Water level data are obtained at preset intervals and stored on a retrievable data cartridge. The data cartridge must be carried to a central base station to retrieve the recorded data.

SONIC TRANSDUCERS

These units measure depth to water in a well by transmission of a sound wave down a well. An electrostatic transducer determines the distance to water level in a well by sending out ultrasonic pulses, and detecting the returning pulses reflected from the water surface. The resultant value must be adjusted to account for the variation of the speed of sound in air with temperature. This requires an independent measurement of temperature. A major advantage of this device is that it does not require inserting tubes, cables, or floats inside the casing. However, cascading water inside the casing may cause inaccurate measurements.

MEASURING FLOW DIRECTIONS, GRADIENTS, AND VELOCITIES

Characterizing groundwater flow in an aquifer system generally requires constructing groundwater maps. These maps are basic tools of hydrogeological interpretation from which one can derive hydraulic gradient and the direction of groundwater flow. To construct these maps, one needs head measurements made in a number of wells, or piezometers that are installed at various depths in an aquifer of interest. Groundwater maps are prepared plotting hydraulic head values of all wells on a base map showing the well locations, and drawing a line connecting the point of equal hydraulic heads (equipotential line). For simplicity one may consider equipotential lines as similar to contours in a topographic map. Hydraulic head values between wells are determined by linear interpolation. When interpolating equipotential lines between data points, one should consider topographic features, such as rivers or lakes in the region. Ignoring such features can cause a significant error in equipotential lines.

In three dimensions, the points with equivalent head values form an equipotential surface called *piezometric*, or *potentiometric*, surface. Potentiometric surface is simply the map of hydraulic head, or the level to which water will rise in a well that is screened within the aquifer. If an aquifer is characterized by primarily horizontal flow, one can draw a potentiometric surface for the entire aquifer. In an unconfined aquifer, the water table represents the potentiometric surface of the aquifer. However, in a confined aquifer, the potentiometric surface is an imaginary surface connecting the water levels in wells. Groundwater maps are actually two-dimensional representations of three-dimensional potentiometric surfaces, and are denoted as water-table maps for an unconfined aquifers, and potentiometric-surface maps for confined aquifers.

Water-table maps or potentiometric-surface maps are graphical representations of hydraulic gradient of water table or potentiometric surface. If the water table or potentiometric surface has a shallow gradient, equipotential lines are spaced well apart. If the gradient is steep, equipotential lines are closer together (Figure 12.20). Groundwater flow occurs in the direction that the water table or potentiometric surface is sloping. The direction of flow is always perpendicular to equipotential lines in a homogeneous and isotropic aquifer.

Measuring Hydraulic Gradients in the Saturated Zone

The hydraulic gradient represents the driving force for groundwater flow. It is defined as the change in the

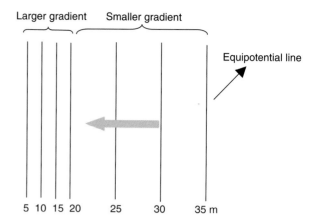

FIGURE 12.20   Schematic of an equipotential map showing change in hydraulic gradient. Gray arrow represents direction of flow.

hydraulic head with a change in the distance in the direction of flow. The hydraulic gradient between two piezometers is:

$$\frac{dh}{dl} = \frac{[(\psi_2 + z_2) - (\psi_1 + z_1)]}{(z_2 - z_1)} = \frac{(h_2 - h_1)}{(z_2 - z_1)} \quad \text{(Eq. 12.29)}$$

where the subscripts represent piezometers at locations 1 and 2. Hydraulic gradient is a vector that is characterized by its magnitude and direction. For two-dimensional flow, hydraulic gradient has both horizontal and vertical components in the direction of flow. To determine the horizontal component of the hydraulic gradient in a given area, one requires water level measurements for three or more wells or piezometers installed to the same level in the ground. A graphical method is available to determine horizontal gradient and the direction of groundwater flow within the area surrounded by three wells. This is known as the "three-point problem." This method is useful especially in sites where there are not many wells to characterize groundwater flow. Application of the graphical method is demonstrated in the following example problem.

The vertical component of hydraulic gradient at a given location can be measured by installing several piezometers to different depths and measuring the hydraulic head at each piezometer. Hydraulic gradient data from nested piezometers (e.g., piezometers in a common borehole) provide clues on vertical flow directions in a profile. Figure 12.21 shows a nest of three piezometers for three hypothetical situations. In the first case, water levels in the three units are at the same elevation, the total hydraulic heads are equal, and the vertical gradient is zero. In the second case, the water levels are at the base of each unit ($\Psi = 0$), and the gravity head is the only driving force. Water flows vertically downward in the

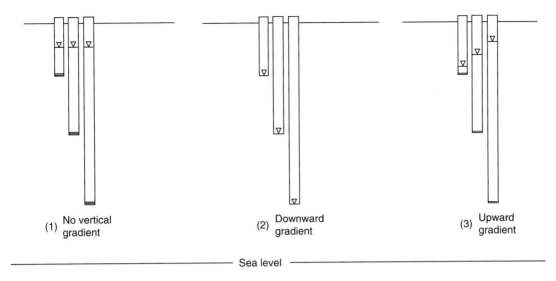

FIGURE 12.21 Three hypothetical situations for vertical hydraulic gradient.

profile. In the third case, the height of water is greatest in the deepest unit and least in the shallow unit. Thus the total hydraulic head decreases vertically upward and flow is in the upward direction.

EXAMPLE PROBLEM 1. This example describes the graphical method for determining horizontal hydraulic gradient within an area surrounded by wells (Figure 12.22). Let's suppose a hypothetical case wherein three wells, denoted as A, B, and C, are drilled to the same level in the aquifer. The wells are fully screened throughout the aquifer. Hydraulic head values noted at wells A, B, and C are 30, 50, and 40 m, respectively. The distance between each well pair is 100 m. Well A is located due west of Well B, and Well C is located due south of the A-B plane. Calculate the hydraulic gradient and direction of groundwater flow.

Steps involved in the solution of three-point problem are summarized in Table 12.6 (see Figure 12.22).

EXAMPLE PROBLEM 2. The following is an example of hypothetical data from three nested piezometers, designated TBW1, TBW2, and TBW3. The units terminate at depths of 30, 50, and 75 m, respectively, from land surface. Depth measurements were taken at the same time from the piezometers. Calculate the total hydraulic head in each piezometer and hydraulic gradients between successive units. Determine the flow directions. Table 12.7 includes the baseline data and results.

Sample calculation for well TBW1 (Figure 12.23):
Distance between baseline (mean sea level where $z = 0$) and piezometer tip = 470 m, or
$z = 470$ m.

Height of water in piezometer = 22 m = ψ
Total hydraulic head = 470 m + 22 m = 492 m
Gradient between well TBW1 and TBW2:

$$\Delta h/\Delta l = (h_{TBW2} - h_{TBW1})/(l_{TBW2} - l_{TBW1})$$
$$= (483 - 492)/20 = -9/20 = -0.45$$

DISCUSSION

1. Data in the Table 12.7 show that the total hydraulic head is simply the elevation of the water level in the piezometers when the datum (reference point) for each piezometer is taken as mean sea level (i.e., where $z = 0$).
2. The negative hydraulic gradient between units TBW1 and TBW2 shows that groundwater is moving vertically down in the profile between these units. In contrast, the positive gradient between units TBW2 and TBW3 shows that groundwater is moving vertically upwards. This suggests that groundwater extraction is occurring at the level of well TBW2, causing a gradient from above and below. In the real world, the relationship between these trends and geological layering would be available from drill cuttings during the installation of the piezometers.

MEASURING HYDRAULIC GRADIENTS IN THE UNSATURATED ZONE

In a previous section, piezometers were described as a technique for measuring hydraulic gradients in saturated media. It is equally important to measure hydraulic gradients in unsaturated media; for example, for irrigation

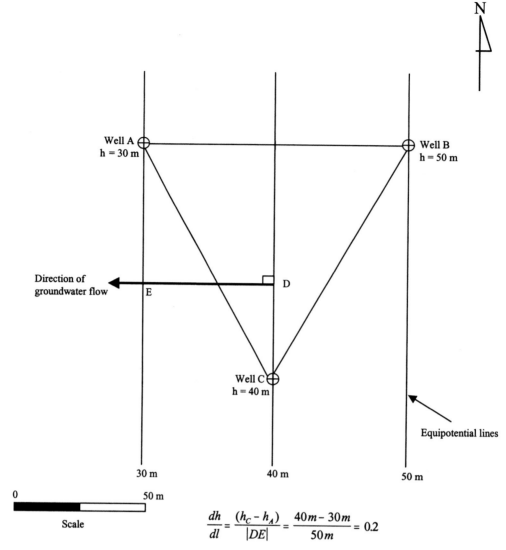

$$\frac{dh}{dl} = \frac{(h_C - h_A)}{|DE|} = \frac{40m - 30m}{50m} = 0.2$$

FIGURE 12.22   Three-point problem for determining hydraulic gradient and direction of groundwater flow.

TABLE 12.6

Steps Involved in Solving "Three-Point Problem" for Determining Hydraulic Gradient and Direction of Groundwater Flow

Steps	Description
1	Make a sketch to scale showing the location of wells and note the head values for each well.
2	Draw a line connecting each well pair.
3	Determine the location of equivalent head values between each well pair by linear interpolation.
4	Draw a line connecting points of equal hydraulic head. Repeat this process for each head value. Equipotential lines must be parallel to each other.
5	The direction of flow is perpendicular to equipotential lines and is from higher head to lower head. Measure the vertical distance between a well pair on the sketch, and then determine the real distance from scale of sketch.
6	Find the head difference between that well pair.
7	The hydraulic gradient is calculated by dividing the distance between a well pair lines by head difference.

(Based on material presented in Fetter, G.W. [1994]. Applied hydrogeology, Prentice Hall, Englewood Cliffs, NJ.)

TABLE 12.7
Base Line Data and Results for Example Problem 2

Piezometer Designation	TBW1	TBW2	TBW3
Surface elevation, m above mean sea level (amsl)	500	500	500
Depth of piezometer below land surface (bls), m	30	50	75
Elevation at point of measurement = z, m amsl	470	450	425
Depth to water level in piezometers, m	8	17	10
Elevation of water level in piezometer bls, m amsl	492	483	490
Height of water in piezometers = ψ	22	33	65
Total hydraulic head = h, m	492	483	490
Gradient		−0.45	0.28

scheduling, monitoring wetting fronts at waste disposal facilities, monitoring at soil aquifer treatment facilities, and conducting basic soil-water research. Piezometers measure the total energy potential at a point in saturated media. The analogous instrument for unsaturated media is the tensiometer.

These devices are simple in design and operation, but require the removal of a soil core followed by insertion (Figures 12.24 and 12.25). A tensiometer device is similar in design and installation to the porous cup soil pore water sampling devices described in the "Soil-Pore Water Samplers (Suction Lysimeters)" section in Chapter 7. Tensiometers consist of an access tube, a porous cup, and a vacuum gage. Water inside the tensiometer responds to changes in the soil matrix potential by moving into or out of the tube through the capillary pores of the porous cup. Since tensiometers are sealed, this gradient creates either a pressure or a vacuum inside the tube that can be measured with a suitable gage.

Data from a nest of tensiometers installed in depth-wise increments are used to calculate hydraulic gradients. The total hydraulic head at each point is calculated using the following equation:

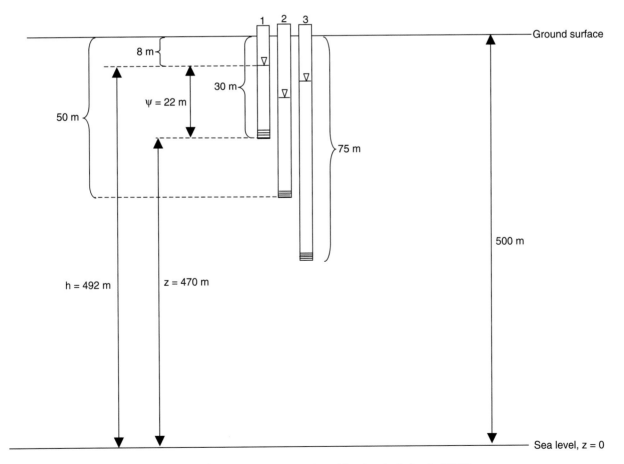

FIGURE 12.23 Schematics of Example Problem 2 and solution for TBW1.

$$\Phi_T = \Phi_m + \Phi_z \qquad \text{(Eq. 12.30)}$$

where

Φ_T is the total soil water potential [L]
Φ_m is the matric potential [L]
Φ_z is the gravitational potential [L].

From Figure 12.24, the basic tensiometer equation for measuring the soil matric potential is:

$$\Phi_m = \psi_g + \psi_h \qquad \text{(Eq. 12.31)}$$

where the term ψ_g is the gage pressure reading from tensiometer; and the term ψ_h accounts for the positive head of water above the tensiometer cup. For example, if the gage reading ψ_g is -150 mbar and the distance from the midpoint of the tensiometer cup and the gage ψ_h is 60 cm, $\Phi_m = -150$ mbar $+ (60)(0.98)$ mbar $= -91$ mbar. See Box 12.1 for a list of equivalent pressure/vacuum units conversion factors.

The hydraulic gradient between two depths is:

$$(\Phi_{T2} - \Phi_{T1})/(z_2 - z_1) = [(\Phi_{m2} + \Phi_{z2}) \\ -(\Phi_{m1} + \Phi_{z1})]/(z_2 - z_1) \qquad \text{(Eq. 12.32)}$$

where $z_2 - z_1$ is the vertical distance between points 1 and 2.

Differences in total hydraulic heads indicate the direction of soil-water flow in the profile. Figure 12.26 shows two hypothetical tensiometers, and Table 12.8 includes three hypothetical cases. In Case 1, the total hydraulic heads in the two tensiometers are equal and there is no detectable movement. The hydraulic gradient $[(\Phi_{T2} - \Phi_{T1})/(z_2 - z_1)]$ is zero. In Case 2, the total hydraulic head in the shallow unit is greater, indicating vertically downward flow. The hydraulic gradient $[(\Phi_{T2} - \Phi_{T1})/(z_2 - z_1)]$ is negative because $\Phi_{T2} - \Phi_{T1} < 0$. In Case 3, the total hydraulic head is greatest in the deeper unit, indicating upward flow (i.e., $\Phi_{T2} - \Phi_{T1} > 0$). The hydraulic gradient is positive because $\Phi_{T2} - \Phi_{T1} > 0$.

FIGURE 12.25 Tensiometer and tensimeter. (Courtesy Soil Measurement Systems, Inc. Tucson, Ariz.)

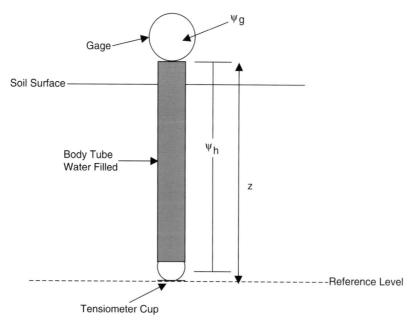

FIGURE 12.24 Basic tensiometer design.

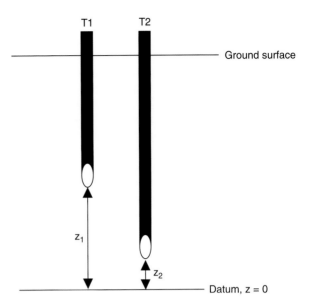

FIGURE 12.26 Use of tensiometers for determining direction of water movement in soil profiles.

EXAMPLE PROBLEM 3. Table 12.9 includes hypothetical data from a nest of three tensiometers installed at depths of 20, 30, and 40 cm, respectively, below land surface. Data were collected using a tensiometer and adjusted for the height of the water column.

1. The tensiometer depth equals the distance from the midpoint of the cup to the water level at the top of the tensiometer.
2. The reference datum is taken at base of tensiometer HAW3.

These values show that the total hydraulic head increases (becomes less negative) with depth. The hydraulic gradients between successive units are positive, suggesting upward flow of soil water; for example, because of evapotranspiration.

MEASURING GROUNDWATER FLOW VELOCITIES

Measurements of groundwater velocity as well as direction of flow are important for hydrogeological studies and environmental monitoring applications, including predicting the rate and direction of contaminant plume movement. There are several techniques available for determining groundwater flow velocities in the field. These include estimation based on Darcy's law, borehole flowmeters, and tracer tests. An estimation of groundwater velocity can be obtained using Darcy's law, rewritten as:

$$v = \frac{K}{n}\frac{dh}{dl} \qquad \text{(Eq. 12.33)}$$

This approach requires knowledge of hydraulic conductivity and the gradient, which can be obtained as described previously.

TABLE 12.8

Summary of Three Hypothetical Total Hydraulic Head Conditions in Two Tensiometer Units

Case Number	Total Hydraulic Head, Φ_T	Flow Direction	Gradient
1	$\Phi_{T2} = \Phi_{T1}$	Equilibrium	0
2	$\Phi_{T2} < \Phi_{T1}$	Downward	Negative
3	$\Phi_{T2} > \Phi_{T1}$	Upward	Positive

TABLE 12.9

Hydraulic Gradient Data for Three Hypothetical Tensiometers[a]

Tensiometer Designation	HAW1	HAW2	HAW3
Tensiometer Depth, ψ_h, cm	20	30	40
Tensimeter Reading, ψ_g, in millibars	−200	−180	−160
Tensimeter Reading, ψ_g, cm	−204	−184	−163
Adjusted Matric Potential, Φ_m, cm	−184	−154	−123
Gravity Potential, Φ_z, cm	20	10	0
Total Hydraulic Head, Φ_T, cm	−164	−144	−123
Hydraulic Gradient	2.1_{1-3}	2_{1-2}	2.1_{2-3}

[a]Subscripts in the hydralic gradient values denote the piezometer locations for the measurement.

Borehole flowmeters are used for many applications including well-screen positioning, recharge zone determination, and estimation of hydraulic conductivity distribution. In addition, they can be used for measurements of horizontal and vertical flow characteristics in a cased well or borehole. The most common borehole flowmeters include the impeller-type flowmeter, heat-pulse flowmeter, and electromagnetic flowmeter.

The impeller flowmeter provides point measurements of flow rate in borehole using low inertia impellers spinning on low-friction jeweled bearings (Figure 12.27A). Whenever the impeller shaft rotates, a magnet in the impeller activates magnetic switches within the probe to detect impeller rotation. The action of these switches indicates both direction and velocity of borehole fluid flow. Conventional impeller flowmeters that are widely used in groundwater studies are useful for measuring higher groundwater velocities, and have a lower measurement limit of about 2 m per minute.

The heat-pulse and electromagnetic flowmeters are high-resolution flowmeters that are used to measure low groundwater flow rates below the threshold limits

of conventional impeller flowmeter ($< 0.03 \, \text{m/min}$). The heat-pulse flowmeter measures the direction and rate of flow based on the thermal migration in an area caused by groundwater flow. This type of flowmeter consists of a heating grid and thermistors (heat sensor) situated above and below it (Figure 12.27B). Through an aperture in the probe, groundwater flow is diverted to the heating grid, where a pulse of electric current is applied to heat the water. Depending on the flow direction, the heated water moves toward any of the thermistors above and below the heating grid, from which one can determine direction of vertical flow. The time required for the warm water front to reach one of the thermistors is used to calculate flow rate.

Electromagnetic borehole flowmeters measure the flow rate based on the Faraday law of induction, which states that the voltage induced across a conductor moving at right angles through a magnetic field is directly proportional to the velocity of the conductor. This flowmeter is comprised of an electromagnet and a pair of electrodes mounted perpendicular to the poles of the electromagnet (Figure 12.27C). The electromagnet creates a magnetic

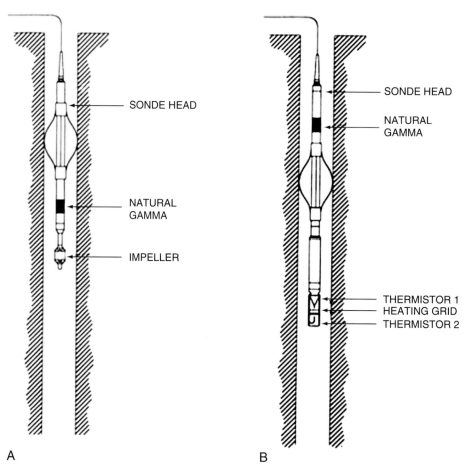

A   B

FIGURE 12.27 Schematics of **(A)** impeller borehole flowmeter and **(B)** heat-pulse borehole flowmeter. (Courtesy of Robertson Geologging Ltd., Cypress, Tex.)

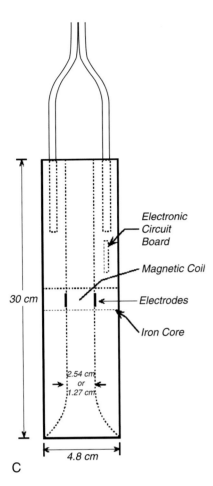

C

FIGURE 12.27—cont'd (C) schematic of electromagnetic borehole flowmeter. (From Young, S.C., Julian, H.E., Pearson, H.S., et al [1998] Electromagnetic borehole flowmeter, EPA/600/SR-98/058.)

field across the flow channel. As water, an electrical conductor, is moving through this magnetic field, a micro-voltage, which is proportional to the average velocity of the water (conductor), is created across the magnetic field.

Tracers may also be used in field studies to determine velocities of groundwater. A tracer test is conducted by injecting the tracer solution into the aquifer and monitoring tracer concentrations at down-gradient locations. The time required for the tracer to travel from the injection point to the monitoring point can be used to calculate groundwater velocity. A suitable tracer must not react physically or chemically with groundwater or aquifer material, and must not undergo transformation reactions. These types of substances are generally called *conservative tracers*. Some of the tracers that have been used in field studies include dyes (fluorescein, uranine, rhodamine B, etc.), salts (sodium chloride, potassium chloride, calcium chloride, lithium chloride), and radioisotopes (^{3}H, ^{131}I, ^{29}Br).

QUESTIONS

1. Determine the texture names for the following soil compositions:

	Sand (%)	Silt (%)	Clay (%)
1	60	20	10
2	30	20	40
3	10	70	20

2. Calculate the bulk density, porosity, and volumetric water content of a sandy soil, based on the given data:

   Mass of saturated soil: 300 g
   Mass of dry soil: 350 g
   Volume of sample container: 150 cm^3

3. A soil sample contains 60 g of water and 700 g of oven dry soil. What is the gravimetric soil-water content of the soil sample?

4. What is the difference between gravimetric water content and volumetric water content?

5. What equation describes flow through a saturated porous medium? Define each term in the equation and associated units.

6. Describe the difference between hydraulic conductivity and intrinsic permeability.

7. Compare unsaturated and saturated soils in terms of their physical properties.

8. Discuss if hydraulic conductivity of sand is always greater than that of clay in unsaturated soil conditions.

9. What is the major difference between unsaturated and saturated hydraulic conductivity?

10. Explain the methods that are used to determine hydraulic conductivity of soils in vadose and saturated zones.

11. Explain the factors determining the porosity of soils.

12. Describe the common methods that are used to measure soil-water content in the field.

13. A tensiometer that is constructed for use with a tensimeter is located 50 cm below the soil surface.

The top of the tensiometer is located 15 cm above the soil surface. If the tensiometer reads −140 kPa, what is the matric potential at that depth?

REFERENCES AND ADDITIONAL READING

American Society of Testing Materials. ASTM D3404, Standard Practices for Measuring Matric Potential in the Vadose Zone Using Tensiometers, Latest Edition Annual Book of ASTM Standards, Section 4, Construction, Volume 04.08 Soil and Rock; Dimension Stone; Geosynthetics. American Society of Testing materials, West Conshohocken, PA.

American Society of Testing Materials. ASTM 5220, Standard Test Method for Water Content of Soil and Rock In-Place by the Neutron Depth Probe Method, Latest Edition Annual Book of ASTM Standards, Section 4, Construction, Volume 04.08 Soil and Rock; Dimension Stone; Geosynthetics. American Society of Testing materials, West Conshohocken, PA.

Batu, V. (1998) Aquifer Hydraulics. A Comprehensive Guide to Hydrogeologic Data Analysis, Wiley Inter Science Publications, New York.

Bedient, P.B., Rifai, H.S., and Newell, C.J. (1999) Ground Water Contamination Transport and Remediation. Prentice Hall PTR, Upper Saddle River, NJ.

Blake, G.R., and Hartge, K.H. (1986) Bulk Density. In Klute, A (ed), *Methods of Soil Analysis. Part I—Physical and Mineralogical Method*, Number 9 (Part 1) in the Series Agronomy, American Society of Agronomy, Inc., Soil Science Society of America, Inc., Madison, WI, 363–375.

Butler, J.J. (1997) The Design, Performance, and Analysis of Slug Test, Lewis Publisher, Boca Raton, FL.

Cassell, D.K., and Klute, A., (1986) Water Potential: Tensiometry. In *Methods of Soil Analysis. Part I-Physical and Mineralogical Method*, Klute, A. edt., Number 9 (Part 1) in the Series Agronomy, American Society of Agronomy, Inc., Soil Science Society of America, Inc., Madison, WI, 563–596.

Charbeneau, R.J. (2000) *Groundwater Hydraulics and Pollutant Transport*, Prentice Hall, NJ.

Danielson, R.E., and Sutherland, P.L. (1986) Porosity. In *Methods of Soil Analysis. Part I-Physical and Mineralogical Method*, Klute, A. edt. Number 9 (Part 1) in the Series Agronomy, American Society of Agronomy, Inc., Soil Science Society of America, Inc., Madison, WI, 443–460.

Domenico, P.A., and Schwartz, F.W. (1990) *Physical and Chemical Hydrogeology*, John Wiley and Sons, New York.

Dorrance, D.W., Wilson L.G., Everett L.G., and Cullen S.J. (1995) A compendium of soil samplers for the vadose zone. In Wilson, L.G., Everett, L.G., and Cullen, S.J. (eds.), *Handbook of Vadose Zone Characterization and Monitoring*, Lewis Press, Chelsea, MI.

Driscoll, F.G. (1986) *Groundwater and Wells*, Johnson Division, St. Paul, MN.

Fetter, C.W. (1994) *Applied Hydrogeology*, 3rd ed. Prentice Hall, Englewood Cliffs, NJ.

Freeze, R.A., and Cherry, J.A. (1979) *Groundwater*, Prentice Hall, NJ.

Gardner (1985) Some study's solutions of the assaturated moisture flow equation to application to evaporation from a water-table. 85, 228–232.

Gee, G.W., and Bauder, J.W. (1986) Partice Size Analysis. In Klute, A. (ed.), *Methods of Soil Analysis. Part I-Physical and Mineralogical Method*, Number 9 (Part 1) in the Series Agronomy, American Society of Agronomy, Inc., Soil Science Society of America, Inc., Madison, WI, 383–409.

Green R.E., Ahuja, L.R., and Chong, S.K. (1986) Hydraulic Conductivity, Diffusivity, and Sorptivity of Unsaturated Soils: Field Methods. In Klute, A. (ed), *Methods of Soil Analysis. Part I-Physical and Mineralogical Method*, Number 9 (Part 1) in the Series Agronomy, American Society of Agronomy, Inc., Soil Science Society of America, Inc., Madison, WI, 771–796.

Haverkamp, R., Ross, P.J., Smettem, K.R.J., Parlange, J.Y. (1994) *Three Dimensional Analysis of Infiltration from Disc Infiltrometer*: 2. Physical based infiltration equation. Water Resources Res., 30, 2931–2935.

Hillel, D. (1998) *Environmental Soil Physics*, Academic Press, New York.

Hubbert, M.K. (1940) The Theory of Groundwater Motion, J. Geol., 785–944.

Klute, A. (1986) *Methods of Soil Analysis. Part I—Physical and Mineralogical Methods*, Soil Science Society of America Inc, Madison, WI.

Klute, A. and Dirksen, C. (1986) Hydraulic Conductivity and Diffusivity: Laboratory Methods. In Klute, A. (ed), *Methods of Soil Analysis. Part I-Physical and Mineralogical Method*, Number 9 (Part 1) in the Series Agronomy, American Society of Agronomy, Inc., Soil Science Society of America, Inc., Madison, WI, 687–732.

Leap, D.I. (1999) Geological Occurence of Groundwater. In Delleur, J. (ed.), CRC Press, Boca Raton, FL.

Leeper, G.W., and Uren, N.C. (1993) *Soil Science. An Introduction*. Melbourne University Press.

Or, D.R., and Wraith, J.M. (1998) *Agricultural and Environmental Soil Physics*, Course Notes Department of Plants, Soils, and Biometeorology, Utah State University, Logan, UT, and Land Resources and Environmental Sciences Department, Montana State University, Bozeman, MT.

Osborne, P.S. (1993) *Suggested Operating Procedures for Aquifer Pumping Tests*, EPA Ground Water Issue, EPA/540/S-93/503.

Reeve, R.C., Water Potential: Piezometry (1986), In Klute, A. (ed), *Methods of Soil Analysis. Part I-Physical and Mineralogical Method*, Number 9 (Part 1) in the Series Agronomy, American Society of Agronomy, Inc., Soil Science Society of America, Inc., Madison, WI, 545–560.

Roscoe Moss Company (1990) *Handbook of Ground Water Development*, John Wiley and Sons, New York.

Selker, J.S., Keller, C.K., and McCord, J.T. (1999) *Vadose Zone Processes*, Lewis Publishers, Boca Raton, FL.

Shepherd, R.G. (1989) *Correlation of Permeability and Grain Size*. Groundwater, 27(5), 633–638.

Smettem, K.R.J., Parlange, J.Y., Ross, P.J., and Haverkamp, R. (1994) *Three Dimensional Analysis of Infiltration from Disc Infiltrometer*: 1. Capillary based theory. Water Resources Res., 30, 2925–2929.

Soil Survey Staff, (1975) *Soil Taxonomy: A basic system of soil classification for making and interpreting soil surveys*. USDA-SCS Agric. Handb., U.S. Government Printing Office, Washington, DC.

Stephens, D.B. (1996) *Vadose Zone Hydrology*. Lewis Publishers, Boca Raton, FL.

Taylor, S.A., and Jackson, R.D. (1986) Temperature. In Klute, A. (ed), *Methods of Soil Analysis. Part I-Physical and Mineralogical Method*, Number 9 (Part 1) in the Series Agronomy, American

Society of Agronomy, Inc., Soil Science Society of America, Inc., Madison, WI, 927–939.

Thien, S. (1979) A flow diagram for teaching texture-by-feel analysis, *J. Agron. Educ.*, 8, 54–55.

Tindall, J.A., and Kunkel, J.R. (1999) *Unsaturated Zone Hydrology for Scientists and Engineers*, Prentice Hall, NJ.

Todd, D.K. (1959) *Groundwater Hydrology*, John Wiley and Sons, New York.

Topp, G.C., Davis, J.L., and Annan, A.P. (1980) *Electromagnetic Determination of Soil Water Content: Measurements in Coaxial Transmission Lines.* Water Resources Res., 16, 574–582.

Wilson, L.G. (1990) Methods for sampling fluids in the vadose zone. In Nielsen, D.M., and Johnson, A.I. (eds), *Ground Water and Vadose Zone Monitoring*, American Society for Testing and Materials, Philadelphia, 7–24.

Wilson, L.G., and Dorrance, D.W. (1995) Sampling from saturated regions of the vadose zone, In Wilson, L.G., Everett, L.G., and Cullen, S.J., (eds), *Handbook of Vadose Zone Characterization and Monitoring*, Lewis Press, Chelsea, MI.

Wilson, L.G., Dorrance, D.W., Bond, W.R., Everett, L.G., and Cullen, S.J. (1995) In situ pore liquid sampling in the vadose zone. In *Handbook of Vadose Zone Characterization and Monitoring*, Wilson, L.G., Everett, L.G., and Cullen, S.J., edts., Lewis Press, Chelsea, MI.

Wooding, R.A. (1968) *Steady Infiltration from a Shallow Circular Pond*, Water Resources Res., 4, 1259.

Yeh, T.-C.J., and Guzman-Guzman, A. (1995) Tensiometry, In *Handbook of Vadose Zone Characterization and Monitoring*, Wilson, L.G., Everett, L.G., and Cullen, S.J., eds., Lewis Publishers, Boca Raton, FL.

Young, S.C., Julian, H.E., Pearson, H.S., Molz, F.J., and Bornan G.K. (1998) *Electromagnetic Borehole Flowmeter*, EPA/600/SR–98/058.

13

ENVIRONMENTAL CHEMICAL PROPERTIES AND PROCESSES

J.F. ARTIOLA

The soil environment presents unique challenges for *in situ* measurements of its chemical properties and processes. The general chemical composition of minerals and other soil components can be estimated by examining color, texture, and reaction to acid. Soil scientists and geologists use their observational skills in the field to identify rocks and minerals, classify soils, and determine environmental factors such as water saturation and redox conditions. However, these observations are no substitute for more accurate and precise laboratory and field methods.

Most soil chemical processes occur in the presence of water, often referred to as the "universal solvent." Therefore water chemistry and soil chemistry are intricately intertwined. It is also very difficult to make direct chemical measurements or to study complex chemical processes in the field because key variables are often impossible to control. Nonetheless, there are several methods available to measure the following parameters: redox potential, pH, elemental composition, and salinity. This chapter presents an overview of the important chemical properties that can be measured directly in the soil environment. Additionally, important interrelated soil chemical processes are also reviewed. After each introductory section, the field methods and equipment necessary to measure these properties and processes are discussed in more detail.

SOIL CHEMICAL CONSTITUENTS

Soils consist of a complex mixture of solid (primary and secondary minerals, organic matter), solution (solvent water and dissolved constituents), and gas (soil atmosphere gases) phase constituents. Soil minerals include those inherited from the rock-parent material (e.g., quartz, feldspars, micas), plus weathered products thereof (e.g., clays, metal oxides, and carbonates). The determination of soil mineral composition is best done in a laboratory using advanced spectroscopic techniques that include X-ray diffraction for crystalline minerals, and infrared spectroscopy for noncrystalline minerals. Refer to standard soil chemical and mineralogical analysis textbooks, such as Sparks (1996) and Klute (1986), for a discussion of laboratory methods for soil chemical analyses. However, the determination of many total and soluble chemical constituents, which is of interest in environmental monitoring, can be accomplished reasonably well using field portable equipment.

SOIL ELEMENTAL COMPOSITION

The soil's elemental composition is similar to that of the underlying geological materials with two notable exceptions: carbon (C) and nitrogen (N) (Figure 13.1). These two elements are significantly more abundant in the soil environment, primarily due to the fixation of atmospheric CO_2 and N_2 gases into soil organic matter during biological activity. Additionally, the weathering processes that result from biological activity and direct contact with the atmosphere control the chemical abundance and forms of several elements. For example, cations such as Na^+, K^+, Ca^{++}, and Mg^{++} are significantly depleted in the soil environment due to dissolution and leaching processes, especially for regions that receive moderate to high precipitation. Selective dissolution

and recrystallization processes change the chemical composition of soils as exemplified by the formation of clay minerals and Fe, Al, and Mn oxides. Trace elements ($< 100 \, mg \, kg^{-1}$) can also be redistributed, enriched, or depleted from the soil environment relative to the parent mineral.

Recent anthropogenic activities that have generated vast quantities of mining, industrial, and municipal wastes have added a new dimension to the elemental cycles that affect the soil environment. Human activities are changing the composition of the soil environment with extensive manipulation, waste disposal, and applications of chemicals for agriculture. These activities have resulted in some "enrichment" of trace elements in soils, and in other cases have exacerbated the loss of elements from the soil environment. Of particular concern are the accumulation of certain trace metals in surface soils. For example, the use of leaded gasoline and lead smelter emissions have resulted in significant increases in lead (Pb) concentrations in urban and industrial soils. Extensive use of arsenic (As)–based defoliants, pesticides, and fungicides, as well as smelter emissions, has resulted in the accumulation of As in soils.

The chemical analysis of soil is inherently difficult, due in no small part to sample preparation requirements. Portable equipment and reagents are available for the measurement of soil elements and chemical species. However, these are mostly limited to water soluble-extractable species that are analyzed using chemical water analysis methods, as described in the "Soil Salinity—Total Dissolved Solids" section and Chapter 9. Elemental chemical composition of soil samples can also be accomplished in the field using portable spectroscopic instruments. Field techniques, however, cannot provide the levels of detection and reliability obtained by well-established laboratory methods that make use of thermal, neutron, proton, or x-ray induced atomic absorption and emission and mass spectroscopy (Sparks, 1996).

X-RAY FLUORESCENCE SPECTROSCOPY

High-energy (short wavelength) X-rays can be used for rapid, direct multi-elemental analysis of dry soil samples without extraction or digestion steps. This method offers significant cost- and time-saving advantages for field monitoring of the soil environment. X-ray fluorescence spectroscopy is based on the principle depicted in Figure 13.2. High-energy X-rays (4–25 kV) are used to irradiate solid samples, and the element-specific fluorescent X-ray emissions that result from K–L electron orbital transitions are collected and quantified using a sophisticated light

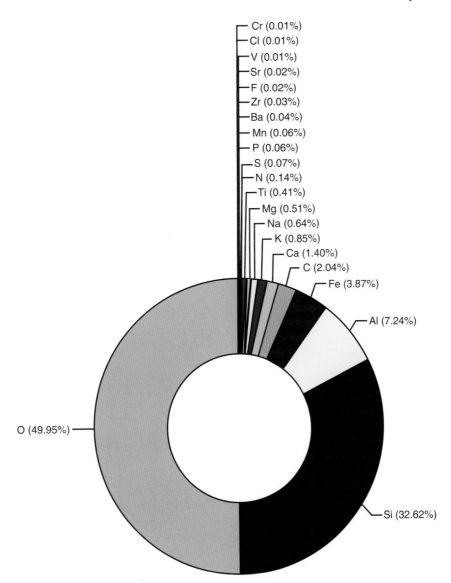

Cr (0.01%)
Cl (0.01%)
V (0.01%)
Sr (0.02%)
F (0.02%)
Zr (0.03%)
Ba (0.04%)
Mn (0.06%)
P (0.06%)
S (0.07%)
N (0.14%)
Ti (0.41%)
Mg (0.51%)
Na (0.64%)
K (0.85%)
Ca (1.40%)
C (2.04%)
Fe (3.87%)
Al (7.24%)
Si (32.62%)
O (49.95%)

FIGURE 13.1   The soil environment elemental composition: 21 most-abundant elements. (Graph by J.F. Artiola, data from Lindsay, 1979, Table 1.1.)

detector. The X-rays emanating from K orbit electrons are typically used for quantitative analysis, because they are more intense and of lower energy than L orbit X-rays. Because of the unique electron structure of each element, characteristic energy X-ray emissions can be used to identify and quantify each element. The range of elements accessible for measurement using field XRF units starts at atomic numbers 8–13 and includes about 80 elements. Unfortunately, complex soil matrixes, emission characteristics, and interferences preclude the use of this technique to quantify elements with atomic numbers <8. Another important limitation of this technique is its quantifiable detection limit, which is usually between 30 and 100 mg kg^{-1} for most elements. However, the upper detection limit range can extend beyond 1%

(10,000 mg kg^{-1}) for some elements. Note that lower levels of detection for elements below atomic number 8 can be achieved only with more sophisticated XRF laboratory units operated by experienced analysts.

Heterogeneous particle-size distributions and chemical compositions have a significant influence on the precision, accuracy, and detection limits of XFR spectrographs. The depth of penetration and scatter of the X-rays into the soil matrix, and the resulting fluorescence emissions, are affected by particle size and composition. Also in soils and sediments, different particle sizes usually do not have the same chemical compositions. Therefore at a minimum, soil sediments should be ground to between less than 50 and 100 μm to obtain a more homogeneous matrix with respect to size and

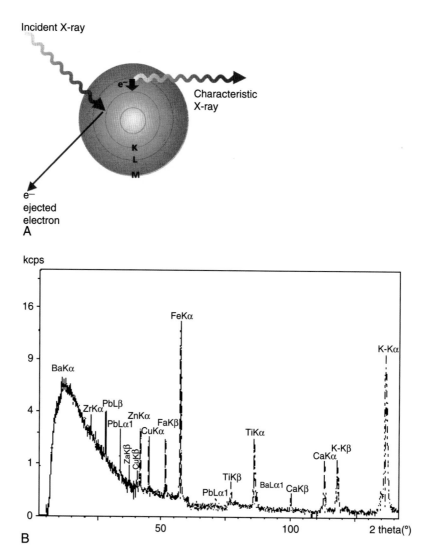

FIGURE 13.2 (**A**) X-ray fluorescence spectroscopy principle (source: company brochure). (**B**) multielement spectrum of a soil sample contaminated with Pb and Ba. (Source: www.pml.tno.nl/en/diac/Layer_2/xrf.html.)

composition. Calibration standards should also have similar particle size and chemical compositions to minimize light scatter errors. Increased sample homogeneity can be obtained by pressing powdered samples into pellets, fused disks, or thin films. However, these sample preparation techniques are not readily implemented in the field. Despite this, the technique can now be used for the routine field analysis of about 26 elements in soils and sediments (Method 6020 of USEPA [1998]). Overall, this method requires minimal sample preparation. Its simplicity, speed of analysis, linear range over four orders of magnitude, and portability make it very attractive for soil chemical analyses. A practical time- and cost-saving application of this technique is the *in situ* delineation of areas contaminated with high concentrations ($>100 \, mg \, kg^{-1}$) of hazardous metals such as Pb, Ni, and Cr.

SOIL SALINITY—TOTAL DISSOLVED SOLIDS

The salinity of the soil environment is controlled by water availability and the presence of soluble soil minerals. Mineral dissolution or formation processes are the indirect result of changes in the total amount and transient states of water within the soil environment. For example, in high-rainfall, humid, and temperate environments, soil minerals are usually dissolving. Soluble minerals (gypsum, calcite) are quickly removed from these environments, leaving behind decreasing amounts of less soluble minerals (feldspars, mica, quartz, and many other minerals). The selective dissolution plus subsequent removal of ions such as $Ca^{++}, Mg^{++}, K^+, Fe^{++}, Al^{3+}$, and Si^{4+} (called weathering) and recombination of some components within their structure can also change the nature of the remaining minerals. This, in turn, leads to the formation of clay minerals and metal oxides such as gibbsite and hematite. In arid environments minerals are seldom completely dissolved because water is present in relatively small quantities. In these soil systems, soluble minerals such as gypsum, calcite, and halite (NaCl) tend to accumulate rather than be removed. For a complete description of soil minerals, refer to Dixon and Weed (1989).

The measurement of total dissolved chemical species in the soil-water environment is very important for characterizing soil degradation and optimizing plant growth. Groundwater and surface water quality degradation are also typically tied to changes in chemical composition of soil-pore water. Several major chemical species that control soil-pore water quality are listed in order of importance in Table 13.1.

The data in Table 13.1 suggest that NaCl (table salt) has the highest potential for degradation of soil-pore water quality because of its high solubility. Thus in saline soil environments it is safe to assume that all the NaCl present will be dissolved in the pore water. Gypsum is a common mineral in soils and, unlike table salt, it has a relatively low solubility. Soils rich in gypsum seldom have dissolved solids much above 2000 mg L^{-1} unless NaCl is also present (see the "Water-Soluble Minerals—Ionic Strength" section). Other much less soluble but common minerals such as calcite, feldspars, and even quartz also add measurable concentrations of dissolved solids to the soil-pore water, but these vary greatly with soil pH and mineral types. Sodium carbonate and nitrate minerals can add dissolved solids in some alkaline-sodic soils, since these minerals are very soluble in water. In acid soils, Al oxides can also add measurable dissolved solids to the soil solution. However, in neutral to alkaline soil environments, total dissolved solid concentrations are usually much higher than in acid soils. This is usually attributed to the presence in these soils of NaCl, sulfate, and bicarbonate minerals that can account for more than 90% of the dissolved solids found in soil solution.

ELECTRICAL CONDUCTIVITY

It is possible to make indirect measurements of the soil-pore water quality using the electrical properties of the dissolved ionic solutes. Electrical conductivity (EC) measurements provide information only on solutes that are in the form of charged cations and anions because neutral soluble species do not conduct electricity in solution. For example, Ca^{++} and $SO_4^=$ ions form ion pairs (loosely attached cation-anion species in soil solutions in equilibrium with gypsum) such that as much as 30% of the dissolved Ca^{++} and $SO_4^=$ ions will not be measured by EC in alkaline soils. Other less prevailing neutral species that are found in soil solutions include $Si(OH)_4$, $MgSO_4$, H_2CO_3, and H_3BO_3.

Electrical conductivity results from the ability of charged dissolved ions to conduct electricity. When two electrodes connected to a battery are immersed in water (Figure 13.3), charged ions will migrate to oppositely charged poles, thereby inducing an electrical current. Resistance of the cell is measured with the flow of current. The total amount of electrical current that can be transferred is proportional to the size and separation of the electrodes, which is characterized as a constant (K) for a given cell. Thus the EC is related to the resistance (R) of a solution as follows:

$$EC = (1/R)K \qquad (Eq.\ 13.1)$$

where

EC = electrical conductivity (dS m^{-1})
R = electrical resistance (ohms)
K = electrode constant

Note that the units for EC are in deciSiemens m^{-1} (dS m^{-1}) or milliSiemens cm^{-1} (mS cm^{-1}) using SI units, but older instruments may still use mmhos cm^{-1}. A cell with 1 cm^2 surfaces that are 1 cm apart (volume = 1 cm^3) is assumed to have a constant (K) of 1.0. Also, EC is temperature dependent because the ionic activity of

TABLE 13.1

Salinity Controlling Minerals in the Soil Environment

Predominant Water Species at pH 7	Mineral(s)	Solubility (mg L^{-1}) @ 25° C
Na^+, Cl^-	Halite (NaCl)	~380,000
Ca^{+2}, SO_4^{-2}	Gypsum ($CaSO_4 * H_2O$)	~2400
$Ca^{+2} Mg^{+2}, CO_3^-$	Calcite ($CaCO_3$)	~15–100 (alkaline envir.)
	Dolomite ($CaMg[CO_3]_2$)	
Na^+, HCO_3^-, NO_3^-	$NaHCO_3, NaNO_3$	High >1000 (alkaline envir.)
$Si(OH)_4^0, Al^{+++}$, other cations	Quartz and other silica minerals (examples: feldspars, olivine, micas, allophane, zeolites, and clays)[a]	~40–>1000 (varies with pH)
H_3BO_3, HPO_4^{3-}	Boron and phosphorus	Variable <1–<10
$Al(OH)_3^0, Al(OH)_4^-$	Al oxides	Variable 100–500 (acid envir.)

[a]Note that silica minerals can release significant amounts of cations such as Ca^+, Mg^+, and K^+.

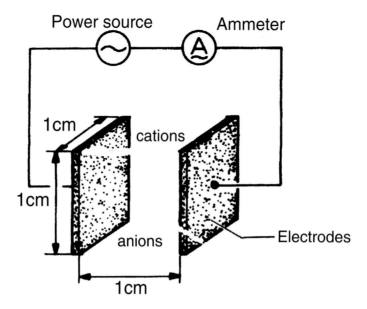

FIGURE 13.3   Electrical conductivity cell: principle of operation and detection of ionic (cations and anions) solutes.

most dissolved species typically increases with temperature. Therefore, by convention, EC is reported at 25° C.

Total dissolved solids (TDSs) for aqueous systems can be estimated using the following conversion factor:

$$TDS = EC * 640 \qquad (Eq. 13.2)$$

where

EC = electrical conductivity (dS m^{-1}) and
TDS = total dissolved solids (mg L^{-1})

The conversion factor is applicable to typical soil systems and natural waters, but the application of this conversion factor to extreme soil values of pH and/or salinity should be done with caution. Note that the electrical conductivity of a solution is not linear with increasing concentration. As explained before, ions form neutral ion pairs. Additionally, each ion has its own equivalent conductance ($\lambda^\circ +$, $\lambda^\circ -$). For example, the $\lambda^\circ +$ or $\lambda^\circ -$ for Cl$^-$, Na$^+$, 1/2SO$_4^=$, 1/2Ca^{2+}, K$^+$, HCO$_3^-$ ions are: 76.4, 50.1, 80, 59.5, 73.5, and 44.5 mho-cm^2 mole$_c^{-1}$, respectively, at 25° C and infinite ion dilution (Clesceri *et al.*, 1998). Note again that individual ion conductances vary with ionic strength (see the "Water-Soluble Materials—Ionic Strength" section) and temperature. However, $\lambda^\circ +$ and $\lambda^\circ -$ values are useful to estimate the conductivity of weak solutions such as natural waters with low concentrations of ions. If the concentration of each ion is known, the electrical conductivity (in mS cm^{-1}) will be the sum of the individual ion $\lambda^\circ +$ or $\lambda^\circ -$ times their respective concentrations (in mmol$_c$ L^{-1}). For example, the EC of a 1-mM KCl solution, assuming infinite dilution (ionic strength = 1),

would have a conductivity of $1(73.5 + 76.4) = 149.9$ mS cm^{-1}. However, if appropriate corrections to account for true ionic strength of the water are made to the ions $\lambda^\circ +$ and $\lambda^\circ -$ (see Clesceri *et al.*, 1998, Method 2510 A), the true electrical conductivity of this solution would be 146.9 mS cm^{-1}. Soil water extract electrical conductivity can be measured using field portable probes, as described in Chapter 9.

SOIL ELECTRICAL RESISTANCE

Direct field measurements of soil resistance can be accomplished by measuring the resistance to current flow between electrodes inserted into the ground. See Wilson *et al.* (1995) for a schematic of a four-electrode setup that can be used to estimate soil salinity with suitable background information. In other words, since resistivity is a function of soil mineralogy, particle-size distribution, water content, and salinity, ER data taken from a similar but salt-free soil system are needed to estimate soil salinity. Soil resistance R is very much dependent on the electrode spacing and geometry. The Wenner array geometry allows for the simplification of a complex relationship between soil-water content and resistance, and its design array eliminates electrode-soil contact resistance. Thus the soil electrical resistance can be estimated from the following equation:

$$R = (2\pi SV)/I \qquad (Eq. 13.3)$$

where

R = electrical resistance (ohms m^{-1})

S = electrode spacing (m)
I = current applied (A)
V = measured voltage (V)

As previously stated, resistance depends not only on water and salinity but also on the types of minerals present. Therefore this technique may be useful to measure relative changes in soil-pore water quality in soil volumes with similar soil physical properties. For further details and an extended discussion on this technique, see Wilson *et al.*, 1995, and LaGrega *et al.*, 1994.

SOIL WATER EXTRACTIONS AND CHEMICAL ANALYSIS

There are numerous field procedures available for quasi *in situ* measurement of water soluble or extractable soil chemical species. These techniques are usually derived from well-established laboratory procedures that have been adapted for field use. Table 13.2 lists the tests that can be conducted in the field using portable kits. Soluble chemical fractions vary depending on the extracting agent used. For example, a 0.01 M $MgCl_2$ solution may extract only the exchangeable ions. In contrast, weak organic acids such as acetic and oxalic acids will extract carbonates and poorly crystalline metal hydroxides, and strong acids such as nitric acid will extract crystalline metal oxides and oxidize most organic matter. Thus many of these kits

provide the user with extracting solutions designed to dissolve, extract, and/or react with chemicals from soil minerals. Included in these kits are premixed, premeasured reagents, dedicated glassware, and precalibrated colorimeters. Often these methods have limited range and accuracy, and only provide semiquantitative data, such as "high," "low," "present," or "absent." Thus it is highly desirable to check the accuracy and precision of each kit against standard laboratory equipment and methods.

Caution should be exercised when using these kits to monitor environmental soil quality, because they often lack the precision and/or detection limits required by regulatory agencies. In addition, they are slow due to lack of automation. The major advantages of these kits are portability and simplicity of the protocols.

MISCELLANEOUS FIELD METHODS

IMMUNOASSAYS

Portable biological tests or immunoassays to detect chemicals in soil or sediment samples are commercially available. They offer portability and speed, making them well suited for *in situ* monitoring of soil contaminants. Some tests have been developed for screening soils contaminated with organic chemicals such as PCBs and hydrocarbons. These portable immunoassays do not

TABLE 13.2
Field Tests Methods for Soil Chemical Analyses

Soil Test	Parameter	Typical Range[b]	Method(s)[c]
pH, total dissolved solids	pH, EC	0–14, 0.2–10 dS m^{-1}	extraction + electrodes
Phosphorus[a]	PO_4-P	variable (0–50 mg k^{-1})	extraction + colorimetry
Nitrate	NO_3-N	variable (0–60 mg k^{-1})	extraction + colorimetry
Ammonia	NH_3-N	variable (0–50 mg k^{-1})	extraction + colorimetry
Potassium[a]	K^+	variable (0–600 mg k^{-1})	extraction + colorimetry
Sulfate[a]	$SO_4^=$	0–100 mg kg^{-1}	extraction + colorimetry
Calcium carbonate	$CaCO_3$	0–25%	titration
Cation exchange capacity and exchangeable acidity base saturation	cmol kg^{-1}	0–?	All: multiple extractions + colorimetry
	%	0–100	
Lime requirement	acidity	cmol kg^{-1}	exchangeable acidity
Gypsum requirement	alkalinity	0–15 tons acre^{-1}	extraction + colorimetry
PCBs, TPHs, pesticides, explosives	see Table 13.3	see Table 13.3	Both: extraction + immunoassay (see "Miscellaneous Field Methods" section) + colorimetry
Organic matter	O.M.	1–5%	digestion + gravimetry
Organic matter		1–16%	digestion + titration
Humus	Low-high	variable	extraction + colorimetry
Cu, Fe, B, Co, Zn, Ca, Mg[a]	mg kg^{-1}	variable (0–25 mg kg^{-1})	extraction + colorimetry

Sources: Hach Company, 1998; LaMotte, 1998.
PCBs, Polychlorinated byphenyls; TPHs, total petroleum hydrocarbons.
[a]Plant available forms only.
[b]Range may vary by manufacturer and method.
[c]For a complete description of extraction and colorimetric procedures, consult the references provided by manufacturers and standard textbooks on methods of soil and water chemical analysis, such as Sparks (1996) and Clesceri *et al.* (1998).

provide the precision, accuracy, and detection limits available with laboratory chemical analytical techniques. Additionally, immunoassays can produce false-positive or false-negative results, depending on the sample matrix. However, immunoassays offer a distinct advantage for screening large numbers of soil samples over short periods.

Immunoassay tests use antibodies to detect a target chemical substance. Typically, a specific antibody molecule binds to the target chemical or antigen. A signal generating or reporter reagent is added to the mixture of antibody and target chemicals. The resulting competitive binding among the three chemicals is related to the concentration of the target chemical present. Because the reporter usually generates a light-producing chemical that changes in intensity, the reaction can be used to detect the chemical of interest. (For further discussion on the biological processes involved in immunoassays, see USEPA SW486, Method 4000.) The most common type of immunoassay is the single point with "positive/negative" results, which is useful for screening large numbers of samples. Some special concerns related to immunoassay reagents include careful adherence to operating temperature ranges and shelf-life. Additionally, immunoassays typically must be done on particulate-free soil solvent extracts. Solvent extraction and filtration often limits the portability and speed of these tests in the field. Table 13.3 lists several contaminants for which there are commercially available immunoassays.

ELECTROMAGNETIC CONDUCTIVITY (EM)

Electromagnetic radiation waves can be used to map the extent of a contaminant plume in the soil environ-

TABLE 13.3
Commercially Available Immunoassay Tests for Screening Pollutants

Target Analyte	Concentration Limits (mg kg^{-1})	Results[a]
Pentachlorophenol	0, 5, 10, or 100	P/N
PCBs	5, 10, or 50	P/N
2,4-D	0.1, 0.5, 1, 5	P/N
Total petroleum	< 40–100, >1000	P/N
Hydrocarbons	5, 25, 100, 500	P/N
PAHs	1	P/N
Chlordane	20, 100, 600	P/N
DDT	0.2, 1, 10	P/N
RDX	0.5	P/N
TNT explosives	0.5	P/N
Toxaphene	>0.5	P/N
Atrazines	—	—

Adapted from USEPA SW486 (1996).
[a]Indicates presence (positive) or absence (negative) of the chemical.

ment with high salt concentrations. Electromagnetic pulses are sent into the ground through an inserted metal probe. These EM waves induce secondary electromagnetic waves that are collected at a distance from the transmitter by another probe, and subsequently amplified and analyzed. A complex mathematical analysis of these EM waves is used to correlate them with soil moisture content and salinity. This process works well to measure soil salinity when other influential factors such as mineral composition and porosity are accounted for. Portable EM equipment offers the distinct advantage of being able to map out soil volumes affected with salts without disturbing the soil profile. This technique affords speed and portability, but lacks precision and sensitivity. Additionally, proper use requires precalibration using uncontaminated soil profiles. (For further details and extended discussion on this technique, see Wilson *et al.* [1995].)

REMOTE SENSING INFRARED SPECTROSCOPY (IR)

The future of soil and water chemical analytical techniques looks promising with the development of new nondestructive techniques that can correlate soil chemical composition with reflected visible and infrared light (see also Chapter 11). These methods usually make use of a passive or active energy source (sun or IR beam) and measure the resulting reflected light that is then subsequently correlated to varying concentrations of chemicals. For example, changes in soil value (color lightness), measured by a handheld color meter, can be correlated to the concentrations of organic matter or iron oxides. Newer methods include the use of IR sources and sensors that can measure the IR absorption spectrum of soil surfaces. The presence of absorbed molecules such as NH_4^+ and NO_3^- ions can be detected by reflected IR because these molecules absorb IR light in different regions of the IR spectrum. Similarly, organic matter and iron oxides absorb IR at different regions, depending on their chemical structures. However, the IR sorption light spectra are affected by other soil chemical and physical properties. So far this technique has been shown to produce acceptable quantitative soil chemical data using calibrations from soils with similar chemical characteristics and particle-size distributions.

SOIL CHEMICAL PROCESSES

The soil environment is formed and maintained through a series of chemical reactions that change the nature and properties of exposed geological materials. Soil chemical transformations obey fundamental chemical reaction principles. A short review of chemical equilibrium principles is presented in the following section. Subsequently, examples of several fundamental processes that are best illustrated with equilibrium chemical reactions will be presented. Finally, methods to measure key variables that affect these processes will be presented and discussed.

CHEMICAL REACTIONS AND EQUILIBRIUM RELATIONSHIPS

All chemical reactions involve the transfer of energy. A reaction constant is defined as the quotient (Q) of products and reactant activities as follows:

$$aA + bB <> cC + dD \qquad \text{(Eq. 13.4)}$$

where

$$Q = (C)^c * (D)^d / (A)^a * (B)^b \qquad \text{(Eq. 13.5)}$$

The brackets denote chemical activities of reactants and products and the Q is equivalent to the equilibrium constant K°, when the reaction is at equilibrium. This equilibrium constant is independent of the ionic strength of the system.

The Standard Gibbs Energy Change (ΔG^0_r) of a reaction, such as the one previously described, is defined by the sum of the standard Gibbs energies of formation of the products minus that of the reactants. The energies of formation of the participating species are known and are tabulated in chemistry references. More specifically then, we can equate the overall change in energy of a reaction under standard conditions as follows:

$$\Delta fG^0_r = \Sigma(\Delta G^0_{f\ prods}) - \sigma(\Delta G^0_{f\ prods}) \qquad \text{(Eq. 13.6)}$$

For a system not at standard state, the Gibbs free energy of reaction is equal to the standard free energy plus the energy of the reaction:

$$\Delta_r = \Delta G^0_r + RT \ln Q \qquad \text{(Eq. 13.7)}$$

However, as the reaction approaches equilibrium the Gibbs free energy $\simeq 0$ and $Q \simeq K^\circ$, as previously shown. Therefore we can simplify the above equation as follows:

$$\Delta G^0_r = -RT \ln K^0 \qquad \text{(Eq. 13.8)}$$

where R = the gas constant and T = absolute temperature.

Furthermore, when nonstandard conditions exist, the aforemention equations become:

$$\Delta G_r = -RT \ln K \qquad \text{(Eq. 13.9)}$$

If the ΔG_r is negative, energy is released and we can assume that the reaction will occur spontaneously, but it is not known how fast. Conversely, reactions with positive ΔG are unlikely to occur without energy inputs. Indeed they are expected to proceed spontaneously in the reverse direction. However, some reactions may also occur even if they require heat energy (ΔH positive), if entropy (randomness) changes make $\Delta G_r < 0$ overall.

Many soil chemical reactions involve electron (e^-) loss (oxidation) or gains (reduction) to and from elements or molecules as well as the loss or gain of protons (H^+). The energy change of redox reactions can be calculated if the e^- potentials and the number of e^- and H^+ ions involved in the reaction are known. At or near equilibrium the energy change of any chemical reaction (Gibbs free energy at standard conditions) is related to the voltage of a reaction by the following equation:

$$\Delta G^0_r = -nFE^0 \qquad \text{(Eq. 13.10)}$$

where
  n = number of e^- involved in the reaction
  E^0 = standard potential of the reaction (volts)
  F = Faraday Constant (23.1 kcal volt-gram-equivalent^{-1})

As was the case with Equation 13.8, Equation 13.10 is also defined for nonstandard conditions as follows:

$$\Delta G_r = -nFE \qquad \text{(Eq. 13.11)}$$

It is also known from laws of thermodynamics that the standard free energy change of any chemical reaction can be expressed in terms of products and reactants, as shown in Equation 13.6. However, at equilibrium $\Delta G_r = 0$; thus combining Equations 13.6 and 13.10, substituting standard state *constants* R (1.987 cal degKelvin-mole^{-1}), T (298.15° *Kelvin* = 25° C) and F, and converting from natural ln to log base 10 values, we obtain:

$$E^0(mv) = (59.2/n) \log K^0 \qquad \text{(Eq. 13.12)}$$

With this relationship chemical redox potentials can be converted to chemical equilibrium constants and vice versa. Note that under nonstandard state conditions Equation 13.12 becomes the Nernst equation as follows, assuming a general reduction reaction:

$$M^{n+} \text{ (oxidized species)} + ne^- \qquad \text{(Eq. 13.13)}$$
$$\leftrightarrow M - \text{ (reduced species)}$$

then

$$E = E^0 - (59.2/n)\log K^0 \qquad \text{(Eq. 13.14)}$$

where

E = electrode potential (mv)

E^0 = standard potential of the reaction (mv)

However, most redox reactions occurring in the soil-water environment involve the gain or loss of hydrogen ions (H^+). Therefore the left side often includes a term to account for the activity of H^+ ions. Now, if we add H^+ ions to Equation 13.4:

$$M^{n+} \text{ (oxidized species)} + ne^- + mH^+ \qquad \text{(Eq. 13.15)}$$
$$\leftrightarrow M^\circ \text{ (reduced species)}$$

which can be written as Equation 13.5 in the following form:

$$K^0 = (M^0)/[(M^{n+})(e^-)^n(H^+)^m] \qquad \text{(Eq. 13.16)}$$

Then Equation 13.14 becomes:

$$E = E^0 - (59.2/n)\log K^0 \qquad \text{(Eq. 13.17)}$$
$$- m(59.2/n)\log H^+$$

This equation relates the electrochemical (redox electrode) potential measurement to equilibrium constants of chemical species and to the pH of the soil-water system. The importance of redox potential and pH, also called *master variables*, is discussed in the following sections.

FUNDAMENTAL SOIL CHEMICAL PROCESSES

There are numerous chemical reactions that can occur in the soil environment. These reactions are at the core of soil chemical processes that control the formation and maintenance of healthy soil chemical systems. Conversely, many of these processes and accompanying chemical reactions, if left unchecked, can act as catalysts for the pollution and degradation of the soil environment. It is beyond the scope of this chapter to present and discuss all known soil chemical reactions and processes. However, those listed in Box 13.1 represent the most important general soil chemical processes.

The most important soil factors that control these processes will be discussed in the following sections.

FACTORS CONTROLLING SOIL CHEMICAL PROCESSES

OXYGEN

Molecular oxygen is the principal electron acceptor species that drives oxidation reactions in the soil-water environment (see Box 13.1, reactions for oxidation-reduction and organic matter transformations). Its presence or absence determines the chemical composition of predominant minerals and the potential for oxidation/reduction reactions in the soil environment. For example, aerobic microorganisms depend on the presence of ample supplies of dissolved oxygen in soil-water to degrade natural organic matter and organic pollutants into carbon dioxide. The rate of biochemical oxidation in the soil environment is partly dependent on the ability of oxygen gas to diffuse from the atmosphere into the soil-pore water. Under water-saturated conditions, aerobic microorganisms including heterotrophic bacteria can, in just a few hours, utilize all dissolved oxygen. If oxygen is not present in the soil environment, microorganisms that can use alternate e^- acceptors become active and start to utilize nitrogen (NO_3^-) iron (Fe^{3+}), sulfur ($SO_4^=$), and even carbon dioxide (CO_2) as electron acceptors to degrade organic materials. This process leads to the formation of gases such as nitrous oxide, dinitrogen, hydrogen sulfide, and methane in the soil environment. Therefore an indirect measurement of O_2 presence in soils can be obtained by measuring soil-water redox potentials (see the "References and Suggested Readings" section in Chapter 15). A direct measurement of oxygen in water can be obtained in the field using portable oxygen probes (see Chapter 9).

MICROBIAL ACTIVITY

Soil microbes, and bacteria in particular, mediate redox biochemical reactions, including organic matter transformations. Bacteria facilitate the transfer of electrons by lowering the energies of activation (E^*) of chemical reactions. Bacteria make use of chemical terminal electron acceptors such as O_2, NO_3^-, and $SO_4^=$ ions and, in the process, catalyze soil chemical reactions. Many soil chemical reactions are thermodynamically favorable, but would proceed very slowly in the absence of bacteria. Without bacteria, organic matter and organic pollutants would experience very slow transformation rates and accumulate in the environment. Thus carbon, nitrogen, and phosphorous nutrients would not proceed significantly through their respective transformation cycles.

BOX 13.1 *Fundamental Soil Chemical Processes and Example Chemical Reactions*

Oxidation-reduction (redox): Electrons are gained or lost from elements/molecules. Example: soluble Iron(II) is oxidized to Fe(III) forming solid iron(III) hydroxides:

$$Fe^{2+} + 1/4O_2 + H^+ >>> Fe^{3+} + 1/2H_2O$$
$$+ 2.5H_2O >>> Fe(OH)_3 + 3H^+$$

Ion substitution and ion exchange: Charged $(+)$ or $(-)$ ions are sorbed onto surfaces of minerals and replaced inside the structures of soil minerals. Examples: Aluminum substitution for a silica (isomorphous substitution) during the formation of clay minerals and sodium for calcium exchange in clay minerals

(A) $(SiO_4)^0 + Al^{+3} >> (SiAlO_4)^- + Si^{4+}$

(B) $2(SiAlO_4)^- 2Na^+ Ca^{2+} >> 2(SiAlO_4)^- Ca^{2+} + 2Na^+$

Note: parenthesis indicates silica-based mineral.

Dissolution and precipitation: Soil minerals and organic matter can dissolved in water and re-precipitate in similar or new forms upon drying. Example: Calcium carbonate (calcite) precipitation in soils.

Water containing Ca^{2+} and $CO_3^=$ ions $>>$

Soil irrigation $>>>$ Water evaporation $>>>$

Soil $+$ calcite mineral

Organic matter transformations: Decaying plant and biological materials decompose into basic components and transform themselves into more resilient soil organic matter (humus). These biologically catalyzed chemical transformations include breakdown (bond cleavage), oxidation (partial), reduction (partial), and polymerization (bond formation) reactions. Example:

Lignin and cellulose $+$ water $+$ nutrients $+$

soil microbes $>>>$ Oxidation, $CO_2 >>$

Breakdown/metabolism $+$ synthesis $>>>>$

Soil humus

Hydrolysis: Aluminum, like all other ions, is quickly surrounded by water molecules (six) when placed in water. If one or more of the water molecules dissociates (due to pH changes), then hydrolysis has occurred leading to the formation of more stable Al-OH$^-$ bonds and the release of H$^+$ ions into the water. The following equilibrium reactions describe this process:

$$Al^{3+} + 6H_2O >> Al(H_2O)_6^{3+} >$$
$$Al(H_2O)_5(OH)^{2+} + H^+ \dots Al(H_2O)$$
$$(OH)_5^{2-} + 5H^+$$

Complexation: The reaction between a cation and one or more anions is very important in soil systems. Metal complexes are stable species that are less likely to participate in sorption, precipitation, and even redox reactions. For example, following hydrolysis reactions when exposed to water a Cd^{++} ion may be quickly adsorbed to clay minerals at neutral pH. However, in the presence of Cl$^-$ ions Cd will shed weakly bonded OH$^-$ ions and react with one or more Cl$^-$ ions forming a much stronger (stable) water soluble complex. This new Cd species may not be readily sorbed by clay minerals:

$$Cd(H_2O)_5(OH)^+ \text{ weak complex } + Cl^- >>$$
$$Cd(H_2O)_5(Cl)^+ \text{ stable complex } + OH^-$$

Surface catalyzed abiotic transformations: Mineral surfaces (in particular metal oxides) can act as a catalyst and facilitate the electron transfer to and from chemical species in solution. Often these reactions are slow but important in pedogenic (soil-forming) processes. However, some can be accelerated with the addition of chemicals. For example, the oxidation of trichloroethene (TCE) pollutant in groundwater can be facilitated with the additions of reactive iron (fillings) to obtain the following reaction:

$$TCE + Fe^0 >> CO_2 + Fe^{+++}$$
$$+ 3H_2O >>> Fe(OH)_3 \text{ (precipitate)}$$

As an example, the overall process of nitrification can be summarized as follows:

$$NH_3 + 3/2O_2 +$$
$$(\textit{Nitrosomonas \& Nitrobacter}) >>> NO_3^- \quad \text{(Eq. 13.18)}$$

These aerobic heterotrophic soil bacteria not only catalyze the reaction but also grow using the energy from this favorable process ($-353\,kJ\,mol^{-1}$). However, under water-saturated conditions and in the absence of O_2, this reaction can be reversed by another group of bacteria that can utilize nitrate as a terminal electron acceptor rather than O_2. Facultative anaerobic bacteria such as *Thiobacillus denitrificans* catalyze the following soil chemical reaction:

$$S + 2NO_3^- >> SO_4^= + N_2 + O_2 \qquad \text{(Eq. 13.19)}$$

This reaction is also energetically favorable (-280 kJ mol^{-1}).

Chapter 14 provides a description of soil microbes and methods of characterization that are available.

WATER

Water, the universal solvent, is at the heart of all soil chemical reactions. It provides all necessary elements to catalyze chemical reactions in the soil environment:

- Water facilitates the supply and removal (mass transfer) of reactants and products. Examples are ions, acid, base and neutral molecules, O_2, and CO_2.
- Water itself (H_2O, H^+, OH^-) is a reactant and is often one of the end-products of soil chemical reactions.
- Water supports life; consequently, all microorganisms involved in biochemical soil reactions only thrive when there is water in the soil environment.

There are numerous techniques available to measure water in the soil environment, discussed in more detail in Chapters 9 and 12.

pH

The activity of hydrogen ions (as measured by pH) is of particular relevance to soil chemical reactions, processes, and soil-water chemical composition. Hydrogen ions are involved in the majority of chemical reactions in the soil environment. The rates of numerous dissolution/pre-cipitation and redox reactions are affected by the presence or absence of H^+ ions. For example, the solubility of most metal hydroxides increases with soil acidity. A simple equilibrium formula for slightly soluble solids may be written as follows:

$$(MX)_{solid} <-> M + X \qquad \text{(Eq. 13.20)}$$

where M and X are water soluble species.

Note that this equation is similar to Equation 13.5. Therefore a solubility product expression can be written as follows:

$$K_{sp} = [M][X] \qquad \text{(Eq. 13.21)}$$

where K_{sp} called a solubility product constant.

If we define the molar solubility of each component as S, then the solubility product expression can be simplified as follows:

$$K_{sp} = S * S = S^2 \qquad \text{(Eq. 13.22)}$$

Note that a more general form of this solubility product expression is needed when the number of moles for each component differs, as in the generalized case of M_aX_b solid. Thus in case of the solid amorphous iron oxides commonly found in soils, their solubilities in water can be summarized by the following equilibrium reaction, which is the sum of Fe oxide dissolution and the dissociation of water:

$$Fe(OH)_3 = Fe^{3+} + 3OH^- \qquad \text{(Eq. 13.23)}$$

Assuming that the activity of the iron oxide solid is $\simeq 1$, then Equation 13.23 can be written in terms of solubility product expression:

$$[Fe^{3+}][OH^-]^3 = K_s^0 \simeq 10^{-38} \qquad \text{(Eq. 13.24)}$$

Also, assuming that the concentration of all species in solution are the same $(Fe^{3+} \simeq OH^-) = nS$, where S is the solubility and n is the number of moles of each species, Equation 13.24 can be simplified to:

$$S * (3S)^3 = 10^{-38} \qquad \text{(Eq. 13.25)}$$

If the activity of OH^- is decreased ten-fold (one pH unit), Equation 13.25 indicates that the activity of Fe^{+++} species in the soil-water environment will increase by 10^3 fold because the right side of the equation cannot change in value (Figure 13.4).

A more general relationship to estimate the solubility of most metal hydroxides and oxides as a function of pH can be derived by combining the equilibrium reaction for metal oxides or hydroxides and acid:

$$M(OH)_n + nH^+ = M^{n+} + nH_2O \qquad \text{(Eq. 13.26)}$$

and metal and water stability constants, K_s and $K_w = 14$, respectively, to yield the general relationship:

$$\log[Me^{z+}] = \log K_s + z(14) - z(pH) \qquad \text{(Eq. 13.27)}$$

where n is also equal to z, which is the charge of the metal ion, and pH is the soil acidity.

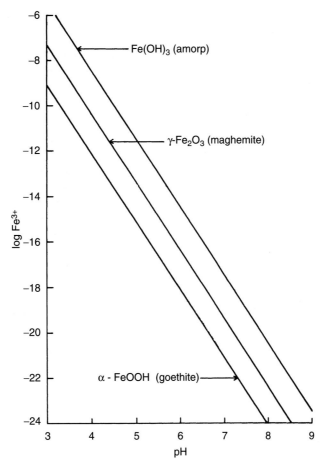

FIGURE 13.4   Solubility (Fe^{3+} activity) of iron oxides as affected by soil pH. (Adapted from Lindsay, 1979, Fig. 10.1.)

Additionally, most redox reactions are also dependent on the activities of H^+ ions. The "Measuring pH" section provides a full description of the principles of pH measurement and a detailed discussion of this measurement.

TEMPERATURE

Heat affects soil chemical processes in two fundamental ways. First, the rate of chemical reactions is dependent on temperature (T) by the following relationship that relates temperature and reaction constants (R):

$$k = Ae^{-E*/RT} \text{ (Arrhenius Equation)} \qquad \text{(Eq. 13.28)}$$

where A is a frequency factor and E^* is the activation energy of the reaction. This is true for both abiotic and biotic chemical reactions. Second, temperature affects the activity of soil organisms involved in chemical reactions. Most soil organisms thrive in moderate soil temperatures. However, the overall chemical reaction rates may depend on the predominance of microbes that are adapted to a given soil temperature (see Chapter

14). The measurement of soil-water temperature in the field can be accomplished with portable thermometers or thermocouple sensor connected to portable dataloggers, (see Chapter 4).

WATER-SOLUBLE MINERALS—IONIC STRENGTH

All minerals, plants, and animal matter dissolve in water to some degree, depending on their ability to interact with water molecules. Chemicals must be dissolved in water to participate in the chemical reactions previously described. The total amount of ions in the soil that are dissolved in water can be measured using individual chemical analyses (see the "Naturally Occurring Particulates" section in Chapter 15). However, an estimate of the total dissolved solids (ions) in solutions can be obtained using electrical conductivity (see the "Electrical Conductivity" section).

Most chemical analyses determine the concentrations of specific chemical species dissolved in water. In studying soil-water chemical reactions, chemical concentrations should not be confused with the activities of ions in the soil solution, as only "active" ions participate in chemical reactions. The activity of a chemical species is related to their concentration in solution by the following equation:

$$\text{Activity } a_i = \text{Concentration } c_i * \text{Activity coefficient } \gamma_i \quad \text{(Eq. 13-29)}$$

and the activity coefficient of an aqueous solution is related to the number of ions present and their charges (Z_i) using the simple Debye-Huckel equation:

$$\log \gamma_i = -A\, Z_i^2\, \mu^{1/2} \quad \text{(Eq. 13.30)}$$

where $\mu = 1/2 \sum c_i Z_i^2$, also known as the ionic strength of a solution. Equation 13.30 shows that as more ions are present in a soil-water system, the activity of these ions decreases. This implies that in saline soil-water the concentrations of species in equilibrium with a solid can increase, because at a constant concentration the activities of ions involved in dissolution reactions are reduced in the presence of other ions. For example, the solubilities of common soil minerals such as calcite and gypsum, expressed in terms of species concentrations, increase significantly in the presence of large concentrations of soluble ions such as NaCl.

RATES OF CHEMICAL PROCESSES IN THE SOIL ENVIRONMENT

The rates of soil chemical reactions are influenced by two major factors, energy and species transport, or mass transfer, processes. In general, exothermic (heat-releasing) reactions proceed faster than endothermic (heat-consuming) reactions in the soil environment. Chemical processes catalyzed by microorganisms are generally exothermic because microbes need energy to grow. Abiotic processes such as mineral dissolution and ion exchange reactions can either release or require heat energy, making their reaction rates more variable. Chemical reactions will not proceed or will proceed slowly if their activation energy (E^*) is very high. Chemical reaction rates generally increase with increasing temperature. In the soil environment many reactions can proceed faster in the presence of microbes as many microbes become more active at moderate to warm temperatures.

The soil environment is a complex system of solids, water, and gases that often limit the movement of reactants and products to and from the reaction sites. Soil particle size and geometry and crystal structure are important in determining the rates of dissolution of soil minerals. In many cases rates of dissolution are controlled by the number of ions that can diffuse from the surface of the crystal to the water solution interface. In other instances the reactions may be controlled by surface reactions limitations. Often the apparent solubilities and rates of mineral dissolution (e.g., SiO_2 based minerals) can be increased significantly by reducing their particle size.

The movement of chemicals via diffusion through all soil media, and the advection or water flux through soil pores (see Chapter 12), often influence soil chemical reaction rates. Therefore the rates of soil chemical reactions are often reported as "apparent" rates and are only applicable to a specific soil system and conditions. There are numerous examples of soil chemical reactions rates that are limited by transport processes. These include oxidation, mineral dissolution, and organic sorption-desorption reactions. For example, the rate of oxidation of soil organic matter is very much dependent on the amount of water present in the soil. Whereas water itself favors chemical reactions, it also acts as a barrier for oxygen molecules to diffuse into the soil and act as an electron acceptor. Oxygen dissolves sparingly in water (<9 mg l^{-1}), and under water saturated conditions, once O_2 is exhausted, it is difficult to replenish it because its diffusion coefficient in water is about four orders of magnitude smaller than in air. In this case the microbial oxidation of organic carbon is said to be oxygen diffusion rate limited.

In summary, the rates of soil chemical reactions vary tremendously, and no soil system is ever completely at equilibrium. Thus cation exchange reactions can take from a few microseconds to hours to reach equilibrium. At the other extreme, clay mineral formation (crystallization) processes may take thousands of years to be completed (Figure 13.5).

SOIL-WATER REDOX POTENTIAL AND pH

OXIDATION-REDUCTION REACTIONS

Electron transfer reactions are chemical reactions involving electrons (oxidation-reduction), and are very common in the soil environment. Most of these reactions are biologically catalyzed, and are an active part of the growth and decay of microbial soil populations. However, the immediate manifestations of these redox reactions are changes in biochemical species that involve the natural cycling of the macroelements (O, C, N, S), and many microelements (Fe, Mn, Cr, Hg, Se, As). Redox reactions are the essence of energy transfer pathways that

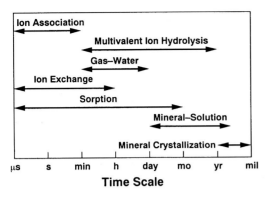

FIGURE 13.5 Soil reaction equilibrium times. Longer times mean slower reaction rates. (From Sparks, D.L. [1995] Environmental Soil Chemistry. Academic Press, San Diego.)

ultimately sustain life. The best illustration, and perhaps the most important pair of redox reactions in nature, are found in photosynthetic processes with O, H, and C elements, as shown in Box 13.2.

In Box 13.2 the oxidation (a) and reduction (b) reactions transfer energy from the sun and convert it into chemical energy by moving electrons from water (oxygen atom) to the carbon atom. Carbon atoms change from an oxidation state of +4 to an oxidation state of 0, thereby storing energy. And O^{2-} atoms (from H_2O) become reduced to O^0 (as O_2).

Electron transfer potentials depend on the chemical species involved in the reaction. For example, the poten-

tial of two electrons leaving metal Fe^0 (to form Fe^{2+}) is lower (+440 mV) than the potential of two electrons leaving Mg^0 (to form Mg^{2+}), which is +2350 mV. Both reactions are favorable in that they will release energy during oxidation. However, to reverse the oxidation process, Mg^0 electrons could be used to reduce Fe^{+2} to Fe^0, but Fe^0 electrons could not be used to reduce Mg^{+2} to Mg^0, because Fe-derived electrons do not have sufficient potential.

The potentials of oxidation or reduction reactions are determined by assuming (by convention) that the electrons involved in the reduction or oxidation of $2H^+$ atoms to form elemental H_2 and vice versa have a potential of 0 mV. All other redox reactions potentials are measured against this baseline. This is the half-cell hydrogen redox reference potential used to measure all other redox potentials, reported as Standard Electrode Potentials. Figure 13.6 shows a diagram of a classical electrochemical cell design, needed to measure a chemical reaction potential. The oxidation of hydrogen gas occurs in the left side (half cell). Note that the platinum wire is not involved in the redox reaction, just e^- transfer and the salt bridge (KCl) is needed to maintain electrical neutrality in each half cell. This figure shows that the electrochemical potential of Cu^{+2} reduction to Cu^0 is +0.34 V. That is, electrons with a potential equal to 340 mV are needed to reduce Cu^{+2}. Note that half-cell reactions can be written as reduction ($M^+ + e^- = M$) (American convention) or as oxidation reactions ($M = M^+ + e^-$) (European convention). Using either convention, electrochemical "series" can be established for various elements. For example, Au, Pt, and Ag are noble metals that are more difficult to oxidize (easier to reduce), and Cu, Fe and Zn are easier to oxidize (more difficult to reduce).

BOX 13.2 *The Photosynthetic Process*

Water disproportionation by sunlight yields:

$$1/2 H_2O + energy(light) \Rightarrow$$
$$1/4 O_2 + H^+ + e^- \quad\quad (a)$$

In the first half of the reaction *hydrogen is oxidized*. The second half of the reaction must therefore involve a reduction. Thus the photosynthesis process is completed by the reduction of carbon dioxide to formaldehyde as follows:

$$1/4 CO_2 + H^+ + e^- \Rightarrow$$
$$1/4 CH_2O + 1/4 H_2O \quad\quad (b)$$

Adding reactions (a) plus (b) and multiplying all sides by 4 (note that H^+ and e^- cancel out):

$$2H_2O + CO_2 + light(energy) \Rightarrow$$
$$CH_2O + O_2 + H_2O$$

The energy required to complete this reaction is about +481 kJ mol^{-1}.

EXAMPLE *What corrodes faster, a Zn or an Fe metal pipe?*   A Zn pipe would corrode faster than an Fe pipe because the $Zn^0 \rightarrow Zn^{+2}$ reaction has a oxidation potential of +760 mV, compared with +447 mV for $Fe^0 \rightarrow Fe^{+2}$. This difference in corrosion tendencies between Zn and Fe metal can be used to protect Fe pipes by coating them with Zn (galvanic process). Zinc slows down the corrosive effects of water by the preferential oxidation of Zn with the subsequent formation of Zn-oxides that act as a barrier to O_2 gas and protons (water acidity). There are numerous chemical reactions involved in the oxidation of metal pipes that could be considered, depending on the water quality parameters, such as pH, alkalinity, O_2 content, and total dissolved solids. As an example, if we consider the reduction of O_2 combined with the formation of hydroxyl (OH^-) ions, and the oxidation of Zn combined with the formation

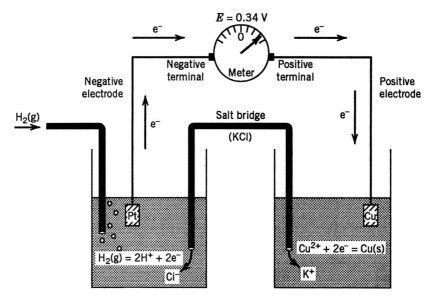

FIGURE 13.6 Electrochemical cell with the standard hydrogen (oxidation side) and copper (reduction side) electrodes. (Adapted from Stumm W., and Morgan, J.J. [1996] Aquatic Chemistry: Chemical Equilibria and Rates in Natural Waters. Wiley, New York. This material is used by permission of John Wiley & Sons, Inc.)

of Zn^{+2}-oxides, two overall reactions can be written as follows:

Oxidation + complex formation $Zn^0 + 2OH^-$

$$\rightarrow Zn(OH)_2 + 2e^- \qquad E = +1249 \, mV$$

O_2 Reduction in water and water dissociation

$$O_2 + H_2O + 2e^- \rightarrow HO_2^- + OH^- \quad E = -76 \, mV$$

the final reaction is the sum of reactions above as follows:

$$Zn^0 + 2OH^- + O_2 + H_2O \rightarrow Zn(OH)_2$$
$$+ HO_2^- + OH^- \qquad E = +1173 \, mV$$

The large positive potential of the combined reactions favors the formation of solid Zn-oxide protective coating on the inside of the pipe.

REDOX MEASURING ELECTRODES

Electrical measurements should be made with probes that will not react with the soil medium. Gold (Au) has a high reduction potential ($\sim +1600 \, mV$), but platinum (Pt) is the preferred metal because of its high reduction potential ($\sim +1200 \, mV$), high electrical conductivity, and hardness. Platinum is also very rare in soil and water environments. Additionally, Pt is much less reactive to the chloride ion that can form stable complexes with noble metals including gold and silver. Figure 13.7 shows the typical design of Pt electrodes used to measure redox conditions in the soil environment.

REDOX MEASUREMENTS

Several considerations are necessary for the proper measurement of redox in soil systems, including moisture content, which impacts the reliability of the redox measurements. As the soil dries, oxygen diffuses into the soil, changing redox conditions. Thus an electrode in contact with an unsaturated soil medium may experience shifts in redox that make it difficult to obtain a "stable" redox reading. In fact, the drier the soil, the more difficult is it to obtain a precise reading. Redox measurements in saturated soils can reach equilibrium within 1 to 2 hr after insertion, whereas it may take several days to obtain a "stable" reading in unsaturated soil systems.

Figure 13.7 also shows a diagram of the installation of redox measurement electrodes in soils. One or more Pt electrodes should be inserted carefully into the soil with firm downward pressure. The electrode(s) must be inserted slowly until the desired depth is reached. Insertion into sandy, gravelly soils must be avoided because they can damage the soft Pt probe (tip). Alternatively, using an access hole, a Pt probe is inserted and sealed with a soil slurry. Here re-establishment of the original redox conditions will take several days, even under saturated conditions.

At least one reference electrode (see the "Reference Electrodes" section) is needed for each Pt electrode or nest of electrodes. These electrodes are usually no more that 50 cm long and are very delicate. The contact with the soil solution is made through a porous tip and must be complete. Accordingly, a reference electrode must be

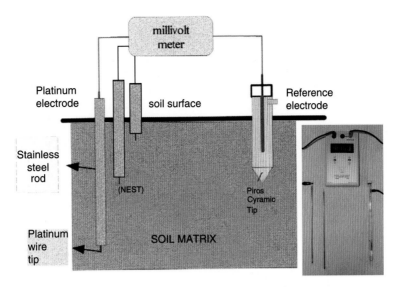

FIGURE 13.7   Soil redox potential measurement; platinum and reference electrodes diagram and picture (*right*).

inserted into the soil using a saturated soil slurry or preferably into small pool of clear water to provide full contact. This installation is applicable even if the soil itself is unsaturated. Periodic soaking with acid to remove metal oxides and salts deposits is recommended to reduce equilibration times and drift.

MEASURING PH

Soil pH affects most reactions in the soil, including redox reactions, precipitation, dissolution, complexation, and exchange reactions in the soil environment. Consequently, changes in soil pH impact the solubility, mobility, and bioavailability of nutrients and pollutants. Similar to redox reactions, the transfer of e^- and H^+ among chemical species is limited by water availability. Therefore accurate and stable *in situ* measurement of soil pH is only possible in near saturated or saturated soil-water systems. The principles of measurement in the field are analogous to those used in the laboratory. Briefly, on exposure of the glass electrode to water solution, a boundary potential develops between the outer (sample) and inner (glass electrode) solution (Figure 13.8). This occurs because H^+ can migrate in and out of the glass matrix in response to their activities in the outer solution. Thus a potential develops that can be measured using a reference electrode. Typically, a rugged form of a glass electrode is inserted into the saturated soil matrix in conjunction with a reference electrode. The measurement process is analogous to that previously described. Thus the electrodes are connected to a mV or pH meter to obtain a direct pH value. Since the soil matrix is disturbed momentarily during the insertion, sufficient time should be allowed for electrode surface equilibration with the bulk

of the soil matrix. In general, low and high values will attain equilibrium faster than near neutral ones. The measurement of pH in poorly buffered neutral soils is difficult because disturbance of the soil matrix will involve O_2 and CO_2 gas exchange that may affect the redox and alkalinity of the soil system.

Measurements of pH are prone to drift and are temperature dependent because the $[H^+]$ changes with temperature. Therefore electrodes used in the field should be calibrated using a two- or three-point standard solution with temperature corrections (usually to $25°\,C$). The working range of pH electrodes should be bracketed using high, medium, and low standards. Extrapolation outside the range is not recommended because pH electrodes have nonlinear responses at the extreme ends of the pH scale (0–14). To reduce drift and equilibration times, periodic soaking with cleansing solutions to remove organic chemicals and salts deposits is recommended. Modern pH electrodes incorporate the reference electrode into their design (see Figure 13.8), making them smaller in size and easier to use. However, care must be taken to protect and maintain the reference solution and to protect the integrity of the fragile glass electrode.

REFERENCE ELECTRODES

Electrochemical measurements using the Standard Hydrogen Electrode require the use of explosive H_2 gas and cumbersome liquid cells (see Figure 13.6). Therefore its use is very impractical in most laboratory and all field conditions. Mixtures of noble metals and their chloride salts have been shown to provide convenient and reliable reference electrodes in place of the hydrogen electrode.

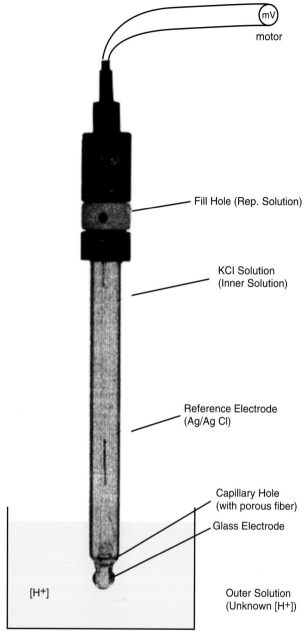

Fill Hole (Rep. Solution)

KCl Solution
(Inner Solution)

Reference Electrode
(Ag/Ag Cl)

Capillary Hole
(with porous fiber)

Glass Electrode

[H+]

Outer Solution
(Unknown [H+])

mV

motor

FIGURE 13.8   Standard pH electrode—glass and reference electrodes combination—principle of operation. (Adapted from Fisher Scientific catalog.)

Two metals are particularly suited for this purpose: silver (Ag) and mercury (Hg). In oxidized regimens, these elements form very insoluble complexes with Cl^- ions. Therefore solid mixtures of either of these two elements, made with Ag or Hg-chloride salts, have surface metal oxidation or reduction without losing ions into solution. The AgCl/Ag reference electrode has a reference potential of +199 mV at 25° C, and the Hg_2Cl_2/Hg (Calomel) electrode has a reference potential of +245 mV, with respect to the H_2 electrode. These potential values are

for electrodes that use an inner solution of saturated KCl. Reference electrodes can experience shift in their reference potentials as a result of interactions of K^+ ions with negatively charged clay surfaces, as well as K^+ and Cl^- ions junction diffusion differences (junction potential). To minimize electrode-soil particle interactions, it is recommended that the reference electrode be placed in a particulate-free soil-water solution in contact with the soil matrix.

OXIDATION-REDUCTION REACTIONS AND pH IN NATURAL SYSTEMS

Electrons are free to flow when reactants are close together, allowing exchanges to occur. Accurate redox measurements should be taken when e^- flow (electric current) is near 0. However, an open environment is dynamic, and equilibrium conditions are seldom attained in the field. Also, redox values measured in natural environments have long been understood to represent the overall (mixed) potential from multiple and simultaneously occurring redox processes. Therefore the isolation of one redox process versus another is impossible, unless a single process dominates, and the inputs and outputs of the system are small or negligible. Since soil is an aqueous system, the redox values in soil systems are constrained by oxidation of water ($O^{-2} \rightarrow O_2$ gas) and the reduction of water ($H^+ \rightarrow H_2$ gas). For the oxidation, a maximum redox value of about +1200 mV is possible for an O_2-saturated system. Conversely, the minimum redox potential is ~ -800 mV for a reduced system limited by a H_2 saturated water. In reality, full saturation (at 1 atm) of either O_2 or H_2 is not possible in natural aqueous systems because other gases are also present in the soil-water atmosphere. Therefore natural systems have redox potentials that range from $\sim +600$ to ~ -500 mV (Figure 13.9).

A useful relationship can be derived between H^+ activity and O_2 as we consider energy involved in the formation of water using electrons, protons, and O_2 in a soil-water system. Consider the following reaction and the energy of formation involved:

$$H^+ + e^- + 1/4O_2 \leftrightarrow 1/2H_2O \qquad \text{(Eq. 13.31)}$$

The standard free energy of this reaction is given by the formation constant $K^0 = 10^{20.78}$. Thus taking the log of K^0, expanding the equilibrium expression (products/reactants), and assuming the activity of water = 1, we obtain:

$$-\log(H^+) - \log(e^-) - \log 1/4O_2 = 20.78 \qquad \text{(Eq. 13.32)}$$

Replacing $-\log(H^+)$ for pH and replacing $-\log(e^-)$ for an equivalent expression called pe, we obtain:

$$pH + pe - \log 1/4O_2 = 20.78 \qquad \text{(Eq. 13.33)}$$

Given that the air contains approximately 20% oxygen, this corresponds to a partial pressure of 0.2 atmospheres. Substitution of this value in Equation 13.33 results in the following expression:

$$pH + pe = 20.61 \qquad \text{(Eq. 13.34)}$$

Thus from Figure 13.9, in theory a soil system at pH $= 0$ could have a pe value of 17, if it was fully saturated with air containing 20% oxygen. In reality this is not possible, because the soil-water environment is seldom this acidic and cannot contain water saturated with 20% oxygen.

Common values for Equation 13.34 are closer to 15 for normal soils, 10 for wet soils, and 5 for waterlogged soils (see Figure 13.9 [inset]).

Mixed redox potentials are very useful to determine the likelihood of occurrence of a given redox process in the soil environment. They should, however, be used with great care for the estimation of activities or concentrations of specific species because it is not possible to isolate individual chemical reactions from the redox measurements; nor should we expect that natural systems are at equilibrium. Several important redox processes in the soil environment affect its overall quality. Figure 13.10 presents a list of important redox processes that may dominate soil and water environments. The likely presence or absence of a given process can be quickly established by measuring the redox potential of the system. For example, is nitrate reduction likely to occur in a soil with a redox potential of $+500\,mV$? According to Figure

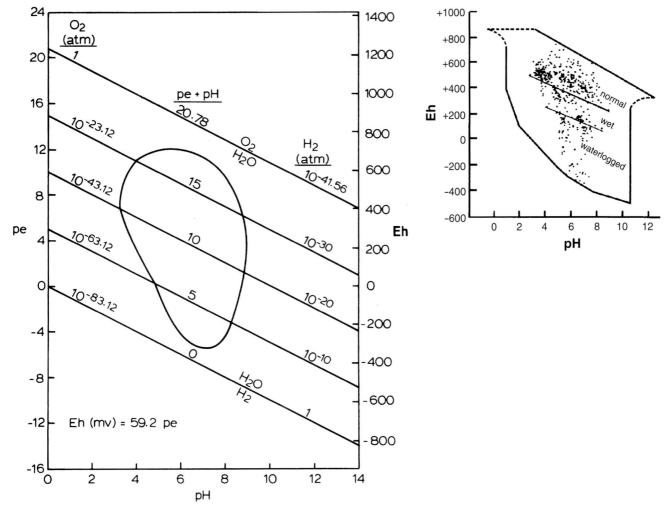

FIGURE 13.9 Soil redox potentials (Eh in mV) and pH ranges *(closed area)*. (From Lindsay, 1979, Fig 2.1.) The approximate soil Eh+pe values are: normal ($\sim$11–16), wet ($\sim$8–11), and waterlogged (1–8) conditions; see smaller graph at right. (From Sparks, D.L. [1995] Environmental Soil Chemistry. Academic Press, San Diego.)

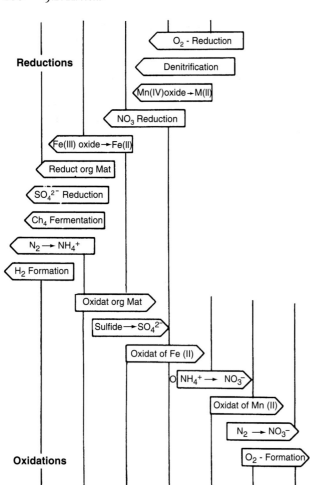

FIGURE 13.10 Reduction and oxidation chemical processes in the soil-water environment. (Adapted from Stumm W., and Morgan, J.J. (1996) *Aquatic Chemistry: Chemical Equilibria and Rates in Natural Waters.* Wiley, New York. This material is used by permission of John Wiley & Sons, Inc.)

13.10, this may not be possible. Conversely, nitrification is likely to occur in the same soil if NH_4^+ is present, although we cannot say anything about its rate of reduction or the amount of NO_3^- produced.

QUESTIONS

1. X-ray fluorescence analysis of a soil sample taken from a car battery recycling site indicates that it contains $1500\,mg\,k^{-1}$ total Pb. Assume that 90% of Pb in this soil is in the form lead carbonate and determine the total (%) amount of $PbCO_3$ in this soil sample. State your assumptions.

2. Estimate the total concentrations of Ca^{++} and $SO_4^=$ ions (in mol L^{-1}) that are present in a gypsum-saturated soil solution. State your assumptions.

3. Calculate the electrical conductivity using λ^0+ and λ^0- values of a solution that contains $1g\,L^{-1}$ of dissolved NaCl and $0.5\,g\,L^{-1}$ of gypsum ($CaSO_4 * 2H_2O$) State your assumptions.

4. The abiotic oxidation of TCE in the soil-water environment is possible using a strong oxidant such as Fe^0 (see Box 13.1). However, $KMnO_4$ is the preferred oxidant because it is soluble in water. The half-cell reaction is:

    $$MnO_4^- + 8H^+ + 5e^- <> Mn^{++} + 4H_2O$$

    (a) Write the other side (oxidation) and then combine and balance the two half-cell reactions into one. Note: for simplicity use acetic acid (HCOOH) instead of TCE.
    (b) Will the oxidation rate of TCE increase in an acid soil or an alkaline soil? Explain your answer.

5. Assuming that the standard cell potential of the reaction from 4a is $+1800\,mV$, estimate the potential of a TCE soil solution treated with permanganate. The final concentrations are:

    $$TCE \sim 0\,\mu g\,L^{-1}, MnO_4^{-1} = 0.2\,mol\,L^{-1},$$
    $$Mn^{++} = 0.05\,mol\,L^{-1}.\text{ State your assumptions.}$$

6. Can $SO_4^=$ be reduced in a soil with a redox eV of $-300\,mV$? Explain your answer.

REFERENCES AND ADDITIONAL READING

Brady, N.C., and Weil, R.R. (2002) *The Nature and Properties of Soils.* 13th edition. Prentice Hall, Upper Saddle River, NJ.

Clesceri, L.S., Greenberg, A.E., and Eaton, A.D. (1998) *Standard Methods for the Examination of Water and Wastewater.* 20th Edition. American Public Health Association. 1015 15th Street, NW. Washington, DC 20005-2605.

Dixon, J.B., and Weed, S.B. (1989) *Minerals in the Soil Environment.* 2nd Edition. Soil Science Society of America, Madison, WI.

Garrels, R.M., and Christ, C.L. (1990) *Solutions, Minerals, and Equilibria.* Jones and Barlett Publishers, Boston.

LaGrega, M.D., Buckingham, P.L., and Evans, J.C. (1994) *Hazardous Waste Management.* McGraw-Hill Inc. Publishers, Boston. Chapter 15.

Lindsay, W.L. (2001) *Chemical Equilibria in Soils*. Blackburn Press, Caldwell, NJ.

Sparks, D.L. (2003) *Environmental Soil Chemistry*. 2nd edition. Academic Press, San Diego.

Sparks, D.L. (1996) *Methods of Soil Analysis, Part 3—Chemical Methods*. Soil Science Society of America, Madison, WI. Chapters 7, 14.

Stumm, W., and Morgan, J.J. (1996) *Aquatic Chemistry: Chemical Equilibria and Rates in Natural Waters*. Wiley, New York.

U.S. EPA. (1998) SW486. Test Methods for Evaluating Solid Waste: Physical and Chemical Methods, Final Update III, January 1998. Method 6020 Revision 0.

Willard, H.H., Merritt, L., Dean, J., and Settle, F. (1988) *Instrumental Methods of Analysis*. 7th edition. Wadsworth Pub, Belmont, CA.

Wilson, L.G., Everett, L.G., and Cullen, S.J. (1995) *Handbook of Vadose Zone Characterization & Monitoring*. Lewis Publishers, Boca Raton, FL. Chapter 22.

14

Environmental Microbial Properties and Processes

I.L. PEPPER, C. RENSING, AND C.P. GERBA

BENEFIT OF ENVIRONMENTAL MICROBES

Microorganisms (bacteria, viruses, algae, fungi, and protozoa) are pervasive and influential throughout every major environment. In fact, any environment that is devoid of microorganisms is certainly the exception and not the rule. Microorganisms are fundamental to all ecosystems, providing the biotic foundation upon which all life exists. Unfortunately, the general public most often perceives microorganisms primarily as the etiological agents of disease (i.e., the causes of disease). In many cases the opposite is true, and many microbes are, in fact, beneficial. Regardless of whether they influence human health and welfare favorably or unfavorably, microorganisms are capable of profound influences on life as we know it. Because of this, methods are needed for the sampling and detection of microbes from a variety of environments including soil, water, food, and municipal wastes (Maier *et al.*, 2000).

This chapter is intended to provide basic information on microbiological properties and processes in the environment. Monitoring processes for pathogenic microbes will be reviewed in Chapter 17. Here we will answer the questions: What are microorganisms? What are some of the methods used to monitor beneficial microbial communities and specific microbes? In addition, specific microbial sampling procedures used in the collection and processing of environmental samples (e.g., air, soil, and water) for microorganisms will be described.

WHAT ARE MICROORGANISMS?

Microorganisms in the environment are ubiquitous and are represented by diverse populations and communities. Microorganisms, which are defined as being individually too small to be seen with the naked eye, include viruses, bacteria, amoeba, fungi, algae, and protozoa. However, there are certain genera of fungi, algae, and protozoa that are macroscopic in nature, and therefore only viruses and bacteria are totally microscopic (Figure 14.1).

Viruses are on average the smallest microorganisms, and are considered to be ultramicroscopic. As defined, viruses are submicroscopic, parasitic, nonfilterable agents consisting primarily and simply of a nucleic acid surrounded by a protein coat. Electron microscopes are required to visualize viruses because they are smaller than the resolution capacity of light microscopes. Ranging in size from 20 to several hundred nm, virus particles are very simple in organization, and normally consist of only an inner nucleic acid genome, an outer protein capsid, and sometimes an additional membrane envelope (Figure 14.2). Unlike other microorganisms, viruses do not necessarily meet the criteria for being considered a live organism because they have no ribosomes and do not metabolize. Viruses must infect and replicate inside living cells. For example, caliciviruses such as the Norwalk virus can be transmitted via ingestion of contaminated water. When a human ingests contaminated water, the Norwalk viruses enter the intestines, attach, infect, and replicate in intestinal cells, ultimately

causing the symptoms of disease, such as gastroenteritis. Because viruses do not metabolize on their own, they are not considered to be living, so the term *microorganism* can be debated.

Bacteria are the very small, relatively simple, single-celled organisms whose genetic material is not enclosed in a nuclear membrane (Figure 14.3). Based on this cellular organization, bacteria are classified as prokaryotes and include the eubacteria and the archea. Although they are classified as bacteria, special mention should be made of the *actinomycetes*. These prokaryotic microbes consist of long chains of single cells, which allow them to exist in a filamentous form. Structurally, from the exterior, actinomycetes resemble miniature fungi, and this is the reason why they are often reported as a specialized subgroup of bacteria. Actinomycetes are important producers of antibiotics such as streptomycin, are important in the biodegradation of complex organics, and also produce geosmin, which is the compound that gives soil its characteristic odor. Geosmin can also result in taste and odor problems in potable waters. Prokaryotes are distinguished from more complex microorganisms such

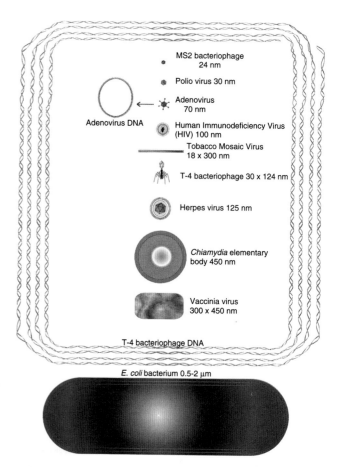

FIGURE 14.1 Comparative sizes of selected bacteria, viruses, and nucleic acids. (Reprinted from *Environmental Microbiology* © 2000, Academic Press, San Diego, p. 11.)

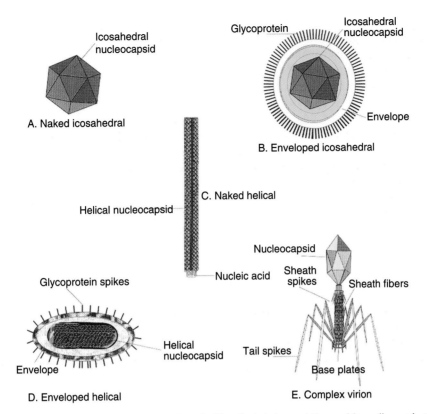

FIGURE 14.2 Simple forms of viruses and their components. The naked icosahedral viruses *(A)* resemble small crystals; the enveloped icosahedral viruses *(B)* are made up of icosahedral nucleocapsids surrounded by the envelope; naked helical viruses *(C)* resemble rods with a fine, regular helical pattern in their surface; enveloped helical viruses *(D)* are helical nucleocapsids surrounded by the envelope; and complex viruses *(E)* are mixtures of helical and icosahedral and other structural shapes. (Reprinted from *Environmental Microbiology* © 2000, Academic Press, San Diego, p. 14.)

as fungi because they lack developed internal structures, especially a membrane-enclosed nucleus, complex internal cell organelles involved in growth, nutrition, or metabolism, and internal cell membranes. Bacteria are the most ubiquitous organisms, and can be found in even the most extreme environments because they have evolved a capacity for rapid growth, metabolism, and reproduction, as well as the ability to use a diverse range of organic and inorganic substances as carbon and energy sources. Bacteria are considered to be the simplest form of life, because viruses are not considered to be alive as previously noted. A single gram of soil can contain up to 10^{10} bacteria. Bacteria are especially vital to life on earth because of their functioning in the major environments. A few of their important activities include biogeochemical processes, nutrient cycling in soils, bioremediation, human and plant diseases, plant-microbe interactions, municipal waste treatment, and the production of important drug agents, including antibiotics.

In contrast to bacteria, *protozoa* are unicellular eukaryotes, meaning that they have characteristic organelles (mitochondria, plasma membrane, nuclear envelope, eukaryotic ribosomal RNA, endoplasmic membranes, chloroplast, and flagella), are large (some are visible with the naked eye) (Figure 14.4), and have independent metabolic pathways. Although they are single-celled organisms, they are by no means simple in structure, and many diverse forms can be observed among the more than 65,000 named species. Morphological variability, evolved over hundreds of millions of years, has enabled protozoan adaptation to a wide variety of environments. Tropism and structure remain the major identifying factors in protist classification; however, molecular and immunological methods promise to increase the database, distinction, and knowledge of community structure. Protozoa may be free living, capable of growth and reproduction outside any host, or parasitic, meaning that they colonize host cell tissues. Some are opportunists, adapting either a free-living or parasitic existence, as their environment dictates. The size of protozoa varies from 2 μm, as with the small ciliates or flagellates, to several cm, as with the amoebae. Therefore the larger protozoa can be seen with the naked eye, but the aid of a microscope is necessary to observe cellular detail. Protozoa can be found in nearly all terrestrial and aquatic environments and are thought to play a valuable role in ecological cycles, in part by controlling bacteria populations. Many species are able to exist in extreme

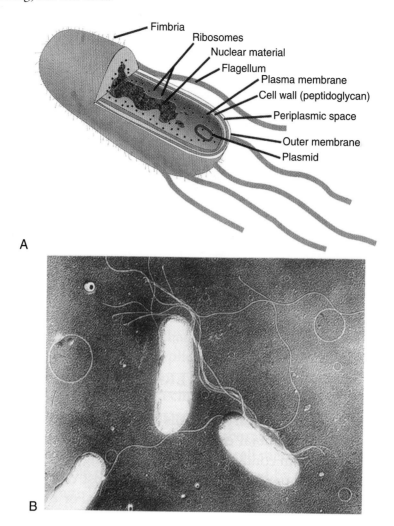

FIGURE 14.3 (**A**) Schematic representation of a typical bacterial cell. (**B**) Scanning electron micrograph of a soil bacterium with multiple flagella. The circles are detached flagella that have spontaneously assumed the shape of a circle. (Reprinted from *Environmental Microbiology* © 2000, Academic Press, San Diego, pp. 20–21).

environments from polar regions to hot springs and desert soils. In recent years protozoan pathogens, such as *Giardia, Cyclospora, Cryptosporidium,* and *Microsporidia,* have emerged to become primary concerns with regard to contamination of drinking water.

Fungi, like protozoa, are also eukaryotic organisms (Figure 14.5). They are very ubiquitous in the environment and also critically affect human health and welfare. For example, the fungi can be both beneficial and harmful to plants, animals, and humans. Certain species of fungi promote the health of many plants through mycorrhizal associations, whereas other species are phytopathogenic and capable of destroying plant tissue and even whole crops. Fungi are also important in the cycling of organics and bioremediation. One of the most important fungi are the yeasts, which are utilized in the fermentation of sugars to alcohol in the brewing and wine industries. Fungi range from microscopic, with a single cell, to macroscopic, with filaments of a single fungal cell being several

centimeters in length. Overall, fungi are heterotrophic in nature, with different genera metabolizing everything from simple sugars to complex aromatic hydrocarbons. Fungi are also important in the degradation of the plant polymers cellulose and lignin. For the most part, fungi are aerobic, though some, such as the yeasts, are capable of fermentation. Fungi are distinguished from algae by their lack of photosynthetic ability.

Algae are a group of photosynthetic organisms that can be macroscopic, as in the case of seaweeds and kelps, or microscopic. Algae are aerobic eukaryotic organisms, and as such, exhibit structural similarities to fungi with the exception of the algal chloroplast (Figure 14.6). Chloroplasts are photosynthetic cell organelles found in algae that are capable of converting the energy of sunlight into chemical energy through a photosynthetic process. Algae are abundant in fresh water and salt water, in soil, and in association with plants. When excessive nutrient conditions occur in aquatic environments, they can cause eu-

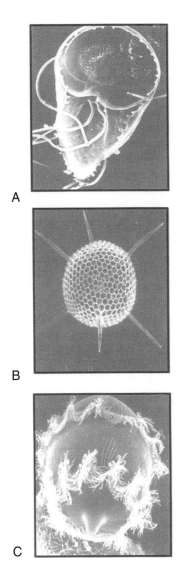

A

B

C

FIGURE 14.4 Basic morphology of protozoa. **(A)** Scanning electron micrograph of a flagella protozoa, *Giardia*. **(B)** Scanning electron micrograph of testate cilia, *Heliosoma*. **(C)** Electron micrograph of ciliated *Didinium*. (Reprinted from *Environmental Microbiology* © 2000, Academic Press, San Diego, p. 35.)

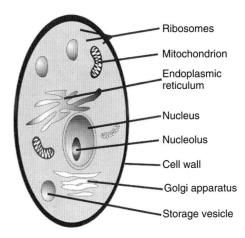

Ribosomes
Mitochondrion
Endoplasmic reticulum
Nucleus
Nucleolus
Cell wall
Golgi apparatus
Storage vesicle

FIGURE 14.5 Structure of a typical fungal cell. (Reprinted from *Environmental Microbiology* © 2000, Academic Press, San Diego, p. 33.)

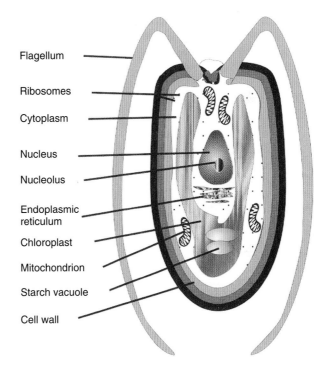

Flagellum
Ribosomes
Cytoplasm
Nucleus
Nucleolus
Endoplasmic reticulum
Chloroplast
Mitochondrion
Starch vacuole
Cell wall

FIGURE 14.6 Structure of a typical algal cell. (Reprinted from *Environmental Microbiology* © 2000, Academic Press, San Diego, p. 34.)

trophication. The capacity of algae to photosynthesize is critical because it forms the basis of aquatic food chains, and can also be important in the initial colonization of disturbed terrestrial environments. Overall, the algae are sometimes known by common names, such as green algae, brown algae, or red algae, based on their predominant color.

Blue-green algae are actually classified as bacteria known as *cyanobacteria*. The *cyanobacteria* are not only photosynthetic but can also fix atmosphere nitrogen. Some species produce toxins in water that are harmful to animals and humans.

Most environmental monitoring of beneficial microbes focuses on bacteria, actinomycetes and fungi, particularly

in the soil environment. Estimates of typical soil microbial numbers and their biomass are shown in Table 14.1. A comparison of their microbial properties is shown in Table 14.2.

MICROORGANISMS IN SOIL

The health of a soil is intimately tied to its microbial populations. When a soil becomes stressed, the microbial populations and microbial community structures reflect

TABLE 14.1

Relative Estimates of the Abundance of Soil Microbes in the Environment

Microbe	Number (per gram of soil)	Biomass Within Root Zone (kg ha^{-1})
Bacteria	10^8	500
Actinomycetes	10^7	500
Fungi	10^6	1500

(Reprinted from *Pollution Science* © 1996 Academic Press, San Diego.)

the stress induced on their environment. The primary factors affecting the microbial ecology of a soil, much like any environment, include natural abiotic factors such as temperature, light, aeration, nutrients, organic matter, pH and moisture content, and biotic factors (e.g., competition, predation, and the ability to utilize available carbon sources). Knowledge of how microbial communities interact with their environment is crucial to a thorough understanding of soils. Indeed, one cannot claim to understand soil without understanding the interactions of the microbial life that exists in that soil.

Soils themselves are classified on the basis of their chemical and physical properties, not their microbiological properties. This is mainly because such chemical and physical properties are more readily defined and measured than their microbiological properties. It is well known that we only have the ability to grow, on agar plates or in broth (i.e., culture), a minute portion of the microorganisms that are actually present in soil. It is estimated that less than 1% of the organisms in any given soil sample have been cultured at this point or even superficially characterized (Amann *et al.*, 1995). Organisms that cannot be cultured are known as "viable but nonculturable" organisms (Roszak and Colwell, 1987). In addition, other obstacles that prevent an accurate characterization of soil microbial environments include the vast diversity of microorganisms that are present at any one time, even within a single gram of soil; their wide range of physiological capacities; the dramatic population variations that can exist even within the same field site; and the rapid fluctuations in a microbial population that result from human influence during sampling. It is also observed in most cases that as the level of stress increases (e.g., stress induced by the addition of a toxic chemical into the soil), the taxonomic diversity of the microbial communities in that soil often changes. This is mainly due to a process of natural selection for those organisms possessing genetic elements conferring some form of advantage in relation to the stress factor. The origination of such organisms may be due to gene exchange (Orgel and Crick, 1980). These new organisms may produce a variety of biodegradative genes allowing for utilization or transformation of the pollutant, or they may develop resistance mechanisms that allow the microbe to survive in the presence of the stress factor. Microbial populations selected by such contamination are often considered beneficial because they may also be responsible for detoxification of the contaminated environment. Note that it is these populations or isolates of microbes that are of current interest in the field of bioremediation. However, though we consider these microorganisms beneficial,

TABLE 14.2

Characteristics of Bacteria, Actinomycetes, and Fungi

Characteristic	Bacteria	Actinomycetes	Fungi
Population	Most numerous	Intermediate	Least numerous
Biomass	Bacteria and actinomycetes have similar biomass		Largest biomass
Degree of branching	Slight	Filamentous, but some fragment to individual cell	Extensive filamentous forms
Aerial mycelium	Absent	Present	Present
Growth in liquid culture	Yes—turbidity	Yes—pellets	Yes—pellets
Growth rate	Exponential	Cubic	Cubic
Cell wall	Murein, teichoic acid, and lipopolysaccharide	Murein, teichoic acid, and lipopolysaccharide	Chitin or cellulose
Complex fruiting bodies	Absent	Simple	Complex
Competitiveness for simple organics	Most competitive	Least competitive	Intermediate
Fix N	Yes	Yes	No
Aerobic	Aerobic, anaerobic	Mostly aerobic	Aerobic except yeast
Moisture stress	Least tolerant	Intermediate	Most tolerant
Optimum pH	6–8	6–8	6–8
Competitive pH	6–8	>8	<5
Competitiveness in soil	All soils	Dominate dry, high-pH soils	Dominate low-pH soils

(Reprinted from *Pollution Science* © 1996, Academic Press, San Diego.)

with the addition of chemical pollutant many naturally beneficial microorganisms that do not possess a genetic advantage, allowing survival in the presence of a contaminant, are no longer able to compete and are eliminated from the microbial community. It is these types of variation in the microbial communities in an environment that make it difficult to accurately characterize the microbiological properties of soil.

AQUATIC ENVIRONMENTS

In reality all microorganisms require an aqueous environment. Aquatic environments occupy more than 70% of the earth's surface, most of this area being occupied by oceans. There are also a broad spectrum of other aquatic environments including estuaries, harbors, major river systems, lakes, wetlands, streams, springs, and aquifers. Microorganisms are the primary producers in that environment (and are responsible for approximately half of all primary production on the earth). They are also the primary consumers. The microbiota that inhabit aquatic environments include bacteria, viruses, fungi, algae, and other microfauna. Identifying the microbiological composition of the aquatic environment and determining the physiological activity of each component are the first steps in understanding the ecosystem as a whole. In general, methods used in the monitoring of microbes in aquatic environments are similar to those used for most other environments. Applied methodologies differ markedly only in how samples are acquired and processed. These sampling principles are discussed later in this chapter.

SAMPLING PROCEDURES FOR MICROBIAL CHARACTERIZATION

REPRESENTATIVE SAMPLING

Two primary issues influence what is considered a representative microbial sampling scheme. The first is the two-dimensional sampling scheme. In two-dimensional sampling each plot is assigned spatial coordinates and set sampling points are chosen according to an established plan. Some typical two-dimensional sampling patterns include random, transect, two-stage, and grid sampling. These topics are covered in Chapter 7, but briefly, random sampling involves choosing random points within the field site; transect sampling involves collection of samples along one direction or path; two-stage sampling involves dividing an area into regular subunits called primary units, which are randomly or systematically

sampled; and finally, in grid sampling, samples are taken systematically at regular intervals, and at fixed spacings covering the entire site. This type of sampling is useful for mapping a large field area when little is known about its variability.

The second issue that must be considered for obtaining representative samples in the environment, especially for microbiological monitoring, is that two-dimensional sampling does not give any information about changes in microbial populations with depth. Therefore three-dimensional sampling is used when information concerning depth is required. Such depth information is critical when evaluating sites that have been contaminated by leaking sewage lines, for instance. Three-dimensional sampling can be as simple as sampling at 50-cm deep increments to a depth of 200 cm, or can involve drilling several hundred meters into the subsurface. It is also useful when collecting water samples, as in oceans, for example.

SAMPLING METHODS FOR SURFACE SOILS

Here we describe methods for processing surface soils before analysis for nonpathogenic microorganisms. Methods for pathogens will be presented in Chapter 17. Bulk soil samples are easily obtained with a shovel or, better yet, a soil auger (see Chapter 7). Soil augers are more precise than shovels because they ensure that samples are taken to exactly the same depth on each occasion. This is important because several soil factors vary considerably with depth, such as oxygen, moisture and organic carbon content, and soil temperature. A simple hand auger is useful for taking shallow soil samples from areas that are unsaturated. Under the appropriate conditions, a hand auger can be used to depths of 2 m, although some soils are simply too compacted or contain too many rocks to allow sampling to this depth. In taking samples for microbial analysis, consideration should be given to contamination that can occur as the auger is pushed into the soil. In this case microbes that stick to the sides of the auger as it is inserted into the soil and pushed downward may contaminate the bottom part of the core. To minimize such contamination, one can use a sterile spatula to scrape away the outer layer of the core and use the inner part of the core for analysis. Contamination can also occur between samples, but this can be avoided by cleaning the auger after each sample is taken. The cleaning procedure involves washing the auger with water, then rinsing it with 75% ethanol or 10% bleach, and giving it a final rinse with sterile water.

SAMPLE PROCESSING AND STORAGE

Microbial analyses should be performed as soon as possible after collection of any type of sample, whether it is soil, water, or air. Once removed from the natural environment, microbial populations within a sample can and will change regardless of the method of storage. Reductions in microbial numbers and microbial activity have been reported even when soil samples were stored in a field moist condition at 4° C for only 3 months (Stotzky *et al.*, 1962). Another factor that influences processing is the type of microbial analysis that will be performed on a given sample. For instance, if soil will be processed for actinomycete populations, the method used to process the sample will be different than if the same soil sample will be processed for protozoan or viral analysis.

Typically, the first step in microbial analysis of a surface soil sample involves sieving through a 2-mm mesh to remove large stones and debris. However, samples must often be air dried to facilitate the sieving. This is acceptable as long as the soil-moisture content does not become so low as to deleteriously impact the microbial populations (Sparkling and Cheshire, 1979). Following sieving, short-term storage should be at 4° C before analysis. If samples are stored, care should be taken to ensure that samples do not dry out and that anaerobic conditions do not develop, because this too can alter microbial populations. Storage up to 21 days appears to leave most soil microbial properties unchanged (Wollum, 1994), but again, time is of the essence with respect to microbial analysis. Note that routine sampling of surface soils does not require sterile procedures. These soils are continually exposed to the atmosphere, so it is assumed that such exposure during sampling and processing will not affect the results significantly.

More care is taken with processing subsurface samples for three reasons. First, they generally have lower cell densities, which means that an exogenous microbial contaminant may significantly affect the cell counts/measurements. Second, subsurface sediments are not routinely exposed to the atmosphere, and microbial contaminants in the atmosphere might substantially contribute to microbial types and numbers counted. Third, it is more expensive to obtain subsurface samples, and often there is no second chance at collection. Subsurface samples obtained by coring are either immediately frozen and sent back to the laboratory as an intact core or processed at the coring site. In either case the outside of the core is usually removed using a sterile spatula, or a subcore is taken using a smaller diameter plastic syringe. The sample is then placed in a sterile plastic bag and analyzed immediately or frozen for future analysis.

STRATEGIES AND METHODS FOR WATER SAMPLING

Sampling water for subsequent microbial analysis is somewhat easier than sampling soils for a variety of reasons. First, because water tends to be more homogeneous than soils, there is less point-to-point variability between two samples collected within the same vicinity. Second, it is often physically easier to collect water samples because it can be done with pumps and hose lines. Thus known volumes of water can be collected from known depths with relative ease. Amounts of water collected depend on the sample being evaluated, but can vary from 100 ml for potable drinking water collected from water utilities to 1000 L for groundwater samples. Sampling strategy is also less complicated for water samples. In many cases, because water is mobile, a set number of bulk samples are simply collected from the same point at various time intervals. Such a strategy would be useful, for example, in sampling a river or a drinking water treatment plant. For marine waters, samples are often collected sequentially in time within the defined area of interest.

Although the collection of a water sample is relatively easy, processing the sample before microbial analysis can be more difficult. Water samples should be cooled to 4° C as soon as possible after collection and analyzed for bacteria. Depending on the type of microbial analysis that will be performed, samples can also be frozen. Microbial populations in water samples change much more rapidly than those in soil samples. The volume of the water sample required for detection of microbes can sometimes become unwieldy, because numbers of microbes tend to be lower in water samples than in soil samples. Therefore strategies have been developed to allow concentration of the microbes within a water sample. For larger microbes, including bacteria and protozoan parasites, samples are often filtered to trap and concentrate the organisms. For bacteria, this often involves filtration using a 0.45-μm membrane filter. For protozoan parasites, coarse, woven fibrous filters have been used. For viruses, water samples are also filtered, but because viral particles are often too small to be physically trapped, collection of the viral particles depends on a combination of electrostatic and hydrophobic interactions of the virus with the filter, which results in their adsorption to the filters. The different requirements for processing water samples for analysis of bacteria are outlined in the "Microbial Numbers" section because these techniques are utilized for beneficial bacteria. Techniques for pathogenic bacteria, viruses, and protozoa are outlined in Chapter 17.

STRATEGIES AND METHODS FOR AIR SAMPLING

Many devices have been designed for the collection of bioaerosols. Choosing an appropriate sampling device is based on many factors, such as availability, cost, volume of air to be sampled, mobility, sampling efficiency (for the particular type of bioaerosol), and the environmental conditions under which sampling will be conducted. Another factor is the biological sampling efficiency of the device. This factor is related to the maintenance of microbial viability during and after sampling. In this section a few of the most common types of samplers are described on the basis of their sampling methods: impingement, impaction, centrifugation, filtration, and deposition. Impingement is the trapping of airborne particles in a liquid matrix; impaction is the forced deposition of airborne particles on a solid surface; centrifugation is the forced deposition of airborne particles enhanced by an increase in the inertial forces of gravity; filtration is the trapping of airborne particles by size exclusion; and deposition is the collection of airborne particles using only natural deposition forces. The two devices most commonly used for microbial air sampling are the all-glass impinger (AGI-30) (Ace Glass, Vineland, NJ) and the Andersen six-stage impaction sampler (6-STG, Andersen Instruments Incorporated, Atlanta).

IMPINGEMENT

The AGI-30 operates by drawing air through an inlet that is similar in shape to the human nasal passage. The air is transmitted through a liquid medium where the air particles become associated with the fluid and are subsequently trapped (Figure 14.7). The AGI-30 impinger is usually run at a flow rate of 12.5 L min^{-1} at a height of 1.5 m, which is the average breathing height for humans. The AGI-30 is easy to use, inexpensive, portable, reliable, easily sterilized, and has good biological sampling efficiency compared with many other sampling devices. The AGI-30 tends to be very efficient for particles in the normal respirable range of 0.8 to 15 μm. The usual volume of collection medium is 30 ml, and the typical sampling duration is approximately 20 minutes, which prevents excessive evaporation during the sampling. Another feature of the impingement process is that the liquid and suspended microorganisms can be concentrated or diluted, depending on the requirements for analysis. Liquid impingement media can also be divided into subsamples to test for a variety of microorganisms by standard methods such as those described by Pillai *et al.*

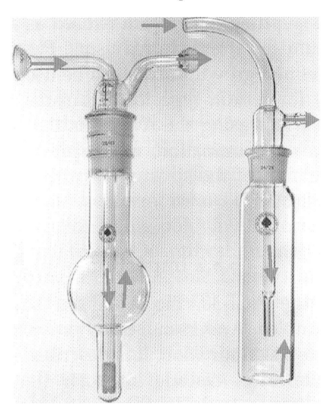

FIGURE 14.7 This is a schematic representation of two all-glass impingers (AGIs). The impinger on the right is the classic AGI-30 impinger. Arrows indicate the direction of air flow. The air enters the impinger drawn by suction. As bioaerosols impinge into the liquid collection medium contained in the bottom of the impinger, the airborne particles are trapped within the liquid matrix. (Photo courtesy Ace Glass Inc., Vineland, NJ.) (Reprinted from *Environmental Microbiology* © 2000, Academic Press, San Diego, p. 103.)

(1996). The impingement medium can also be optimized to increase the relative biological recovery efficiency. This is important because during sampling, the airborne microorganisms, which are already in a stressed state because of various environmental stresses such as ultraviolet (UV) radiation and desiccation, can be further stressed, if a suitable medium is not used for recovery. Examples of sampling media can range from 0.85% NaCl, which is a basic, osmotically balanced, sampling medium that would prevent osmotic shock of recovered organisms, to 0.1% peptone, which is used as a resuscitation medium for stressed organisms. The major drawback when using the AGI-30 is that there is no particle-size discrimination, which prevents accurate characterization of the nature of the airborne particles that are collected.

IMPACTION

Unlike the AGI-30, the Andersen six-stage impaction sampler provides accurate particle-size discrimination. It is described as a multilevel, multiorifice, cascade impactor (Jensen *et al.*, 1992). The Andersen 6-STG was developed by Andersen in 1958 and operates at an input flow rate of 28.3 L min^{-1}. The general operating principle is that air is sucked through the sampling port and strikes agar plates (Figure 14.8). Larger particles are collected on the first layer, and each successive stage collects smaller and smaller particles, which are selected by increasing the flow velocity and, consequently, the impaction potential. The shape of the Andersen sampler does not conform to the shape of the human respiratory tract, but the particle-size distribution can be directly related to the particle-size distribution that occurs naturally in the lungs of animals. The lower stages correspond to the alveoli and the upper stages to the upper respiratory tract. The Andersen sampler is constructed of stainless steel with glass petri dishes, allowing sterilization, ease of transport, and reliability. It is useful over the same particle-size range as the AGI-30 (0.8 to more than 10 μm), corresponding to the respirable range of particles. It is more expensive than the AGI-30, and the biological sampling efficiency is somewhat lower because of the method of collection, which is impaction on an agar surface. Analysis of viruses collected by impaction is also somewhat difficult, because after impaction, the viruses must be washed off the surface of the impaction medium and collected before assay. Bacteria or other microorganisms, with the exception of viruses, can be grown directly on the agar surface or, as with viruses, they can be washed off the surface and assayed using other standard methodologies. The biggest single advantage of the Andersen 6-STG sampler is that particle-size determinations can be obtained. Thus the two reference samples (AGI-30 and Andersen 6-STG) complement each other's deficiencies.

CENTRIFUGATION

Centrifugal samplers use circular flow patterns to increase the gravitational pull within the sampling device to deposit particles. The Cyclone, a tangential inlet and returned flow sampling device, is the most common type. These samples are able to sample a wide range of air volumes (1 to 400 L min^{-1}), depending on the size of the unit. The unit operates by applying suction to the outlet tube, which causes air to enter the upper chamber of the unit at an angle. The flow of air falls into a characteristic tangential flow pattern, which effectively circulates air around and down along the inner surface of the conical glass housing. As a result of the increased centrifugal forces imposed on particles in the airstream, the particles are sedimented out. The conical shape of the upper chamber opens into a larger bottom chamber, where most of this particle deposition occurs. Although these units are able to capture some respirable-size particles, in order to trap microorganisms efficiently, the device must be combined with some type of metered fluid flow that acts as a trapping medium. This unit, when used by someone proficient, can be very effective for microbiological air sampling. It is relatively inexpensive, easily sterilized, and portable, but it lacks high biological sampling efficiency and particle sizing capabilities. Rinsing the sampler with an elluent medium, collection of the elluent, and subsequent assay by standard methodologies is the procedure for processing.

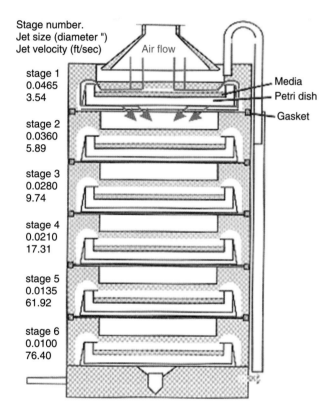

FIGURE 14.8 This is a schematic representation of the Andersen six-stage impaction air sampler. Air enters through the top of the sampler and larger particles are impacted upon the surface of the petri dish on stage 1. Smaller particles, which lack sufficient impaction potential, follow the air stream to the subsequent levels. As the air stream passes through each stage the air velocity increases, thus increasing the impaction potential so that particles are trapped on each level based on their size. Therefore larger particles are trapped efficiently on stage 1 and slightly smaller particles on stage 2, and so on until even very small particles are trapped on stage 6. The Andersen six stage thus separates particles based on their size. (Reprinted from *Environmental Microbiology* © 2000, Academic Press, San Diego, p. 104.)

FILTRATION

Filtration and deposition methods are both widely used for microbial sampling for cost and portability reasons. Filter sampling involves passage of air through a filter, where the particles are trapped. Filtration requires a vacuum source, as in the preceding three methods, making it a less portable methodology. Filter sampling involves using a filter holder containing an appropriate membrane filter, which is then attached to the vacuum source. Membrane filters can have variable pore sizes, although smaller pore sizes restrict flow rates. Filtration sampling for microorganisms is not highly recommended, however, because of its low overall sampling efficiency and desiccation of microbial cells. After collection, the filter is washed to remove the organisms before analysis.

DEPOSITION

Deposition sampling is by far the easiest and most cost-effective method of sampling. Merely opening an agar plate and exposing it to the atmosphere, which results in direct impaction, gravity settling, and other depositional forces, can accomplish deposition sampling. Problems with this method of sampling include low overall sampling efficiency, because it relies on natural deposition, no defined sampling rates or particle sizing, and an intrinsic difficulty in testing for multiple microorganisms with varied growth conditions. Analysis of microorganisms collected by depositional sampling is similar to impaction sample analysis.

Air samples are unique in that they are collected in liquid or solid media. Thus processing is dependent on how the air was collected. In general, air samples should be stored immediately at 4° C for transportation to the laboratory. Once in the laboratory, the type of microbial analysis to be performed, again, governs how the sample should be processed. For instance, if the sample is being processed for bacterial pathogens or indicators, it can be treated as any other water sample. If the sample is being used for the monitoring of viruses, it can be considered as a concentrated and eluted water sample, as described in the following sections on virus analysis. If the sample is being processed for protozoa, then similar to virus analysis, it can also be considered a concentrated and purified water sample, as described in the following sections on protozoan analysis. If the air sample was collected onto a solid media, an extra step is added to the analysis in which the solid media (usually agar) is washed with sterile water to remove collected organisms. At this point, it too can be considered as a water sample.

METHODS FOR CHARACTERIZING MICROORGANISMS AND MICROBIAL PROPERTIES IN WATER AND SOIL

Typically, microbial analyses include determination of either microbial numbers or microbial biomass. If estimates of these parameters can be made, the microbial health (or impairment) of a particular environment can be ascertained. This, in turn, will allow for the assessment of whether or not a particular beneficial microbial process such as *in situ* bioremediation will occur. In the following sections we will briefly describe various methodologies that can be used for the general or specific characterization of environments. Unfortunately most of these methods are laboratory oriented and require specialized and/or sophisticated equipment. Because of this, few methods are available that allow for real-time microbial analysis in the field.

MICROBIAL NUMBERS

HETEROTROPHIC PLATE COUNTS

Heterotrophic plate counts (HPCs) are obtained through a methodology that involves dilution and plating of a sample to allow specific microbes to grow, resulting in the formation of macroscopic colonies that can be seen with the naked eye and enumerated. These procedures are generally utilized for bacteria, and can be modified to give estimates of culturable fungi. The "dilution" of the environmental sample (typically soil or water) is necessary because of the vast numbers of microbes associated with such samples. The "plating" of the diluted environmental sample allows discrete colonies to grow on a solid culture media. Thus if the assumption is made that each colony arises from the presence of a single cell, then the number of colonies on the plate can be related back to the original number of microbes in the environmental sample, by taking into account the dilution factor (see Box 14.1 for sample calculation).

Heterotrophic plate counts give an indication of the general "health" of the soil, as well as an indication of the availability of organic nutrients within the soil. This concept is based on the assumption that the higher the concentration of culturable heterotrophic microorganisms the better the health of the soil. Similarly, the availability of organic nutrients in a soil can often be directly

BOX 14.1 *Dilution and Plating Calculations*

A 10-g sample of soil with a moisture content of 20% on a dry weight basis is analyzed for viable culturable bacteria via dilution and plating techniques. The dilutions were made as follows:

Step	Dilution
10 g soil→	95 ml saline (solution A) 10^{-1} (weight/volume)
1 ml solution A→	9 ml saline (solution B) 10^{-2} (volume/volume)
1 ml solution B→	9 ml saline (solution C) 10^{-3} (volume/volume)
1 ml solution C →	9 ml saline (solution D) 10^{-4} (volume/volume)
1 ml solution D →	9 ml saline (solution E) 10^{-5} (volume/volume)

1 ml of solution E is pour plated onto an appropriate medium and results in 200 bacterial colonies.

Number of CFU

$$= \frac{1}{\text{dilution factor}} \times \text{number of colonies}$$

$$= \frac{1}{10^{-5}} \times 200 \text{ CFU per g moist soil}$$

$$= 2.00 \times 10^7 \text{ CFU per g moist soil}$$

But, for 10 g of moist soil,

$$\text{Moisture content} = \frac{\text{moist weight} - \text{dry weight (D)}}{\text{dry weight (D)}}$$

Therefore,

$$0.20 = \frac{10 - D}{D} \text{ and}$$

$$D = 8.33 \text{ g}$$

Number of CFU per g dry soil =

$$2.00 \times 10^7 \times \frac{1}{8.33} = 2.4 \times 10^7$$

related to HPCs. Two basic types of media can be used for HPC: nutrient-rich and nutrient-poor media. Examples of nutrient-rich media are nutrient agar (Difco), peptone–yeast agar (Atlas, 1993), and soil extract agar (Atlas, 1993). These media contain high concentrations of peptone, yeast, and/or extracts from beef or soil. Nutrient-poor media are often called minimal media and contain as much as 75% less of these ingredients, often with substitutions such as casein, glucose, glycerol, or gelatin. Examples of minimal media are R2A agar (BBL, Difco) and m-HPC agar (Difco). In many cases higher colony counts are obtained with a nutrient-poor medium because many stressed, starved, or injured organisms in the environment cannot be cultured on a rich medium. This is especially true for water samples, and the nutrient-poor media, primarily R2A, are used in this case.

The *spread plate* procedure is relatively straightforward. A basic procedure would be addition of 10 g of soil to 90 mL of saline (0.85% NaCl). This soil suspension is then shaken or vortexed for up to 1 minute to break up soil aggregates and dislodge the soil microorganisms from the soil particles. This cell extraction step is followed by serial dilution of the sample to separate the microorganisms into individual reproductive units known as colony forming units (CFUs). Saline, buffered peptone, or phosphate solutions are commonly used for this step.

MICROSCOPIC DIRECT COUNTS

Microscopic direct counts (DCs) are used to visually count and provide quantitation of the concentrations of microorganisms in a sample. Concurrent with HPC, the use of direct microscopic enumeration of direct counts of soil microorganisms can provide added information on the health of an environment. Direct counts in environmental samples usually provide numbers that are higher than culturable counts because direct counts include viable organisms, viable but nonculturable organisms, and even nonviable but intact or nondegraded organisms. Direct microscopy of soil microorganisms involves separating the organisms from soil particles. Soils are first suspended in NaCl solution, usually 10 g of soil into 90 g of saline, and vortexed or shaken vigorously to separate organisms from soil particles and to disrupt soil aggregates. A known volume of the resulting soil suspension (usually 0.1 ml) is then placed on a glass microscope slide or filtered onto a membrane, stained to provide contrast, and examined with the microscope (Jones and Mollison, 1948).

To enhance the efficacy of direct microscopy, organisms must be stained to provide contrast. Perhaps the most widely used stain for direct microscopy of bacteria is acridine orange (AO), used in obtaining acridine orange direct counts (AODCs) (Figure 14.9). Common examples of fluorochromes, with their excitation and emission wavelengths, are shown in Table 14.3.

The classical methods for characterization of microbial populations and the microbial related health of a soil were by means of HPCs and direct microscopic enumerations

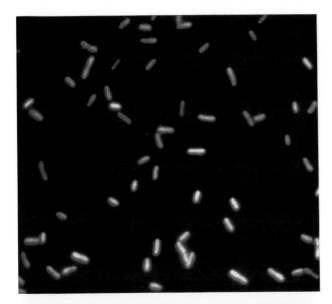

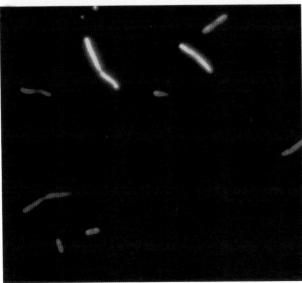

FIGURE 14.9 Acridine orange direct counts of bacteria. (Photo courtesy K.L. Josephson.) (Reprinted from *Environmental Microbiology* © 2000, Academic Press, San Diego, p. 203.)

of total organism using methods, such as AODCs. A useful concept is that the ratio of HPC to AODC is an indication of the health and/or the levels of stress on a given microbial community. A low HPC in relation to the AODC suggests low diversity and high environmental stress, whereas an HPC closer to the enumeration pro-

vided by the AODC indicates a healthy soil, often with high diversity and high microbial numbers.

MICROBIAL BIOMASS

The most common assay performed to give an indication of microbial biomass is chloroform fumigation. The principle behind the use of fumigation is that chloroform addition to a soil sample kills the indigenous microorganisms. The dead microbial biomass becomes available for microbial consumption. After fumigation of a soil sample, a starter culture from a similar sample of nonfumigated soil is added and mixed. During this microbial consumption of the dead biomass, CO_2 is produced. The amount of CO_2 generated is then used to calculate the quantity of microbial biomass that was originally present in the soil. The amount of CO_2 related by mineralization of dead microbial biomass is quantified using an alkaline trap and titration or by gas chromatography (GC) analysis of headspace gases. A control soil sample from the same source that is not fumigated is also incubated, along with the fumigated sample and the CO_2 generation used as a baseline correction when calculating biomass. In other words, the amount of respiration in the fumigated sample, corrected for the basal rate in the unfumigated sample, is a measure of the amount of microbial biomass (Maier *et al.,* 2000). The calculation of the amount of carbon held in microbial biomass is:

Biomass C =

$$\frac{F_c - U_{fc}}{K_c} \qquad \text{(Eq. 14.1)}$$

where

Biomass C = the amount of carbon trapped in microbial biomass;
F_c = CO_2 produced from the fumigated soil sample;
U_{fc} = CO_2 produced by the nonfumigated soil sample; and
K_c = fraction of biomass C mineralized to CO_2.
The value of K_c is often listed as 0.41 to 0.45.

CHARACTERIZATION AND IDENTIFICATION OF MICROORGANISMS

Individual isolates of soil bacteria are normally obtained by dilution or plating techniques described in the "Heterotrophic Plate Counts" section. Pure culture isolates can now be identified fairly routinely using molecular methods of analyses. This usually entails polymerase chain reaction (PCR) amplification of 16S ribosomal DNA sequences, and automated DNA sequencing of

TABLE 14.3
Excitation and Emission Wavelengths for Common Fluorochromes

Fluorochrome	Excitation Wavelength (nm)	Emission Wavelength (nm)
Acridine orange (bound to DNA)	502	526
Fluorescein isothiocyanate (FITC)	494	520

the resulting PCR product. Following this, sequence comparisons with known bacteria can be conducted using computer software sequence analysis, such as BLAST (Altuschul *et al.*, 1990).

APPLICATION OF MICROORGANISMS FOR TOXICITY TESTING

BIOSENSOR APPROACHES TO TOXICITY TESTING

Among the many challenges the 21st century brings are questions concerning how: (1) to maintain agricultural soil fertility to ensure food production to a still growing world population, (2) to prevent further environmental damage, and (3) to invent cost-effective ways to remediate existing contamination of soil, sediments, and water. Improved knowledge of how microorganisms interact with their environment and contribute to geochemical transformations will be critical to these endeavors. Of the three domains of life, the majority of microorganisms reported to date from soils and sediments are bacteria. Microbes in both aquatic and terrestrial ecosystems are responsible for many fundamental ecological processes, such as the biogeochemical cycling of chemical elements or the decomposition of plant and animal residues. Consequently, the potential impact of pollutants on such microbe-mediated processes may greatly affect the quality of the biosphere. Toxicity of pollutants to soil organisms in terms of potential impact on number, diversity, and microbial activity is of primary importance. It must be emphasized that for microbes in soil, total pollutant concentration does not necessarily reflect the amount of pollutant that is toxic. Rather, toxicity depends on the "bioavailable" concentration of the pollutant. Bioreporter genes signal the activity of an associated gene of interest. The theory behind a reporter gene is that the reporter gene product is more easily detected than the product of the gene of interest. Bioreporters provide a unique analytical capability because contaminants are quantified relative to the concentrations experienced by the reporter organism, as opposed to being relative to the extraction technique that is used for traditional analysis. Microbial whole-cell biosensors, which are specific for certain metals, measure only the bioavailable metal concentration that is taken up by the cells. Although quantification of bioavailable metal does not replace traditional chemical analysis, the comparison of metal bioavailability with its toxicity and total metal content will allow better definition of the biologically active portion of the metal.

Numerous, nonspecific microbial whole-cell sensors that react to nearly any kind of toxic substance have been developed. A novel approach for a microbial whole-cell sensor is to use recombinant DNA technology to construct a plasmid or other vector system in which a strictly regulated promoter is connected to a sensitive reporter gene. Reporter genes code for proteins that produce a signal that allows the protein to be determined in a complex mixture of other proteins and enzymes (Figure 14.10). The *lacZ* reporter system has been used extensively for pure culture analysis of microbial genetic expression. The product of the *lacZ* gene is a stable enzyme, β-galactosidase, which is easily detected and quantified by colorimetric assays using substrates such as X-Gal (5-bromo-4-chloro-3-indolyl-b-D-galactopyranoside) or ONPG (*o*-nitrophenol-b-D-galactopyranoside) (Snyder and Champness, 1977). Reporter genes that do not require substrate addition include those that produce luminescence or fluorescence proteins, such as the green fluorescent protein or the luciferase enzymes (Burlage and Kuo, 1994). The luciferase enzymes encoded by the *lux* and *luc* reporter genes are widely used reporter genes in prokaryotic as well as eukaryotic systems because they provide simple and sensitive detection of gene expression and regulation (Wood and Gruber, 1996). The luciferase enzymes produce a photon emission when the necessary substrates are present. The quantification of light emission (i.e., bioluminescence) is one of the most sensitive means of detection, and it can be measured with a liquid scintillation counter, a luminometer, or even with x-ray film. This makes it very suitable for environmental monitoring. The selection of the optimal bioreporter will depend on the nature of the application and the potential interference from samples being analyzed.

Inducible bioreporters contain reporter genes that are fused to a gene that is regulated by the concentration of the contaminant of interest. The most interesting promoters for environmental analyses are found in bacteria that survive in extreme environments contaminated by, for example, heavy metals or organic compounds. The ability of bacteria to survive in a contaminated environment is usually based on a genetically encoded resistance system, the expression of which is precisely regulated. The best studied example is the mercury resistance (*mer*) operon (Silver and Phung, 1996). The *mer* promoter is activated when Hg(II) binds to the regulatory protein MerR. Indicator bacteria that contain gene fusions between the promoter of the *mer* operon and a reporter gene are able to detect Hg(II). Sensor bacteria in which this promoter-reporter gene concept is operable have been developed to detect mercury, arsenic, cadmium, zinc and lead ions, and xenobiotic

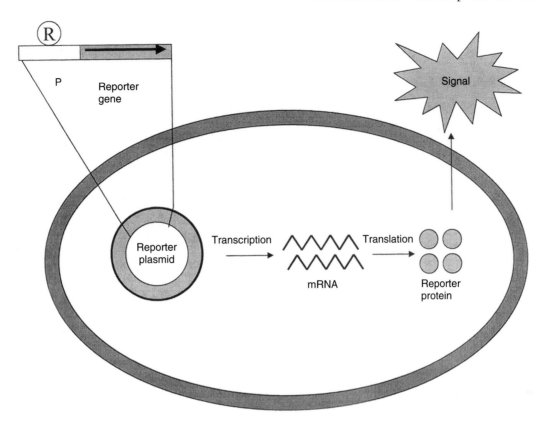

FIGURE 14.10 Regulation of a reporter gene by a regulatory protein. Binding of the regulatory protein *(R)* to the promoter *(P)* controls transcription, followed by translation of the mRNA to produce the protein. Both of these steps produce multiple copies of the reporter protein, leading to an increased protein concentration.

compounds (Selifnova *et al.*, 1993; Tauriainen *et al.*, 1998).

FIELD-BASED MICROBIAL BIOASSAYS FOR TOXICITY TESTING (SEE ALSO CHAPTER 17)

Microorganisms are useful ecotoxicity test organisms because a large number of independent organisms can be evaluated in a short time and they often occur at high levels where bioaccumulation and/or bioconcentration are potential problems. There are numerous bacterial toxicity tests that measure a wide variety of endpoints, including mutagenicity (Ames *et al.*, 1973), population growth (Nendza and Seydel, 1988), CO_2 production (Jardin *et al.*, 1990), enzyme biosynthesis (Dutton *et al.*, 1990), glucose mineralization (Reteuna *et al.*, 1989), and the inhibition of bioluminescence (Ribo and Kaiser, 1987). Although each of these studies provide useful data on the relative toxicity toward a particular species of bacteria, there are often discrepancies when comparing data to different test systems or to the same organism with different endpoints. These discrepancies are probably due to differences in protocol or toxicokinetics. Among the bacterial toxicity databases, the most extensive is the bioluminescence inhibition in *Vibrio fisheri*, or the "Microtox" assay. However, applications of the standard Microtox assay to environmental samples have not always been successful (Paton *et al.*, 1995). This is partially because *V. fisheri* is a marine bacterium that is sensitive to changes in pH and osmotic conditions. To obtain environmentally relevant test organisms, genetically engineered bioluminescent bacteria were constructed for terrestrial, wastewater, and brewery environments (Boyd *et al.*, 1997).

IMPORTANT MICROBIAL PROCESSES

Environmental microorganisms consist of large, diverse populations that are capable of biological transformations and processes that can be beneficial or detrimental to human health and welfare. As shown in Box 14.2, beneficial soil microorganisms are those that can fix atmospheric nitrogen, decompose organic wastes and residues, detoxify pesticides, suppress plant diseases and soilborne pathogens, enhance nutrient cycling, and produce bioactive compounds such as vitamins, hormones, and enzymes that stimulate plant growth. Harmful soil microorganisms are those that can induce plant disease,

BOX 14.2 *Beneficial and Harmful Soil Microbial Processes*

Beneficial Soil Microbial Processes

- Fixation of atmospheric nitrogen
- Decomposition of organic wastes and residues
- Suppression of soilborne pathogens
- Recycling and increased availability of plant nutrients
- Degradation of toxicants including pesticides
- Production of antibiotics and other bioactive compounds
- Production of simple organic molecules for plant uptake
- Complexation of heavy metals to limit plant uptake
- Solubilization of insoluble nutrient sources
- Production of polysaccharides to improve soil aggregation

Functions of Harmful Soil Microorganisms

- Induction of plant diseases
- Stimulation of soilborne pathogens
- Immobilization of plant nutrients
- Inhibition of seed germination
- Inhibition of plant growth and development
- Production of phytotoxic substances

TABLE 14.4

Examples of Important Autotrophic Soil Bacteria

Organism	Characteristics	Function
Nitrosomonas	Gram negative, aerobe	Converts $NH_4^+ \rightarrow NO_2^-$ (first step of nitrification)
Nitrobacter	Gram negative, aerobe	Converts $NO_2^- \rightarrow NO_3^-$ (second step of nitrification)
Thiobacillus	Gram negative, aerobe	Oxidizes $S \rightarrow SO_4^{2-}$ (sulfur oxidation)
Thiobacillus denitrificans	Gram negative, facultative anaerobe	Oxidizes $S \rightarrow SO_4^{2-}$; functions as a denitrifier
Thiobacillus ferrooxidans	Gram negative, aerobe	Oxidizes $Fe^{2+} \rightarrow Fe^{3+}$

(Reprinted from *Environment Microbiology* © 2000, Academic Press, San Diego.)

stimulate soilborne pathogens, immobilize nutrients, and produce toxic and putrescent substances that adversely affect plant growth and health. Obvious detrimental impacts also occur via human pathogenic microbes, which will be covered in Chapter 17. Processes can be carried out by autotrophic microbes (Table 14.4) or heterotrophic microbes (Table 14.5). Any one process can be beneficial or detrimental, depending on the particular environment. For example, nitrification within

a soil can lead to decreased plant nitrogen use efficiency, as well as health impairment via methemoglobinemia from nitrate pollution of water. However, in sewage treatment plants, nitrification is a prerequisite for denitrification, which is the major mechanism for nitrogen removal from sewage. In all of these processes the product of the microbial transformation can often be measured or monitored. Thus nitrate can be a measure of nitrification, whereas measurement of sulfate concentrations can indicate the degree of sulfur oxidations. Often the measurements involve some kind of soil extraction, followed by sophisticated analyses (e.g., atomic absorption or UV spectroscopy). The autotrophic bacteria can be important in causing pollution via nitrification or sulfur oxidation, which results in reduced soil pH values that solubilize metals. In contrast, heterotrophic bacteria are often used beneficially for bioremediation of organics. Obviously the "black box" approach, wherein the product of a microbial transformation is monitored, can be applied to any compound of interest, provided specific sensitive methods of analysis are available.

TABLE 14.5

Examples of Important Heterotrophic Soil Bacteria

Organism	Characteristics	Function
Actinomycetes (e.g., *Streptomyces*)	Gram positive, aerobic, filamentous	Produces geosmin's "earthy odor," and antibiotics
Bacillus	Gram positive, aerobic, spore former	Carbon cycling, production of insecticides and antibiotics
Clostridium	Gram positive, anaerobic, spore former	Carbon cycling (fermentation), toxin production
Methanotrophis (e.g., *Methylosinus*)	Aerobic	MethylosinusMethane oxidizers that can cometabolize trichloroethene (TCE) using methane monooxygenase
Alcaligenes eutrophus	Gram negative, aerobic	2,4-D degradation via plasmid pJP4
Rhizobium	Gram negative, aerobic	Fixes nitrogen symbiotically with legumes
Frankia	Gram positive, aerobic	Fixes nitrogen symbiotically with nonlegumes
Agrobacterium	Gram negative, aerobic	Important plant pathogen, causes crown gall disease

(Reprinted from *Environmental Microbiology* © 2000, Academic Press, San Diego.)

TABLE 14.6
State-of-the-Art Approaches for Characterizing Microbial Properties and Activities

Technique	Function
I. Microbial	
– epifluorescent microscopy	Detection of specific microbes
– electron microscopy	Magnification up to $10^6 \times$ 3-D images
– confocal scanning microscopy	
II. Physiological	
– carbon respiration	Heterotrophic activity
– radiolabeled tracers	Degradation of specific compounds
– enzyme assays	Specific microbial transformations
III. Immunological	
– immunoassays	Detection of specific microbes
– immunocytochemical assays	Location of specific antigens
– immunoprecipitation assays	Semi-quantitative determination of antigens
IV. Nucleic Acid Based	
– polymerase chain reaction (PCR)	Detection of genes or specific microbes
– reverse transcriptase (RT-PCR)	Detection of mRNA or viruses
– gene probes	Detection of genes
– denaturing gradient gel electrophoresis	Microbial diversity changes
– plasmid profile analysis	Unique microbial functions

STATE-OF-THE-ART APPROACHES FOR MONITORING MICROBIAL ACTIVITIES

Many sophisticated analyses are currently available that allow for enhanced environmental monitoring of microbial activities. Although they fall outside the scope of this book and are generally unavailable for routine environmental monitoring, it is important to be aware that such analyses do exist. Table 14.6 illustrates some of these important new techniques. For details on these techniques, the reader should review specific texts such as *Environmental Microbiology* (Maier *et al.*, 2000).

QUESTIONS

1. Briefly discuss the statement: "Soil is a favorable habitat for microorganisms."

2. Student "A" performs a soil extraction with peptone as the extracting solution. A dilution and plating technique is performed using the spread plate method. The following results are obtained:

   a. 150 colonies are counted on an R2A plate at the 10^5 dilution.
   b. 30 colonies are counted on a nutrient agar plate at 10^4 dilution.

   Calculate the number of viable bacteria per gram of soil for R2A and nutrient agar. Why are the counts for nutrient agar lower than for the R2A media?

3. Student "B" performs a soil extraction with peptone as the extracting solution. A dilution and plating technique is performed using the pour plate method. The number of bacteria per gram of soil is calculated to be 8×10^8. What dilution plate contained 80 colonies?

4. What are the major differences in sampling and storage protocols when soils are sampled for subsequent microbial as opposed to chemical analysis?

5. What are some of the major problems associated with "air sampling" and the collection of bioaerosols?

6. a. Give at least two examples of aerobic heterotrophic bacteria commonly found in surface soils.
   b. Give at least two examples of anaerobic heterotrophic bacteria commonly found in surface soils.
   c. Give at least two examples of aerobic autotrophic bacteria commonly found in surface soils.
   d. Give at least two examples of anaerobic autotrophic bacteria commonly found in surface soils.

REFERENCES AND ADDITIONAL READING

Altuschul, S.F., Gish, W., Miller, W., Myers, E.W., and Lipman, D.J. (1990) Basic local alignment search tool. *J. Mol. Biol.* **215**, 403–410.

Amann, R.I., Ludwig, W., and Schleifer, K.H. (1995) Phylogenetic identification and *in situ* detection of individual microbial cells without cultivation. *Microbiol. Rev.* **59**, 143–169.

Ames, B.N., Durston, W.E., Yamasaki, E., and Lee, F.D. (1973) Carcinogens are mutagens: A simple test system combining liver homogenates for activation and bacteria for detection. *Proc. Natl. Acad. Sci. USA* **70**, 2281–2285.

Atlas, R.M. (1993) "Handbook of Microbiological Media." CRC Press, Boca Raton, FL.

Boyd, E.M., Meharg, A.A., Wright, J., and Killham, K. (1997) Assessment of toxicological interactions of benzene and its primary degradation products (catechol and phenol) using a *lux*-modified bacterial bioassay. *Environ. Toxicol. Chem.* **16**, 849–856.

Burlage, R.S., and Kuo, C. (1994) Living biosensors for the management and manipulation of microbial consortia. *Ann. Rev. Microbiol.* **48**, 291–309.

Dutton, R.J., Britton, G., Koopman, B., and Agami, O. (1990) Effects of environmental toxicants on enzyme biosynthesis: A comparison of β-galactosidase, α-glucosidase, and tryptophanase. *Arch. Environ. Contam. Toxicol.* **19**, 395–398.

Jardin, W.F., Pasquini, C., Guimaraes, J.R., and deFaria, L.D. (1990) Short-term toxicity test using *Escherichia coli*: Monitoring CO_2 production by flow injection analysis. *Wat. Environ. Res.* **24**, 351–354.

Jensen, P.A., Todd, W.F., Davis, G.N., and Scarpino, P.V. (1992) Evaluation of eight bioaerosol samplers challenged with aerosols of free bacteria. *J. Ind. Hyg. Assoc.* **53**, 660–667.

Jones, P.C.T., and Mollison, J.E. (1948) The technique for the quantitative estimation of soil microorganisms. *J. Gen. Microbiol.* **2**, 54–64.

Maier, R.M., Pepper, I.L., and Gerba, C.P. (2000) *Environmental Microbiology. A Textbook.* Academic Press: San Diego.

Nendza, M., and Seydel, J.K. (1988) Quantitative structure-toxicity relationships for ecotoxicologically relevant biotest systems and chemicals. *Chemosphere* **17**, 1585–1602.

Orgel, L.E., and Crick, F.H.C. (1980) Selfish DNA. *Nature* **288**, 601–603.

Paton, G.I., Palmer, G., Kindnes, A., Campbell, C., Glover, L.A., and Killham, K. (1995) Use of luminescence-marked bacteria to assess copper bioavailability in malt whiskey distillery effluent. *Chemosphere* **31**, 3217–3224.

Pepper, I.L., Gerba, C.P., and Brusseau, M.L., eds. (1996) *Pollution Science*, Academic Press, San Diego, CA.

Pillai, S.D., Widmer, K., Dowd, S.E., and Ricke, S.L. (1996) Occurrence of airborne bacteria and pathogen indicators during land application of sewage sludge. *Appl. Environ. Microbiol.* **62**, 296–299.

Reteuna, C., Vasseur, P., and Cabridenc, R. (1989) Performances of three bacterial assays in toxicity assessment. *Hydrobiologia* **188**, 149–153.

Ribo, R.M., and Kaiser, K.L.E. (1987) *Photobacterium phosphoreum* toxicity bioassay: I. Test *Proc. Appl. Toxicol. Assess.* **2**, 305–323.

Roszak, D.B., and Colwell, R.R. (1987) Survival strategies of bacteria in the natural environment. *Microbiol. Rev.* **51**, 365–379.

Selifnova, O., Burlage, R., and Barkay, T. (1993) Bioluminescent sensors for detection of bioavailable Hg(II) in the environment. *Appl. Environ. Microbiol.* **59**, 3083–3090.

Silver, S., and Phung, L.T. (1996) Bacterial heavy metal resistance: new surprises. *Annu. Rev. Microbiol.* **50**, 753–789.

Snyder, L., and Champness, W. (1977) Molecular Genetics of Bacteria. American Society for Microbiology, Washington, D.C.

Sparkling, G.P., and Cheshire, M.V. (1979) Effects of soil drying and storage on subsequent microbial growth. *Soil Biol. Biochem.* **11**, 317–319.

Stotzky, G., Goos, R.D., and Timonin, M.I. (1962) Microbial changes occurring in soil as a result of storage. *Plant Soil* **16**, 11–18.

Tauriainen, S., Karp, M., Chang, W., and Virta, M. (1998) Luminescent bacterial sensor for cadmium and lead. *Biosens. Bioelectron.* **13**, 931–938.

Wollum, A.G. (1994) Soil sampling for microbiological analysis. In *Methods of Soil Analysis, Part 2, Microbiology and Biochemical Properties.* SSSA Book Series No. 5. Soil Science Society of America, Madison, WI, pp. 2–13.

Wood, K.V., and Gruber, M.G. (1996) Transduction in microbial biosensors using multiplexed bioluminescence. *Biosens. Bioelectronics* **11**, 204–214.

15

PHYSICAL CONTAMINANTS

J.L. WALWORTH

Small particles can pollute air and water supplies. These particles pose a hazard to human health and to the environment in a variety of ways. Sources of particulates are divided into those arising from a single, well-defined emission source, which is called *point-source pollution*, and those originating from a wide area, which is called *nonpoint source pollution*. Particulate emissions can be human-made or natural. Naturally-occurring particulates may become a threat when aggravated by human activ-

ities, such as agriculture, logging, and construction, or they may originate from completely natural processes, such as volcanoes or soil erosion. Human-made particulates include those created by industrial processes, combustion in power plants, wood stoves, fireplaces, internal combustion engines, and rubber particles from tire wear. In this chapter we will explore the properties of particulate pollutants, the health threats they present, where they come from, their fate in the environment, and how these pollutants can be controlled.

NATURALLY OCCURRING PARTICULATES

SOIL PARTICLES

Soil particles are natural contaminants, but are not considered to be pollutants unless they move into the atmosphere or surface waters. Soil particles shown in Box 15.1 are divided into three size categories in the U.S. Department of Agriculture (USDA) classification system.

To give some perspective to these size ranges, blonde human hair ranges from 0.02 to 0.05 mm (20 to 50 μm) in diameter (dark hair is somewhat thicker), so its diameter is similar to that of a silt particle. An individual silt particle is too small to be seen with the naked eye, but is visible through a light microscope. Clay particles, on the other hand, are generally too small to be seen with a light microscope. A wide range of particle sizes are shown in Figure 15.1.

Stokes' Law (Box 15.2) is used to describe the fall of particles through a fluid medium, such as air or water, and is shown in the following equation:

$$V = [D^2 * (\rho_p - \rho_1) * g]/18\eta \qquad \text{(Eq. 15.1)}$$

where

V = velocity of fall (cm s^{-1})

g = acceleration of gravity (980 cm s^{-2})

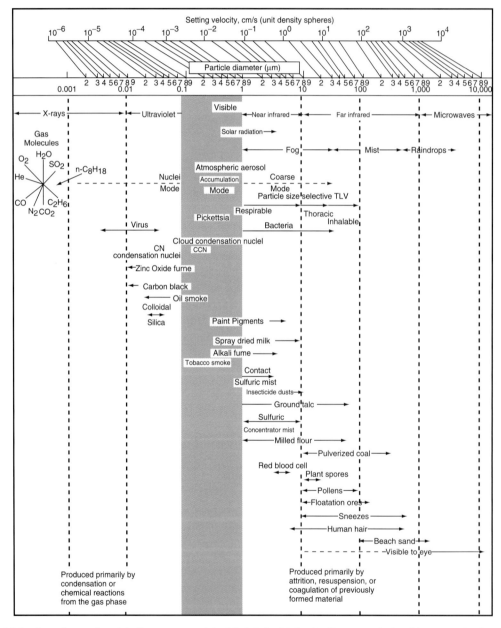

FIGURE 15.1   Molecular and aerosol particle diameters, copyright © P.C. Reist. Molecular diameters calculated from viscosity data. See Bird, Stewart, and Lightfoot, 1960, *Transport phenomena*, Wiley. (From Liu, D.H.F, and Liptak, B.G. [2000] *Air Pollution*. Lewis Publishers, Boca Raton, FL, an imprint of CRC Press.)

D = diameter of particle (cm)
ρ_p = density of particle (density of quartz particles is 2.65 g cm^{-3})
ρ_1 = density of dispersion medium (air has a density of about 0.001213 g cm^{-3}; water has a density of about 1 g cm^{-3})
η = viscosity of the dispersion medium (about 1.83×10^{-4} poise or g cm s^{-1} for air; 1.002×10^{-2} poise for water)

Small particles, once suspended in either air or water, can remain suspended for a long time and can be transported great distances. In fact, sedimentation ("settling out" of solution) is not an important factor for particles less than 0.0005 mm in diameter. Clay size particles are small enough that they may act as *colloids*, which are so small that once suspended in a dispersion medium such as air or water, they do not settle out through the force of gravity, which means they can remain in suspension indefinitely. Large particles, such as those in the sand size category, remain suspended for only a short time, and their range of movement is limited. Turbulence (disorderly flow or mixing of the dispersion medium) can increase suspension times considerably, so the calculated settling rates are valid only in the absence of turbulence. Actual settling rates are generally greater.

Sand and silt size particles are dominated by primary minerals such as quartz, feldspar, and mica formed directly from molten magma. Clay particles, on the other hand, are composed largely of secondary layered silicate clay minerals that are weathering products of primary minerals, hydrous oxides of aluminum and iron, and organic materials. Most of these particles have a net negative electric charge (although some can have a net positive charge; others have no charge at all), either from characteristics of their crystalline structure, as in the case of layered silicate clay minerals, or due to surface chemical reactions. Positively charged cations surrounding clay particles balance the surface charge of the particles. Negatively charged clay particles are always surrounded by a cloud of cations. Additionally, certain anions and organic chemicals can be sorbed to clay particles and transported by moving particles. Thus clay particles can act as a carrier in the distribution and transport of associated molecules.

PARTICLES IN AIR

Health and environmental concerns associated with naturally occurring particles can be categorized based on the contaminated medium. Particles suspended in air are known as *aerosols*. Aerosols pose a threat to human health mainly through respiratory intake and deposition in nasal and bronchial airways. Smaller aerosols travel farther into the respiratory system and generally cause more health problems than larger particles. For this reason the U.S. Environmental Protection Agency (EPA) has divided airborne particulates into two size categories: PM$_{10}$, which refers to particles with diameters less than or equal to 10 μm, and PM$_{2.5}$, which are particles less than or equal to 2.5 μm in diameter. See Chapter 10 for a description of air particulate monitoring equipment. For this classification, the diameter of aerosols is defined as the *aerodynamic diameter*:

$$d_a = d_p * \left(\rho_p / \rho_w \right)^{1/2} \qquad (Eq. \ 15.2)$$

where
d_a = aerodynamic particle diameter (μm)
d_p = actual particle diameter (μm)
ρ_p = particle density (g cm^{-3})
ρ_w = density of water (g cm^{-3})

Atmospheric particulate concentration is expressed in micrograms of particles per cubic meter of air (μg m^{-3}). (Sidebar 15.1)

Symptoms of particulate matter inhalation include decreased pulmonary function, chronic coughs, bronchitis, and asthmatic attacks. The specific causal mechanisms are poorly understood.

One well-documented episode of aerosol pollution occurred in London in 1952, when levels of smoke and sulfur dioxide aerosols, largely associated with coal combustion, reached elevated levels due to local weather conditions. Approximately 4000 deaths occurring over a 10-day period were attributable to cardiovascular and lung disorders brought on or aggravated by these aerosols (Koenig, 1999).

Some pollen, fungal, and plant spores, most bacteria, and viruses fall into the PM_{10} and $PM_{2.5}$ categories. Suspended pathogenic spores and microbial organisms, in addition to the detrimental respiratory and pulmonary health effects caused by other types of particles, can cause diseases. Coccidioidomycosis (also known as Valley Fever) is one such disease caused by inhalation of spores of the fungus *Coccidioides immitis*, which is indigenous to hot, arid regions, including the Southwestern United States. The fungus can travel from the respiratory tract to the skin, bones, and central nervous system and can result in systemic infection and death.

Some sources of naturally occurring PM_{10} in the United States are shown in Table 15.1. In addition to the sources shown in Table 15.1, volcanoes and breaking waves also generate airborne particles, sometimes in very large quantities.

Airborne particles can travel great distances. Intense dust storms during 1998 and 2001 in the Gobi desert of Western China and Mongolia (Figure 15.2) elevated aerosol levels to concentrations near the health standard in Western North America several thousand miles away!

PARTICLES IN WATER

In water, suspended particulates pose quite different risks. Inorganic particles cause an increase in the turbidity of affected water, and the particles themselves can create problems through sedimentation that can fill lakes, dams, reservoirs, and waterways. In the United States, waterborne soil particles fill more than 123 million cubic meters of reservoir capacity each year, reducing water storage capacity and necessitating expensive dredging operations. Suspended particles increase wear on pumps, hydroelectric generators, and related equipment. Also, lands from which suspended particles are derived suffer damage that can reduce agricultural productivity and land values. Severe soil erosion can threaten buildings, roads, and other structures.

Naturally occurring organic particles are composed of plant and animal residues, which consist of carbohydrates, proteins, and more complex compounds. Under favorable conditions, these compounds can be degraded via microbially mediated oxidation processes (see Chapter 14). They may also be subject to abiotic reactions (see Chapter 13).

The oxidation of waterborne organic particulates consumes dissolved oxygen and produces CO_2. The use of oxygen in biological oxidation reactions is called the *biological oxygen demand* (BOD). Increasing BOD and resulting oxygen depletion, called *hypoxia*, can have a profound effect on aquatic animals, such as fish, that depend on the dissolved oxygen supply. (See Chapter 9

TABLE 15.1

Annual Aerosol Production from Human Activities

Source	PM_{10} (millions of metric tons)
Industrial processes	
Chemical industries	0.064
Metals processing	0.200
Petroleum industries	0.037
Other industries	0.481
Solvent utilization	0.005
Storage and transport	0.103
Waste and disposal and recycling	0.269
Fuel combustion	
Electric utilities	0.263
Industrial	0.285
On-road vehicles	0.243
Non-road sources	0.423
Other	
Agricultural and forestry	4.270
Fire and other combustion	0.921
Unpaved roads	11.163
Paved roads	2.282
Construction	3.649
Wind erosion	4.823

(From Council on Environmental Quality, 1997.)

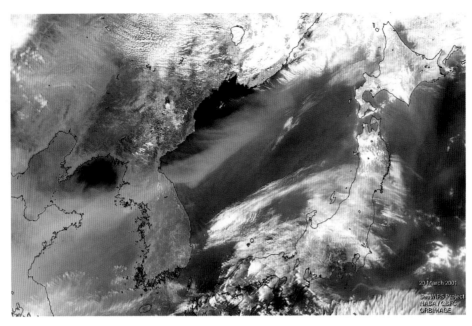

FIGURE 15.2 Satellite image of Mongolian dust cover over Sea of Japan. (Provided by NASA.)

for a description of methods to collect and measure particulates in water.)

CHEMICAL TRANSPORT BY PARTICLES

Both organic and inorganic particles can act as carriers of other contaminants that are sorbed to particle surfaces. One of the more important of these contaminants is phosphorus. When aquatic phosphorus levels are elevated, the increase in nutrient levels can cause excessive growth of algae and other aquatic plants. Although phosphorus is not the sole factor in aquatic nutrient and oxygen status, it limits growth in many fresh waters, and increases in dissolved phosphorus are related to accelerated aquatic algal and plant growth. Following algal blooms or other increased plant growth, aerobic decomposition of the plant tissue increases BOD and depletes dissolved oxygen. As the dissolved oxygen levels decrease, oligotrophic waters (deficient in plant nutrients, with abundant dissolved oxygen) change to mesotrophic, and finally to eutrophic waters (Figure 15.3). This process is known as *eutrophication*.

A dramatic example of this has been documented in the Gulf of Mexico, where a large hypoxic zone is known to have existed for decades. In 1993 flood waters from the Midwest, carrying nutrients and suspended particulates, moved down the Mississippi River and into estuaries in the Gulf of Mexico. The hypoxic zone doubled in size, reaching an area approximately twice the size of New Jersey. In this area dissolved oxygen levels dropped below two parts per million, resulting in widespread death of fish and other aquatic life (minimum levels of four to five parts per million are required to support healthy populations of most fish species).

In aquatic and soil systems, dissolved phosphate is present mainly as phosphate anions $H_2PO_4^-$ (pH 2 to 7) and HPO_4^{2-} (pH 7 to 12). Although these negatively charged ions are electrostatically repelled by negatively charged clay particles, they are strongly sorbed by clay,

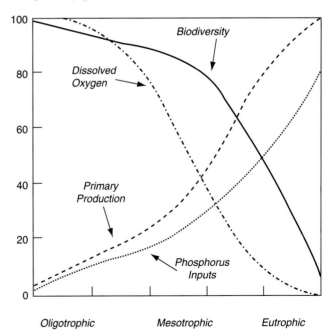

FIGURE 15.3 Changes that occur as fresh waters become eutrophic. (Adapted from Pierzynski, G.M., Sims, J.T. and Vance, G.F. [2000] *Soils and Environmental Quality.* CRC Press, Boca Raton, FL.)

particularly in acid soils. Phosphate ions can replace water or hydroxy groups on the surfaces of metal oxides, such as those of hydrous aluminum and iron oxide particles or the metal oxide surfaces found at the edges of layer silicate minerals (Figure 15.4). Particulates composed of or coated with iron and aluminum oxides have a greater capacity to sorb phosphorus than other types of clay particles. The bond between the phosphate ion and the metal oxide surface is initially monodentate (attached at one point) but, over time, becomes bidentate (attached at two points) and more stable. When clay particles with sorbed phosphorus are suspended in water, or deposited in sediments following suspension, sorbed phosphorus equilibrates with phosphorus dissolved in the water, raising the water's phosphate content and buffering the system, with respect to phosphorus. Once waters are contaminated with phosphate-laden particulates, it is difficult to reduce aquatic phosphate levels because of the buffering provided by the phosphorus-clay complex. The amount of buffering provided by soil particles can be illustrated through phosphorus adsorption isotherms, which relate phosphorus sorption and desorption to the aqueous-phase phosphorus concentration in equilibrium with the particles (Figure 15.5). The equilibrium phos-

phorus concentration at the point where there is zero adsorption or desorption is the EPC_0. The higher the EPC_0 value, the greater the ability of the soil particles to desorb phosphorus into affected waters.

Organic chemicals, including herbicides, insecticides, fuels, solvents, preservatives, and other industrial and agricultural chemicals, can also adsorb and desorb from waterborne particulates. The health threats associated with this very diverse group of chemicals are similarly broad, as discussed in Chapter 16. Aquatic life also can be influenced by pesticide contamination and through increased BOD and hypoxia caused by oxidation of organic chemicals, as previously described. Nonpolar organic chemicals tend to associate with organic soil particles, which also exhibit nonpolar properties (although there may also be considerable sorption onto clay particles). For this reason, organic chemical sorption on soil materials is modeled by describing how the organic chemical partitions between water (a polar solvent) and octanol (a nonpolar solvent) with the *octanol-water partition coefficient* (K_{ow}). Organic chemicals with a high K_{ow} bind to organic soil particles more than those with low K_{ow} values, and are more likely to be transported with waterborne particles. This relationship is shown in Figure 15.6. Adsorption isotherms similar to those shown for phosphorus adsorption/desorption can be constructed for organic contaminants.

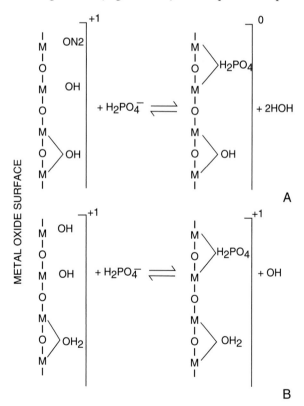

FIGURE 15.4 Displacement of (A) aquo and (B) hydroxy group from a metal oxide surface by phosphate sorption. (Adapted from Uehara, G. and Gilman, G. [1981] *The Mineralogy, Chemistry, and Physics of Tropical Soils with Variable Charge Clays.* Westview Tropical Agriculture Series, Westview Press, Boulder, CO.)

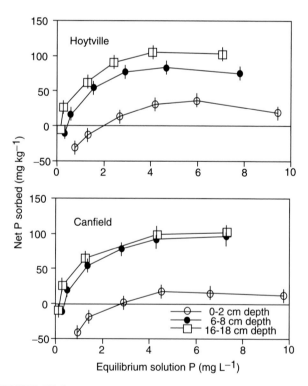

FIGURE 15.5 Phosphorus sorption isotherms for Hoytville and Canfield soils at three depths. (From Guertal, E.A., Eckert, D.J., Traina, S.J., and Logan, T.J. [1991] *Soil Sci. Soc. Am. J.* 55, 410–413.)

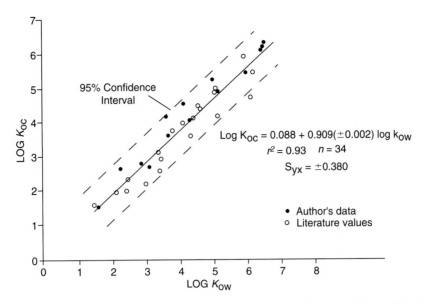

FIGURE 15.6 Log K_{oc}-log K_{ow} relationship (sorption on soil organic fraction versus octonol-water partition coefficient). (From Sawney, B.L. and Brown, K. [eds] [1989] *Reactions and Movement of Organic Chemicals in Soils.* SSSA Special Pub. No. 22, Madison, WI.)

PARTICLE FLOCCULATION

As previously indicated, the settling rates of particles suspended in water depends on, among other things, their effective diameter. This can be greatly increased when particles adhere to one another, or *flocculate*, to form aggregates made up of groups of soil particles. Particles with like charges repel each other, so the forces of repulsion must be overcome for particles to aggregate. The negative charge on clay particles is balanced by a *diffuse double layer*, which is a cloud of counterions (cations) surrounding the clay particle. In this double layer the concentration of cations increases as the clay surface is approached, while concomitantly, the concentration of anions decreases. The thickness of this diffuse double layer depends on several factors, the most important of which are counterion valence and concentration.

$$1/\chi = \left(\frac{K}{z^2 n}\right)^{1/2} \qquad \text{(Eq. 15.4)}$$

where

$1/\chi$ = the relative effective thickness of the double layer

z = valence of the counterions (cations)

K = a constant dependent on temperature and the dielectric constant of the solvent

n = electrolyte concentration in the equilibrium solution

The thickness of the double layer decreases as the valence of counterions increase, or as the electrolyte concentration in the solvent increases (Figure 15.7). In

some clay systems, particularly those with hydrous aluminum and iron oxides, the charge on the clay surfaces can be altered by solvent pH, but the relationship in Equation 15.4 can still be used to describe the nature of the double layer. As the double layer thickness decreases, attractive energy between particles becomes equal to repulsive forces, and the particles flocculate. The electrolyte concentration at which this occurs is called the *critical*

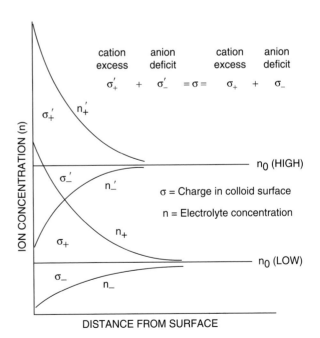

FIGURE 15.7 Thickness of the electrical double layer and ion distribution at two electrolyte concentrations. (From Sumner M.E. and Stewart B.A. [eds] [1992] *Soil Crusting: Chemical and Physical Processes.* Lewis Publishers, an imprint of CRC Press, Boca Raton, FL.)

flocculation concentration (CFC), which is described by the Schulze-Hardy Rule:

$$CFC = \left(\frac{1}{z^6}\right) \qquad \text{(Eq. 15.5)}$$

The Schulze-Hardy Rule indicates the CFC is a function of $1/z^6$. Although this relationship does not completely explain the variable ability of different cations to flocculate soil particles, we can see that the effect of counterion valence is extremely important. The relative flocculating power of common soil cations is shown in Table 15.2.

Soil particles in suspension in water will tend to be dispersed if the electrolyte concentration is low, and the dominant counterions are potassium or sodium. In contrast, flocculation will be more prone to occur with a high electrolyte concentration or if the dominant counterions are higher valence. Where rivers flow into the ocean, sediment-laden river water mixes with saline ocean water. Suspended particles flocculate and settle out to form river deltas (Figure 15.8). To flocculate and settle particles out of wastewater, high-valence flocculating agents such as aluminum, iron, and copper sulfates and chlorides are added as clarifiers. Organic polymers and synthetic polyelectrolytes that have anionic or cationic functional groups are also used to flocculate particles suspended in water.

Depending on the properties of the suspended particles and the impacted water, particles may flocculate into aggregates and quickly settle out of suspension, or they may remain dispersed. As previously indicated, dispersed colloidal particles can remain in suspension indefinitely.

HUMAN-MADE PARTICULATES

Particulate matter in the atmosphere can be from direct emissions, pollution that enters the atmosphere as previously formed particles. These are called *primary particles*. Alternatively, *secondary particles* are formed in the atmosphere from precursor components, such as ammonia, volatile organics, or oxides of nitrogen (NO_x) and sulfur (SO_x). Primary particles may fall into the $PM_{2.5}$ or the PM_{10} size range, whereas secondary particles generally fall into the $PM_{2.5}$ category.

Industrially generated primary particles arise from incomplete combustion processes and high-temperature metallurgical processes. Secondary particles, on the other hand, are produced from gases emitted from industrial processing, and various combustion processes (automobiles, power plants, wood burning, incinerators) that undergo gas-to-particle conversion followed by growth and coagulation. In the atmosphere sulfur oxides are oxidized to form sulfuric acid and fine sulfate particles. Gases condense to form ultrafine particles (less than $0.01~\mu m$), either from supersaturated vapor produced in high-temperature combustion processes or through photochemical reactions. These particles grow in size through condensation and coagulation to form larger particles (0.1 to $2.5~\mu m$). The principal sources of SO_x in the United States include coal power plants, petroleum refineries, paper mills, and smelters. NO_x, on the other hand, is largely produced by industrial and automotive combustion processes.

Health threats from $PM_{2.5}$ and PM_{10} generated by human activities are much like those from naturally occurring particulates. Adverse health effects are most severe in senior citizens, and those with pre-existing heart or lung problems (Sidebar 15.2).

Particles formed from incomplete combustion, such as those from wood-burning and diesel engines, contain organic substances that may have different health effects. Diesel exhaust has been shown to increase lung tumors in rats and mice, and long-term human exposure to diesel exhaust may be responsible for a 20% to 50% increase in the risk of lung cancer (Koenig, 1999).

In addition to human health concerns, $PM_{2.5}$ associated with wood burning, automobile exhaust, and industrial activities is responsible for much of the atmospheric haze in the United States. Aerosols (particularly $PM_{2.5}$) absorb and scatter light, producing haze and

TABLE 15.2

Relative Flocculating Power of Common Monovalent and Divalent Cations

Ion	Relative Flocculating Power (Relative to Na^+)
Na^+	1.00
K^+	1.70
Mg^{2+}	27.00
Ca^{2+}	43.00

(From Sposito, G. [1984] *The Surface Chemistry of Soils.* Oxford University Press, New York.)

SIDEBAR 15.2 Recent studies estimate that with each $10~\mu g^{-3}$ increase in PM_{10} above a base level of $20~\mu g m^{-3}$, daily respiratory mortality increases by 3.4%, cardiac mortality increases 1.4%, hospitalizations increase 0.8%, emergency room visits for respiratory illnesses increase by 1.0%, days of restricted activity due to respiratory symptoms increase 9.5%, and school absenteeism increases 4.1% (Vedal, 1995).

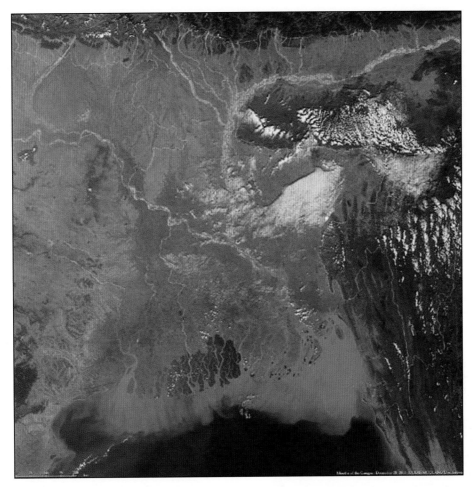

FIGURE 15.8   Sediment-laden water from the Ganges and Brahmaputra rivers flows into the Bay of Bengal. (Courtesy of NASA.)

reducing visibility. When severe, this interferes with automobile and aviation navigation, posing safety threats. Atmospheric particulates can also be a nuisance by settling on and in cars, homes, and other buildings.

MECHANISMS AND CONTROL OF PARTICULATE POLLUTION

As noted previously, naturally occurring particulate pollution comes from many sources, including agricultural operations, logging, construction-related activities, mining and quarrying, unpaved roads, and from wind erosion. Soil erosion is a natural process that occurs continuously but is often accelerated by human activities. Several factors are required for soil material to become dislodged and transported into air or water. Soil must be susceptible to erosional processes, which generally requires that the soil be exposed to erosional forces. In the first phase of soil erosion, soil particles become dislodged. Various soil properties determine the suscepti-

bility of soil particles to dislodgement. In the second phase the particles are transported.

WATER-INDUCED SOIL EROSION AND CONTROL

We will first consider particle movement caused by water. As indicated previously, soil particles are usually formed into aggregates, which vary considerably in size, shape, and stability. Organic materials and certain soil cations are the primary agents of soil aggregation. Individual particles are dispersed or separated from aggregates in the detachment phase. The source of the energy responsible for detaching soil particles is either raindrop impact or the flow of water. When raindrops, which travel at approximately $900 \, cm \, s^{-1}$, hit bare soil, the kinetic energy of the raindrops is transferred to the soil particles, breaking apart aggregates and dislodging particles. Dislodged particles can be moved more than 1 m in the splash from raindrop impact. They are moved larger distances by flowing water, which can dislodge additional particles through a scouring action. Smaller particles are transported more easily than larger ones, and faster flowing water can carry a heavier particulate load than slower moving water. When uniform shallow layers of soil are eroded off areas of land, this is called *sheet erosion*. Directed water flow cuts channels into the soil forming small channels or *rills*, or large channels known as *gullies* (Figure 15.9).

The process of water erosion has been described by the USDA Universal Soil Loss Equation (USLE), later modified to the Revised Universal Soil Loss Equation (RUSLE). These equations, originally developed to predict soil loss or movement from agricultural fields, but since expanded to include forestlands and rangelands, are instructional because they enumerate the factors important in the transformation of soil into a water pollutant regardless of soil use.

The RUSLE equation is:

$$A = 2.24R * K * LS * C * P \qquad \text{(Eq. 15.6)}$$

where

A = the estimated average annual soil loss (metric tons hectare^{-1})

R = the rainfall and runoff erosivity index. This describes intensity and duration of rainfall in a given geographical area. It is the product of the kinetic energy of raindrops and the maximum 30-minute intensity (Figure 15.10).

K = the soil erodibility factor. K is related to soil physical and chemical properties that determine how easily soil particles can be dislodged. It is related to soil texture, aggregate stability, and soil permeability or ability to absorb water. It ranges from 1 (very easily eroded) to 0.01 (very stable soil) (Table 15.3).

TABLE 15.3
Values of the Erodibility Factor (K) for Selected Soils

| Textural Class | Soil Erodibility Factor (K)^a | | |
| | Organic Matter Content | | |
	<0.5%	2%	4%
Sand	0.05	0.03	0.02
Loam	0.38	0.34	0.29
Silt loam	0.48	0.42	0.33
Silt	0.60	0.52	0.43
Clay	0.13–0.29	0.13–0.29	0.13–0.29

^{a}This is an approximation of K based primarily on soil texture.

(Adapted from Pepper, I.L., Gerba, C.P., Brusseau, M.L. [eds] [1996] *Pollution Science*. Academic Press, San Diego.)

FIGURE 15.9   Gully erosion. (Photo by Walworth.)

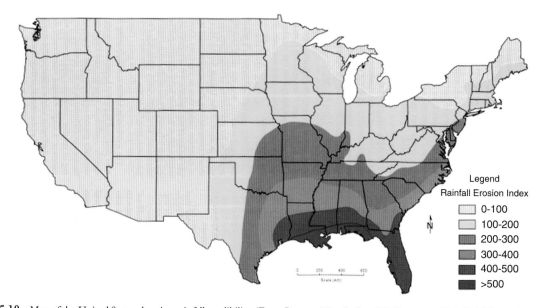

FIGURE 15.10   Map of the United States showing rainfall erodibility. (From Pepper, I.L., Gerba, C.P., Brusseau, M.L. [eds] [1996] *Pollution Science*. Academic Press, San Diego.)

LS = a dimensionless topography factor determined by length and steepness of a slope. The LS factor is related to the velocity of runoff water. Water moves faster on a steep slope than a more level one, and it picks up speed as it moves down a slope, so the steeper and longer the slope, the faster runoff water will flow. The faster water flows, the more kinetic energy it can impart to the soil surface (kinetic energy = mass * velocity^2) (Table 15.4).

C = the cover and management factor. Cover of any kind can help protect the soil surface from rain-drop impact and can force runoff water to take a longer, more tortuous path as it moves down-slope, slowing the water and reducing its kinetic energy. This factor ranges from 0.001 for a well-protected soil to 1 for a bare, exposed soil (Table 15.5).

P = the factor for supporting practices. This factor takes into account specific erosion control measures. A value of 1 represents erosion in the absence of erosion control practices. Erosion control practices reduce the P factor (Table 15.6).

We can use the RUSLE to determine the effects of various factors and practices on soil particle movement

TABLE 15.4
Selected Values of the LS (Topographic) Factor

	Slope Length (m)			
Slope %	7.6	22.9	45.7	91.4
1	0.09	0.12	0.15	0.18
5	0.27	0.46	0.66	0.93
12	0.90	1.6	2.2	3.1
30	4.0	6.9	9.7	14.0

(Adapted from Pepper, I.L., Gerba, C.P., Brusseau, M.L. [eds] [1996] *Pollution Science*. Academic Press, San Diego.)

TABLE 15.5
Values of C (the Cover and Management) Factor for Selected Management Practices

Vegetative condition	C value
Cotton after cotton, seedbed period up to 80% cover	0.60
Corn mulch, 40% of soil covered, no-till cultivation	0.21
Corn mulch, 90% of soil covered, no-till cultivation	0.05
Undisturbed forest, 90%–100% duff cover, 70%–100% of area has canopy	0.001–0.0001
Continuous pasture, 95% grass cover	0.003

(From Pepper, I.L., Gerba, C.P., Brusseau, M.L. [eds] [1996] *Pollution Science*. Academic Press, San Diego.)

TABLE 15.6
Selected Values for P (Erosion Control Practices) Factor for RUSLE Using Contour Strip Cropping as an Erosion Control Practice

Slope %	Maximum Length of Contour (m)	P Value
1–2	122	0.60
3–5	91.4	0.50
6–8	61.0	0.50
9–12	36.6	0.60
13–16	24.4	0.70
21–25	15.2	0.90

(From Pepper, I.L., Gerba, C.P., Brusseau, M.L. [eds] [1996] *Pollution Science*. Academic Press, San Diego.)

via water. Some of the factors, such as the rainfall intensity factor (R) are not manageable, whereas others such as the cover and management factor (C) can be easily manipulated. Table 15.5 that shows providing soil cover is a major factor in reducing particulate movement.

Agriculturally, there has been a move over the past few decades to reduce tillage operations that leave soil bare, and to minimize the amount of time that the soil surface is exposed to raindrops. These practices are collectively known as *reduced tillage* or *minimum tillage* systems. There are many alternatives in these systems, but the overall objective is to leave as much crop residue as possible on the soil surface and to minimize disturbance of the soil. Often, crops are planted directly into residue from the previous crop with no tillage, or with minimal tillage between crops (Figure 15.11). Selection of crops can also be an important management tool. Perennial crops provide greater crop cover than annual crops, and are used to protect soils. Note that in Table 15.5 continuous pasture has a value of 0.003, compared with cotton, which has a value of 0.60, indicating that other things being equal, pasture can have 200 times less erosion than fields planted with cotton. Another control measure aimed at reducing the amount of time that ground remains bare includes the use of cover crops. These are crops whose primary function is to provide soil cover, and that are often plowed into the soil or chemically killed, rather than being harvested. On highly erodible lands, specific erosion control practices including contour planting, strip cropping, or terracing can effectively reduce erosion. The primary mechanism of these practices is to alter the LS factor by breaking up long slopes into smaller segments, and increasing the length of the path that runoff water must take. These both reduce the velocity and erosive power of runoff water. On construction sites and similarly disturbed lands, erosion control measures are also directed at providing soil cover and reducing runoff water velocity. Five measures used to

FIGURE 15.11 No-till cotton planted into crop residue to protect the soil surface. (From USDA, image number K8550-8.)

reduce soil movement off these sites include (Brady and Weil, 2000):

1. Schedule soil-disturbing operations for low rainfall seasons.
2. Clear small areas of soil at one time so the amount of exposed bare ground is minimized.
3. Cover bare soil with vegetation, mulch, or other cover.
4. Control flow of runoff water to prevent channeling or formation of gullies.
5. Trap sediment before it can move off-site.

Bare soil at construction sites and along road cuts is often covered with synthetic fabrics called *geotextiles*. These coarse woven materials provide immediate protection and may be used in conjunction with seeding of cover crops that can provide long-term cover. Grass or other plants can be seeded with a hydroseeder that sprays a mixture of seed, fertilizer, mulch, and polymers, which cement the mixture into a cohesive soil covering that gives temporary protection to the soil surface until the seeds can germinate and plants cover the soil. Permeable

barriers made of straw bales or woven fabrics can be used to slow water and reduce its ability to carry sediments. Runoff water can be trapped in settling ponds in which water velocity is eliminated or greatly reduced, allowing suspended particles to settle out, and reducing sediment loads before overflow water is released. If suspended colloids are in a dispersed condition, flocculating agents may be added to aggregate particles into larger assemblages that rapidly settle out of suspension.

On a larger scale, numerous engineering solutions are available to control water erosion. For example, gabions (Figure 15.12), which are wire mesh containers filled with rocks that act as porous retaining structures, allow water to flow through them, but reduce its velocity. Fixed solid structures are often used to stop gully erosion. They can be designed to slow water at the foot or bottom of a gully, allowing sediment to fill in above-stream from the structure, slowly filling the gully (Figure 15.13). Alternatively,

FIGURE 15.12 A gabion erosion control structure.

FIGURE 15.13 Erosion control structure at the foot or bottom of a large gully. Sediment has filled in on the upstream *(left)* side of the structure, filling in the gully above the structure.

FIGURE 15.14 Erosion control structure at the head cut of a large gully. This structure has stopped growth of the gully by providing a stable channel to drop the water into the bottom of the gully.

they can be placed at the head cut or top of a gully, acting as a plug to allow water to fall without causing further erosion (Figure 15.14).

WIND-INDUCED SOIL EROSION AND CONTROL

Like water erosion, wind erosion has two phases: detachment and movement. As the wind blows, soil particles are dislodged and begin to roll or bounce along the soil surface in a process called *saltation*. Larger soil particles can move relatively short distances in this way, but more importantly, as the large particles bounce and strike smaller particles and aggregates, they provide the energy necessary to break aggregates apart and suspend smaller particles into the air. As discussed previously, the smaller the suspended particle, the longer it will take to settle out, with $PM_{2.5}$-size particles remaining in suspension for long periods and traveling great distances. Similar to water erosion, models developed to predict wind erosion are useful for examining the factors important in wind erosion. One such wind erosion equation (Saxton *et al.*, 2000) is:

$$E = f(W, EI, SC, K, WC) \qquad (Eq.\ 15.7)$$

where
E = estimated annual erosion (kg m^{-2})
W = erosive wind energy (g s^{-2})
EI = soil erodibility index (defined below)
SC = surface cover (%)
K = surface roughness factor (cm)
WC = soil moisture and crusting (ranges from 0 to 1): wet, crusted soils will have values close to 0, whereas dry, disturbed, and noncrusted soils have values closer to 1.

The combination of these factors can be used to predict wind erosion, which is indirectly related to airborne particulate generation. Soil erodibility index (EI) measures the susceptibility of a bare, tilled soil to wind erosion in the absence of any crusting or soil moisture. This can be calculated as the ratio of erosion to wind energy:

$$EI = \text{measured erosion/wind energy} \qquad (Eq.\ 15.8)$$

Soils with similar erodibility properties are grouped into wind erodibility groups (Table 15.7).

Wind energy (W) is related to the cubic power of the wind velocity above a threshold velocity required to start suspending particles (generally considered to be 10 to 15 miles per hour), so the eroding and particle suspending power of the wind increases dramatically as wind speed increases.

$$W_N(U_w - V_T)^3 \qquad (Eq.\ 15.9)$$

where
U_w = wind velocity (m s^{-1})
V_T = threshold velocity (m s^{-1})

Percent surface cover (SC) and soil surface roughness act in a synergistic manner to protect soil from wind

TABLE 15.7
Wind Erodibility Grouping of Various Soils

WEG	Properties of Soil Surface Layer	Wind Erodibility Index (T ha^{-1} yr^{-1})
1	Very fine sand, fine sand, sand, or coarse sand	682–352
2	Loamy sands, and sapric organic soil materials	295
3	Sandy loams	189
4	Clay, silty clay, noncalcareous clay loam, noncalcareous silty clay loam with >35% clay, calcareous loam, silt loam, clay loam, and silty clay loam	189
5	Noncalcareous loam and silt loam with >20% clay or sandy clay loam, sandy clay, and hemic organic soil materials	123
6	Noncalcareous loam and silt loam with >20% clay or noncalcareous clay loam with <35% clay	106
7	Silt, noncalcareous silty clay loam with <35% clay, and fibric organic soil material	84
8	Soils not susceptible to wind erosion due to coarse fragments at the surface or wetness	0

erosion. Either maintaining soil cover or keeping the soil surface rough are effective methods of control.

Wetness and crusting (WC) also reduce wind erosion. Moist soil is more cohesive than dry soil, so particles are less likely to become suspended for moist soil than for dry soil. Crusting occurs when surface soil particles are dispersed (i.e., when aggregates are broken down), usually by raindrop impact, and small particles migrate into surface pores filling soil void space and sealing the soil surface. When the soil dries, it solidifies into a solid crust that is resistant to wind erosion.

Table 15.7 illustrates that finer textured soils (those with more silt and clay size particles) are less erodible and are grouped in higher WEGs than sandy soils. This reflects the ability of soil aggregates to hold the soil in place during high wind events. On the other hand, PM_{10} consists largely of silt size particles and $PM_{2.5}$ is mainly clay. Therefore soil wind erosion and particulate matter production are not directly related. Saxton *et al.* (2000) developed a "Soil Dustiness Index" that describes the amount of PM_{10} a soil can produce compared with the total amount of suspended material produced from that soil:

$$D = m_{sp}/m_s \qquad (Eq.\ 15.10)$$

where

D = soil dustiness index

m_{sp} = mass of suspended PM_{10} particles

m_s = mass of soil sample <2 mm suspended

Thus the Soil Dustiness Index relates soil wind erosion to potential PM_{10} production.

Multiplying soil wind erosion times the soil dustiness index gives a good estimate of PM_{10} generation:

$$PM_{10} = E * D \qquad (Eq.\ 15.11)$$

An examination of the factors included in the soil erosion in Equation 15.7 gives a good indication of methods of controlling particulate matter generation. Wind velocity can be decreased with windbreaks of planted trees, shrubs, or grasses, or they can be constructed from material such as fences or screens. Windbreaks are most effective when placed perpendicular to the direction of the prevailing wind. Effects of windbreaks extend to as much as 40 to 50 times the height of the windbreak, although the area adequately protected by the windbreak is somewhat less. Effective control is usually considered to extend to about 10 times the height of the windbreak.

Surface roughness can be controlled by creating ridges or a rough surface with tillage implements. Ridges 5 to 10 cm in height are most effective for controlling wind erosion. Soil surface can also be protected by providing vegetative or other surface cover, including straw, hay, animal manure, or municipal sludges.

FIGURE 15.15 Spreading water to control aerosol production from an unpaved road.

Soil-wind erosion and particulate matter suspension can be controlled by maintaining an adequate level of soil moisture. For example, bare soil surfaces at construction sites and unpaved road surfaces are often watered to reduce production of suspended particulate matter (Figure 15.15). For agricultural applications, field operations can sometimes be timed to coincide with periods of high soil moisture, or soil can be irrigated prior to field operations.

Wind is not always the source of energy that suspends particles in the air. Airborne particulates from natural sources such as volcanoes and ocean spray are completely unrelated to soil erosion. In agricultural and construction operations, equipment provides suspension energy. On unpaved roads vehicular traffic is a major energy source, even when wind speeds are minimal. Reducing these energy sources, by reducing vehicle speed for example, can effectively cut dust emissions. Simply restricting access to unpaved roads is a very effective means of control. Various amendments that bind soil particles together include: calcium chloride ($CaCl_2$), soybean feedstock processing byproducts, calcium lignosulfate, polyvinyl acrylic polymer emulsion, polyacrylamide, and emulsified petroleum resin. These can be applied to unpaved roads. Alternatively, unpaved roads can be covered in gravel or similar nonerodible surfacing materials.

Agricultural field operations can be combined to reduce the number of passes through the field with soil-disturbing equipment. In Table 15.8 the results of a study on reduced tillage in cotton production are shown (Coates, 1996). In this study the reduced tillage systems utilized three or four operations compared with the eight operations used in the conventional tillage system. Some of the reduced tillage systems reduced aerosol production between 30% and 40%; however, others were ineffective, or even increased the production of airborne particulates. Equipment can sometimes be engineered to reduce fugitive dust through the use of deflectors that direct exhaust sideways or upward, or by using dust shrouds to contain

TABLE 15.8
Aerosol Production from Conventional and Reduced Tillage Systems in Cotton

Tillage System	Number of Operations	Total Suspended Particulates (kg ha^{-1})
Sundance	4	202.2
Conventional	8	119.5
Modified conventional	4	107.0
Uprooter-Shredder-Mulcher	3	78.3
Puller	4	66.6

(From Coates, W. [1996] *Trans. Am. Soc. Agric. Eng.* 39: 1593-1598.)

airborne particles. Limiting soil-disturbing operations during high-wind events will not decrease the amount of dust produced by the operation, but may decrease the distance that suspended dust travels before settling out.

INDUSTRIAL PARTICULATE CONTROL

Methods of preventing or reducing industrial or man-made airborne particulates fall into several categories:

1. *Material substitution*—using process materials that do not contain pollutants or their precursors. Examples include use of alternative fuels (e.g., replacing high sulfur coal with low sulfur coal or natural gas to reduce SO_2 production), or using oxygenated fuels to power vehicles.
2. *Process modification*—changing industrial processes to reduce pollutant output by optimizing process parameters, replacing a polluting process with a cleaner one, or removing or decreasing a pollutant used in a process. Reducing excess oxygen in combustion processes is an example of process modification that can reduce SO_2 production.
3. *Behavior modification*—changing public behavior to decrease particulate output. Examples include restricting use of two-cycle engines on watercraft, lawnmowers, etc.
4. *Cleaning up industrial output streams*—using dry collectors, cyclones, filters, or electrostatic precipitators to reduce particulates in exhaust streams before they are emitted to the atmosphere.

Dry collectors operate by using either gravity or centrifugal force to settle relatively large particles out of air. These may be as simple as gravity settling chambers that reduce air velocity, allowing particles to settle out according to Stokes' law, described previously. These are equivalent to the settling ponds used for reducing particulate load in flowing water. Preliminary screening may be used to separate out larger particles. Gravitational settling is only effective for particles larger than 50 μm. Accelerated settling can be produced with a cyclone, using centrifugal force (Figure 15.16). A cyclone creates two vortexes, an outer vortex that moves downward, and an inner vortex that moves upward and carries the waste stream out. The outer vortex throws the particles outward and down, dropping them out of the bottom of the cyclone. This can effectively remove particles larger than 5 μm. Cyclones are used in grain elevators, grain mills, coal dryers, industrial kilns, etc.

Filters can be constructed out of paper or fabrics woven or pressed out of dacron, orlon, nylon, wool, cotton, teflon, glass, or other fibers. Pressed or felted fabrics are used for furnace, air conditioner, and automobile air filters. They are generally replaced when dirty. Woven fabric filters are usually formed into a bag, through which contaminated air is passed. These are used to collect building material dust, such as that generated by saws, sanders and grinders, grain and other industrial processing dust, and dry chemical powders. Paper filter bags are used to filter out suspended particles in many vacuum cleaners. Particles are attracted to the fiber

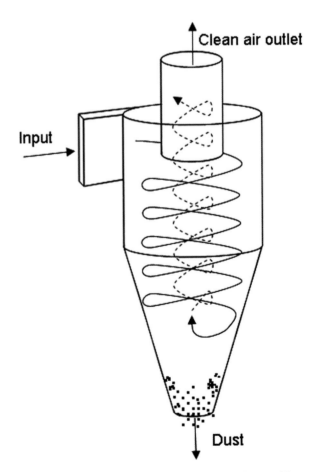

FIGURE 15.16 Diagram of a cyclone-gas particulate scrubber.

material, and attach to fibers when they move within approximately one particle radius of the fiber surface. Particles must be removed when the bags are filled to capacity by shaking or by reversing air flow. Filters are effective for removing particles as small as 1 μm or less.

Electrostatic precipitators operate by electrically charging suspended particles, then passing the particulate contaminated air through an electrical field, which attracts particles to an oppositely charged collection electrode. The charging electrodes are usually wires charged with high-voltage direct current, which ionizes the gas near the electrode. Charged gas molecules adhere to suspended solid or liquid particles, giving the particles an electrical charge. Collection electrodes, with an opposite electrical charge, can be either plates or wires. Particles that accumulate on the collection electrodes are removed via gravity or mechanical action. Electrostatic precipitators are highly efficient for removing small particles, and can remove particles as small as 0.01 μm and as large as 1000 μm. They can handle large volumes of gas, but have relatively high initial costs. Electrostatic precipitators are used in various industrial processes including coal power plant combustion to remove fly ash, in smelters, cement kilns, paper mills, etc.

Wet collectors or scrubbers use water or some other liquid to entrap suspended particles. Liquid can be sprayed into the contaminated air stream, where droplets collide with particles that adhere to the droplet. The droplets settle out with the attached particles. This process is similar to the natural processes of particle removal from the atmosphere by rain, snow, or mist. Settling can be accelerated through the use of centrifugal force, as in cyclones. Alternative configurations can be used. Particles can be trapped on wetted surfaces, such as wetted plates, fibrous steel, plastic, or glass, or crushed rock or similar packing materials. Fluidized bed scrubbers use packing materials light enough to float in the gas stream, improving mixing of gas and scrubbing liquid. Wet collectors are used to remove particles as small as 1 μm in a wide range of industrial processes. Some internal combustion engines use wet collectors to remove particulates by drawing engine intake air through an oil bath or oil-soaked packing.

QUESTIONS

1. Define the terms PM_{10} and $PM_{2.5}$.

2. Based on the data in Table 15.1 quantifying PM_{10} production from various sources in the United States, construct a coherent strategy to reduce PM_{10}.

3. How do waterborne particles reduce dissolved O_2 in surface water? What are the resulting effects on aquatic biota?

4. Explain how flocculation of particles suspended in water changes the time required for the particles to settle out. Use Stokes' Law to justify your answer.

5. Explain why sediments carried in freshwater rivers quickly settle out when these sediment-laden waters combine with ocean water.

6. Explain the differences between primary and secondary particles in terms of size, formation, and composition.

7. Describe and contrast the mechanisms by which soil particles become pollutants in air and in water.

8. What are reduced tillage and minimum tillage? How do they help reduce the amount of soil particles entering surface waters and the atmosphere?

9. Wind is the major source of energy for suspending soil particles in the atmosphere. What other sources of energy are important in this process?

10. Discuss the principles of operation of an electrostatic precipitator.

REFERENCES AND ADDITIONAL READING

Brady, N.C., and Weil, R.R. (2000) *Elements of the Nature and Properties of Soils*. Prentice-Hall, Inc. Upper Saddle River, NJ.

Cha, S.S., Chuang, K.T., Liu, D.H.F., Ramachandran, G., Reist, P.C., and Sanger, A.R. (2000) In Liu, D.H.F. and B.G. Liptak (Eds.) *Air Pollution*. Lewis Publishers, Boca Raton, FL.

Coates, W. (1995) Particulates Generated by Five Cotton Tillage Systems. *Trans. Am. Soc. Agric. Eng.* **39**, 1593–1598.

Council on Environmental Quality. (1997) *Environmental Quality: The 1997 Report of the Council on Environmental Quality*. U.S. Government Printing Office. Washington, DC.

Guertal, E.A., Echert, D.J., Traina, S.J., and Logan, T.J. (1991) Differential Phosphorus Retention in Soil Profiles under No-Till Crop Production. *Soil Sci. Soc. Am. J.* **55**, 410–413.

Hassett, J.J., and Banwart, W.L. (eds) (1989) The Sorption of Nonpolar Organics by Soils and Sediments. *In* B.L. Sawney and K. Brown (Eds.) *Reactions and Movement of Organic Chemicals in Soils*, SSSA Special Publication no. 22. Soil Science Society of America, Madison, WI.

Koenig, J.Q. (1999) *Health Effects of Ambient Air Pollution*. Kluwer Academic Publishers, Boston.

Liu, D.H.F. and Liptak, B.G. (2000) *Air Pollution*. Lewis Publishers, Boca Raton, FL.

National Research Council (U.S.). Committee on Particulate Control Technology. (1980) *Controlling airborne particles*. National Academy of Sciences, Washington, DC.

Natural Resources Conservation Service. (2001) *National Soil Survey Handbook*. United States Department of Agriculture, Washington, DC.

Pierzynski, G.M., Sims, J.T., and Vance, G.F. (2000) *Soils and Environmental Quality*, 2nd Edition. CRC Press, Boca Raton, FL.

Post, D.F. (1996) Sediments (Soil Erosion) as a Source of Pollution. *In* I.L. Pepper, C.P. Gerba, M.L. Brusseau (Eds.) *Pollution Science*. Academic Press, San Diego.

Rao, S.S. (1993) *Particulate Matter and Aquatic Contaminants*. Lewis Publishers, Boca Raton, FL.

Rengasamy, P., and Sumner, M.E. (1998) Processes Involved in Sodic Behavior. *In* M.E. Sumner and R. Naidu (Eds.) *Sodic Soils: Distribution, Properties, Management, and Environmental Consequences*. Oxford University Press, New York.

Ritter, W.F., and Shirmohammadi, A. (Eds.) (2001) *Agricultural Nonpoint Source Pollution*. Lewis Publishers, Boca Raton, FL.

Saxton, K., Chandler, D., Stetler, L., Lamb, B., Claiborn, C., and Lee, B.-H. (2000) Wind Erosion and Fugitive Dust Fluxes on Agricultural Lands in the Pacific Northwest. 2000 *Trans. ASAE*. **43**, 623–630.

Sumner, M.E. (1992) The Electrical Double Layer and Clay Dispersion. *In* M.E. Sumner and B.A. Stewart (Eds.) *Soil Crusting: Chemical and Physical Processes*. Lewis Publishers, Boca Raton, FL.

Uehara, G., and Gillman, G. *The Mineralogy, Chemistry, and Physics of Tropical Soils with Variable Charge Clays*. Westeview Tropical Agriculture Series, No. 4. Westview Press, Boulder, CO.

Vedal, S. (1995) *Health Effects of Inhalable Particles: Implications for British Columbia*. Air Resources Branch, Ministry of Environment, Lands and Parks.

Yu, M-H. (2001) *Environmental Toxicology*. Lewis Publishers, Boca Raton, FL.

16

CHEMICAL CONTAMINANTS

M.L. BRUSSEAU, G.B. FAMISAN, AND J.F. ARTIOLA

It can be argued that all matter in one form or another can become a contaminant when found out of place or at concentrations above normal. However, chemical contaminants become pollutants when accumulations are sufficient to adversely affect the environment or to pose a risk to living organisms. On average, nine new chemicals are registered every second with the Chemical Abstracts Service (CAS). Today there are more than 20,000 industrial chemicals that can be dangerous to human beings and the environment. Fortunately, the vast majority of these chemicals are not produced in large enough quantities to be a human or environmental threat. However, there are more than 3000 natural and manmade chemicals that are toxic enough, and are produced in sufficient quantities, to be a potential environmental hazard. Thus the production, transport, storage, and disposal of these chemicals are regulated by government agencies. There are numerous sources of chemical contaminants released to the environment, but these generally fall into a few general categories. This chapter will present an overview of the various types of chemical contaminants, their sources, and the characteristics that make them potential contaminants and pollutants.

TYPES OF CONTAMINANTS

There are three basic categories of contaminants: organic, inorganic, and radioactive. In turn, there are several classes of contaminants within each of these categories. Major classes of contaminants are listed in Box 16.1.

Thousands of chemicals are released into the environment every day. Thus when conducting site characterization studies, it is important to prioritize the suite of chemicals under investigation. For most sites this is done by focusing on so-called priority pollutants, those which are regulated by federal, state, or local governments. The primary list of priority pollutants, which is governed by the National Primary Drinking Water Regulations, has legally enforceable standards that apply to all

BOX 16.1 *Examples of Organic, Inorganic, and Radioactive Contaminants*

Organic Contaminants

Petroleum hydrocarbons (fuels)—benzene, toluene, xylene, polycyclic aromatics, MTBE
Chlorinated solvents—trichloroethene, tetrachloroethene, trichloroethane, carbon tetrachloride
Pesticides—DDT, toxaphene, atrazine
Polychlorinated biphenyls—PCBs
Coal tar—polycyclic aromatics
Pharmaceuticals/food additives/cosmetics—drugs, surfactants, dyes
Gaseous compounds—CFCs (freon), HCFCs

Inorganic Contaminants

Inorganic "salts"—sodium, calcium, nitrate, sulfate, chloride
Heavy/trace metals—lead, zinc, cadmium, mercury, arsenic, selenium

Radioactive Contaminants

Solid elements—uranium, strontium, cobalt, plutonium
Gaseous elements—radon

public water systems. These standards protect public health by limiting the levels of contaminants allowed to exist in drinking water. The organic and inorganic contaminants on this list are presented in Table 16.1. Note that the full list also includes microorganisms, radionuclides, and water disinfection by-products.

The frequency of occurrence of the contaminants listed in Table 16.1, as well as other chemicals, differs greatly for each specific contaminated site. The contaminants that are most frequently encountered at U. S. Environmental Protection Agency (EPA) designated Superfund sites are presented in Box 16.2. It is quite likely that one or more of these contaminants will be present at most hazardous waste sites.

SOURCES OF CONTAMINANTS

AGRICULTURAL ACTIVITIES

Agricultural systems comprise highly controlled tracts of land that generally receive large inputs of *fertilizers* in the

forms of chemicals, biosolids, and animal wastes. The ultimate goal of these chemical additions is to generate optimum amounts of food and fiber. However, fertilizers are often applied in excess of the crop needs or are in chemical forms that make them very mobile in soil and water environments. For example, nitrate pollution of groundwater is often caused by excessive nitrogen fertilizer applications, which results in leaching below the root zone. Agricultural contaminants fall into two major categories: inorganic and organic.

Inorganic fertilizers are routinely applied at least once a year and include, in order of decreasing amounts, N, P, K, and metals. The annual applications of these chemicals range from 50 to 200 kg ha^{-1}, as N, P, or K. Micronutrient (e.g., Fe, Zn, Cu, B, Mo) fertilizer additions are also applied regularly to agricultural fields but with less frequency because of lower crop requirements. These chemicals are applied to agricultural lands at average rates of 0.5 to 2 kg ha^{-1}, in their respective elemental forms, every 2 to 5 years. A third type of inorganic chemicals applied to agricultural land are called *soil amendments*. These materials are applied to agricultural fields with some frequency for two reasons: (1) to decrease or increase soil pH, decrease soil salinity, and improve soil structure; and (2) to replenish macronutrients like Ca^{++}, Mg^{++}, K^+, and $SO_4^=$. To control macronutrient deficiencies, the application rates of these chemicals range from 50 to 500 kg ha^{-1}. To control soil pH and salinity, applications typically range from 2000 to 10,000 kg ha^{-1}. The common forms of these chemicals are listed in Table 16.2 in order of decreasing probable impact to the environment.

Table 16.2 illustrates that the aforementioned inorganic chemicals can act as a nutrient and as a pollutant, depending on the amounts applied, the location of these chemicals, and soil-plant-water dynamics. For example, Figure 16.1 shows the soil-nitrogen cycle, which illustrates the transformations, sinks, and sources of this element. Plants and some soil minerals can act as sinks for the two major forms of N. But some plants, animals, the atmosphere, and human beings (fertilizer additions) can sometimes contribute to excessive N concentrations that lead to groundwater pollution.

Most *pesticides* are organic compounds, and are often applied in agricultural systems at least once a year, albeit in much smaller quantities than fertilizers. However, synthetic pesticides, designed to be very toxic to plants and pests, may have deleterious effects at very low concentrations. Most synthetic pesticides are broadly classified as insecticides, herbicides, and fungicides. Although most pesticides are solids, they are usually dissolved in water or oil to facilitate their handling and application. Fumigants are gaseous pesticides typically used to control insects. A list of common organic pesti-

TABLE 16.1
National Primary Drinking Water Regulations

	MCL or TT^a (mg L^{-1})	Potential Health Effects from Ingestion of Water
Inorganic Chemicals		
Antimony	0,006	Increase in blood cholesterol; decrease in blood glucose
Arsenic	0,01	Skin damage; circulatory system problems; increased risk of cancer
Asbestos (fiber > 10 micrometers)	7 MFL	Increased risk of developing benign intestinal polyps
Barium	2	Increase in blood pressure
Beryllium	0,004	Intestinal lesions
Cadmium	0,005	Kidney damage
Chromium (total)	0,1	Some people who use water containing chromium well in excess of the MCL over many years could experience allergic dermatitis
Copper	TT^8; Action Level = 1.3	Short-term exposure: Gastrointestinal distress. Long-term exposure: Liver or kidney damage. People with Wilson's disease should consult their personal doctor if their water systems exceed the copper action level
Cyanide (as free cyanide)	0,2	Nerve damage or thyroid problems
Fluoride	4	Bone disease (pain and tenderness of the bones); children may get mottled teeth
Lead	TT^8; Action Level = 0.015	Infants and children: Delays in physical or mental development. Adults: Kidney problems; high blood pressure
Mercury (inorganic)	0,002	Kidney damage
Nitrate (measured as nitrogen)	10	"Blue baby syndrome" in infants less than 6 months—life threatening without immediate medical attention. Symptoms: Infant looks blue and has shortness of breath
Nitrite (measured as nitrogen)	1	"Blue baby syndrome" in infants less than 6 months—life threatening without immediate medical attention. Symptoms: Infant looks blue and has shortness of breath
Selenium	0,05	Hair or fingernail loss; numbness in fingers or toes; circulatory problems
Thallium	0,002	Hair loss; changes in blood; kidney, intestine, or liver problems
Organic Chemicals		
Acrylamide	TT^9	Nervous system or blood problems; increased risk of cancer
Alachlor	0,002	Eye, liver, kidney or spleen problems; anemia; increased risk of cancer
Atrazine	0,003	Cardiovascular system problems; reproductive difficulties
Benzene	0,005	Anemia; decrease in blood platelets; increased risk of cancer
Benzo(a)pyrene (PAHs)	0,0002	Reproductive difficulties; increased risk of cancer
Carbofuran	0,04	Problems with blood or nervous system; reproductive difficulties
Carbon tetrachloride	0,005	Liver problems; increased risk of cancer
Chlordane	0,002	Liver or nervous system problems; increased risk of cancer
Chlorobenzene	0,1	Liver or kidney problems
2,4-D	0,07	Kidney, liver, or adrenal gland problems
Dalapon	0,2	Minor kidney changes
1,2-Dibromo-3-chloropropane (DBCP)	0,0002	Reproductive difficulties; increased risk of cancer
o-Dichlorobenzene	0,6	Liver, kidney, or circulatory system problems
p-Dichlorobenzene	0,075	Anemia; liver, kidney or spleen damage; changes in blood
1,2-Dichloroethane	0,005	Increased risk of cancer
1,1-Dichloroethylene	0,007	Liver problems
cis-1,2-Dichloroethylene	0,07	Liver problems
trans-1,2-Dichloroethylene	0,1	Liver problems
Dichloromethane	0,005	Liver problems; increased risk of cancer
1,2-Dichloropropane	0,005	Increased risk of cancer
Di(2-ethylhexyl)adipate	0,4	General toxic effects or reproductive difficulties
Di(2-ethylhexyl)phthalate	0,006	Reproductive difficulties; liver problems; increased risk of cancer
Dinoseb	0,007	Reproductive difficulties
Dioxin (2,3,7,8-TCDD)	0	Reproductive difficulties; increased risk of cancer
Diquat	0,02	Cataracts
Endothall	0,1	Stomach and intestinal problems
Endrin	0,002	Nervous system effects
Epichlorohydrin	TT^9	Stomach problems; reproductive difficulties; increased risk of cancer
Ethylbenzene	0,7	Liver or kidney problems
Ethylene dibromide	0,00005	Stomach problems; reproductive difficulties; increased risk of cancer
Glyphosate	0,7	Kidney problems; reproductive difficulties
Heptachlor	0,0004	Liver damage; increased risk of cancer
Heptachlor epoxide	0,0002	Liver damage; increased risk of cancer

(Continued)

TABLE 16.1 (Continued)

	MCL or TT[a] (mg L^{-1})	Potential Health Effects from Ingestion of Water
Organic Chemicals (Continued)		
Hexachlorobenzene	0,001	Liver or kidney problems; reproductive difficulties; increased risk of cancer
Hexachlorocyclopentadiene	0,05	Kidney or stomach problems
Lindane	0,0002	Liver or kidney problems
Methoxychlor	0,04	Reproductive difficulties
Oxamyl (Vydate)	0,2	Slight nervous system effects
Polychlorinated biphenyls (PCBs)	0,0005	Skin changes; thymus gland problems; immune deficiencies; reproductive or nervous system difficulties; increased risk of cancer
Pentachlorophenol	0,001	Liver or kidney problems; increased risk of cancer
Picloram	0,5	Liver problems
Simazine	0,004	Problems with blood
Styrene	0,1	Liver, kidney, and circulatory problems
Tetrachloroethylene	0,005	Liver problems; increased risk of cancer
Toluene	1	Nervous system, kidney, or liver problems
Toxaphene	0,003	Kidney, liver, or thyroid problems; increased risk of cancer
2,4,5-TP (Silvex)	0,05	Liver problems
1,2,4-Trichlorobenzene	0,07	Changes in adrenal glands
1,1,1-Trichloroethane	0,2	Liver, nervous system, or circulatory problems
1,1,2-Trichloroethane	0,005	Liver, kidney, or immune system problems
Trichloroethylene	0,005	Liver problems; increased risk of cancer
Vinyl chloride	0,002	Increased risk of cancer
Xylenes (total)	10	Nervous system damage

From: www.epa.gov/safewater/standards.html.

[a]Maximum Contaminant Level, the highest level of a contaminant that is allowed in drinking water.

> **BOX 16.2** *Common Pollutants Found in Superfund Sites*
>
> - Acetone
> - Aldrin/Dieldrin
> - Arsenic
> - Barium
> - Benzene
> - 2-Butanone
> - Cadmium
> - Carbon tetrachloride
> - Chlordane
> - Chloroform
> - Chromium
> - Cyanide
> - DDT, DDE, DDD
> - 1,1-Dichloroethene
> - 1,2-Dichloroethane
> - Lead
> - Mercury
> - Methylene chloride
> - Naphthalene
> - Nickel
> - Pentachlorophenol
> - Polychlorinated biphenyls (PCBs)
> - Polycyclic aromatic hydrocarbons (PAHs)
> - Tetrachloroethylene
> - Toluene
> - Trichloroethylene
> - Vinyl chloride
> - Xylene
> - Zinc
>
> Source: www.epa.gov/superfund/resources/chemicals.htm.

cides is presented in Table 16.3. Less common forms of inorganic pesticides are used to control roaches and rats. These chemicals, which have all too often been used in close proximity to human beings, have, as their primary

acting agent, toxic forms of arsenic (AsO_4^{3-}), boron (H_3BO_3), and S (SO_2).

The chemical structure of organic pesticides controls their water solubility, mobility, environmental persistence, and toxicity. The first generation of organic pesticides had multiple chlorine groups inserted into their structures to give them a broad spectrum of biotoxic effects. However, the chlorine groups also made them very difficult to degrade, thus making them very persistent. The next step in pesticide development sought a compromise between persistence and toxicity with chemical structures that were moderately soluble in water, and more targeted toxicity effects. The next generation of pesticides again sought to decrease the persistence of these chemicals in the environment by making them even more water soluble, and continued to focus their toxic effects. This class of pesticides seldom bioaccumulates in human beings or animals and has short life spans (days) in the environment. However, when misused, these chemicals can be found in water sources. For example, today the members of the triazine family are the most commonly found pesticides in surface and groundwater resources. Conversely, chlorinated pesticides are seldom found in water but can still be found in soils and sediments.

Animals generate significant amounts of residues that are benign to the environment in open systems with low concentrations of animals. However, in the last 100 years,

TABLE 16.2
Common Fertilizer and Soil Amendment Materials and Potential Contaminant Forms

Fertilizers	Nutrient form	Pollutant properties
NH_4NO_3, KNO_3	NO_3^-	Very mobile, promotes plant and microbial growth
NH_3 (gas), $CO(NH_2)_2$ (urea)	NH_4^+	Toxic gas and cation toxic to aquatic life; converts to NO_3^-
Superphosphate, triple	PO_4^{3-}	Variable mobility, promotes microbial growth (eutrophication)
Ammonium phosphate	$NO_3^-, NH_4^+, PO_4^{3-}$	See above
Calcite ($CaCO_3$)	$Ca^{++}, CO_3^=$	Increases soil-water alkalinity
Gypsum ($CaSO_4 2H_2O$)	$Ca^{++}, SO_4^=$	Increases sulfate ion concentrations in water
Micronutrients: salt forms, chelates	$Fe^{++}, Mn^{++}, Zn^{++}, Cu^{++}, MoO_4^=, H_3BO_3$	Cations are mobile in acid soils, acid forming Anions are mobile in alkaline soils; plant, animal toxicity

large-scale animal production systems have created concentrated sources of animal-derived contaminants. Large-scale animal feeding operations include feedlots for beef, swine, and poultry production, dairies, and fish farms. These operations act as point sources for the common chemicals listed in Table 16.1. Nitrate-N, Ammonium-N, and Phosphate-P are the three most common contaminants derived from unregulated animal waste disposal practices. These three chemicals are usually found at concentrations ranging from 1000 to 50,000 mg kg^{-1} (elemental form) in animal wastes. Nitrates are very mobile in the environment, and can only be controlled by plant and microorganism uptake, or by the process of denitrification. Large releases of ammonium can have several detrimental effects in the environment. First, the ammonium ion is unstable, volatilizes in alkaline water, and oxidizes to nitrate, increasing the pool of this anion. Second, the ammonium ion is very toxic to fish. And third, the process of ammonium oxidation to nitrate (nitrification) releases acidity into the environment.

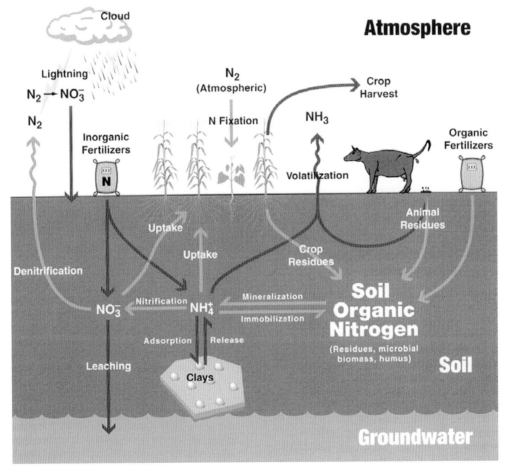

FIGURE 16.1 Soil-nitrogen transformations.

TABLE 16.3

Major Classes of Organic Pesticides and their Potential Pollutant Properties

Class/Elemental	Common	Pollutant Properties
Organochlorines	DDT	Resistant to degradation (persistent)
Organophosphates	Chlorpyrifos	Mobile in the soil environment
Carbamates	Carbaryl	Very mobile in the soil environment
Triazines	Atrazine	Very mobile in the soil environment
Plant insecticides	Pyrethroids	Some toxic to fish
Fumigants	Dichloropropene	Toxic to animals, volatile

Note: *All* of these chemicals have some degree of toxicity (acute and/or chronic) to humans.

$$NH_4^+ + 2O_2 \rightarrow 2H^+ + NO_3^- + H_2O \qquad \text{(Eq. 16.1)}$$

Phosphates are much less mobile in the environment. However, small quantities ($>1\,mg\,L^{-1}$) can be extremely deleterious to stagnant water bodies because phosphates can trigger excessive microbial growth, which leads to eutrophication. Box 16.3 shows a list of contaminants, in addition to N and P, that concentrated animal wastes can introduce in significant amounts into the environment.

MUNICIPAL WASTEWATER TREATMENT

Wastewater treatment plants produce wastes that contain many potential contaminants. Reclaimed wastewater is usually clean enough to be used for irrigation, but usually contains higher ($\sim$1.5 times) concentrations of dissolved solids than the source water. Also, chlorine-disinfected reclaimed water can contain significant trace amounts of disinfection by-products such as trihalomethanes and haloacetic acids.

The solid residues of wastewater treatment plants, called *biosolids*, typically contain common fertilizers such as those listed in Table 16.4 and may also contain heavy metals and synthetic organic compounds found in household products (Table 16.4). Because biosolids usually contain macronutrients, micronutrients, and organic matter, they are routinely applied to agricultural lands as fertilizer and soil amendments. The application of biosolids to land is limited by the concentrations of potential pollutants, such as heavy metals that vary from one treatment plant to another. More specifically, maximum pollutant loading rates are set by federal (40 CFR, Part 503, promulgated by U.S. EPA under the Clean Water Act, Section 405), and state regulations. These regulations set

---

> **BOX 16.3** *Pollutants Released from Animal Wastes.*
>
> - Total dissolved solids (Na, Cl, Ca, Mg, K, soluble N and P forms): Most animal wastes are high ($>10,000\,mg\,L^{-1}$) in total dissolved solids (TDSs).
> - Organic carbon: Excessive amounts of soluble carbon, together with soluble P, can quickly reduce O_2 availability in water by raising the biochemical oxygen demand.
> - Residual pesticides: Used to control pests in animal facilities.
> - Residual metals: Cu, As; from animal diets and pesticides.
> - Medicines: Antibiotics, growth regulators.
> - Gases: From waste-storage facilities and waste-disposal activities, greenhouse—(CO_2, N_2O), toxic (NH_3, H_2S), odors—H_2S, mercaptans, indoles, org-sulfides, NH_3.

annual and lifetime application rates for pollutants such as heavy metals; some of which are also plant and animal micronutrients. Restrictions also exist on maximum allowable concentrations of certain organic pollutants in biosolids, such as pesticides, and solvents (see Table 16.4). Less specific guidelines exist usually at the state level for the application of biosolids as an NPK fertilizer source. In Arizona for example, up to 8 tons of dry biosolids may be applied annually to agricultural fields to meet the N demands of crops. In this way the land application of biosolids completes the natural cycles of C, N, and P in the environment. Repeated applications of biosolids often increases the concentrations of metals, P, and some salts in the soil environment. In addition, excessive, concentrated, or uneven applications of biosolids can result in surface and groundwater pollution.

TABLE 16.4

Major Classes of Potential Contaminants Found in Municipal Wastes

Common Name/Group	Common Examples	Pollutant Properties
Fertilizers	NO_3^-, NH_4^+, PO_4^{3-}	See Table 16.2
Salts	Na^+, Cl^-	Increase salinity
Metals	Cu, Zn, Ni, Pb, Cr, Mo, Cd	See Table 16.7
Solvents	Chlorinated solvents	See Table 16.1, 16.7
Pesticides	Chlorinated aromatic	See Table 16.1, 16.7
Surfactants	Alkylphenols	Potential endocrine disruptors
Pharmacueticals	Antibiotics	Similar to other organics
Plastics and Plasticizers	Vinyl chloride, phthalates	Toxic at low concentrations
Gases	NH_3, H_2S, mercaptans	Irritants, odorous

Table 16.4 illustrates the fact that municipal residues can contain many types of chemical contaminants. The concentrations of these chemicals vary by several orders of magnitude, ranging from less than $100\,mg\,L^{-1}$ for most synthetic organic chemicals, to 1000 or more $mg\,L^{-1}$ for most metals and salts, and more than $30,000\,mg\,L^{-1}$ (3%) for N and P. It should be noted that the pollution-potential effects of some of the chemicals found in biosolids are well known, particularly for N, P, and salts. However, concerns over the potential accumulation and environmental effects of residual pesticides, surfactants, and pharmaceuticals found in biosolids are prompting new regulations and research into the fate of these chemicals in the environment.

INDUSTRIAL AND MANUFACTURING ACTIVITIES

There are numerous sources of industrial chemical contaminants, the result of controlled or uncontrolled waste disposal and releases into the environment. Industrial wastes may contain contaminants classified by the federal government as hazardous and nonhazardous. However, this classification primarily separates wastes containing high concentrations of pollutants versus wastes that contain low concentrations. For example, metal-plated industrial wastes contain high concentrations of toxic metals, such as Cr, Ni, and Cd, and are usually classified as hazardous. However, municipal wastes, classified as nonhazardous, also contain these metals and many others, but at much lower concentrations. Most industrial contaminants originate from a few general categories of industrial wastes. These are summarized in Box 16.4 with examples of industries and their common classes of contaminants.

SERVICE-RELATED ACTIVITIES

There are many service activities that produce waste materials that are potential sources of contamination. The service industries that produce substantial amounts of waste include dry cleaners and laundry plants, automotive service and repair shops, and fuel stations. These facilities would be subject to regulation under the Resource, Conservation and Recovery Act (RCRA) if they generate wastes that fall under RCRA's definition of a hazardous waste.

Dry-cleaning, a service industry involved in the cleaning of textiles, uses solvents in the cleaning process that are considered a hazardous waste. These solvents include tetrachloroethene, petroleum solvents, CFC-113 (1,1,2-trichloro-1,2,2-trifluoroethane, valclene), and

---

> **BOX 16.4** *Industrial Wastes and Sources of Contaminants*
>
> **Solid, liquid and slurry wastes with high concentrations of metals, salts, and solvents**
>
> Industries: metal plating, painting
> Types of pollutants: metals, solvents, toxic aromatic and nonaromatic hydrocarbons
>
> **Liquid wastes with high concentrations of hydrocarbons and solvents**
>
> Industries: chemical manufacturing, electronics manufacturing, plastics manufacturing
> Types of pollutants: chlorinated solvents, hydrocarbons, plastics, plasticizers, metals, catalysts, cyanides, sulfides
>
> **Wastewaters that contain organic chemicals**
>
> Industries: paper processing, tanneries, food processing, industrial wastewater treatment plants, pharmaceuticals
> Types of pollutants: various organic chemicals

1,1-trichloroethane (1,1,1-TCA). Along with spent solvents, other wastes produced are solvent containers, spent filter cartridges, residues from solvent distillation, and solvent-contaminated wastewater.

Petroleum-based fuels, such as gasoline, diesel fuel, and aviation fuels, are ubiquitous sources of contamination at automotive, train, and aviation fuel stations. The lower molecular weight, more soluble constituents such as benzene and toluene are of special concern, with respect to groundwater contamination potential. In addition, some fuel additives may also be of concern. For example, methyl-tertiary-butylether (MTBE) is a hydrocarbon derivative that has been added to gasoline for the past several years to boost the oxygen content of the fuel. This was done in accordance with federal regulations formulated to improve air quality. However, MTBE is a very soluble compound that is also resistant to biodegradation. Thus it is a very mobile and persistent compound, and this nature has lead to widespread groundwater contamination. Low levels of MTBE can make water supplies undrinkable because of its offensive taste and odor. The use of MTBE in gasoline is now being phased out because of its mobility and persistence in the environment.

Automotive service and repair shops can be a source of numerous contaminants. Various types of solvents are used to degrease and clean engine parts. Metal contaminants can originate from batteries, circuit boards, and other vehicle components. Fuel-based contaminants are also typically present.

RESOURCE EXTRACTION/PRODUCTION

Mineral extraction (mining) and petroleum and gas production are major resource extraction activities that provide the raw materials to support our economic infrastructure. An enormous amount of pollution is generated from the extraction of natural resources from the environment. The EPA's Toxic Releases Inventory report lists mining as the single largest source of toxic waste of all industries in the United States. Mineral extraction sites, which include strip mines, quarries, and underground mines, contribute to surface and groundwater pollution, erosion, and sedimentation. The mining process involves the excavation of large amounts of waste rock to remove the desired mineral ore. The ore is then crushed into finely ground tailings for chemical processing and separation to extract the target minerals. After the minerals are processed, the waste rock and mine tailings are stored in large, above-ground piles and containment areas. These waste piles, along with the bedrock walls exposed from mining, pose a huge environmental problem due to metals pollution associated primarily with acid mine drainage. Acid mine drainage is caused when water draining through surface mines, deep mines, and waste piles comes in contact with exposed rocks that contain pyrite, an iron sulfide, causing a chemical reaction. The resulting water is high in sulfuric acid and contains elevated levels of dissolved iron. This acid runoff also dissolves heavy metals such as lead, copper, and mercury, resulting in surface and groundwater contamination.

Petroleum and natural gas extraction pose environmental threats such as leaks and spills occurring during drilling and extraction from wells, and air pollution as natural gas is burned off at oil wells. The petroleum and natural gas extraction process generates production wastes, including drilling cuttings and muds, produced water, and drilling fluids. Drilling fluids, which contain many different components, can be oil based, consisting of crude oil or other mixtures of organic substances such as diesel oil and paraffin oils, or water based, consisting of freshwater or seawater with bentonite (clay) and barite ($BaSO_4$). Each component of a drilling fluid has a different chemical function. For example, barite is used to regulate hydrostatic pressure in drilling wells. As a result of being exposed to these drilling fluids, drilling cuttings and muds contain hundreds of different substances. This waste is usually stored in waste pits, and if the pits are unlined, the toxic chemicals in the spent waste cut-

tings and muds, such as hydrocarbon-based lubricating fluids, can pollute soil, surface, and groundwater systems. Produced water is the wastewater created when water is injected into oil and gas reservoirs to force the oil to the surface, mixing with formation water (the layer of water naturally residing under the hydrocarbons). At the surface, produced water is treated to remove as much oil as possible before it is reinjected, and eventually when the oil field is depleted, the well fills with the produced water. Even after treatment, produced water can still contain oil, low-molecular-weight hydrocarbons, inorganic salts, and chemicals used to increase hydrocarbon extraction.

RADIOACTIVE CONTAMINANTS

Radioactive waste primarily originates from nuclear fuel production and reprocessing, nuclear power generation, military weapons development, and biomedical and industrial activities. The largest quantities of radioactive waste, in terms of both radioactivity and volume, are generated by commercial nuclear power and military nuclear weapons production industries, and by activities that support these industries, such as uranium mining and processing. However, radioactive material can also originate from natural sources.

Naturally occurring sources of radioactive materials, including soil, rocks, and minerals that contain radionuclides, can be concentrated and exposed by human industrial activity, such as uranium mining, oil and gas production, and phosphate fertilizer production. For example, when uranium is mined using *in situ* leaching or surface methods, bulk waste material is generated from excavated top soil, uranium waste rock, and subgrade ores, all of which can contain radionuclides of radium, thorium, and uranium. Other extraction and processing practices that can generate and accumulate radioactive wastes similar to that of uranium mining are aluminum and copper mining, titanium ore extraction, and petroleum production. According to EPA reports, the total amount of naturally occurring radioactive waste that is enhanced by industrial practices number in excess of 1 billion tons annually. Sometimes the levels of radiation are relatively low in comparison to the large volume of material that contains the radioactive waste. This causes a problem because of the high cost of disposing of radioactive waste compared with the relatively low value of the product from which the radioactive waste is separated (i.e., fertilizers). Additionally, relatively few landfills or other licensed disposal locations can accept radioactive waste.

Selected examples of radioisotopes are presented in Table 16.5. There are three main types of ionizing radiation found in both natural and anthropogenic sources: alpha, beta, and gamma X-ray radiation. Alpha particles are subatomic fragments consisting of two neutrons and two protons. Alpha radiation occurs when the nucleus of an atom becomes unstable (the ratio of neutrons to protons is too low) and alpha particles are emitted to restore balance. Alpha decay occurs in elements with high atomic numbers, such as uranium, radium, and thorium. The nuclei of these elements are rich in neutrons, which makes alpha particle emission possible. Alpha particles are relatively heavy and slow, and therefore they interact readily with materials into which they come in contact. As a result, these particles have low penetrating power and can be blocked with a sheet of paper. Beta radiation occurs when an electron is emitted from the nucleus of a radioactive atom. Beta decay also occurs in elements that are rich in neutrons. Just like electrons found in the orbital of an atom, beta particles have a negative charge and weigh significantly less than a neutron or proton. Therefore beta particles can travel farther than alpha particles, because they interact less readily with materials they encounter. Beta particles can be blocked by a sheet of metal or plastic and are typically produced in nuclear reactors. Gamma or X-ray radiation is produced during a nucleus' excited state after a decay reaction. Instead of releasing another alpha or beta particle, it purges the excess energy by emitting a pulse of electromagnetic radiation called a gamma ray. Gamma rays are similar in nature to light and radio waves except that they have very high energy. Gamma rays have no mass or charge, but they penetrate materials by colliding with electrons in the atomic shells. They can travel for long distances and thus pose external and internal hazards for people. An example of a gamma emitter is cesium-137, which is used to calibrate nuclear instruments.

Radioactive wastes are classified for disposal according to their physical and chemical properties, along with the source from which the waste originated. The half-life of the radionuclide and the chemical form in which it exists are the most influential of the physical properties that determines waste management. The United States divides its radioactive waste into the following categories:

low-level waste, high-level waste, and *transuranic waste.* High-level wastes consist of spent irradiated nuclear fuel from commercial reactors, and the liquid waste from solvent extraction cycles, along with the solids that liquid wastes have been converted into from reprocessing. Transuranic wastes are alpha-emitting residues that contain elements with atomic numbers greater than 92, which is the atomic number of uranium. Wastes are considered transuranic when the elements have half-lives greater than 20 years and concentrations exceeding $100\,nCi\,g^{-1}$. Wastes in this category originate primarily from military manufacturing, with plutonium and americium being the principal elements of concern. Low-level waste defines the radioactive waste that is not classified under the aforementioned categories. Low-level wastes are separated into subcategories: Classes A, B, C, and greater-than Class-C (GTCC), with Class A being the least hazardous and GTCC being the most hazardous. Commercial low-level waste is generated by industry, medical facilities, research institutions and universities, and a few government facilities.

In some commercial and military activities radioactive wastes are mixed with hazardous waste, creating a complex environmental problem. Mixed waste is dually regulated by the EPA and the U.S. Nuclear Regulatory Commission, and waste handlers must comply with both the Atomic Energy Act and Resource, Conservation and Recovery Act statutes and regulations once a waste is deemed a mixed waste. Military sources are regulated by the Department of Energy and comply with the Atomic Energy Act, in regard to radiation safety.

Radon, a naturally occurring radioactive gas that is produced by the radioactive decay of uranium in rock, soil, and water, is of great concern because of the potential for the gas to become concentrated in buildings and homes. The higher the uranium levels in the rocks, the greater the chances that a home or building may have radon gas contamination. Once the parent material decays into radon, it dissolves into the water contained in the pore spaces between soil grains. A fraction of the radon in the pore water volatilizes into the soil atmosphere gas, rendering it more mobile via gas-phase diffusion.

Exposure of human beings to radon occurs in several ways. Decay products of radon are electrically charged when formed, so they tend to attach themselves to atmospheric dust particles that are normally present in the air. This dust can be inhaled, and although the inert gases are mostly exhaled immediately, a fraction of the dust particles deposit on the lungs, building up with every breath. Radon dissolved in groundwater is another source of human exposure, mainly because radon gas is released into the home atmosphere from water as it exits the tap. Another source of human exposure in home and building

TABLE 16.5
Selected Natural and Anthropogenic Radioisotopes

Element	Radioisotope	Origin	Activity
Uranium	^{238}U	Natural, enriched	Uranium mining
Radium	^{226}Ra	Natural, enriched	Uranium mining
Radon	^{222}Rn	Natural, enriched	Uranium mining, construction
Strontium	^{90}Sr	Fission product	Reactors, weapons
Cesium	^{137}Cs	Natural, fission product	Reactors, weapons

settings is the tendency for radon gas to enter structures via diffusion through their foundations and from certain construction materials. Radon gas availability in structures is mainly associated with the concentration of radon in the rock fractures and soil pores surrounding the structure and the permeability of the ground to gases. Slight pressure differentials between structure and soil foundations, which can be caused by barometric changes, winds, and temperature differentials, creates a gradient for radon gas to move from soil gas, through the foundations, and into the structural atmosphere.

Natural Sources of Contaminants

The contaminant sources just presented are associated with human activities involving the production, use, and disposal of chemicals and products. It is important to realize that there are also "natural" sources of contaminants. A major source of such contaminants is drinking water pumped from aquifers composed of sediments and rocks containing naturally occurring elements that dissolve into the groundwater. One example, that of radioactive contaminants such as radon, was discussed in the previous section. Another major example is

arsenic, which has become of great concern in recent years. (See Case Study 16.1.)

CONTAMINANT TRANSPORT AND FATE

The potential effects of hazardous compounds released into the environment on human health and the environment are often difficult to define. Understanding the transport and fate of contaminants in all environmental media (soil/sediment, atmosphere, and hydrologic systems) is critical for proper monitoring and assessment of contaminated sites (see Chapters 18 and 19). The term *fate* refers to the disposition of a contaminant as it is transported through the environment. There are many physical, chemical, and biological processes that influence the ultimate fate of a contaminant in the environment. Some of the major processes will be briefly reviewed in the next section, based on material presented in a recent report published by NRC (2000). Additional information regarding the factors and processes that influence the transport and fate of contaminants in the environment can be found in Pepper *et al.* (1996).

CASE STUDY 16.1

Arsenic (As) occurs naturally in aquifers of the country of Bangladesh. As a result, perhaps as many as 50% of the 125 million people living in this country may be exposed to abnormally high (from 50 to $< 1000\,\mu\text{g L}^{-1}$) As concentrations found in their drinking water. Long-term chronic exposure to As promotes several skin diseases (from dermititis to depigmentation). More advanced stages of As exposure produce gastroenteritis, gangrene, and cancer, among other diseases. More than 2 million people in Bangladesh suffer from one or more of these As-induced deseases. High As concentrations in the groundwater have been associated with As-rich sediments from the Holocene period. These sediments are primarily found in the flood and delta plains of Bangladesh. In these areas $> 60\%$ of the wells have elevated As concentrations.

Arsenic exists in two oxidation states, arsenate, As(III), and arsenite, As(V), both of which are anions. Although both forms are toxic, arsenite is much more toxic, and is also very soluble and mobile in water environments. The exact mechanism of As

enrichment in the groundwater of Bangladesh is not known but likely may be related to the presence of arsenite-bearing minerals and the reductive dissolution of arsenate to the much more soluble form of arsenite. Iron reacts with As anions and can form insoluble and eventually very stable Fe-As complexes that remove As from water. In fact, amorphous Fe oxide is commonly used by water utilities to decontaminate drinking water. Another possible means of treating As-contaminated water includes the use of natural soil material (as filtering devices) that contain high concentrations of iron minerals, such as goethite and hematite that can adsorb and chemically bond As.

No country is immune to the effect of this natural pollutant. In the United States, the Drinking Water Standard has recently been lowered to $10\,\mu\text{g L}^{-1}$ by the EPA. The annual added costs for drinking-water source treatment needed to comply with the new As standard are placed in the billions of dollars. The states most likely to have groundwater sources with elevated As concentrations include Arizona, New Mexico, Nevada, Utah, and California.

TRANSPORT AND FATE PROCESSES

The four general processes that control the transport and fate of contaminants in the environment are advection, dispersion, interphase mass transfer, and transformation reactions. Advection is the transport of matter by the movement of a fluid responding to a gradient of fluid potential. For example, a contaminant molecule dissolved in water will be carried along by the water as it flows through (infiltration) or above (run-off) the soil. Similarly, a molecule residing in air will be carried along as the air flows. Dispersion represents spreading of matter about the center of the contaminant mass. Spreading is caused by molecular diffusion and nonuniform flow fields. Contaminant molecules can reside in several phases in the environment, such as in the atmosphere, in water, and associated with soil particles. So-called mass transfer processes, such as sorption, liquid-liquid partitioning, and volatilization, involve the transfer of matter between phases in response to gradients of chemical potential or, more simply, to concentration gradients. Transformation reactions include any process by which the physicochemical

nature of a chemical is altered. Examples include biotransformation (metabolism by organisms) and hydrolysis (interaction with water molecules). Many of the processes influencing contaminant transport and fate are illustrated in Figure 16.2, which shows the disposition of an organic liquid spilled into the subsurface.

The fate of a specific contaminant in the environment is a function of the combined influences of these four general processes. The combined impact of the four processes determines the "pollution potential" and "persistence" of a contaminant in the environment. The pollution potential characterizes, in essence, the "ability" of the chemical to contaminate the medium of interest (soil, water, air). Compounds that are transported readily (e.g., high aqueous solubility, low sorption), and that are not transformed to any great extent (i.e., are persistent), generally have larger pollution potentials. Large rates of transport means that the contaminant readily moves away from the site where it first entered the environment. A low transformation potential means that the chemical will persist, and thus remain hazardous, for longer periods.

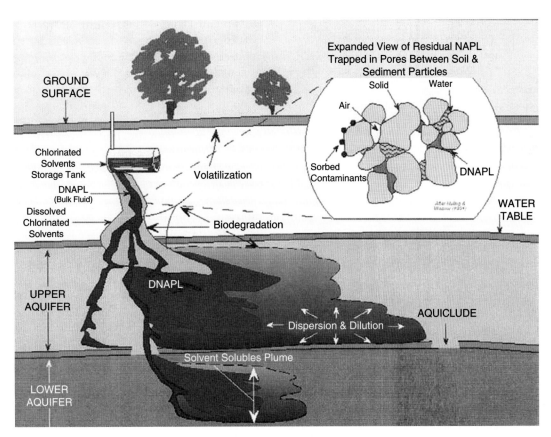

FIGURE 16.2 Schematic of chlorinated solvent pollution as dense nonaqueous phase liquids migrating downward in an aquifer and serving as a source for a solvent soluble plume. Also shown are natural attenuation processes. *DNAPL*, Dense nonaqueous phase liquid; *NAPL*, nonaqueous phase liquid. (U.S. EPA, 1999.)

The health risk posed by a specific contaminant to humans or other receptors is, of course, a function of its toxicological characteristics, as well as its pollution potential, as discussed in Chapter 20. Thus it is important to understand both types of properties. For example, the greatest potential health risk will generally be associated with contaminants that are persistent and highly toxic. However, actual harmful effects will occur only if the receptor is exposed to the contaminant. Thus, the pathways of exposure are critical to assessing risks posed by contaminants. The processes influencing the transport and fate of contaminants in the environment have a significant impact on pathways and levels of exposure.

The EPA has developed special reporting rules for certain chemicals of concern under the Toxic Release Inventory program. These chemicals, listed in Table 16.6, are classified as persistent, bioaccumulative, and toxic (PBT) chemicals. These compounds pose increased risk to human health, not only because they are toxic, but

also because they remain in the environment for long periods, are not readily destroyed, and build up or accumulate in body tissue.

In a related development, an international treaty was recently enacted to control the future production of a class of chemicals termed *persistent organic pollutants (POPs)*. The Stockholm Convention is a global treaty to protect human health and the environment from POPs, which are chemicals that remain intact for long periods, become widely distributed geographically, accumulate in the fatty tissue of living organisms, and are toxic. There are 12 chemicals currently on the POP list: aldrin, chlordane, chlordecone, DDT, dieldrin, endrin, HCH, heptachlor, hexabromobiphenyl, hexachlorobenzene, Mirex, and toxaphene. These chemicals are pesticides, and inspection of Table 16.6 shows some of them are also listed as PBTs.

Once a chemical is applied to (or spilled onto) the land surface, it may remain in place, or it may transfer to the air, surface runoff, or to soil-pore water. Chemicals with moderate to large vapor pressures may evaporate or volatilize into the gas phase, thus becoming subject to atmospheric transport and fate processes, including advection (carried along by wind), dispersion, and transformation reactions (see Chapter 10). Breathing contaminated air is one potential source of exposure. However, atmospheric concentrations of most toxic chemicals are relatively low due to dilution effects.

Transfer of contaminants to surface runoff during precipitation or irrigation events is a major concern associated with the so-called nonpoint source pollution issue. Once entrained into surface runoff, the contaminant may then be transported to surface water bodies. Consumption of contaminated surface water is one major potential route of exposure to toxic chemicals.

Another major potential route of toxic-chemical exposure for human beings is consumption of contaminated groundwater. Once applied to land surface, a contaminant can partition to the soil-pore water. The contaminant then has the potential to move downward to a saturated zone (aquifer), thereby contaminating the groundwater. Whether or not this will occur, as well as the time it will take and the resulting magnitude of contamination, depends on numerous factors, including the magnitude and rate of infiltration/recharge, the soil type, the depth to the aquifer, and the quantity of contaminant and its physicochemical properties (e.g., solubility, degree of sorption, and transformation potential). Volatile contaminants can move by advection and diffusion in the soil-gas phase, in addition to the pore-water phase, which provides an additional means to travel to an aquifer.

TABLE 16.6
Persistent, Bioaccumulative, and Toxic Chemicals[a]

Chemical Categories	Sources
Dioxin and dioxin-like compounds	Chemical manufacturing and processing by-product, waste combustion
Lead compounds	Mining, manufacturing, leaded fuels
Mercury compounds	Mining, manufacturing
Polycyclic aromatic compounds	Petroleum production, combustion, coal tar

Chemicals	Sources
Aldrin	Pesticide
Benzo(g,h,i)perylene	Petroleum refining, fuel combustion
Chlordane	Pesticide
Heptachlor	Pesticide
Hexachlorobenzene	Pesticide
Isodrin	Pesticide
Lead	See above
Mercury	See above
Methoxychlor	Pesticide
Octachlorostyrene	Chemical manufacturing by-product, combustion
Pendimethalin	Pesticide
Pentachlorobenzene	Pesticide production, combustion
Polychlorinated biphenyls (PCBs)	Transformer fluids, lubricants, flame retardants, water-proofing agents
Tetrabromobisphenol A	Flame retardant
Toxaphene	Pesticide
Trifluralin	Pesticide

[a]From www.epa.gov/triinter/lawsandregs/pbt/pbtrule.htm.

PROPERTIES OF THE CONTAMINANT

The physicochemical properties of the contaminant have a major influence on its transport and fate behavior. The first property to consider is the phase state of the contaminant. Under "natural" conditions (temperature $[T] = 25°C$, pressure $[P] = 1$ atm), chemicals in their pure form exist as solids, liquids, or gases (Table 16.7). Clearly, the mobility of a chemical in the environment will depend in part on the phase in which it occurs, with gases being most mobile and solids least mobile.

Many of the organic contaminants of greatest concern happen to exist as liquids in their pure state under natural conditions. These organic compounds are referred to as immiscible or nonaqueous phase liquids (NAPLs). Examples of NAPLs include fuel components (benzene, toluene), chlorinated solvents, and polychlorinated biphenyls. The presence of NAPLs in the subsurface at a contaminated site greatly complicates remediation efforts. Once released into the subsurface, the NAPL becomes trapped in pore spaces, after which it is very difficult to physically remove. Hence they serve as long-term sources of contamination as the molecules transfer to other phases (as will be discussed in a coming paragraph). An additional complicating factor is that many NAPLs are composed of multiple consitutents. Examples of such multicomponent NAPLs include fuels (gasoline, diesel fuel, aviation fuel), coal tar, and creosote that contain hundreds of organic compounds. These multicomponent NAPLs can contain individual compounds such as naphthalene and anthracene that normally occur as solids, but which are "dissolved" in the organic liquids.

Most inorganic contaminants of concern occur as solids in their elemental state. One notable exception is mercury, which is a liquid under standard conditions. An important factor for inorganic contaminants is their "speciation." For example, many inorganics occur primarily in ionic form in the environment (e.g., Pb^{+2}, Cd^{+2}, Hg^+). Speciation can greatly influence aqueous solubility and sorption potential. In addition, many inorganics may combine with other inorganics to form complexes whose transport behavior differs from that of the parent ions.

As noted in the previous section, the ability of a contaminant to transfer from its original state to other states and phases is an important aspect of contaminant behavior. The primary properties that govern these phase transfers are solubility, vapor pressure, volatility, and sorption potential. The extent to which contaminant molecules or ions will transfer from their pure forms into water is the aqueous solubility. The process of transferring from the pure state to water is called dissolution. The solubilities of different contaminants vary by orders of magnitude. A critical aspect of solubility is how it compares to action levels for the contaminant. For example, the aqueous solubility of trichloroethene is approximately $1\,g\,L^{-1}$. This is very low compared to many other compounds. However the maximum contaminant level for trichloroethene is $5\,\mu g\,L^{-1}$, which is 200,000 times

TABLE 16.7

Contaminant Properties

	Representative Contaminants	Solubility	Vapor Pressure	Volatility	Sorption	Biodegradation
Solids						
Organic	naphthalene	low	medium	medium	medium	medium
	pentachlorophenol	low	medium	low	high	medium
	DDT	low	low	low	high	low
Inorganic	lead	low	low	low	medium	—
	chromium	high	low	low	low	—
	arsenic	medium	low	low	low	—
	cadmium	low	low	low	medium	—
Liquids						
Organic	trichloroethene	medium	high	medium	low	low
	benzene	medium	high	medium	low	medium
Inorganic	mercury	low	medium	low	medium	—
Gases						
Inorganic	carbon dioxide	medium	very high	very high	low	none
	carbon monoxide	low	very high	very high	low	none
	sulfur dioxide	medium	very high	very high	low	none
Organic	methane	medium	very high	very high	low	low

—, Not applicable.

smaller. Thus it does not take very much trichloroethene to pollute water.

Evaporation of a compound involves transfer from the pure liquid or pure solid phase to the gas phase. The *vapor pressure* of a pollutant, then, is the pressure of its gas phase in equilibrium with the solid or liquid phase, and is an index of the degree to which the compound will evaporate. In other words, we can think of the vapor pressure of a compound as its "solubility" in air. Evaporation can be an important transfer process when pure-phase pollutant is present in the vadose zone, such that contaminant molecules evaporate into the soil atmosphere.

Volatilization, which is the transfer of pollutant between water and gas phases, is an important component of the transport of many organic compounds in the vadose zone. Volatilization is different from evaporation, which specifies a transfer of pollutant molecules from their pure phase to the gas phase. For example, the transfer of benzene molecules from a pool of gasoline to the atmosphere is evaporation, whereas the transfer of benzene molecules from water (where they are dissolved) to the atmosphere is volatilization. The vapor pressure of a compound gives us a rough idea of the extent to which a compound will volatilize, but volatilization also depends on the solubility of the compound as well as environmental factors.

Sorption is a major process that influences the transport of many contaminants in soils. The broadest definition of *sorption* (or *retention*) is the association of contaminant molecules with the solid phase of the soil (soil particles). Sorption can occur by numerous mechanisms, depending on the properties of the contaminant and of the sorbent.

As discussed in the previous section, the transport and fate of many contaminants is influenced by biodegradation and abiotic reactions. The susceptibility of the contaminant to such transformation reactions is very dependent on their molecular structure (organics) or speciation (inorganics). For example, chlorinated hydrocarbons are generally more resistant to biodegradation than are nonchlorinated hydrocarbons.

QUESTIONS

1. Why are mining processes responsible for the greatest amount of radioactive pollution?

2. What makes gamma radiation so much greater of a health risk than alpha or beta radiation?

3. What are the processes that lead to acid mine drainage?

4. What are the major properties of the contaminant that influences its transport and fate behavior?

5. What are "POPs," and what types of chemicals are they?

REFERENCES AND ADDITIONAL READING

Eisenbud, M, Gesell, T.F. (1997) *Environmental radioactivity: from natural, industrial, and military sources.* San Diego, Academic Press.

Harun-ur-Rashid, Mridna, A.K. (1998) Arsenic Contamination of Groundwater in Bangladesh. 24th WEDC conference, Islamabad, Pakistan, 1998.

Kathren, R.L. (1984) *Radioactivity in the environment: sources, distribution, and surveillance.* New York, Harwood Academic Publishers.

17

Microbial Contaminants

C.P. GERBA AND I.L. PEPPER

Although there are very few organisms that cause disease in humans, the World Health Organization (WHO) has indicated that infectious disease, or disease caused by microorganisms, is the number one killer of human beings in the world today. The majority of organisms that result in infectious diseases are transmitted via the environment (Pepper *et al.*, 1996) (i.e., water, food, or air). For this reason monitoring the environment for pathogens has become both necessary and fundamental to protection of human health.

ENVIRONMENTALLY TRANSMITTED PATHOGENS

A small portion of microorganisms are capable of causing infection and disease in human beings. Such microorganisms are termed pathogens. Pathogens are classified into two general categories based on their potential to cause disease. The first group is termed frank pathogens, which are microorganisms capable of producing disease in healthy and immunocompromised individuals.

The second group is the opportunistic pathogens, which are usually only capable of causing infections when a host is immunocompromised. Examples of such immunocompromised individuals include burn patients; patients taking antibiotics; individuals infected with human immunodeficiency virus (HIV) who are showing acquired immune deficiency syndrome (AIDS); infants; the elderly; and people with uncontrolled diabetes.

Opportunistic and frank pathogens can occur naturally in the environment, or be deposited there by natural processes. Those deposited by natural processes may be present in the guts of animals for instance, and be deposited into the environment through defecation. The concentrations of organisms released into the environment in feces of an infected human being or animal vary with the type of microorganism (Table 17.1). Many human pathogens, such as *Neisseria gonorrhoeae* (gonorrhea) and HIV, can only be transmitted by direct or close contact with an infected person. This is because their survival time in the environment is limited. Others, such as the viral pathogen hepatitis, may survive hours or even years in the environment.

There are several routes of transmission for environmental pathogens. The primary routes, are fecal-oral, waterborne, and airborne transmission. Microorganisms transmitted by the fecal-oral route are usually referred to as enteric pathogens because they infect the gastrointestinal tract. They are characteristically stable in water and food, and in the case of enteric bacteria are capable of growth outside the host, under appropriate environmental conditions. Waterborne diseases are those transmitted through the ingestion of contaminated water that serves as the passive carrier of the infectious agent (Table 17.2). The classic waterborne diseases cholera and typhoid fever,

which frequently ravaged densely populated areas throughout human history, have been effectively controlled by the protection of water sources, and by treatment of contaminated water supplies. In fact, the control of these diseases gave water supply treatment its reputation and played an important role in the reduction of infectious diseases. It is important to remember that many waterborne diseases are transmitted through the fecal-oral route, from human being to human being or animal to human being, so that drinking water is only one of several possible sources of infection. Airborne diseases are those transmitted through the atmosphere, by air currents, and deposited on or in a host where they can begin a new infection. Influenza virus (flu) is the classic example of an airborne disease-causing agent (Table 17.3).

SPECIFIC BACTERIAL PATHOGENS

Detection of specific pathogens is often complex and time consuming. Methods can be based on culturable technologies or nucleic-acid–based methods, such as the polymerase chain reaction (PCR). Detailed procedures for their detection can be found in *Standard Methods for the Examination of Water and Wastewater* (APHA, 1998).

CONCEPT OF INDICATOR ORGANISMS

The routine examination of environmental samples for the presence of intestinal pathogens is often an expensive, tedious, difficult, and time-consuming task. One way to overcome the problems inherent in actual pathogen monitoring is through the use of indicator microorganisms. Indicator organisms are those whose presence indicates that pathogenic microorganisms may also be present. As far back as 1914, the U.S. Public Health Service had adopted the coliform group as an indicator of fecal contamination of drinking water. Coliform bacteria occur naturally in the intestines of all warm-blooded animals and are excreted in great numbers in feces. In polluted water, coliform bacteria are found in densities that can be correlated to the degree of fecal pollution. It has been suggested that the organism must meet several criteria (Box 17.1) to be considered an ideal indicator of fecal contamination.

Unfortunately, no one indicator meets all these criteria. Thus depending on the goal of the monitoring project, various groups of microorganisms have been suggested and used as indicator organisms. Concentra-

TABLE 17.1
Concentrations of Enteric Pathogens in Feces

Organism	Organism/gram feces
Protozoan Parasites	10^6-10^7
Enteric Viruses	
Enteroviruses	10^3-10^7
Rotavirus	10^{10}
Adenovirus	10^{12}
Enteric Bacteria	
Salmonella spp.	10^4-10^{11}
Shigella	10^5-10^9
Worms	
Ascaris	10^4-10^5
Indicator Bacteria	
Coliform	10^7-10^9
Fecal coliform	10^6-10^9

TABLE 17.2
Examples of Waterborne Pathogens

Agent	Disease or Symptoms
Bacteria	
Salmonella typhi	Typhoid fever, fever, headache, nausea, abdominal pain
Shigellae	Dysentery, fever
Escherichia coli 0157:H7	Hemolytic uremic syndrome, diarrhea, vomiting
Campylobacter jejuni	Acute gastroenteritis, diarrhea
Vibrio cholerae 01	Cholera; profuse, watery diarrhea; vomiting
Yersinia enterocolitica	Diarrhea, abdominal pain
Viruses	
Poliovirus	Paralytic poliomyelitis, meningitis, febrile illness
Echoviruses	Meningitis, encephalitis, paralysis, otitis, common cold, pneumonia, diarrhea, exanthema
Coxsackie B	Meningitis, myocarditis, congenital heart defects, pleurodynia, otitis, common cold, pneumonia, others
Enteroviruses 69–71	Meningitis, paralysis, conjunctivitis, pneumonia, bronchiolitis, common cold
Protozoa	
Giardia lamblia	Abdominal pain, diarrhea, bloating
Cryptosporidium parvum	Profuse, watery diarrhea
Entamoeba histolytica	Abdominal pain, diarrhea
Naegleria fowleri	Meningoencephalitis, fever, nausea, usually fatal
Cyclospora cayetanesis	Diarrhea
Enterocytozoon bieneusi	Profuse, nonself-limiting diarrhea
Encephalitozoon intestinalis	Nonself-limiting diarrhea

TABLE 17.3
Examples of Airborne Pathogens

Pathogens	Human Diseases
Bacterial Diseases	
Mycobacterium tuberculosis	Pulmonary tuberculosis
Chlamydia psittaci	Pneumonia
Bacillus anthracis	Pulmonary anthrax
Staphylococcus aureus	Staph, respiratory infection
Streptococcus pyogenes	Strep, respiratory infection
Legionella spp.	Legionellosis
Neisseria meningitidis	Meningococcal infection
Yersinia pestis	Pneumonic plague
Bordetella pertussis	Pneumonic plague
Bordetella pertussis	Whooping cough
Corynebacterium diptheriae	Diphtheria
Fungal Diseases	
Aspergillus fumigatus	Aspergillosis
Blastomyces dermatiridi	Blastomycosis
Coccidioides immitis	Coccidioidomycosis
Cryptococcus neoformans	Cryptococcosis
Histoplasma capuslatum	Histoplasmosis
Nocardia asteriodes	Nocardiosis
Viral Diseases	
Influenza virus	Influenza
Hantavirus	Hantavirus pulmonary syndrome
Coxsackievirus, Echovirus	Pleurodynia
Rubivirus	Rubella
Morbillivurus	Measles
Rhinovirus	Common cold
Protozoan Diseases	
Pneumocystis carinii	Pneumocystosis

BOX 17.1 *Criteria for an Ideal Indicator Organism*

1. The organism should be able to be used as an indicator for all types of water.
2. The organism should be present whenever enteric pathogens are present.
3. The organism should have a reasonably longer survival time than the hardiest enteric pathogen.
4. The organism should not grow in water.
5. The testing method should be easy to perform.
6. The density of the indicator organisms should have some direct relationship to the degree of fecal pollution.
7. It is a member of the natural intestinal microflora of animals, particularly mammals.

tions of indicator bacteria found in wastewater are shown in Table 17.4.

TOTAL COLIFORMS

The coliform group, which includes *Escherichia, Citrobacter, Enterobacter,* and *Klebsiella* species, is relatively

TABLE 17.4
Estimated Levels of Indicator Organisms in Raw Sewage

Organism	Concentration/100 ml
Total coliforms	10^7-10^9
Fecal coliforms	10^6-10^7
Fecal *Streptococci*	10^5-10^6
Enterococci	10^4-10^5
Clostridium perfringens	10^4
Staphylococcus (coagulase positive)	10^3
Pseudomonas aeruginosa	10^5
Acid-fast bacteria	10^2
Coliphage	10^2-10^3
Bacteroides	10^7-10^{10}

easy to detect. Specifically, this group includes all aerobic and facultatively anaerobic, gram-negative, nonspore-forming, rod-shaped bacteria that produce gas upon lactose fermentation in prescribed culture media within 48 hours at 35° C. This group of bacteria has been used as the standard for assessing fecal contamination of recreational and drinking waters since 1914. It has been learned that absence of these organisms in 100 ml of drinking water ensures the prevention of bacterial waterborne disease outbreaks. However, it has also been learned that there are a number of deficiencies in the use of these indicators. These deficiencies include the potential for them to actually proliferate in aquatic environments and distribution systems, and the potential for them to be suppressed (during assay) by high background bacterial growth, and they are not always indicative of a health threat, especially for the more environmentally resistant enteric protozoan and viral pathogens. Still, the coliform group of bacteria has proved its merit in assessing the bacterial quality of water. Three methods are commonly used to identify coliforms in water. These methods are: the most probable number (MPN), the membrane filter (MF), and the presence–absence (P–A) tests.

MOST PROBABLE NUMBER TEST

The MPN test is used for detection and estimation of coliform concentration in environmental samples. This test consists of three steps: a presumptive test, a confirmed test, and a completed test (Figure 17.1). In the *presumptive test*, lauryl sulfate–tryptose–lactose broth is placed in a set of test tubes with different dilutions of the water to be tested. Usually, three to five test tubes are prepared per dilution. These test tubes are incubated at 35° C for 24 to 48 hours before examination for gas and acid production that indicates a positive test. Once the positive tubes have been

identified and recorded, it is possible to estimate the total number of coliforms in the original sample by using an MPN table that gives numbers of coliforms per 100 ml, or by using the following equation (Table 17.5):

$$\frac{\text{\# of positive tubes} \times 100}{(\text{ml of sample in negative tubes}) \times (\text{ml of sample in all tubes})} \quad \text{(Eq. 17.1)}$$

In the confirmation test, positive presumptive tests are verified using selective bacteriological agars such as Levine's eosin–methylene blue (EMB) agar or Endo agar, inoculated with a small amount of culture from the positive tubes. Confirmation is indicated by the formation of colonies showing a green sheen or a dark center. In some cases a completed test is performed in which colonies from the agar are inoculated back into lauryl sulfate–tryptose–lactose broth to demonstrate the production of acid and gas.

MEMBRANE FILTER TEST

The MF test is easier to perform than the MPN test because it requires fewer test tubes and less handling. Usually about 100 ml of water is passed through a membrane filter (pore size 0.45 μm). This membrane is then placed on a thin, absorbent pad saturated with differential or selective medium (Figure 17.2). For example, if total coliform organisms are sought, a modified Endo medium is used. For coliform bacteria, the filter is incubated at 35° C for 18 to 24 hours. The success of the method depends on using the correct media to facilitate identification of the bacterial colonies growing on the membrane filter surface. To continue with the previous example, enumerating colonies that have a green sheen indicates the number of coliform bacteria in a water sample.

PRESENCE–ABSENCE TEST

P–A tests answer the question: "Are there indicators in this water sample?" It should be noted, however, that they are nonquantitative. In P–A tests a single tube of lauryl sulfate–tryptose–lactose broth can be used to indicate the presence of coliforms. However, in recent years enzymatic assays have been developed that allow the detection of both total coliform bacteria and *E. coli* in drinking water at the same time. This test is primarily used for drinking water in which 100-ml volumes are tested.

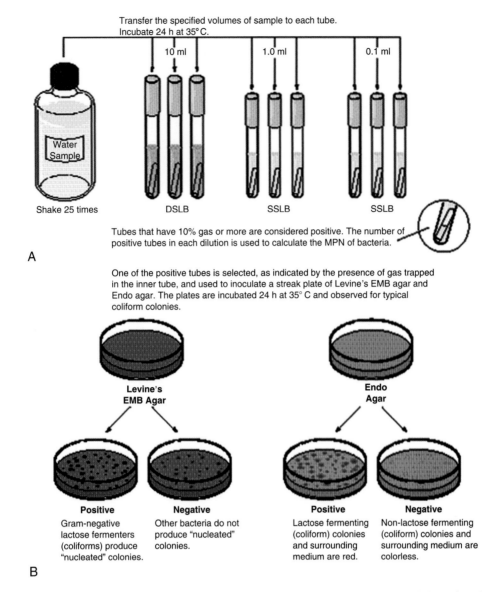

Transfer the specified volumes of sample to each tube.
Incubate 24 h at 35°C.

10 ml   1.0 ml   0.1 ml

Water Sample

Shake 25 times   DSLB   SSLB   SSLB

Tubes that have 10% gas or more are considered positive. The number of positive tubes in each dilution is used to calculate the MPN of bacteria.

A

One of the positive tubes is selected, as indicated by the presence of gas trapped in the inner tube, and used to inoculate a streak plate of Levine's EMB agar and Endo agar. The plates are incubated 24 h at 35° C and observed for typical coliform colonies.

Levine's EMB Agar   **Endo Agar**

Positive
Gram-negative lactose fermenters (coliforms) produce "nucleated" colonies.

Negative
Other bacteria do not produce "nucleated" colonies.

Positive
Lactose fermenting (coliform) colonies and surrounding medium are red.

Negative
Non-lactose fermenting (coliform) colonies and surrounding medium are colorless.

B

FIGURE 17.1 Procedure for performing an MPN test for coliforms on water samples: (**A**) Presumptive test. (**B**) Confirmed test. *DSLB,* Double-strength lactose broth; *SSLB,* single-strength lactose broth. (From *Pollution Science* © 1996, Academic Press, San Diego.)

Enzymatic Assay (Colilert)

The use of Colilert assay can be a simple P–A test (Figure 17.3) or an MPN assay using a bubble tray (Figure 17.4). In this test total coliform bacteria produce the enzyme β-galactosidase, which hydrolyzes the substrate *o*-nitrophenyl-β-D-galactopyranoside (ONPG) to yellow nitrophenol. *E. coli* is detected at the same time by incorporation of a substrate, 4-methylumbelliferone glucuronide (MUG) (Figure 17.5), which produces a fluorescent end product after cleavage with the enzyme β-glucuronidase found specifically in *E. coli*. The end product can be detected with a long-wave ultraviolet (UV) lamp. As mentioned, this test can be performed by adding the sample to a single bottle (P–A test) or MPN and can be done routinely in the field. After 24 hours of incubation, samples positive for total coliforms turn yellow, whereas *E. coli*–positive samples fluoresce under long-wave UV illumination in the dark.

Fecal Coliforms

Although the total coliform group has served as the main indicator of water pollution for many years, many of the organisms in this group are not limited to fecal sources. Thus methods have been developed to restrict the enumeration to coliforms that are more clearly of fecal

TABLE 17.5

Most Probable Number (MPN) Table Used for Evaluation of the Data in this Experiment, Using Three Tubes in Each Dilution

Number of Positive Tubes in Dilutions				Number of Positive Tubes in Dilutions			
10 ml	1 ml	0.1 ml	MPN /100 ml	10 ml	1 ml	0.1 ml	MPN/100 ml
0	0	0	<3	2	0	0	9.1
0	1	0	3	2	0	1	14
0	0	2	6	2	0	2	20
0	0	3	9	2	0	3	26
0	1	0	3	2	1	0	15
0	1	1	6.1	2	1	1	20
0	1	2	9.2	2	1	2	27
0	1	3	12	2	1	3	34
0	2	0	6.2	2	2	0	21
0	2	1	9.3	2	2	1	28
0	2	2	12	2	2	2	35
0	2	3	16	2	2	3	42
0	3	0	9.4	2	3	0	29
0	3	1	13	2	3	1	36
0	3	2	16	2	3	2	44
0	3	3	19	2	3	3	53
1	0	0	3.6	3	0	0	23
1	0	1	7.2	3	0	1	39
1	0	2	11	3	0	2	64
1	0	3	15	3	0	3	95
1	1	0	7.3	3	1	0	43
1	1	1	11	3	1	1	75
1	1	2	15	3	1	2	120
1	1	3	19	3	1	3	160
1	2	0	11	3	2	0	93
1	2	1	15	3	2	1	150
1	2	2	20	3	2	2	210
1	2	3	24	3	2	3	290
1	3	0	16	3	3	0	240
1	3	1	20	3	3	1	460
1	3	2	24	3	3	2	1100
1	3	3	29	–	–	–	–

original—that is, the fecal coliforms. These organisms, which include the genera *Escherichia* and *Klebsiella*, are differentiated in the laboratory by their ability to ferment lactose with the production of acid and gas at 44.5° C within 24 hours. Fecal coliforms may be detected by methods similar to those used for coliform bacteria. For the MPN method a growth media known as EC broth (Difco, Detroit) is used, and for the membrane filter method another growth media known as m-FC agar is used for water analysis. A medium known as m-T7 agar has been proposed for use in the recovery of injured fecal coliforms from water (LeChevallier *et al.*, 1983), and results in greater recovery from water.

FECAL *STREPTOCOCCI*

The fecal *streptococci* are a group of gram-positive *streptococci* belonging to the genera *Enterococcus* and *Streptococcus* (Gleeson and Gray, 1997). The genus *Enterococcus*

includes all *streptococci* that share certain biochemical properties and have a wide range of tolerance for adverse growth conditions. They are differentiated from other *streptococci* by their ability to grow in 6.5% sodium chloride, pH 9.6 and at 10, and 45° C. Fecal *streptococci* have been used to assess whether the fecal contamination in a sample is of human or animal origin, by determining the ratio of fecal coliforms to fecal *streptococci*. A fecal coliform/fecal *streptococci* (FC/FS) ratio of 4 or more indicates a contamination of human origin, whereas a ratio below 0.7 is indicative of animal pollution (Geldreich and Kenner, 1969) (Table 17.6). This ratio is valid only for recent (24 hours) fecal pollution. It should be noted that the validity of this test has been questioned considerably, though it is still considered one of the few methods capable of this type of differential assay.

The filtration and MPN methods are acceptable methods for monitoring fecal *streptococci*. The MF method uses fecal *Streptococcus* agar with incubation at 37° C for 24 hours. Red, maroon, and pink

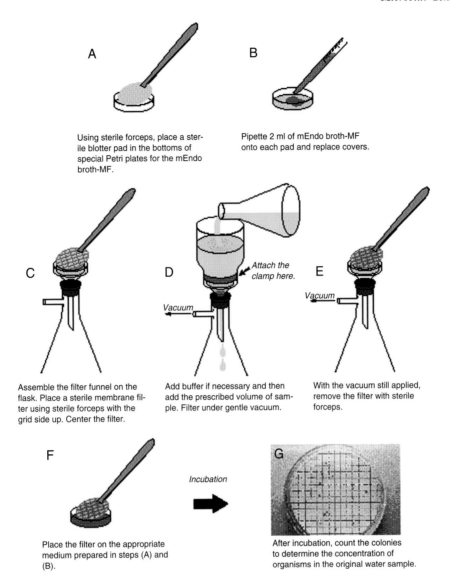

A

Using sterile forceps, place a sterile blotter pad in the bottoms of special Petri plates for the mEndo broth-MF.

B

Pipette 2 ml of mEndo broth-MF onto each pad and replace covers.

C

Assemble the filter funnel on the flask. Place a sterile membrane filter using sterile forceps with the grid side up. Center the filter.

D

Attach the clamp here.

Vacuum

Add buffer if necessary and then add the prescribed volume of sample. Filter under gentle vacuum.

E

Vacuum

With the vacuum still applied, remove the filter with sterile forceps.

F

Incubation

Place the filter on the appropriate medium prepared in steps (A) and (B).

G

After incubation, count the colonies to determine the concentration of organisms in the original water sample.

FIGURE 17.2 Membrane filtration for determining the coliform count in a water sample, using vacuum filtration. (From *Pollution Science* © 1996, Academic Press, San Diego.)

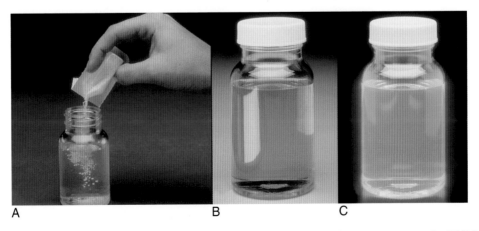

A B C

FIGURE 17.3 Detection of indicator bacteria with Colilert. (**A**) Addition of salts and enzyme substrates to water sample. (**B**) Yellow color indicating the presence of coliform bacteria. (**C**) Fluorescence under long-wave ultraviolet light indicating the presence of *E. coli*. (Courtesy of IDEXX, Westbrook, ME.)

FIGURE 17.4   Colitray.

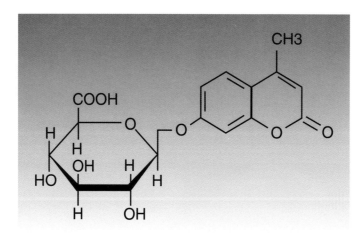

FIGURE 17.5   The structure of 4-methylumbelliferyl-β-D-glucuronide (MUG).

colonies are counted as presumptive fecal *streptococci* (Figure 17.6). Confirmation of fecal *streptococci* is by subculture on bile aesulin agar and incubation for 18 hours at 44° C. Discrete colonies surrounded by a brown or black halo are considered positive.

CLOSTRIDIUM PERFRINGENS

Clostridium perfringens is a sulfite-reducing anaerobic spore former; it is gram positive, rod shaped, and exclusively of fecal origin. The spores are very heat resistant,

TABLE 17.6
The FC/FS Ratio

FC/FS Ratio	Source of Pollution
>4.0	Strong evidence that pollution is of human origin
2.0–4.0	Good evidence of the predominance of human wastes in mixed pollution
0.7–2.0	Good evidence of the predominance of domestic animal wastes in mixed pollution
<0.7	Strong evidence that pollution is of animal origin

persist for long periods in the environment, and are very resistant to disinfectants. The hardy spores of this organism limit its usefulness as an indicator. However, it has been suggested that it could be an indicator of past pollution, a tracer of less hardy indicators, and an indicator of removal of protozoan parasites or viruses during drinking water and wastewater treatment (Payment and Franco, 1993). One method described by Dowd *et al.* (1997) involves incubation of samples at 75° C for 15 minutes, followed by anaerobic incubation on Clostricel media at 37° C for 72 hours.

BACTERIOPHAGE

Bacteriophage, which are viruses that infect bacteria, have been proposed as indicators of fecal pollution.

These organisms have also been suggested as indicators of the presence of human pathogenic virus. Basically the structure, morphology, size, and behavior in the environment of many bacteriophages closely resemble those of enteric viruses. For these reasons, they have also been used extensively to evaluate virus resistance to disinfectants during water and wastewater treatment, and as surface and groundwater tracers. The presence of bacteriophage in an environmental sample is an indication that bacteria capable of supporting their replication are present. Two groups of phage, in particular, have been studied: the somatic coliphage, which infect *E. coli* host strains through cell wall receptors, and the F-specific RNA coliphage, which infect strains of *E. coli* and related bacteria through conjugation tubes. A significant advantage of using coliphage is that they can be detected by simple and inexpensive techniques that yield results in 8 to 18 hours. Both a plating method (the agar overlay method) (Figure 17.7) and the MPN method can be used to detecut coliphage in volumes ranging from 1 to 100 ml. Because F-specific phage are infrequently detected in human fecal matter and show no direct relationship to the fecal pollution level, they cannot be considered indicators of fecal pollution (Havelaar *et al.*, 1990). However, their presence in high numbers in wastewater, and their relatively high resistance to chlorination, contribute to their consideration

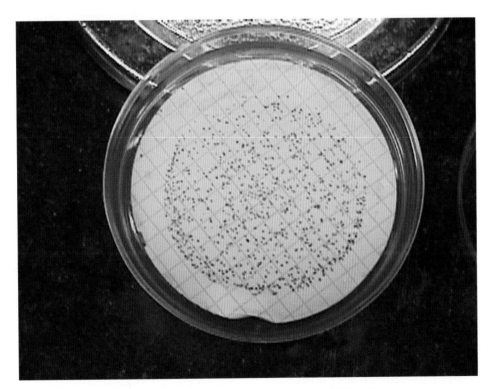

FIGURE 17.6   Colonies of fecal *streptococcus* on K–F media.

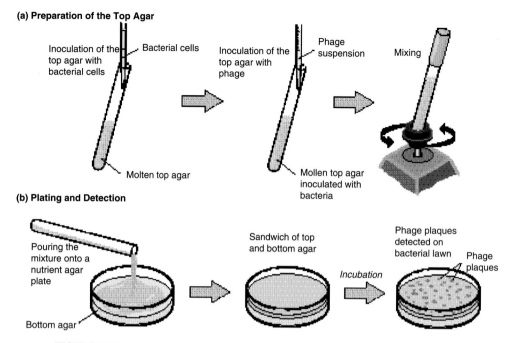

(a) Preparation of the Top Agar

Inoculation of the top agar with bacterial cells — Bacterial cells

Molten top agar

Inoculation of the top agar with phage — Phage suspension

Mollen top agar inoculated with bacteria

Mixing

(b) Plating and Detection

Pouring the mixture onto a nutrient agar plate

Bottom agar

Sandwich of top and bottom agar

Incubation

Phage plaques detected on bacterial lawn — Phage plaques

FIGURE 17.7   Technique for performing a bacteriophage assay. (From Pepper *et al.*, 1995.)

as an index of wastewater contamination and potential indicator of enteric viruses.

OTHER INDICATOR ORGANISMS

Table 17.7 provides a list indicator organisms and their potential applications in environmental monitoring.

STANDARDS AND CRITERIA FOR INDICATORS

The U.S. Environmental Protection Agency (EPA) states that no detectable coliforms per 100 ml of drinking water shall occur in drinking water. This standard is legally enforceable in the United States. If this level is exceeded by water industries, they are required to take corrective action or they face stiff fines by the state or federal government. Microbial standards set by various government bodies in the United States are shown in Table 17.8.

The use of microbial standards also requires the development of standard methods and quality assurance or quality control plans for the laboratories that will do the monitoring. Knowledge of how to sample and how often to sample is also important. All of this information is usually defined in the regulations when a standard is set. For example, frequency of sampling may be determined by the size (number of customers) of the utility providing the water. Sampling must proceed

in some random fashion so that the entire system is characterized. Because of the wide variability in numbers of indicators in water, some positive samples may be allowed or tolerance levels or averages may be allowed. Usually, *geometric averages* are used in standard setting because of the often-skewed distribution of bacterial numbers. This prevents one or two high values from giving overestimates of high levels of contamination, which would appear to be the case with *arithmetric averages* (Table 17.9).

TABLE 17.7

Indicators and Their Potential Uses

Indicator	Possible Uses
Coliforms	Sewage
Fecal coliforms	Sewage, pathogen, fecal indicators
Enterococci	Fecal indicators, sewage indicator, separation of human from lower animal sources, proximity to fecal source
Clostridium perfringens	Fecal indicators, sewage indicator, proximity to fecal source
Candida albicans	Pathogen, fecal indicators, sewage indicator
Bifidobacteria	Fecal indicators, sewage indicator, proximity to fecal source
Coliphage	Sewage indicator
Pseudomonas aeruginosa	Pathogen, sewage indicator, indicator of nutrient pollution
Aeromonas hydrophila	Pathogen, sewage indicator, indicator of nutrient pollution

TABLE 17.8
Federal and State Standards for Microorganisms

Authority	Standards
U.S. Environmental Protection Agency Safe Drinking Water Act	0 coliforms/100 ml
Clean Water Act	
– wastewater discharges	200 fecal coliforms/100 ml
– sewage sludge (Class A)	<1000 fecal coliforms/4 g
	<3 *Salmonella*/4 g
	<1 enteric virus/4 g
	<1 helminth ova/4 g
California Wastewater reclamation for irrigation	<2.2 MPN coliforms/100 ml
Arizona Wastewater reclamation for irrigation water with public access	25 fecal coliforms/100 ml
	125 enteric virus/40 L
	No detectable *Giardia*/40 L
U.S. Environmental Protection Agency contact recreation recommended	*Freshwater*
	33 *Enterococci*/100 ml
	126 fecal coliforms/100 ml
	Marine water
	35 *Enterococci*/100 ml

TABLE 17.9
Arithmetic and Geometric Averages of Bacterial Numbers in Water

MPN	Log
2	0.30
110	2.04
4	0.60
150	2.18
1100	3.04
10	1.00
<u>12</u>	<u>1.08</u>
198 = arithmetic average	1.46 = log $\bar{x}$ antilog $\bar{x}$ = 29
	29 = geometric average

MPN, Most probable number.

Geometric means ($\bar{x}$) are determined by the following equation:

$$\text{Log } \bar{x} = \frac{\Sigma_i^N (\log x_i)}{N} \qquad \text{(Eq. 17.2)}$$

$$\bar{x} = \text{antilog } (\log \bar{x}) \qquad \text{(Eq. 17.3)}$$

where N is the number of samples $\bar{x}$, and x_i is the number of organisms per sample volume for sample i.

SAMPLE PROCESSING AND STORAGE

Sample processing and storage for soil, water, and air were described in Chapter 14.

PROCESSING ENVIRONMENTAL SAMPLES FOR VIRUSES

PROCESSING SLUDGE, SOIL, AND SEDIMENT FOR VIRUSES

The EPA currently has guidelines for the monitoring of biosolids for enteroviruses for certain types of land application. In addition, soil and sediment are also of interest when monitoring the environment for pathogens. For detection of viruses on solids, the first step is to desorb or detach them from the solid. As with microporous filters, viruses are thought to bind to these solids by a combination of electrostatic and hydrophobic forces. Thus eluting solutions, or eluents, are used. These solutions are able to desorb the virus from the solid surface, or alternatively to lessen these attractive forces, which allows the virus to be recovered in the eluting fluid (Berg, 1987; Gerba and Goyal, 1982).

The most common procedure for biosolids involves collecting 500 to 1000 ml of material and adding $AlCl_3$ and HCl to adjust the pH to 3.5 (Figure 17.8). This causes virus binding to the sludge solids and allows concentration of the solids containing the virus by centrifugation. After this procedure the solids are resuspended in a beef extract solution at neutral pH to elute the virus (Straub *et al.*, 1994). The desorbed viruses can then be reconcentrated by protein flocculation, which occurs when the solution's pH is lowered to 3.5, resuspended in 20 to 50 ml, and neutralized. This same type of elution procedure is also used for the recovery of viruses from soils and aquatic sediments (Hurst *et al.*, 1997). A sample can then be enumerated for viruses in cell culture in the "Enumeration of Human Pathogenic Viruses" section.

PROCESSING OF WATER SAMPLES FOR VIRUSES

WATER SAMPLE COLLECTION

The suitability of a virus concentration method depends on the probable virus density, the volume limitations of the concentration method for the type of water, and the presence of interfering substances. A sample volume of less than 1 L may suffice for recovery of viruses from raw and primary sewage. For drinking water and relatively nonpolluted waters, the virus levels are likely to be so low that hundreds or perhaps thousands of liters must be sampled to increase the probability of virus detection.

Most methods employed for virus concentration depend on adsorption of the virus to a surface, such as a filter or mineral precipitate, although hydroextraction and ultrafiltration have been used (Gerba, 1987). For

Recovery and concentration of viruses from sludge

Procedure	Purpose
500-2000 ml sludge	
Adjust to pH 3.5 0.005 M AlCl₃	Adsorb viruses to solids
Centrifuge to pellet solids	
Discard supernatant	
Resuspend pellet in 10% beef extract	Elute (desorb) viruses from solids
Centrifuge to pellet solids	
Discard pellet and filter through 0.22-μm filler	Remove bacteria viruses are in supernatant
Assay using cell culture	

FIGURE 17.8   Procedure for recovery and concentration of viruses from sludge. (Reprinted from *Environmental Microbiology* © 2000, Academic Press, San Diego.)

large volumes of water, pleated cartridge filters are used (Figure 17.9), which can process the water at flow rates of 30 to 40 L per minute. The entire system can usually be contained in a 20-L capacity ice chest (Figure 17.10).

The class of filters most commonly used for virus collection from large volumes of water is adsorption–elution microporous filters (APHA, 1998) (Table 17.10). This procedure involves passing the water through a filter to which the viruses adsorb. The pore size of the filters is much larger than the viruses, and adsorption takes place by a combination of electrostatic and hydrophobic interactions (Gerba, 1984). Two general types of filters are available: electronegative (negative surface charge) and electropositive (positive surface charge). Electronegative filters are composed of either cellulose esters or fiberglass with organic resin binders. Because the filters are negatively charged, cationic salts ($MgCl_2$ or $AlCl_3$) must be added, in addition to lowering the pH to 3.5. This reduces the net negative charge usually associated with viruses, allowing adsorption to be maximized. Such pH adjustment can be cumbersome, because it requires modifying the water before filtering and use of additional materials and equipment, such as pH meters. The most commonly used electronegative filter is the Filterite. Generally it is used as a 10-inch (25.4 cm) pleated cartridge with either a 0.22- or 0.45-μm pore-size rating.

Electropositive filters are ideal when concentrating viruses from waters with high amounts of organic matter and turbidity. Electropositive filters may be composed

FIGURE 17.9 (A) Field Viradel system for concentrating viruses from water. (B) Elution of virus from filter with beef extract. (Reprinted from *Environmental Microbiology* © 2000, Academic Press, San Diego.)

viruses for adsorption sites. Finally, the concentration efficiency varies depending on the type of virus, presumably because of differences in the isoelectric point of the virus, which influences the net surface charge of the virus at any pH.

SAMPLE ELUTION AND RECONCENTRATION

Adsorbed viruses are usually eluted from the filter surfaces by pressure filtering a small volume (1 to 2 L) of an eluting solution through the filter (see Figure 17.9). The eluent is usually a slightly alkaline proteinaceous fluid, such as 1.5% beef extract adjusted to pH 9.5. The elevated pH increases the negative charge on both the virus and the filter surfaces, which results in desorption of the virus from the filter. The organic matter in the beef extract also competes with the virus for adsorption on the filter, further aiding desorption. The 1- to 2-L volume of the eluent is still too large to allow sensitive virus analysis, and therefore a second concentration step (reconcentration) is used to reduce the volume to 20 to 30 ml before assay. The elution–reconcentration process is shown in detail in Figure 17.10. Overall, these methods can recover enteroviruses with an efficiency of 30% to 50% from 400- to 1000-L volumes of water.

ENUMERATION OF HUMAN PATHOGENIC VIRUSES

Living cells are necessary for virus replication, as previously mentioned. To assay for viruses, we use human cell lines that have been isolated and developed to grow *in vitro* (in culture). These cells may be placed in flasks, tubes, or Petri dishes, depending on the needs of the laboratory. Plastic flasks and multiwell plates especially prepared for cell culture use are currently the most commonly used type of containers. The cells attach to the surface of the vessel and begin replicating. Replication ceases when a single layer or monolayer of cells occupies all the available surface of the vessel. The most commonly used cell line for enterovirus detection in water is the BGM cell line. In recent years the $CaCO_2$ cell line originating from a human colon carcinoma has been found to grow more types of enteric viruses (hepatitis A, astroviruses, adenoviruses, enteroviruses, rotaviruses) than any cell line and is seeing increased use in environmental virology.

The two most common methods for the detection and quantification of viruses in cell culture are the cytopathogenic effect (CPE) method and the plaque-forming unit (PFU) method (Figure 17.11). Cytopathic effects are observable changes that take place in the host cells as a result of virus replication (Figures 17.12 and 17.13).

of fiberglass or cellulose containing a positively charged organic polymeric resin, which creates a net positive surface charge to enhance absorption of the negatively charged virus. These filters adsorbed viruses efficiently over a wide pH range without a need for polyvalent salts. However, they cannot be used with seawater or water with a pH exceeding 8.0 to 8.5 (Sobsey and Glass, 1980). The electropositive 1MDS Virozorb is specifically manufactured for virus concentration from water. Filter methods suffer from a number of limitations. Suspended matter in water tends to clog the filters, thereby limiting the volume that can be processed and interfering with the elution process. Dissolved and colloidal organic matter in some waters can interfere with virus adsorption to filters, presumably by competing with

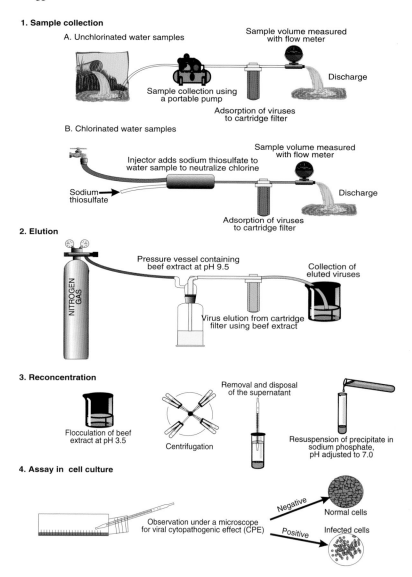

FIGURE 17.10   Procedures for sampling and detection of viruses from water. (Reprinted from *Environmental Microbiology* © 2000, Academic Press, San Diego.)

TABLE 17.10

Methods Used for Concentrating Viruses from Water

Method	Initial Volume of Water	Applications	Remarks
Filter adsorption–elution			
Negatively charged filters	Large	All but the most turbid waters	Only system shown useful for concentrating viruses from large volumes of tap water, sewage, seawater, and other natural waters; cationic salt concentration and pH must be adjusted before processing
Positively charged filters	Large	Tap water, sewage	No preconditioning of water necessary at neutral or acidic pH levels
Adsorption to metal salt precipitate, aluminum hydroxide, ferric hydroxide	Small	Tap water, sewage	Have been useful as reconcentration
Charged filter aid	Small	Tap water, sewage	40-L volumes tested, low cost; used as a sandwich between prefilters

(*Continued*)

TABLE 17.10 (Continued)

Method	Initial Volume of Water	Applications	Remarks
Polyelectrolyte PE60	Large	Tap water, lake water, sewage	Because of its unstable nature and lot-to-lot variation in efficiency for concentrating viruses, method has not been used in recent years
Bentonite	Small	Tap water, sewage	
Iron oxide	Small	Tap water, sewage	
Glass powder	Large	Tap water, seawater	Columns containing glass powder have been made that are capable of processing 400-L volumes
Positively charged glass wool	Small to large	Tap water	Positively charged glass wool is inexpensive; used in pipes or columns
Protamine sulfate	Small	Sewage	Very efficient method for concentrating reoviruses and adenoviruses from small volumes of sewage
Hydroextraction	Small	Sewage	Often used as a method for reconcentrating viruses from primary eluates
Ultrafiltration			
Soluble filters	Small	Clean waters	Clogs rapidly even with low turbidity
Flat membranes	Small	Clean waters	Clogs rapidly even with low turbidity
Hollow fiber or capillary	Large	Tap water, lake water, seawater	Up to 100 to 1000 L may be processed, but water must often be prefiltered
Reverse osmosis	Small	Clean waters	Also concentrates cytotoxic compounds that adversely affect assay methods

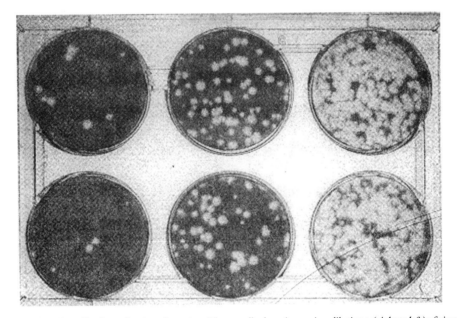

FIGURE 17.11 Multiwell plates for cell culture for virus detection. These wells show increasing dilutions *(right to left)* of virus on a monolayer. Each clear zone, or plaque, theoretically arises from a single infectious virus particle, that is, a plaque-forming unit. (Photo courtesy C.P. Gerba.)

Such changes may be observed as changes in morphology, including rounding or formation of giant cells, or formation of a hole in the monolayer due to localized lysis of virus-infected cells. Different viruses may produce very individual and distinctive CPE. For example, adenoviruses cause the formation of grapelike clusters, and enteroviruses cause rounding of the cells. The production of CPE may take as little time as 1 day to 2 to 3 weeks, depending on the original concentration and type of virus. Not all viruses may produce CPE during replication in a given cell line. In addition, the virus may grow to high numbers, but no visible CPE is observed. In these instances other techniques for virus detection must be used. Alternatives include the use of an immunoassay that detects viral antigens or the use of PCR to detect viral nucleic acid.

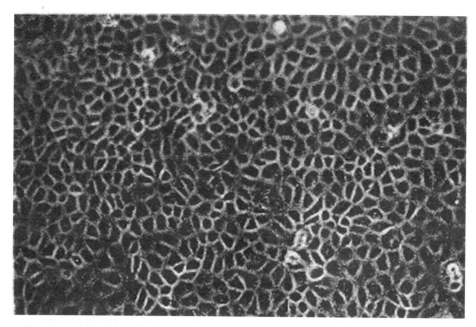

FIGURE 17.12   Normal uninfected cell culture monolayer. Compare with Figure 17.13. (Photo courtesy M. Abbaszadegan.)

FIGURE 17.13   Infected cell culture monolayer exhibiting CPE. Compare with Figure 17.12. (Photo courtesy M. Abbaszadegan.)

ENUMERATION OF BACTERIOPHAGES

Bacteriophages in water are usually assayed by addition of a sample (usually 0.1 to 2 ml) to soft or overlay agar, along with a culture of host bacteria (*E. coli* if coliphages are being assayed) in the log phase of growth. The phage attach to the bacterial cell and lyse the bacteria.

The bacteria produce a confluent lawn of growth, except for areas where the phage has grown and lysed the bacteria (wee Figure 17.7). The resulting clear areas are known as plaques. The soft agar overlay is used to enhance the physical spread of viruses between the bacterial cells. Plaques will appear within 12 to 24 hours.

PROCESSING ENVIRONMENTAL SAMPLES FOR PROTOZOAN PARASITES

WATER SAMPLE COLLECTION

As with enteric viruses, ingestion of only a few protozoan parasites causes infection in human beings. As a result, sensitive methods for analysis of protozoa are required. Current methods have been developed primarily for the concentration and detection of *Giardia* cysts and *Cryptosporidium* oocysts (Hurst *et al.*, 1997). The first step usually involves collection and filtration of large volumes of water (often hundreds of liters). During filtration, the cysts or oocysts are entrapped in a 1-μm nominal porosity spun-fiber filter (25 mm in length) made of Orlon- or polypropylene-pleated cartridge filters, or foam filters (Simmons *et al.*, 2001). The precise volume of water collected depends on the investigation. For surface waters it is common to collect 100 to 400 L, and volumes in excess of 1000 L have been collected for drinking water analysis. Usually, a pump running at a flow rate of 4 to 12 L per minute is used to collect a sample. The filter cartridge is placed in a plastic bag, sealed, stored on ice, and sent to the laboratory to be processed within 72 hours.

SAMPLE ELUTION AND PURIFICATION

In the laboratory the cysts and oocysts are extracted by cutting the filter in half, unwinding the yarn, and placing the fibers in an eluting solution of Tween 80, sodium dodecyl sulfate, phosphate-buffered saline (Figure 17.14). The filters are either hand washed in large beakers or washed in a mechanical stomacher. The 3- to 4-liter resulting volume is then centrifuged to concentrate the oocysts, which are then resuspended in 10% formalin or 2.5% buffered potassium dichromate (for *Cryptosporidium*). Cysts and oocysts are stable for many weeks when stored in formalin or potassium dichromate. A great deal of particulate matter is often concentrated, along with the cysts and oocysts, and the pellet is further purified by density gradient centrifugation with Percoll–sucrose, sucrose, or potassium citrate solutions (Figure 17.15).

ENUMERATION OF HUMAN PATHOGEN PROTOZOA

Aliquots of the concentrated and purified water samples are filtered through a membrane cellulose acetate filter, which collects the cysts and oocysts in its surface. The cysts and oocysts are then stained with fluorescent monoclonal antibodies and the filter placed under an

FIGURE 17.14   Spun fiber filter for concentrating *Giardia* cysts and *Cryptosporidium* oocysts. (Reprinted from *Environmental Microbiology* © 2000, Academic Press, San Diego.)

1. Sample collection

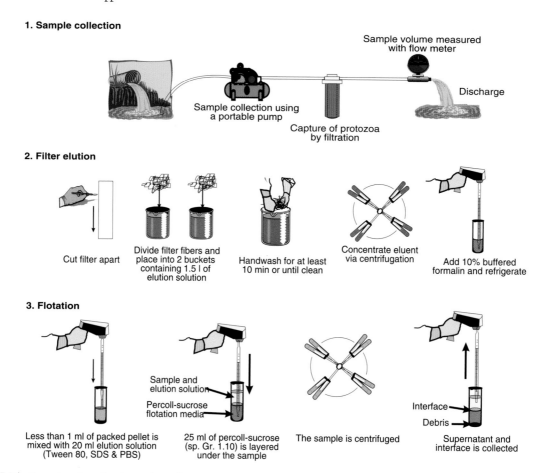

2. Filter elution

Cut filter apart

Divide filter fibers and place into 2 buckets containing 1.5 l of elution solution

Handwash for at least 10 min or until clean

Concentrate eluent via centrifugation

Add 10% buffered formalin and refrigerate

3. Flotation

Less than 1 ml of packed pellet is mixed with 20 ml elution solution (Tween 80, SDS & PBS)

Sample and elution solution
Percoll-sucrose flotation media

25 ml of percoll-sucrose (sp. Gr. 1.10) is layered under the sample

The sample is centrifuged

Interface
Debris

Supernatant and interface is collected

FIGURE 17.15   Procedure for collection and sampling protozoa from water. (Reprinted from *Environmental Microbiology* © 2000, Academic Press, San Diego.)

epifluorescence microscope. Fluorescent bodies of the correct sizes and shapes are identified and examined by differential interference contrast microscopy for the presence of internal bodies (i.e., trophozites or sporozites) (Figure 17.16). This method is time consuming and cumbersome, but has been successful in isolation of *Giardia* and *Cryptosporidium* from most of the surface waters in the United States (Rose *et al.*, 1997). There are several disadvantages associated with this method of analysis. These include frequently low recovery efficiencies (5% to 25%), a 1- to 2-day processing time, interference caused by fluorescent algae, inability to determine viability, and finally, the limitation in sample volume that can be collected from some sites because of high turbidity and filter clogging. It is for these reasons that other methods (e.g., calcium carbonate flocculation, membrane filtration, ultrafiltration, and immunomagnetic separation) are under development to improve the efficiency of concentration.

FIELD ANALYSIS

Many of the microbiological assays described in this chapter must be performed in the laboratory. However, it may be necessary to obtain real-time monitoring information while at remote field locations. With careful planning several of the indicator assays can be performed in the field, including the Colilert analysis. In addition, many microbial monitoring systems are currently available for use in the field. Millipore Corporation (Bedford, MA), for instance, offers a complete field kit for total coliform detection. This kit comes with a water-sampling apparatus, a filter holder, forceps, valved syringe, sterile Millipore filters, culture media, Petri dishes/pads, sealing tape, and even an alcohol bottle. This field test kit is capable of performing 24 tests. Another kit manufactured by Millipore also offers an incubator for onsite testing. The incubator can be plugged into an automobile cigarette lighter or powered with a battery. Finally,

4. Indirect antibody staining

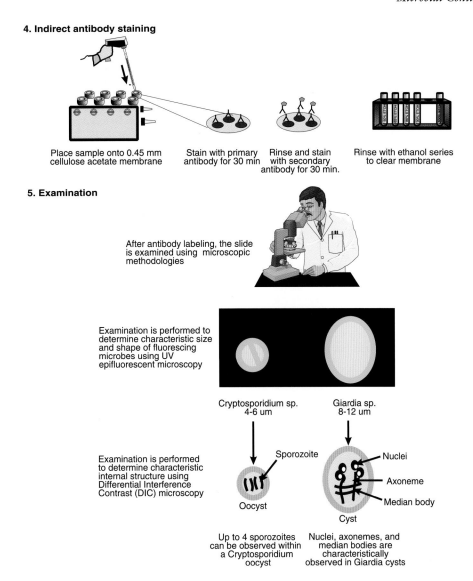

Place sample onto 0.45 mm
cellulose acetate membrane

Stain with primary
antibody for 30 min

Rinse and stain
with secondary
antibody for 30 min.

Rinse with ethanol series
to clear membrane

5. Examination

After antibody labeling, the slide
is examined using microscopic
methodologies

Examination is performed to
determine characteristic size
and shape of fluorescing
microbes using UV
epifluorescent microscopy

Cryptosporidium sp.
4-6 um

Giardia sp.
8-12 um

Examination is performed
to determine characteristic
internal structure using
Differential Interference
Contrast (DIC) microscopy

Sporozoite

Nuclei

Axoneme

Median body

Oocyst

Cyst

Up to 4 sporozoites
can be observed within
a Cryptosporidium
oocyst

Nuclei, axonemes, and
median bodies are
characteristically
observed in Giardia cysts

FIGURE 17.16   Procedure for antibody staining and examination of *Giardia* and *Cryptosporidium*. (Reprinted from *Environmental Microbiology* © 2000, Academic Press, San Diego.)

they also offer the complete field laboratory Potable Water Laboratory, which includes a portable incubator, a filter holder, a handheld vacuum syringe, pipettes, Petri dishes, sterile Millipore filters, m-Endo ampouled culture medium for 100 tests, a colony counter, and a two-stage normal Flash-O-Lens Magnifier. These test kits are available for indicator organisms because of the relative ease of testing. Field testing for actual pathogens would be much more difficult.

With the advent and acceptance of molecular methodologies, however, it may soon become possible to perform such pathogen screening in less than 2 hours in the field. One company, at least, is even marketing a portable thermocycler that could be easily used for onsite molecular testing of samples. Cepheid (Sunnyvale, CA) has released MicroBE, which is a portable, handheld instrument. Essentially, samples of blood, feces, tissue homogenates, food, or water can be deposited into a specialized cartridge, and the cartridge loaded into the handheld instrument. The system then chemically processes the sample, extracts and concentrates nucleic acid, performs the nucleic acid amplification process (PCR), and analyzes the product. Each disposable cartridge contains all necessary reagents, reactions chambers, waste chambers, and disposable microfluidic components. Fluid handling takes place within the cartridge only, eliminating contact of specimen fluids with the handheld instrument.

QUESTIONS

1. What are some of the criteria for indicator bacteria?

2. What is the difference between standards and criteria?

3. Why are geometric means used to report average concentrations of indicator organisms?

4. Calculate the arithmetic and geometric averages for the following data set: Fecal coliforms/100 ml on different days on a bathing beach were reported as 2, 3, 1000, 15, 150, and 3000.

5. Define coliform and fecal coliform bacteria. Why are they not ideal indicators?

6. Why have coliphage been suggested as indicator organisms?

7. What are two methods that can be used to detect indicator bacteria in water?

8. What are MUG and ONPG, and how are they used to detect coliform and fecal coliform bacteria?

9. What types of filters are used to concentrate viruses from water, and how do they work?

REFERENCES AND ADDITIONAL READING

Alonso, J.L., Batella, M.S., Amoros, I., Rambach, A. (1992) *Salmonella* detection in marine waters using a short standard method. *Water Res.* **26**, 973–978.

APHA. (1998) *Standard Methods for the Examination of Water and Wastewater,* 20th ed. American Public Health Association, Washington, DC.

Benton, W.H., Ward, R.L. (1982) Induction of cytopathogenicity in mammalian cell lines challenged with culturable enteric viruses and its enhancement by 5-iododeoxyuridine. *Appl. Environ. Microbiol.* **43**, 861–868.

Berg, G. (1987) *Methods for Recovering Viruses from the Environment.* CRC Press, Boca Raton, FL.

Chapelle, F.H. (1992) *Ground-Water Microbiology and Geochemistry.* John Wiley & Sons, New York.

Dowd, S.E., Widmer, K.W., Pillai, S.D. (1997) Thermotolerant clostridia as an airborne pathogen indicator during land application of biosolids. *J. Environ. Qual.* **26**, 194–199.

England, B.L. (1984) Detection of viruses on fomites. In *Methods in Environmental Virology* (C.P. Gerba, S.M. Goyal, eds.) Marcel Dekker, New York, pp. 179–220.

Geldreich, E.E. (1978) Bacterial populations and indicator concepts in feces, sewage, storm water and solid wastes. In *Indicators of Viruses in Water and Food* (G. Berg, ed.) Ann Arbor Science, MI. pp. 51–97.

Geldreich, E.E., Kenner, B.A. (1969) Comments of fecal streptococci in stream pollution. *J. Wat. Poll. Contr. Fed.* **41**, R336–R341.

Gerba, C.P. (1984) Applied and theoretical aspects of virus adsorption to surfaces. *Adv. Appl. Microbiol.* **30**, 33–168.

Gerba, C.P. (1987) Recovering viruses from sewage, effluents, and water. In *Methods for Recovering Viruses from the Environment,* (G. Berg, ed.) CRC Press, Boca Raton, FL, pp. 1–23.

Gerba, C.P., Goyal, S.M. (1982) *Methods in Environmental Virology.* Marcel Dekker, New York.

Gleeson, C., Gray, N. (1997) *The Coliform Index and Waterborne Disease.* E. and FN Spon. London, UK.

Havelaar, A.H., Hogeboon, W.M., Furuse, K., Pot, R., Horman, M.P. (1990) F-specific RNA bacteriophages and sensitive host strains in faeces and wastewater of human and animal origin. *J. Appl. Bacteriol.* **69**, 30–37.

Hurst, C.J., Knudsen, G.R., McInerney, M.J., Stetzenbach, L.D., Watler, M.V. (1997) *Manual of Environmental Microbiology.* ASM Press, Washington, DC.

Jensen, P.A., Todd, W.F., Davis, G.N., Scarpino, P.V. (1992) Evaluation of eight bioaerosol samplers challenged with aerosols of free bacteria. *J. Ind. Hyg. Assoc.* **53**, 660–667.

LeChevallier, M.W., Cameron, S.C., McFeters, G.A. (1983) New medium for improved recovery of coliform bacteria from drinking water. *Appl. Environ. Microbiol.* **45**, 484–492.

Pattison, C.P., Boyer, K.M., Maynard, J.E., Kelley, P.C. (1974) Epidemic hepatitis in a clinical laboratory: Possible association with computer card handling. *J. Am. Med. Assoc.* **230**, 854–857.

Payment, P., Franco, E. (1993) *Clostridium perfringens* and somatic coilphage as indicators of the efficiency of drinking water treatment for viruses and protozoan cysts. *Appl. Environ. Microbiol.* **59**, 2418–2424.

Pepper, I.L., Gerba, C.P., Brusseau, M.L. eds. (1996) *Pollution Science.* Academic Press, San Diego.

Pillai, S.D., Widmer, K.W., Dowd, S.E., Ricke, S.C. (1996) Occurrence of airborne bacteria and pathogen indicators during land application of sewage sludge. *Appl. Environ. Microbiol.* **62**, 296–299.

Rose, J.B., Lisle, J.T., LeChevallier, M. (1997) Waterborne cryptosporidiosis: Incidence, outbreaks, and treatment strategies. In *Cryptosporidium and cryptosporidiosis* (R. Fayer, ed.) CRC Press, Boca Raton, FL, pp. 93–109.

Simmons, O.D., Sobsey, M.D., Heaney, C.D., Schaefer, F.W., Francy, D.S. (2001) Concentration and detection of *Cryptosporidium* oocysts in surface water samples by method 1622 using ultrafiltration and capsule filtration. *Appl. Environ. Microbiol.* **65**, 1123–1127.

Sobsey, M.D., Glass, J.S. (1980) Poliovirus concentration from tap water with electropositive adsorbent filters. *Appl. Environ. Microbiol.* **40**, 201–210.

Straub, T.M., Pepper, I.L., Abbaszadegan, M., Gerba, C.P. (1994) A method to detect enteroviruses in sewage-sludge amended soil using the PCR. *Appl. Environ. Microbiol.* **60**, 1014–1017.

U.S. EPA, United States Environmental Protection Agency. (1992) Control of pathogens and vector attraction in sewage sludge. EPA/625/R–92/013. Washington, DC.

Ward, R.L., Knowlton, D.R., Pierce, M.J. (1984) Efficiency of human rotavirus propagation in cell culture. *J. Clin. Microbiol.* **19**, 748–753.

18

SOIL AND GROUNDWATER REMEDIATION

M.L. BRUSSEAU AND R.M. MAIER

Public concern with polluted soil and groundwater has encouraged the development of government programs designed to control and remediate this contamination, as well as to prevent further contamination. In the United States the first major piece of federal legislation that dealt with remediation of contaminated environments was the *Water Pollution Control Act* of 1972. This Act authorized funds for the remediation of hazardous substances released into navigable waters, but did not provide for remediation of contaminated land. Subsequently, in 1976 the first legislation to directly address the problem of contaminated land was passed—the *Resource Conservation and Recovery Act (RCRA)*, which authorized the federal government to order operators to clean up hazardous waste emitted at the site of operation. This Act did not, however, cover abandoned or previously contaminated hazardous waste sites.

The *Comprehensive Environmental Response, Compensation, and Liability Act (CERCLA)* of 1980—otherwise known as the *Superfund* program—explicitly addressed the remediation of hazardous waste sites. Together with a later amendment, the *Superfund Amendments and Reauthorization Act (SARA)* of 1986, CERCLA is the major federal act governing activities associated with the remediation of soil, sediment, and groundwater contamination. In addition, the Department of Defense has a separate remediation program (Installation Restoration Program) for military sites, and the Department of Energy has a program designed specifically for remediation of radioactive waste sites associated with the production of materials used for nuclear weapons and nuclear reactor cores. Below the federal level, other governing bodies have also enacted legislation to regulate and control pollution. For example, some states have passed their own versions of the federal Superfund program.

The institution of Superfund and other regulatory programs mentioned previously has generated an entire industry focused on the characterization and remediation of hazardous waste sites. This industry is composed of regulators working for environmental agencies at all levels of government, research scientists and engineers developing and testing new technologies, scientists and engineers working for consulting firms contracted to characterize and remediate specific sites, and attorneys involved in myriad legal activities. In the following sections, basic concepts associated with site characterization and remediation will be presented.

SUPERFUND PROCESS

Because of its importance, we will briefly discuss the major components of Superfund. Its purpose is twofold: to respond to releases of hazardous substances on land and in navigable waters, and to remediate contaminated sites. The former deals with future releases, whereas the latter deals with sites of existing contamination. There are two types of responses available within Superfund: (1) removal actions, which are responses to immediate threats, such as leaking drums; and (2) remedial actions, which involve cleanup of hazardous sites. The Superfund provisions can be used either when a hazardous substance is actually released or when the threat of such a release is substantial. They may also be used when the release of a contaminant or threat thereof poses imminent and substantial endangerment to public health and welfare. The process by which Superfund is applied to a site is illustrated in Figure 18.1.

The first step is to place the potential site in the *Superfund Site Inventory*, which is a list of sites that are candidates for investigation. The site is then subjected to a preliminary assessment and site inspection, which may be performed by a variety of local, state, federal, or even private agencies. The results of this preliminary investigation determine whether the site qualifies for the *National Priorities List (NPL)*, which is a list of sites deemed to require remedial action by the Environmental Protection Agency (EPA). [*Note*: non-NPL sites may also need to be cleaned up, but their remediation is frequently handled by nonfederal agencies, with or without the help of the EPA.]

The two-component *remedial investigation/feasibility study (RI/FS)* is the next step in the process. The purpose of the RI/FS is to characterize the nature and extent of risk posed by contamination, and to evaluate potential remediation options. The investigation and feasibility study components of the RI/FS are performed concurrently, using a "phased" approach that allows feedback

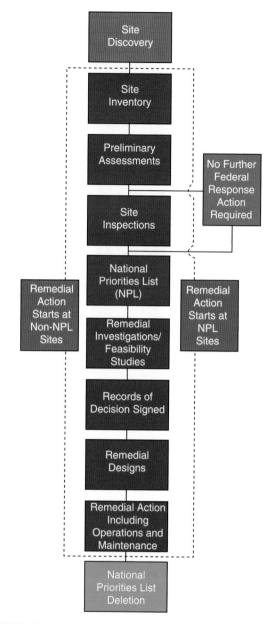

FIGURE 18.1   The Superfund process for treatment of a hazardous waste site. (Adapted from U. S. EPA, 1998.)

between the two components. A diagram of the RI/FS procedure is shown in Figure 18.2.

The selection of the specific remedial action to be used at a particular site is a very complex process. The goals of the remedial action are to protect human health and the environment, to maintain protection over time, and to maximize waste treatment (as opposed to waste containment or removal). Specifically, Section 121 of CERCLA mandates a set of three categories of criteria to be used for evaluating and selecting the preferred alternative: threshold criteria, primary balancing criteria, and modifying criteria. These criteria are detailed in Box 18.1. The *threshold criteria* ensure that the remedy protects human

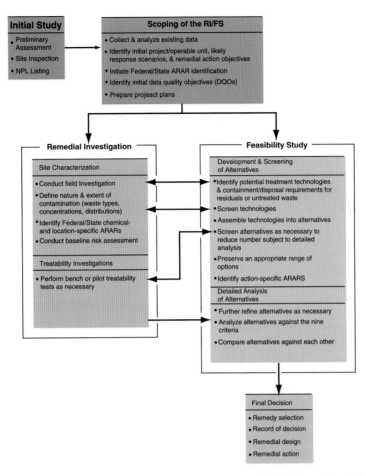

FIGURE 18.2 The remedial investigation/feasibility study process (RI/FS). (Adapted from U.S. EPA, 1988.)

health and the environment and is in compliance with applicable or relevant requirements (ARARs); the *balancing criteria* ensure that such trade-off factors as cost and feasibility are considered; and the *modifying criteria* ensure that the remedy meets state and community expectations.

After a remedial action has been selected, it is designed and put into action. Some sites may require relatively simple actions, such as removal of waste-storage drums and surrounding soil. However, the sites placed on the NPL generally have complex contamination problems, and are therefore much more difficult to clean up. Because of this, it may take a long time to completely clean up a site and to have the site removed from the NPL.

An important component of Superfund and all other cleanup programs is the problem of deciding the target level of cleanup. Related to this problem is the question: "How clean is clean?" If, for example, the goal is to lower contamination concentration, how low does that concentration level have to be before it is "acceptable"? As can be imagined, the stricter the cleanup provision, the greater will be the attendant cleanup costs. It may require

tens to hundreds of millions of dollars to return large, complex hazardous waste sites to their original conditions. In fact, it may be impossible to completely clean up many sites. However, a site need not be totally clean to be usable for some purposes. It is very important therefore that the technical feasibility of cleanup and the degree of potential risk posed by the contamination be weighed against the economic impact and the future use of the site. Consideration of the risk posed by the contamination and the future use of the site allows scarce resources to be allocated to those sites that pose the greatest current and future risk. The great difficulty of completely remediating contaminated sites highlights the importance of pollution prevention.

SITE CHARACTERIZATION

Site characterization is a critical component of hazardous waste site remediation. Site characterization provides information that is required for conducting risk assessments and for designing and implementing remediation systems. The primary objectives of site characterization

BOX 18.1 *Criteria for Evaluating Remedial Action Alternatives*

The EPA has developed criteria for evaluating remedial alternatives to ensure that all important considerations are factored into remedy selection decisions. These criteria are derived from the statutory requirements of CERCLA Section 121, particularly the long-term effectiveness and related considerations specified in Section 121 (bX1), as well as other technical and policy considerations that have proved important for selecting among remedial alternatives.

Threshold Criteria

The two most important criteria are statutory requirements that any alternative must meet before it is eligible for selection.

1. *Overall protection of human health and the environment* addresses whether or not a remedy provides adequate protection. It describes how risks posed through each exposure pathway (assuming a reasonable maximum exposure) are eliminated, reduced, or controlled through treatment, engineering controls, or institutional controls.
2. *Compliance with applicable or relevant and appropriate requirements (ARARs)* addresses whether a remedy meets all of the applicable or relevant and appropriate requirements of other federal and state environmental laws or whether a waiver can be justified.

Primary Balancing Criteria

Five primary balancing criteria are used to identify major trade-offs between remedial alternatives. These trade-offs are ultimately balanced to identify the preferred alternative and to select the final remedy.

1. *Long-term effectiveness and permanence* addresses the ability of a remedy to maintain reliable protection of human health and the environment over time once cleanup goals have been met.
2. *Reduction of toxicity, mobility, or volume through treatment* addresses the anticipated performance of the treatment technologies employed by the remedy.
3. *Short-term effectiveness* addresses the time needed to achieve protection. It also assesses any adverse impacts on human health and the environment that may be posed during the construction and implementation period, until cleanup goals are achieved.
4. *Implementability* addresses the technical and administrative feasibility of a remedy, including the availability of materials and services needed to implement a particular option.
5. *Cost* addresses the estimated capital and operation and maintenance costs, and net present worth costs.

Modifying Criteria

These criteria may not be considered fully until after the formal public comment period on the Proposed Plan and RI/FS report is complete, although EPA works with the state and community throughout the project.

1. *State acceptance* addresses the support agency's comments. Where states or other federal agencies are the lead agencies, EPA's acceptance of the selected remedy should be addressed under this criterion. State views on compliance with state ARARs are especially important.
2. *Community acceptance* addresses the public's general response to the alternatives described in the Proposed Plan and the RI/FS report.

are to identify the nature and extent of contamination. This includes identifying the types of contaminants present, the amount and location of contamination, and the phases in which it is occurring (e.g., is it associated with the groundwater only, or is it also sorbed to the aquifer solids?).

Generally, the first step in identifying the contaminants present at the site involves inspection of material records, if available. Useful information regarding the types of contaminants potentially present at the site can be obtained from chemical-stock purchasing records, deliv-ery records, storage records, and from other types of documents. Important information may also be obtained from records of specific chemical spills or leaks, as well as waste disposal records. Unfortunately, such information is incomplete for many sites, especially for older or defunct sites.

After examining available records, site inspections are conducted to identify potential sources of contamination. This includes searching the site for drums, leaking storage tanks, and abandoned disposal pits or injection wells. Once located, actions are taken, if necessary, to prevent further release of contamination. For example, leaking tanks, along with the surrounding porous media, are excavated and removed.

The next step in site characterization generally involves a "survey" of groundwater contamination. This is accomplished through collecting groundwater samples from wells in and adjacent to the site. This component is often implemented in a phased approach. For example, in the initial characterization phase, when little is known about the type or extent of groundwater contaminants present, it is often necessary to analyze samples for a broad suite of possible contaminants. As noted in Chapter 16, this is typically done using the priority pollutants list associated with the National Primary Drinking Water Standards (Table 16.1). After several rounds of sampling, the list of analytes is often shortened to those that are most commonly found at the site or that pose the greatest risk. In addition, the initial phase typically makes use of existing wells, and a few new wells located to obtain a complete (but sparse) coverage of the site. Once the specific zones of contamination are identified, additional wells are often drilled in those areas to increase the density of sampling locations. Site characterization programs are generally focused on obtaining information on the areal (x–y plane) distribution of contamination. However, once this information is available, it is important to characterize the vertical distribution of contamination, if possible. Specific methods for developing and implementing a groundwater sampling program are discussed in Chapter 8.

Groundwater sampling provides information about the types and concentrations of contaminants present in groundwater. This is a major focus of characterization at most sites because groundwater contamination is often the primary risk driver. However, sampling programs can also be conducted to characterize contamination distribution for other phases. For example, as discussed in Chapter 5, core material can be collected and analyzed to measure the amount of contamination associated with the porous-media grains. In addition, various methods can be used to collect gas-phase samples in the vadose zone to evaluate vapor-phase contamination.

One component of site characterization that is often critical for sites contaminated by organic chemicals is determining if immiscible liquids are present in the subsurface. As discussed in Chapter 16, many organic contaminants are liquids, such as chlorinated solvents and petroleum derivatives. Their presence in the subsurface serves as a significant source of long-term contamination, and greatly complicates remediation. In some cases it is relatively straightforward to determine the presence of immiscible-liquid contamination at a site. For example, when large quantities of gasoline or other fuels are

present, a layer of "free product" may form on top of the water table because the fuels are generally less dense than water. This floating free product can be readily observed by collecting samples from a well intersecting the contamination.

Conversely, in other cases it may be almost impossible to directly observe immiscible-liquid contamination. For example, chlorinated solvents such as trichloroethene are denser than water. Thus large bodies of floating free product are not formed as they are for fuels (see Figure 18.2). Therefore evidence of immiscible-liquid contamination must be obtained through the examination of core samples, and not from well sampling. Unfortunately, it is very difficult to identify immiscible-liquid contamination with the use of core samples. When these denser-than-water liquids enter the subsurface, localized zones of contamination will usually form due to the presence of subsurface heterogeneities (e.g., permeability variability). Within these zones, the immiscible liquid is trapped within pores, and may form small pools above capillary barriers. Given the small diameter of cores (5 to 20 cm), the chances of a borehole intersecting a localized zone of contamination are relatively small. Thus it may require a large, cost-prohibitive number of boreholes to characterize immiscible-liquid distribution at a site. Recognizing these constraints, alternative methods for characterizing subsurface immiscible-liquid contamination are being developed and tested. These include methods based on geophysical and tracer-test techniques, some of which provide a larger scale of measurement.

Although identifying the nature and extent of contamination is the primary site-characterization objective, other objectives exist. For example, another common objective involves determining the physical properties of the subsurface environment. For example, pumping tests are routinely conducted to determine the hydraulic conductivity distribution for the site (see Chapter 12). This information is useful for determining the potential location of contaminants, their rates and direction of movement, and for evaluating the feasibility and effectiveness of proposed remediation systems. In some cases core samples will be analyzed to characterize chemical and biological properties and processes pertinent to contaminant transport and remediation (see Chapters 13 and 14). For example, assessing the potential of the microbial community associated with the porous medium to biodegrade the target contaminants is critical for evaluating the feasibility of employing bioremediation at the site.

In summary, site characterization is an involved, complex process comprising many components and activities. Generally, the more information available, the better informed are the site evaluations and thus the greater chance of success for the planned remediation system. However, site characterization activities are expensive and often time consuming. Thus different levels of site characterization may be carried out at a particular site depending on goals and available resources. Examples of three levels of site characterization are given in Box 18.2. Clearly the standard approach is least costly, whereas the state-of-the-science approach is much more costly. However, the state-of-the-science approach provides significantly more information about the site, compared with the standard approach.

The use of mathematical modeling has become an increasingly important component of site characterization and remediation activities. Mathematical models

can be use to characterize site-specific contaminant transport and fate processes, to predict the potential spread of contaminants, to help conduct risk assessments, and to assist in the design and evaluation of remediation activities. As such, it is critical that models are developed and applied in such a manner to provide an accurate and site-specific representation of contaminant transport and fate. Employing advanced characterization and modeling efforts that are integrative and iterative in nature enhances the success of remediation projects.

Numerous approaches are available for modeling transport and fate of contaminants at the field scale. These can

---

BOX 18.2 *Various Approaches to Site Characterization*

Standard Approach

Activities

- Use existing wells; install several fully screened monitor wells
- Sample and analyze for priority pollutants
- Construct geological cross sections using driller's log and cuttings
- Develop water level contour maps

Advantages

- Rapid screening of site
- Moderate costs involved
- Standardized techniques

Disadvantages

- True nature and extent of problem not identified

State-of-the-Art Approach

Activities

- Conduct standard site-characterization activities
- Install depth-specific monitor well clusters
- Refine geological cross sections using cores
- Conduct pumping tests
- Characterize basic properties of porous media (grain size, organic matter, clay content)
- Conduct solvent-extraction of core samples to evaluate sorbed- and NAPL-phase contamination
- Conduct soil-gas surveys
- Conduct limited geophysical surveys (resistivity soundings)
- Conduct limited mathematical modeling

Advantages

- Better understanding of nature/extent of contamination
- Improved conceptual understanding of problem

Disadvantages

- Greater costs
- Detailed understanding of problem still limited
- Demand for specialists increased

State-of-the-Science Approach

Activities

- Employ state-of-the-art approach
- Conduct tracer tests and borehole geophysical surveys
- Characterize geochemical properties of porous media (oxides, mineralogy)
- Measure redox potential, pH, dissolved oxygen, etc., of subsurface
- Evaluate sorption-desorption and NAPL dissolution behavior using selected cores
- Assess potential for biotransformation using selected cores
- Conduct advanced mathematical modeling

Advantages

- Information set as complete as generally possible
- Enhanced conceptual understanding of problem

Disadvantages

- Characterization cost significantly higher
- Field and laboratory techniques not yet standardized
- Demand for specialists greatly increased

be grouped into three general types of approaches, differentiated primarily by the level of complexity (Box 18.3). The "standard" approach is just that. It is the standard approach used almost exclusively in the analysis of Superfund and other hazardous wastes sites. The primary advantage of this approach is the relative simplicity of the model, and the relatively minimal data requirements. Unfortunately, the usefulness of such modeling is very limited. Except for the simplest of systems (e.g., a nonreactive, conservative solute), this type of modeling cannot be used to characterize the contribution of specific processes or factors influencing the transport and fate of contaminants. In addition, because there is no real mechanistic basis to the model, its use for generating predictions is severely constrained. The "state-of-the-science" approach is based on implementing cutting-edge understanding of transport and fate processes into fully three-dimensional models that incorporate spatial distributions of all pertinent properties. Clearly this type of modeling,

although desirable, is ultimately impractical for all but the most highly characterized research site. The "state-of-the-art" approach is an intermediate-level approach that is process based, but also is developed with recognition of the data-availability limitations associated with most sites. Because models based on this approach are process based, this type of modeling can be used to characterize the contribution of specific processes or factors influencing the transport and fate of contaminants. Thus this modeling approach can be used to effectively evaluate the impact of the key factors influencing the contamination potential of waste sites.

REMEDIATION TECHNOLOGIES

There are three major categories or types of remedial actions: (1) *containment*, where the contaminant is restricted to a specified domain to prevent further

---

BOX 18.3 *Various Approaches for Mathematical Modeling*

Standard Approach

- "One-layer" model with areal hydraulic-conductivity distribution
- No spatial variability of chemical/biological properties
- Lumped macrodispersion term (lumps all contributions to spreading, mixing, dilution)
- Dissolution of immiscible liquid simulated using a source term function
- Lumped retardation, with mass transfer processes treated as linear and instantaneous
- Lumped dispersion term for systems with gas-phase transport
- Lumped first-order transformation term

State-of-the-Art Approach

- "Multilayer" model with vertical and areal hydraulic conductivity distribution
- Layer/areal distribution of relevant chemical and biological properties
- Layer/areal distribution of immiscible liquid
- Rate-limited immiscible liquid dissolution
- Multiple-component immiscible liquid partitioning (using Raoult's Law)
- Rate-limited, nonlinear sorption/desorption
- Gas-phase transport of volatile organic contaminants (gas-phase advection, diffusion)

- Biodegradation processes coupled to mass-transfer processes
- Geochemical zone-dependent (layer/areal distributed) first-order rate coefficients

State-of-the-Science Approach

- Three-dimensional distributed hydraulic conductivity field
- Three-dimensional distribution of relevant chemical and biological properties
- Three-dimensional distribution of immiscible liquid
- Rate-limited immiscible liquid dissolution
- Immiscible liquid composition effects (multiple-component behavior)
- Rate-limited, nonlinear sorption/desorption
- Gas-phase transport of volatile organic contaminants (gas-phase advection, diffusion)
- Gas-phase retention/mass transfer (air-water mass transfer, adsorption at air-water interface, vapor-phase adsorption, immiscible liquid evaporation)
- Biodegradation coupled to mass-transfer processes
- Microbial dynamics: population growth, death, cell transport
- Biogeochemical properties: multiple electron acceptors

spreading; (2) *removal*, where the contaminant is transferred from an open to a controlled environment; and (3) *treatment*, where the contaminant is transformed to a nonhazardous substance. Since the inherent toxicity of a contaminant is eliminated only by treatment, this is the preferred approach of the three. Containment and removal techniques are very important, however, when it is not feasible to treat the contaminant. Although we will focus on each of the three types of remedial actions in turn, it is important to understand that remedial actions often consist of a combination of containment, removal, and treatment.

CONTAINMENT TECHNOLOGIES

Containment can be accomplished by controlling the flow of the fluid that carries the contaminant, or by directly immobilizing the contaminant. Here, we will briefly discuss the use of physical and hydraulic barriers for containing contaminated water, which are the two primary containment methods. We will also briefly mention other containment methods.

PHYSICAL BARRIERS

The principle of a physical barrier is to control the flow of water to prevent the spread of contamination. Usually the barrier is installed in front (down gradient) of the contaminated zone (Figure 18.3); however, barriers can also be placed up gradient, or both up and down gradient of the contamination. Physical barriers are primarily used in unconsolidated materials such as soil or sand, but they may also be used in consolidated media such as rock if special techniques are employed. In general, physical barriers may be placed to depths of about 50 m. The horizontal extent of the barriers can vary widely, depending on the size of the site, from tens to hundreds of meters.

One important consideration in the employment of physical barriers is the presence of a zone of low permeability, into which the physical barrier can be seated. Without such a seating into a low permeability zone, the contaminated water could flow underneath the barrier. Another criterion for physical barriers is the permeability of the barrier itself. Since the goal of a physical barrier is to minimize fluid flow through the target zone, the permeability of the barrier should be as low as practically possible.

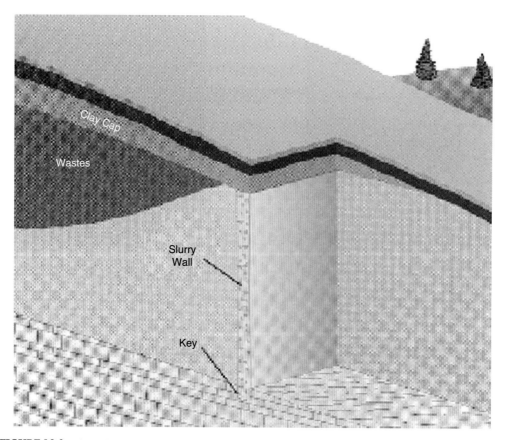

FIGURE 18.3   Physical containment of a contaminant by the use of a slurry wall. (Adapted from U.S. EPA, 1985.)

Another factor to consider is the potential of the contaminants to interact with the components of the barrier and degrade its performance. The properties of the barrier material should be matched to the properties of the contaminant to minimize failure of the barrier.

There are three major types of physical barriers: slurry walls, grout curtains, and sheet piling. *Slurry walls* are trenches filled with slurries of clay or mixtures of clay and soil (see Figure 18.3). *Grout curtains* are hardened matrices formed by cement-like chemicals that are injected into the ground. *Sheet piling* consists of large sheets of iron that are driven into the ground. Slurry walls are the least expensive and most widely used type of physical barrier, and they are the simplest to install. Grout curtains, which can be fairly expensive, are limited primarily to sites having consolidated subsurface environments. Sheet piles have essentially zero permeability and are generally of low reactivity. They can leak, however, because it is difficult to obtain perfect seals between individual sheets. Moreover, sheet piling is generally more expensive than slurry walls, and it is difficult to drive sheet piles into rocky ground. Thus they are used primarily for smaller-scale applications at sites composed of unconsolidated materials.

HYDRAULIC BARRIERS

The principle behind hydraulic barriers is similar to that behind physical barriers—to manipulate and control water flow. But unlike physical barriers, which are composed of solid material, hydraulic barriers are based on the manipulation of water pressures. They are generated by the pressure differentials arising from the extraction or injection of water. The key performance factor of this approach is its capacity to capture the *contaminant plume* (i.e., to limit the spread of the zone of contamination). Plume capture is a function of the number and placement of the wells or drains, as well as the rate of water flow through the wells or drains. Often, attempts are made to optimize the design and operation of the containment system so that plume capture is maximized while the volume of contaminated water removed is minimized.

The simplest hydraulic barrier is that established by a drain system. Such a system is constructed by installing a perforated pipe horizontally in a trench dug in the subsurface and placed to allow maximum capture of the contaminated water. Water can then be collected and removed by the use of gravity or active pumping. The use of drain systems is generally limited to relatively small, shallow contaminated zones.

Well-field systems are more complicated—and more versatile—than drain systems. Both extraction and injection wells can be used in a containment system, as illustrated in Figure 18.4. An extraction well removes the water entering the zone of influence of the well, creating a cone of depression. Conversely, an injection well creates a pressure ridge, or mound of water under higher pressure than the surrounding water, which prevents flow past the mound. One major advantage of using wells to control contaminant movement is that this is essentially the only containment technique that can be used for deep ($>50\,\text{m}$) systems. In addition, wells can be used on contaminant zones of any size; the number of wells is simply increased to handle larger problems. For example, some large sites have contaminant plumes that are several kilometers long. For these and other reasons, wells are the most widely used method for containment, despite disadvantages that include the cost of long-term operation and maintenance, and the need to store, treat, and dispose of the large quantities of contaminated water pumped to the surface.

OTHER CONTAINMENT METHODS

A variety of techniques have been used to attempt to immobilize subsurface contaminants by fixing them in an impermeable, immobile solid matrix. These techniques are referred to alternatively as solidification, stabilization, encapsulation, and immobilization. They are generally based on injecting a solution containing a compound that will cause immobilization or encapsulation of the contamination. For example, cement or a polymer solution can be added, which converts the contaminated zone into a relatively impermeable mass encapsulating the contaminant. In another approach a reagent can be injected to alter the pH or redox conditions of the subsurface, thus causing the target contaminant to "solidify" *in situ*. For example, promoting reducing conditions will induce chromium to change its predominant speciation from hexavalent, which is water soluble and thus "bioavailable," to trivalent, which has low solubility and thus precipitates on the porous-media grains (and is therefore no longer readily available).

In general, it is difficult to obtain uniform immobilization due to the natural heterogeneity of the subsurface. Thus containment may not be completely effective. In addition, a major factor to consider for these techniques is the long-term durability of the solid matrix, and the potential for leaching of contaminants from the matrix.

Vitrification is another type of containment, wherein the contaminated matrix is heated to high temperatures to "melt" the porous media (i.e., the silica components), which subsequently cools to form a glassy, impermeable

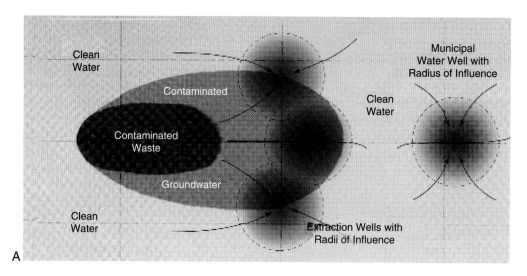

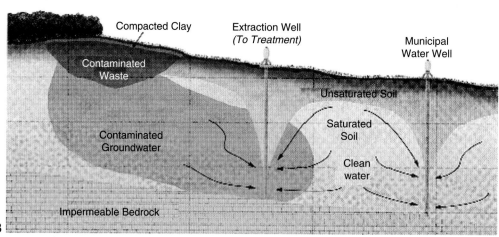

FIGURE 18.4 Containment of a contaminant plume by a hydraulic barrier. **(A)** Overhead view of the remediation site showing groundwater surfaces, flow directions, and contaminated waste (soil material removed). **(B)** Cut-away view of the above site showing lateral movement of leached contaminants and groundwater. (Adapted from U.S. EPA, 1985.)

block. Vitrification may be applied either *in situ* or *ex situ*. For treatment of typical hazardous waste sites, potential applications would primarily be *in situ*. Above-ground applications are being investigated for use in dealing with radioactive waste. Vitrification is an energy-intensive, disruptive, and relatively expensive technology, and its use would generally be reserved for smaller-scale sites for which other methods are not viable.

REMOVAL

EXCAVATION

A very common, widely used method for removing contaminants is *excavation* of the soil in which the contaminants reside. This technique has been used at many sites and is highly successful. There are, however, some disadvantages associated with excavation. First,

excavation can expose site workers to hazardous compounds. Second, the contaminated soil requires treatment and/or disposal, which can be expensive. Third, excavation is usually feasible only for relatively small, shallow areas. Excavation is most often used to remediate shallow, localized, highly contaminated source zones.

PUMP AND TREAT

For the *pump-and-treat method,* currently the most widely used remediation technique for contaminated groundwater, one or more extraction wells are used to remove contaminated water from the subsurface. Furthermore, clean water brought into the contaminated region by the flow associated with pumping removes, or "flushes," additional contamination by inducing desorption from the porous-media grains. The contaminated water pumped from the subsurface is directed to some type of treatment operation, which may consist of air stripping, carbon adsorption, or perhaps an above-ground biological treatment system. An illustration of a pump-and-treat system is provided in Figure 18.5.

Usually discussed in terms of its use for such saturated subsurface systems as aquifers, the pump-and-treat method can also be used to remove contaminants from the vadose zone. In this case it is generally referred to as *in situ soil washing*. For this application, infiltration galleries, in addition to wells, can be used to introduce water to the contaminated zone.

When using water flushing for contaminant removal (as in pump-and-treat), contaminant-plume capture and the effectiveness of contaminant removal are the major performance criteria. Recent studies of operating pump-and-treat systems have shown that the technique is very successful at containing contaminant plumes and, in some cases, shrinking them. However, it appears that pump and treat is frequently ineffective for completely removing contaminants from the subsurface. There are many factors that can limit the effectiveness of water flushing for contaminant removal. Selected major factors include the following:

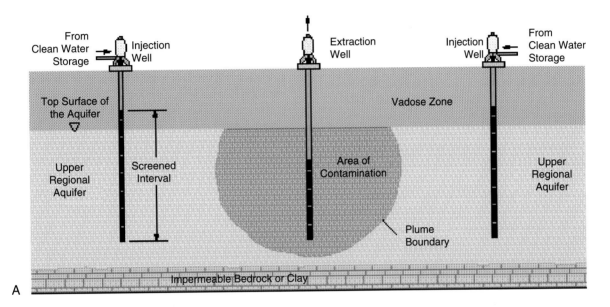

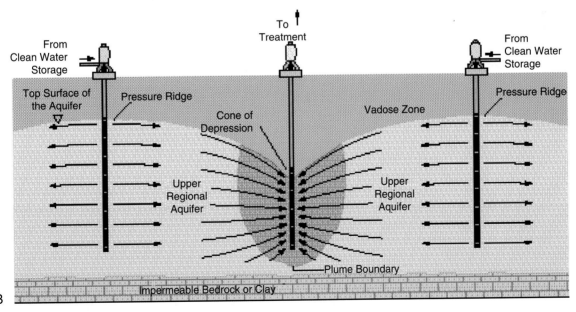

FIGURE 18.5 Removal of a contaminant by the pump-and-treat processes. **(A)** Before treatment. **(B)** After flow initiation. (From Pepper, I.L., Gerba, C.P., Brusseau, M.L. *Pollution Science*, Academic Press, 1996.)

1. *Presence of low-permeability zones.* When low-permeability zones (e.g., silt/clay lenses) are present within a sandy subsurface, they create domains through which advective flow and transport are minimal in comparison to the surrounding sand. The groundwater flows preferentially around the clay/silt lenses, rather than through them. Thus contaminant located within the silt/clay lenses is released to the flowing water primarily by pore-water diffusion, which can be a relatively slow process. Thus the mass of contaminant being removed diminishes with each volume of water pumped, thereby increasing the time required to completely remove the contaminant.
2. *Rate-limited desorption.* Research has revealed that adsorption/desorption of many solutes by porous media can be significantly rate-limited. When the rate of desorption is slow enough, the concentration of contaminant in the groundwater is lower than the concentrations obtained under conditions of rapid desorption. Thus less contaminant is removed per volume of water, and removal by flushing will therefore take longer.
3. *Presence of immiscible liquid.* In many cases immiscible organic liquid contaminants may be trapped in portions of the contaminated subsurface. Because it is very difficult to displace or push out this trapped contamination with water, the primary means of removal will be dissolution into water and evaporation into the soil atmosphere. It can take a very long time to completely dissolve immiscible liquid, thus greatly delaying removal. The immiscible liquid therefore serves as a long-term source of contamination.

Because pump and treat is a major remedial action technique, methods are being tested to enhance its effectiveness. One way to improve the effectiveness of pump and treat is to contain or remove the *contaminant source zone* (i.e., the area in which contaminants were disposed or spilled). If the source zone remains untreated or uncontrolled, it will serve as a continual source of contaminant requiring removal. Thus failure to control or treat the source zone can greatly extend the time required to achieve site cleanup. It is important therefore that the source zone at a site be delineated and addressed in the early stages of a remedial action response. This might be done by using a physical or hydraulic barrier to confine the source zone. Other methods of treating the source zone involve enhancing the rate of contaminant removal, as detailed in the following section.

ENHANCED FLUSHING

Contaminant removal can be difficult because of such factors as low solubility, high degree of sorption, and the presence of immiscible-liquid phases, all of which limit the amount of contaminant that can be flushed by a given volume of water. Approaches are being developed to enhance the removal of low-solubility, high-sorption contaminants. One such approach is to inject a reagent solution into the source zone, such as a *surfactant* (e.g., detergent molecule), which will promote dissolution and desorption of the contaminant, thus enhancing removal effectiveness. Such surfactants work like industrial and household detergents, which are used to remove oily residues from machinery, clothing, or dishes: individual contaminant molecules are "solubilized" inside of surfactant micelles, which are groups of individual surfactant molecules ranging from 5 to 10 nm in diameter. Alternatively, surfactant molecules can coat oil droplets and emulsify them into solution. Other enhanced-removal reagents are also being tested, such as alcohols and sugar compounds.

In laboratory tests these enhanced-removal reagents have been successfully used to increase the apparent aqueous solubility of organic contaminants and to enhance the rate of removal. However, only a few field tests have been attempted, and results have been mixed. A key factor controlling the success of this approach in the field is the ability to deliver the reagent solution to the places that contain the contaminant. This would depend, in part, on potential interactions between the reagent and the soil (e.g., sorption) and on properties of water flow in the subsurface. An important factor concerning regulatory and community acceptance of this approach is that the reagent should be of low toxicity. It would clearly be undesirable to replace one contaminant with another. Another important factor, especially with regard to cost effectiveness of this approach, is the potential for recovery and reuse of the reagent.

SOIL VAPOR EXTRACTION

The principle of *soil vapor extraction*, or *soil venting*, is very similar to that of pump and treat—a fluid is pumped through a contaminated domain to enhance removal. In the case of soil venting, however, the fluid is air rather than water. Because air is much less viscous than water, much less energy is required to pump air. Thus it is usually cheaper and more effective to use soil venting rather than soil washing to remove volatile contaminants from the vadose zone. There are two key conditions for using soil venting. First, the soil must contain a gas phase through which the contaminated air can travel. This condition generally limits the use of soil venting to the vadose zone. In some cases groundwater is pumped to lower the water table, thus allowing the use of soil venting for zones that were formerly water saturated. Second, contaminants must be capable of transfer from other phases (solid, water, immiscible liquid) to the gas phase. This requirement limits soil venting to volatile and semivolatile contaminants. Fortunately, many of the organic contaminants of greatest concern, such as chlorinated solvents (trichloroethene, tetrachloroethene) and certain components of fuels (benzene, toluene), are volatile or semivolatile. Soil venting is the most widely used method for removing volatile contaminants from contaminated vadose-zone systems.

A blower is generally used to extract contaminated air from the subsurface. In some cases passive or active air injection is used to increase air circulation. Passive air injection simply involves drilling boreholes through which air can then move, as opposed to the use of a blower for actively injecting air. A cap made of plastic or asphalt is often placed on the ground surface to increase venting effectiveness and to prevent water from infiltrating into the subsurface.

Once the contaminant is removed from the vadose zone, it is either released to the atmosphere or placed into a treatment system. The major performance criteria for soil venting are the effectiveness of capturing and removing the contaminant. The effectiveness of contaminant removal by soil venting can be limited by many of the same factors that limit removal by water flushing.

AIR SPARGING

In situ air sparging is a means by which to enhance the rate of mass removal from contaminated saturated-zone systems. Air sparging involves injecting air into the target contaminated zone, with the expectation that volatile and semivolatile contaminants will undergo mass transfer (volatilization) from the groundwater to the air bubbles (Figure 18.6C). Because of buoyancy, the air bubbles generally move upward toward the vadose zone, where a soil-venting system is usually employed to capture the contaminated air stream.

Recent laboratory and pilot-scale research has shown that the effectiveness of air sparging is often limited by a number of factors in practice. One major constraint is the impact of "channeling" on air movement during sparging. Studies have shown that air injected into water-saturated porous media often moves in discrete channels that comprise only a fraction of the entire cross section of the zone, rather than passing through the entire medium as bubbles (as proposed in theory). This channeling phenomenon greatly reduces the "stripping efficiency" of air sparging.

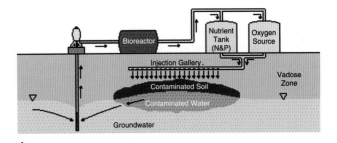

A

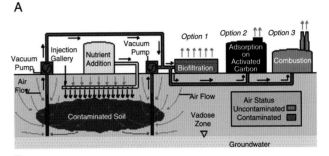

B

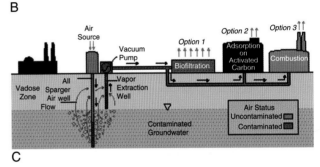

C

FIGURE 18.6 (**A**) *In situ* bioremediation of a contaminated aquifer. Contaminated groundwater pumped out of the site may be treated using an above-ground bioreactor, as shown, or by other methods including air stripping, activated carbon, oil/water separation, and oxidation. After treatment the groundwater is pumped back into the aquifer via an injection well. The water may have oxygen or nutrients added to stimulate *in situ* biodegradation, but if the level of contamination is very low, these additions may not be required. (**B**) Contaminant bioremediation in the vadose (unsaturated) zone using bioventing and biofiltration. Air is drawn through the contaminated site, stimulating aerobic degradation. In addition, volatile contaminants are removed with the flow of air. These vapor phase contaminants can be treated above ground by piping them through a soil bed, also called a biofilter. Alternative treatments include incineration of extracted vapors or adsorption on activated carbon. (**C**) Contaminant bioremediation in the saturated zone using air sparging. This process is similar to bioventing, but takes place in the saturated zone. Air is pumped into the contaminated site to stimulate aerobic degradation. Volatile contaminants are carried to the surface and can be treated by biofiltration, incineration, or activated carbon. (Adapted from National Research Council, 1993.)

Another significant limitation to air sparging applications is the presence of low-permeability zones overlying the target zone. As previously noted, air-sparging systems are designed to operate in tandem with a soil-venting system, so that the contaminated air can be collected and treated. The presence of a low-permeability zone overlying the target zone can prevent the air from passing into the vadose zone, preventing capture by the soil-venting system. In such cases, the air-sparging operation may act to spread the contaminant. Although limited by such constraints, air sparging may be of potential use for specific conditions, such as for targeting localized zones of contamination.

THERMAL METHODS

As noted previously, contaminant removal can be difficult because of such factors as low solubility, large magnitude of sorption, and the presence of immiscible-liquid phases, all of which limit the amount of contaminant that can be flushed by a given volume of water or air. One means by which to enhance desorption, volatilization, and evaporation, and thus improve the rate of contaminant removal is to raise the temperature in the contaminated zone. This can be done in several ways. One method is to inject hot air or water during standard soil-venting or pump-and-treat operations. This approach may increase temperatures by several degrees. Another approach involves injecting steam into the subsurface, which can enhance contaminant removal through a number of complex processes. A third approach is based on the application of electrical heating methods (e.g., directing electrical current or radio waves into the subsurface) to heat the target zone. This last approach can generate higher temperatures than the other methods (e.g., 100° to 150° C) but is much more energy intensive and thus expensive. However, electrical heating may be particularly useful for remediating low-permeability zones, which are difficult to address with many other methods.

ELECTROKINETIC METHODS

All electrokinetic processes depend on creation of an electrical field in the subsurface. This is done by placement of a pair or series of electrodes around the area to be treated, to which direct current (typically between about 50 and 150 V) is applied. Electrokinetic treatment encompasses several different processes that individually or in combination can be used for contaminant removal, depending on the particular characteristics of the contaminants present. Electromigration, electrophoresis, and electro-osmosis all enhance mobilization of the target contaminant to a location for extraction or treatment. The first two mobilization processes act on contaminants that are charged (ionic) or highly polar. Thus these methods can be used for metals, radionuclides, and selected ionic or ionizable organic contaminants. Electro-osmosis, the movement of water in response to an electrical gradient, can be used for uncharged organic contaminants because the dissolved molecules will be carried along with the water. The presence of water is required for these methods to work.

One potentially promising application for electrokinetic methods is for removing contaminants from low-permeability zones. As previously noted, such zones are difficult to remediate with existing methods, most of which are based on water or air flow. By placing the electrodes directly into the low-permeability zone, contaminant movement can be generated within that zone. The feasibility and effectiveness of electrokinetic techniques are still being tested.

In Situ Treatment

In situ treatment technologies are methods that allow in-place cleanup of contaminated field sites. There is great interest in these technologies because they are in some cases cheaper than other methods. Second, *in situ* treatment can in some cases eliminate the risk associated with the hazardous form of the contaminant. The two major types of *in situ* treatment are biological (*in situ* bioremediation) and chemical.

BIOREMEDIATION

The objective of bioremediation is to exploit the activity of naturally occurring microorganisms to clean up contaminated sites. There are several types of bioremediation: *in situ bioremediation* is the in-place treatment of a contaminated site; *ex situ bioremediation* is the treatment of contaminated soil or water that is removed from a contaminated site; and *intrinsic bioremediation* is the indigenous level of contaminant biodegradation that occurs without any stimulation or treatment. All of these types of bioremediation are receiving increasing attention as viable remediation alternatives for several reasons. These include generally good public acceptance and support, good success rates for some applications, and the comparatively low cost of bioremediation when it is successful. As with any technology, there are also drawbacks. First, success can be unpredictable because biological systems are themselves complex and unpredictable. A second consideration is that bioremediation rarely restores an environment to its original condition. Often the residual contamination left after treatment is strongly sorbed and not available to microorganisms for degradation. Over a long period (years), these residuals can be slowly released, generating additional pollution. There is little research concerning the fate and potential toxicity of such released residuals; therefore both the public and regulatory agencies continue to be concerned about the possible deleterious effects of residual contamination.

Domestic sewage waste, which has been treated biologically for many years, is one of bioremediation's major successes. From this application, it is clear that biodegradation is dependent on pollutant structure and bioavailability. Therefore application of bioremediation to other pollutants depends on the type of pollutant or pollutant mixtures present, and the type of microorganisms present. The first successful application of bioremediation outside of sewage treatment was in the cleanup of oil spills, where aerobic heterotrophic bacteria were used to biodegrade hydrocarbon products. In the past few years many new bioremediation technologies have emerged; these are listed in Box 18.4.

Several key factors are critical to successful application of bioremediation: environmental conditions, contaminant and nutrient availability, and the presence of microorganisms capable of degrading the contaminant. Thus when biodegradation fails to occur, it is important to isolate the "culprit"—the limiting factors of bioremediation. This task can be very complex. Initial laboratory tests on soil or water from a polluted site can usually determine the presence or absence of degrading micro-

organisms; such tests may also reveal an obvious environmental factor that limits biodegradation, such as limited oxygen, or extremely low or high pH. Often, however, the limiting factor is not easy to identify. For example, pollutants are often present as mixtures, and one component of the pollutant mixture can have toxic effects on the growth and activity of degrading microorganisms. Similarly, low bioavailability, which is another factor that can limit bioremediation, can be very difficult to evaluate in the environment.

Most of the developed bioremediation technologies are based on two standard practices: the addition of oxygen and the addition of other nutrients. These are briefly discussed in the next section.

ADDITION OF OXYGEN OR OTHER GASES

One of the most common limiting factors in bioremediation is the availability of oxygen, which is required for aerobic biodegradation. Moreover, oxygen is sparingly soluble in water and has a low rate of diffusion (movement) through water. The combination of these

---

BOX 18.4 *Description of Bioremediation Technologies*

In Situ Source Treatment Processes

- Bioventing: Oxygen is delivered to contaminated unsaturated soils by movement of forced air (either extraction or injection of air) to increase concentrations of oxygen and stimulate biodegradation.
- Slurry-phase lagoon aeration: Air and soil are brought into contact with each other in a lagoon to promote biological degradation of the contaminants in the soil.

In Situ Groundwater Processes

- Biosparging: Air is injected into groundwater to enhance biodegradation and volatilization of contaminants; biodegradation occurs aerobically.
- Aerobic: Air, oxygen, and/or nutrients are injected into groundwater to enhance biodegradation of contaminants. Systems include direct injections of oxygen release compound (ORC) or hydrogen peroxide, or groundwater recirculation systems.
- Anaerobic: Carbon sources such as molasses, lactic acid, or hydrogen release compound (HRC) are injected into groundwater to enhance biodegradation of contaminants using

direct injection or groundwater recirculation systems.

Ex Situ Source Treatment Processes

- Land treatment: Contaminated soil, sediment, or sludge is excavated, applied to lined beds, and periodically turned over or tilled to aerate the contaminated media. Amendments can be added to the contaminated media in the beds.
- Composting: Contaminated soil is excavated and mixed with bulking agents such as wood chips and organic amendments such as hay, manure, and vegetable wastes. The types of amendments used depends on the porosity of the soil and the balance of carbon and nitrogen needed to promote microbial activity.
- Biopiles: Excavated soils are mixed with soil amendments and placed in aboveground enclosures. The process occurs in an aerated static pile in which compost is formed into piles and aerated with blowers or vacuum pumps.
- Slurry-phase treatment: An aqueous slurry is created by combining soil, sediment, or sludge with water and other additives. The slurry is mixed to keep solids suspended and microorganisms in contact with the contaminants. Treatment usually occurs in a series of tanks.

Source: U.S. EPA 2001.

three factors makes it easy to understand why inadequate oxygen supplies often limit bioremediation.

Several technologies have been developed to overcome a lack of oxygen. Consider the typical bioremediation system shown in Figure 18.6A. This system is used to treat a contaminated aquifer, together with the contaminated zone above the water table. It contains a series of injection wells, or galleries, and a series of recovery wells, thus providing a two-pronged approach to bioremediation. First, the recovery wells remove contaminated groundwater, which is treated above ground, in this case using a *bioreactor* containing microorganisms that are acclimated to the contaminant. After bioreactor treatment, the clean water is supplied with oxygen and nutrients, and then reinjected into the subsurface. The reinjected water provides oxygen and nutrients to stimulate *in situ* biodegradation. In addition, the reinjected water flushes the vadose zone to aid in removal of the contaminant for above-ground bioreactor treatment. This remediation scheme is a very good example of a combination of physical/chemical/biological treatments that can be used to maximize the effectiveness of the remediation treatment.

Bioventing is a technique used to add oxygen directly to a site of contamination in the vadose zone (unsaturated zone). Bioventing is a combination of soil-venting technology and bioremediation. The bioventing zone, which is highlighted in red in Figure 18.6B, includes the vadose zone and contaminated regions just below the water table. As shown in this figure, a series of wells has been constructed around the zone of contamination. To initiate bioventing, a vacuum is drawn on these wells to force accelerated air movement through the contamination zone. This effectively increases the supply of oxygen throughout the site, and hence the rate of contaminant biodegradation. In the case of volatile pollutants, some of the pollutants are removed as air is forced through this system (i.e., soil venting). This contaminated air can also be treated biologically by passing the air through above-ground soil beds in a process called *biofiltration*, as shown in Figure 18.6B. Air sparging (see the "Air Sparging" section) can be used to add oxygen to the saturated zone (Figure 18.6C).

Methane is another gas that can be added with oxygen into extracted groundwater and reinjected into the saturated zone. Methane is used specifically to stimulate methanotrophic activity and cometabolic degradation of chlorinated solvents. Methanotrophic organisms produce the enzyme methane monooxygenase to degrade methane, and this enzyme also cometabolically degrades several chlorinated solvents, such as trichloroethene. Cometabolic degradation of chlorinated solvents is currently being tested in field trials to determine the usefulness of this technology.

NUTRIENT ADDITION

Second only to the addition of oxygen in bioremediation treatment is the addition of nutrients, nitrogen and phosphorus in particular. Many contaminated sites contain organic wastes that are rich in carbon, but poor in nitrogen and phosphorus. Nutrient addition is illustrated in the bioremediation schemes shown in Figures 18.6A and 18.6B. Injection of nutrient solutions takes place from an above-ground batch-feed system. The goal of nutrient injection is to optimize the carbon/nitrogen/phosphorus ratio (C:N:P) in the site to approximately 100:10:1. However, sorption of added nutrients can make it difficult to achieve the optimal ratio.

STIMULATION OF ANAEROBIC DEGRADATION USING ALTERNATIVE ELECTRON ACCEPTORS

Until recently, anaerobic degradation of many organic compounds was not considered feasible. Now, however, it is being proposed as an alternative bioremediation strategy, even though aerobic degradation is generally a much more rapid process. Because it is difficult to establish and maintain aerobic conditions in some saturated subsurface systems, several alternative electron acceptors have been proposed for use in anaerobic degradation. These acceptors include nitrate, sulfate, and iron (Fe^{3+}) ions, as well as carbon dioxide. Although limited in number, field trials using nitrate have shown promise. Another approach under investigation for mixed-waste sites is the use of sequential anaerobic/aerobic degradation processes. This relatively new area in bioremediation will undoubtedly receive increased attention in the next few years.

ADDITION OF MICROORGANISMS

If appropriate biodegrading microorganisms are not present in soil, or if microbial populations have been reduced because of contaminant toxicity, specific microorganisms can be added as "introduced organisms" to enhance the existing populations. This process is known as *bioaugmentation*. Scientists are now capable of creating "engineered" microorganisms that can degrade pollutants at extremely rapid rates. Such organisms can be developed through successive adaptations under laboratory conditions, or they can be genetically engineered. In terms of biodegradation, these engineered microorganisms are often far superior to organisms found in the environment. A major problem with use of engineered microorganisms is that introduction of a microorganism to a contaminated site may fail for two reasons. First, the introduced microbe often cannot establish a niche in the environment; in fact, these introduced organisms rarely survive in a new environment beyond a few weeks. Second, microorganisms, like contaminants, can be strongly sorbed by solid surfaces; so there are difficulties in delivering the introduced organisms to the site of contamination. Currently, very little is known about the establishment of environmental niches or about microbial transport; these are areas of active research.

Another constraint to this approach is the resistance of regulators and the general public to the release of genetically engineered organisms into the environment. Perhaps in the next few years scientists will gain further understanding of microbial behavior in soil ecosystems. However, until we discover how to successfully deliver and establish introduced microorganisms, and their ultimate fate, their addition to contaminated sites will not be a feasible method of bioremediation. However, one way to take advantage of these microorganisms is to use them in above-ground bioreactor systems under controlled conditions.

METAL CONTAMINANTS

Current approaches to bioremediation of metals are based on the complexation, oxidation–reduction (redox), and alkylation reactions introduced in Chapter 14. Microbial leaching, microbial surfactants (biosurfactants), volatilization, and bioaccumulation/complexation are all strategies that have been suggested for removal of metals from contaminated environments. Unfortunately, the number of accompanying field-based studies has thus far been small.

Bioleaching: *Ex situ* removal of metals from soil can sometimes be accomplished by microbial leaching, or *bioleaching*. This technique has been used in mining to remove metals such as copper, lead, and zinc from low-grade ores. In bioleaching, metals are solubilized as a result of acid production by specific microorganisms, such as *Thiobacillus ferrooxidans* and *T. thiooxidans*. Analogously, in bioremediation, this process has been used to leach uranium from nuclear-waste–contaminated soils and to remove copper from copper tailings. Another potential application is the treatment of sewage sludge earmarked for disposal in soil. Sludge-amended soils exhibit improved productivity, but also show increased metal content. The use of *T. ferrooxidans* and *T. thiooxidans* has been demonstrated on the laboratory scale for leaching metals from contaminated sludge before soil application. Bacterial surfactants can also be used for removal of metals from contaminated soils and water.

Volatilization: Although alkylation is not a desirable reaction for most metals, particularly arsenic or mercury, alkylation with subsequent volatilization has been proposed as a technique for remediation of selenium-contaminated soils and sediments. For example, selen-

ium-contaminated soils from San Joaquin, Calif., were remediated using selenium volatilization stimulated by the addition of pectin in the form of orange peels. It was found that the addition of pectin enhanced the rate of selenium alkylation, ranging from 11.3% to 51.4% of added selenium; this result suggests that volatilization is a feasible approach to treating selenium-contaminated soils.

Bioaccumulation/Complexation: Technologies for metal removal from solution are based on several specific microbial–metal interactions: the binding of metal ions to microbial cell surfaces, the intracellular uptake of metals, and the precipitation of metals via complexation with microbially produced ligands. These interactions lead to the *bioaccumulation* of metals within cells or on the outside of the cell. Treatment of metal-containing waste streams generally involves the use of *biofilms*, which are concentrated films of microorganisms that may be composed of bacteria, fungi, or algae. Some systems utilize viable organisms, whereas other systems utilize processed, nonviable cells. In either case the biofilm essentially traps metals as the contaminated water is pumped through. Bioaccumulation has been used to treat acid mine drainage, mining effluent waters, and waste streams from nuclear processing. Metals removed include zinc, copper, iron, manganese, lead, cadmium, arsenic, and uranium.

PHYTOREMEDIATION

A relatively new area in bioremediation of hazardous waste sites is the application of plants to aid in cleanup, a process called phytoremediation. Phytoremediation can be used to treat both organic and inorganic contaminants. Plants interact with contaminants in several ways. *Accumulation* is the uptake of contaminants into root and shoot tissues. Accumulation is being studied for treatment of metal contaminants. Some plants, referred to as *hyperaccumulators*, are extremely efficient at metal accumulation, achieving up to several percent of the plant tissue dry weight as metal contaminant. The advantage of this approach is that metal contaminants are actually removed from a site by harvesting the plants after metals are accumulated. Metal *stabilization* is a second phytoremediation approach and involves metal stabilization in the root zone through complexation with plant root exudates and the plant root surface. This approach is attractive from two perspectives. First, metals are not accumulated in above-ground plant tissues, avoiding any potential risk of wildlife exposure to metal-containing shoot material. Second, this approach does not require plant harvesting. A third approach is the use of plants to aid in the *mineralization* of organic contaminants. In this approach plants can merely stimulate the growth and activity of contaminant-degrading microorganisms in the root zone by providing root exudate materials. However, there is a great deal of interest in using plants themselves to mineralize organic contaminants. The construction of "designer" contaminant-degrading plants through genetic engineering has great potential. Contaminants shown to be amenable to phytoremediation include petroleum, pentachlorophenol, TCE, polychlorinate biphenyls (PCBs), and pesticides.

IN SITU CHEMICAL TREATMENT

In situ chemical remediation is a process in which the contaminant is degraded by promoting a transformation reaction, such as hydrolysis, oxidation, or reduction, within the subsurface. Although this approach has been used much less frequently than *in situ* bioremediation, it has begun to receive increasing attention in the past few years. It can be accomplished by using active methods such as injecting a reagent into the contaminated zone of the subsurface, or by using passive methods such as placing a permeable treatment barrier down gradient of the contamination. An advantage of these methods is that they promote *in situ* destruction of the contamination, thus reducing or eliminating the associated hazard potential.

As an alternative to *in situ* bioremediation, *in situ* chemical oxidation has recently become a popular focus of research. Three oxidizing reagents in particular, potassium permanganate ($KMnO_4$), Fenton's reagent ($Fe[II]$ and H_2O_2), and ozone (O_3), are being tested. The organic contaminants are oxidized to carbon dioxide and other oxidant-specific by-products. Both ozone and potassium permanganate have been used for decades as oxidants for above-ground disinfection and purification of drinking water. As previously noted, chlorinated compounds are, in general, relatively resistant to biodegradation, which complicates or limits the use of bioremediation methods. There is thus great interest in using *in situ* chemical oxidation to remediate sites contaminated by chlorinated compounds.

All of the oxidizing reagents used for treatment of organic subsurface contamination are relatively nonselective and will thus react with most organics present in the subsurface. Therefore soils and groundwaters containing high natural organic carbon concentrations can exhibit a high oxidant demand, competing with the demand from the actual organic contaminant. Given that this approach is based on injection (flow) of the oxidant solution, the presence of low-permeability zones can reduce the overall effectiveness of *in situ* oxidation techniques. Since treatment depends on the chemical oxidant being uniformly distributed throughout the zone of contamination, significant heterogeneity will cause preferential channeling of the oxidant solution to the high conductivity zones with poor transport to the low conductivity zones.

In situ treatment walls or barriers are of particular interest for chlorinated solvents, such as trichloroethene. This approach generally involves digging a trench down gradient of the contaminant plume and filling that trench with a wall of permeable, reactive material that can degrade the contaminant to nontoxic by-products. The wall is permeable so that the water from which the contaminant has been removed can pass through. For example, iron filings can degrade compounds such as trichloroethene via reduction reactions while permitting water to pass through. One potential promising application of treatment walls is to use them to prevent the down-gradient spread of contamination. For example, the barrier can be emplaced at the boundary of a site property, on the down-gradient side of the contaminant zone, with the objective being that the barrier would prevent movement of contamination off the site property (Case Study 18.1).

MONITORED NATURAL ATTENUATION

Recently, monitored natural attenuation (MNA) has become of great interest as a low-cost, low-tech approach for site remediation. Monitored natural attenuation is based on natural transformation and retention processes reducing contaminant concentrations *in situ*, thereby containing and shrinking groundwater contaminant plumes. The application of MNA requires strict monitoring to ensure that it is working; thus the name "monitored" natural attenuation. Extensive information has been reported concerning the role of natural attenuation processes for remediation of contaminated sites. This information has been summarized recently in a report released by the NRC (2000). Two of the major conclusions reported in this document are that (1) MNA has the potential to be used successfully at many sites, and (2) MNA should be accepted as a formal remedy only when the attenuation processes are documented to be working and that they are sustainable. MNA has proved to be a success for sites contaminated by petroleum hydrocarbons (e.g., gasoline). However, its use for chlorinated-solvent contaminated sites generally has not been as successful to date. There are three major questions to address when evaluating the feasibility and viability of applying MNA to a particular site: (1) Are natural attenuation processes occurring at the site? (2) Is the magnitude and rate of natural attenuation sufficient to accomplish the remediation goal (e.g., plume contain-

CASE STUDY 18.1 *Bioremediation—The Exxon Valdez*

In March 1989, 10.8 million gallons of crude oil was spilled from a large tanker—the Exxon *Valdez*—in Prince William Sound, Alaska. Subsequently spread by a storm, this oil eventually coated onto the shores of the islands in the Sound. Initially, more than 1500 miles of Alaska coastline was polluted. A resultant court settlement included $900 million, to be paid by the Exxon Corporation for damage assessment and restoration of natural resources. A good portion of this sum was spent on conventional cleanup, primarily by physical methods. But these methods failed to remove all of the oil on the beaches, particularly that under rocks and in the beach sediments. In late May 1989, Exxon reached a cooperative agreement with the U.S. Environmental Protection Agency (EPA) to test bioremediation as a cleanup strategy. The approach followed was to capitalize on existing conditions as much as possible. For example, scientists quickly realized that the beaches most likely contained indigenous oil-degrading organisms that were already adapted to the cold climate. Preliminary studies revealed that, indeed, the polluted beaches contained such organisms, together with a plentiful carbon source (spilled oil) and sufficient oxygen. In fact, natural biodegradation, also called *intrinsic bioremediation*, was clearly already occurring in the area. Further studies showed that intrinsic degradation rates were limited by the availability of nutrients, such as nitrogen, phosphorus, and certain trace elements. At this point, time was of the essence in treating the contaminated sites because the relatively temperate Alaskan season was rapidly advancing. Therefore it was decided to attempt to enhance the intrinsic rates of biodegradation by amendment of the contaminated areas with nitrogen and phosphorus.

In early June 1989, field demonstration work was started. Several different fertilizer nutrient formulations and application procedures were tested. One problem was the tidal action, which tended to quickly wash away added nutrients. Therefore a liquid oleophilic fertilizer (Inipol EAP-22), which adhered to the oil-covered surfaces, and a slow-release water-soluble fertilizer (Customblen) were tested as nutrient sources. Within about 2 weeks after application of the fertilizers, rock surfaces treated with Inipol EAP-22 showed a visible decrease in the amount of oil. Subsequent independent scientific studies on the effectiveness of bioremediation in Prince William Sound concluded that bioremediation enhanced removal from threefold to eightfold over the intrinsic rate of biodegradation without any adverse effects to the environment.

Bioremediation in Prince William Sound was considered a success not only because contaminated areas were restored, but because the rapid rate of removal of the petroleum contamination prevented further spread of the petroleum to uncontaminated areas. This well-documented and highly visible case has helped gain attention for the potential of bioremediation.

CASE STUDY 18.2 *Pilot-Scale Tests of In Situ Chemical Oxidation of TCE Using Potassium Permanganate at a Superfund Site*

Pilot-scale tests were performed as part of a collaborative project between academia, industry, consulting firms, and government agencies to assess the efficacy of potassium permanganate for *in situ* chemical oxidation (ISCO) of trichloroethene (TCE) contamination at Air Force Plant 44 (AFP44), a Superfund site in Tucson, Ariz. The target contamination resides in the upper zone of the regional aquifer. Remediation of the site is constrained by the difficulty of removing contamination from the source zones. Extensive characterization indicates high concentrations of TCE reside in layers of relatively low permeability. In addition, evidence suggests the presence of immiscible-phase liquids in the saturated zone. Experiments were conducted to evaluate the potential of *in situ* chemical oxidation using potassium permanganate for directly treating source-zone contamination.

AFP44 is part of the Tucson International Airport Area (TIAA) Superfund Site, located in southwest Arizona. From the early 1950s and 1960s to approximately 1978, industrial solvents including TCE, trichloroethane (TCA), and methylene chloride, machine lubricants and coolants, paint thinners and sludges, and other chemicals used were disposed of in open, unlined pits and channels. In addition, wastewater containing chromium from plating operations flowed through a drainage ditch that crosses part of the site. A 1981–1982 investigation determined that TCE had moved down through overlying sediments into the regional aquifer. In 1983 the TIAA site was added to the national priorities list as a Superfund site.

(Continued)

CASE STUDY 18.2 *Continued*

A groundwater pump-and-treat remediation system was started in 1987, and a soil-venting system was initiated in 1998. The pump-and-treat system has successfully controlled and reduced the extent of contamination, and the soil-venting system has removed large quantities of contaminant mass from the vadose zone. Unfortunately, the cleanup target has not yet been attained. The results of advanced characterization studies indicated that contaminant removal is constrained by site heterogeneities, physical and chemical rate limitations, and the presence of immiscible-liquid contamination within the source zones.

ISCO is a technique that shows promise for directly treating source-zone contaminants. ISCO pilot-tests with permanganate were conducted at two source-zone sites on the AFP44 facility, Site 2 and Site 3, from June through November 2000. These tests were designed to affect a relatively small area of the aquifer and provide useful information for the design of a full-scale application at AFP44. The ISCO pilot-scale test was performed with the collaboration of Raytheon Systems Environmental section, The University of Arizona, Errol L. Montgomery and Associates, Inc., and IT Corporation.

Potassium permanganate application in the subsurface is dependent on the effective delivery in the aqueous phase. Fortunately, potassium permanganate is quite soluble at 20° C ($64\,g\,L^{-1}$) and even more so at higher temperatures ($250\,g\,L^{-1}$ at 65° C). Application concentrations have generally ranged from 1% to 6%. Successful subsurface treatment is also dependent on the oxidizing solution remaining reactive in the subsurface for extended periods. Although permanganate is not thermodynamically stable, the time span for significant degradation is fairly long. In field experiments, permanganate has been observed to persist for periods longer than 5 months. Other oxidizing agents, such as hydrogen peroxide and ozone, are much less stable in the subsurface.

The simplified proposed reaction mechanism for potassium permanganate ($KMnO_4$) with chlorinated compounds is as follows:

$$C_2Cl_3H(TCE) + 2MnO_4^-$$
$$(Permanganate) \rightarrow \qquad (Eq.\ 18.1)$$
$$2CO_2 + 2MnO_2(s) + 3Cl^- + H^+$$

with 0.81 kg of chloride (Cl^-) produced and 2.38 kg manganese oxide (MnO_2) reduced per kilogram of TCE oxidized.

Potassium permanganate is fairly inexpensive; for example, the cost for the industrial grade potassium permanganate purchased for the field project was approximately $4.40 per kilogram. Since extraction of groundwater is not required, disposal costs are minimized. Mobilization, demobilization, and costs associated with the delivery method are the primary costs for ISCO treatment. Since oxidation occurs rapidly with adequate delivery, the time frame of projects can be dramatically reduced in comparison with traditional treatment technologies, such as pump and treat.

In addition to limitations associated with stability of the oxidant solution, transport to and contact with the contaminant, *in situ* oxidation agents have a significant potential for geochemical interactions with the matrix and groundwater and for reactions with co-contaminants. Production of MnO_2 precipitate has the potential to clog the aquifer, and production of hydrochloric acid (HCl) has the potential to dissolve aquifer material, especially carbonates. Additionally, interaction with co-contaminants is possible. Chromium is a fairly common contaminant in the industrial waste sites for which ISCO is a potential remediation strategy. When oxidized, chromium converts to the hexavalent form, which is more toxic, as well as more mobile, and is therefore a health concern. *In situ* oxidation for remediation of chlorinated compounds temporarily creates an oxidizing environment in the subsurface. However, when the oxidizing solution leaves the treatment zone, conditions generally return to pre-treatment conditions. The potential for the mobilization of chromium and other metals, and its subsequent potential for reduction and stabilization (natural attenuation), is an important consideration for this technology.

Batch and column experiments were conducted to assess the potential for *in situ* chemical treatment of TCE contamination at AFP44, Tucson, Ariz. Additional batch studies were performed to investigate aquifer demand for the oxidant and calculate injection concentrations. Results of the laboratory studies showed rapid oxidation of TCE. The presence of the aquifer material had minimal impact on reaction rates, suggesting that the aquifer material does not have a significant oxidative demand. The amount of chromium mobilized was less than 16% of the total initial chromium content of the aquifer material. These results suggested that *in situ* oxidation using potassium permanganate may be an effective method for remediating TCE contamination at this site.

Two pilot-scale tests were conducted, both of which were designed to attempt to address the contamination associated with the low-permeability zones. At one location at the site, a horizontal natural-gradient flush was performed in the saturated zone. A vertical percolation flush wherein permanganate solution was injected above the capillary fringe, which coincides with a low-permeability source zone, was performed at another location at the site. The impact of the pilot tests on system properties was evaluated by a number of methods. Aquifer pumping tests were conducted before and after the oxidant flush to examine effects on aquifer properties. Contaminant concentrations, oxidizing solution concentrations, and a number of geochemical parameters were monitored. Bromide was employed as a nonreactive tracer to differentiate between hydraulic effects on contaminant concentrations and the effects caused by the oxidation reaction. Groundwater monitoring was performed for three periods: baseline sampling, during the test, and post-test sampling. Parameters evaluated by groundwater monitoring included: TCE, chromium, permanganate, carbonate/alkalinity, cations (K^+, Ca^{2+}, Mg^{2+}, Na^+, $^{3+} + Fe^{2+}$), anions (Br^-, Cl^-, F^-), field parameters (color, conductivity, oxidation-reduction potential [ORP], temperature, and pH) and groundwater elevation. The success of the pilot-scale field test of ISCO can be most directly evaluated by examining TCE concentrations before, during, and after the treatment test. The baseline was represented by four rounds of sampling, which occurred over a 2-month period before the injection of the potassium permanganate solution.

Groundwater samples from all wells in which permanganate was observed showed a decrease in TCE concentrations coincident with the arrival of the potassium permanganate in concentrations above approximately 50 mg L^{-1}. During the operational period, when permanganate solution was active in the treatment area, the average TCE concentration reduction was 84% for site 3 and 64% for site 2. As permanganate left the treatment area, TCE concentrations in all wells experienced some rebounding. Possible causes of rebounding include movement of contaminated groundwater into the treatment zone from up gradient, and mass transfer of untreated TCE from the aquifer material by desorption, diffusion from dead-end pores or areas of lower permeability, and dissolution from nonaqueous phase liquid (NAPL) contamination. Rebounding to an average of 98% of pretest concentrations was observed in water-flushing studies conducted previously at site 3. However, in the potassium permanganate study, rebound concentrations averaged only 50% of the pretest concentrations. The significant difference in rebound likely reflects a decrease in mass of TCE associated with nonaqueous phases within the source zones due to permanganate oxidation. These results indicate that *in situ* oxidation using potassium permanganate was successful in removing TCE associated with the source zones at two locations at the AFP44 site.

The presence of chromium as a co-contaminant at the site was an important aspect of this pilot project. Chromium concentrations increased as expected. However, concentrations quickly returned to low levels. Further attenuation is expected with time, based on the history of the site and review of other ISCO projects. The observed increase in soluble chromium was anticipated and may be due more to the presence of chromium as an impurity in the injection solution than the mobilization of chromium associated with the aquifer solids.

ment, plume reduction) and be protective of human health and the environment? (3) Will natural attenuation be sustainable over the long timeframes generally required for MNA to be successful? Successful implementation of MNA requires an accurate assessment of these three questions.

As noted above, the first step in evaluating the feasibility and viability of applying MNA to a particular site is to determine whether or not natural attenuation processes are occurring at the site. Evaluating and documenting the potential occurrence of natural attenuation involves the following steps: (1) develop a general conceptual model of the site that is used to provide a framework for evaluating the predominant transport and transformation processes at the site; (2) evaluate contaminant and geochemical data for the presence of known "footprints" indicative of active transformation-processes. This includes analyzing contaminant concentration histories, the temporal and spatial variability of transformation products and reactants (e.g., O_2, CO_2, etc.), and isotope ratios for relevant products and reactants; and (3) screen groundwater and soil samples for microbial populations capable of degrading the target contaminants.

The second question to address for MNA application is the magnitude and rate of natural attenuation at the site. Characterizing the magnitudes and rates of attenuation at the field-scale is a complex task. For example, the initial mass of contaminant released into the subsurface is not known at most sites. This means it is difficult to accurately quantify the magnitude and rate of attenuation. Thus we can generally obtain estimates of such rates at best. One method is based on characterizing temporal changes in contaminant concentration profiles, electron acceptor concentrations (e.g., O_2, NO_3^-, SO_4^{2-}, Fe^{3+}), or contaminant transformation products. However, the complexity of typical field sites makes it difficult to accurately determine attenuation rates from temporal changes in these parameters. Another approach often used to characterize potential attenuation processes is to conduct bench-scale studies in the laboratory using core samples collected from the field. However, results obtained from laboratory tests may not accurately represent field-scale behavior. *In situ* microcosms, in which tests are conducted in a small volume of the aquifer, can be used to minimize this problem. Given the complexity of most field sites, including spatial variability of physical and chemical properties and of microbial populations, a prohibitive number of sampling points may often be required to fully characterize a field site using either of these methods. Tracer tests have recently been proposed as an alternative method for field-scale characterization of the *in situ* attenuation potential associated with subsurface environments.

QUESTIONS

1. What are the essential components of a typical hazardous waste site characterization program? Why are these the essential components?

2. How do so-called LNAPLs and DNAPLs behave when spilled into the ground? How do these behaviors influence our ability to find them during site characterization projects?

3. Compare and contrast the use of physical barriers versus well fields to contain a groundwater contaminant plume.

4. What is a major limitation of the pump-and-treat method? What are some of the methods that can be used to enhance the performance of pump and treat?

5. For which of the following contaminants would you propose to use air sparging as a possible remediation method—trichloroethene, phenol, nitrate, benzene, pyrene, cadmium? Why or why not?

6. What are some of the potential benefits of using *in situ* bioremediation? What are some of the major potential limitations to its use?

7. What constraints were discovered during the remediation of AFP44? How was potassium-permanganate injection used to help overcome these constraints?

8. What criteria are used to evaluate proposed remedial actions?

9. What are the major advantages and disadvantages of using excavation?

REFERENCES AND ADDITIONAL READING

Alternatives for Groundwater Cleanup. (1994) Natural Research Council, National Academy Press, Washington, D.C.

Groundwater and Soil Cleanup. (1999) Natural Research Council. National Academy Press, Washington, D.C.

In Situ Bioremediation, When Does It Work? (1993) National Research Council. National Academy Press, Washington, DC.

National Research Council. (2000) *National Attentuation for Groundwater Remediation.* National Academy Press, Washington, D.C.

Practical Handbook of Soil, Vadose Zone, and Ground-Water Contamination (1995) Boulding, J.R., Lewis Publishers, Boca Raton, FL.

U.S. Environmental Protection Agency. (2001) *Use of Bioremediation at Superfund Sites.* EPA 542-R-01-019, Washington, D.C.

19

ECOLOGICAL RESTORATION

V.J. GERHART, W.J. WAUGH, E.P. GLENN, AND I.L. PEPPER

Human beings have greatly disturbed most of the world's natural ecosystems. Socioeconomic pressures, land-use patterns and recently, the wide scale removal of natural resources, such as forests, have devastated vast tracts of land across the globe. Often the original species' composition of the area is lost, but the essential biophysical resources may remain intact. The primeval forests of Europe were cleared for farms and settlements thousands of years ago, and although there is little chance that they will return to their former state, they remain productive albeit in a different form. Activities that result in degraded lands that are candidates for restoration include: deforestation; overgrazing; secondary salinization from poor irrigation management; wetland clearing and draining; oil production; mining; and toxic spills. In fragile ecosystems such as the desert and semidesert regions of the world, overuse of land can lead to the irreversible loss of fertile top soil, vegetation, and nutrient cycling, a process called *desertification*. According to the United Nations Atlas of Desertification (Middleton and Thomas, 1997), over half of the world's arid and semi-arid lands have been affected by desertification. In wet regions of the world deforestation and other unsustainable land-use practices have left large tracts of land with unusable, unfertile soils prone to water and wind erosion. In any ecosystem there is a threshold for self-repair, but once that threshold has been crossed, severe degradation occurs. It has become apparent that land, and particularly soil, are finite resources that need to be preserved and restored whenever possible.

Increasingly strict regulations have been promulgated to reduce potentially destructive land use. Today, for example, nearly all mining activities in developed countries require a *closure plan* that describes how the land will be restored to a productive state after mining ceases. The United Nations has an antidesertification program that aims to return millions of hectares of arid lands around the world to productivity. In the United States the Endangered Species Act has mandated that key ecosystems, such as riparian corridors and wetlands that have become degraded, must be restored or recreated to provide habitat for threatened species. In return for permits to build new factories and power plants, developers are now often required to provide *environmental offsets*, in which they restore abandoned farmland, create wetlands, or plant trees on logged-over property. The science of *ecological restoration* has developed, rather recently, to find ways of repairing damage to disturbed ecosystems.

In 1996 The Society for Ecological Restoration defined restoration as "the process of assisting the recovery and management of ecological integrity. Ecological integrity includes a critical range of variability in biodiversity, ecological processes and structures, regional and historical context, and sustainable cultural practices." Technically, *rehabilitation*, *revegetation*, and *reclamation* fall under the umbrella of *restoration*, each referring to specific goals within a restoration project. However, these terms have been widely used by different land-management agencies, and their definitions tend to be interchangeable. Rehabilitation means repairing some or most of the damage done to land so that it can serve some productive function. For example, salinized farmland, unable to support native plants, can be planted with salt-tolerant plants (*halophytes*) to prevent erosion and provide wildlife habitat. Revegetation involves planting or seeding an area that has received minor damage. In contrast, reclamation is often used as a synonym for rehabilitation, although it generally refers to restoring biotic function and productivity to the most severely degraded land, such as an Environmental Protection Agency (EPA) Superfund site (see Chapter 18). However, by another definition reclamation means "...making land available for human use by changing natural conditions" (Merriam-Webster, 1993). This is the definition adopted by the U.S. Bureau of Reclamation, which has sponsored programs to convert desert land in the southwestern United States into irrigated farmland. By this definition, the original human inhabitants of Europe could be said to have engaged in a massive reclamation project by converting the forests to farms and towns. In this chapter we are concerned with methods to repair human-caused damage to natural ecosystems, and like many restoration ecologists, we tend to use the aforementioned terms interchangeably.

In the following sections we describe the aims and methodology of restoration ecology. The process usually starts with a *site characterization*. This includes a *conceptual plan* and a *site assessment*. These two components are often combined to provide a complete picture of a project site. The conceptual plan summarizes the restoration potential of a particular site, whereas the site assessment details current conditions. Essentially the questions asked at this point include: What was the land like before human intervention? What is it like now? What changes in topography, soil properties, surface and subsurface hydrology, and vegetative cover have taken place? The information garnered during this stage enhances the development of realistic restoration objectives. Can the land be reasonably restored to its original state, or has it been so altered that it must be converted into a different type of habitat? Once the objectives are set, a *site design* and *implementation* follows, which includes a schedule and detailed protocols for repairing the land. Abiotic components such as soils must be replaced, stabilized, or amended so they can once again support plants, and biodiversity must be restored to the site. Restoration even under favorable circumstances can take many years. The final stage of the process is *monitoring and evaluation*. Did it work? What more must be done? There must be feedback loops built in to the plan so necessary changes can be made along the way. The costs associated with restoration can reach millions of dollars for the most severely disturbed sites, but if the plan is not cost effective, it is not likely to be implemented despite its benefits to the environment.

In reading this chapter, keep in mind that restoration ecology is a new science. There is no cookbook method for restoring a damaged ecosystem, and each restoration site has its own unique characteristics and problems. As the field of restoration ecology moves out of its infancy, practitioners commonly define specific output goals, which in turn provide the blueprint for input and management decisions. Clearly the best restoration project is the one that is not required because safeguards minimizing land degradation were built into the original land-use plan.

SITE CHARACTERIZATION

In an ideal world the restoration objectives for a site will be chosen in advance, even before the site is disturbed. For example, a new electric power plant might have a useful life of 50 years. A closure plan will be submitted along with the application for a construction permit for review by local, state, and federal agencies and other interested parties (*stakeholders*). Thus when the plant is closed, the original site conditions will have been documented, and a procedure for mitigating any damage to the land will already be in place. In reality most current restoration work is done after the fact. Land that decades ago was converted to some use, such as mining or farming, then abandoned, may only recently have become a candidate for restoration through stricter environmental laws. Often state or federal agencies assume responsibility for restoration, even if the land was damaged by private owners. Restoration of these sites starts with a conceptual plan and site assessment that can proceed like a detective story. Figure 19.1 shows the main points to be considered during each step of the process, from conception to final evaluation. As you read through the following sections, refer to the case study on a former uranium mill site on the Navajo Reservation in Monument Valley, Ariz., which details the formal procedure followed by the U.S. Department of Energy as they began the restoration process.

CONCEPTUAL PLAN AND SITE ASSESSMENT

A conceptual plan often begins with a detailed history of the site where information is collected to determine the anthropogenic changes that have lead to the current state of degradation. This information is cross referenced against historical and current topographical, geological, and vegetation maps to determine what changes have occurred spatially and temporally. Information on the physiochemical soil properties and the water quality of the site, before disturbance, can show the restoration potential of the area, and perturbations or stressors that enable the degradation process are identified. Before deciding to proceed with a project, social and cultural values of the neighboring residents need to be assessed to ensure that the proposed restoration objectives are compatible with the local socioeconomic needs of the public.

Once a complete history of the site has been compiled, and it has been determined that the restoration project is

feasible, a thorough site assessment of abiotic and biotic conditions takes place. This step is probably one of the most important in any restoration project, since it not only provides baseline measurements on such parameters as hydrologic features, soil conditions, and biological information, but it also serves as the benchmark on which to evaluate the project through time. This step involves placing the site in the context of the regional landscape with respect to habitat fragmentation, disconnected surface and subsurface hydrological flows, water quality issues, physical and chemical properties of the soil, and finally plant and sometimes animal inventories. Depending on the nature of degradation, additional data may be collected on the presence or absence of toxic chemicals, such as organophosphates, heavy metals, and radioactive waste, as discussed in Chapter 16. In assessing a site it is important to determine which basic ecological functions are damaged or fragmented because this often sets the restoration priorities, and ultimately the success of the project.

It is worth noting that the soil conditions and water quality in a given area generally dictate the type of vegetation cover and thus the biodiversity of biotic components, so extra care should be taken in analyzing and describing these two elements both horizontally and vertically across the landscape. Figures 19.2 to 19.4 show the soil moisture and soil salinity gradients, plus the depth to the impeding layer at a restoration site in western Maricopa County, Ariz. This 11,000-acre site was a cotton farm until it was abandoned in the late 1980s due to salinity problems and decreases in the price of cotton. The climatic conditions of this site are harsh, with annual precipitation ranging from 150 to 200 mm per year and summer temperatures rising to 44°C. During preliminary surveys taken in the fall of 2000, it was found that although some areas had extensive stands of mesquite trees, and other areas were covered with saltbush, most of the ground was bare and subject to wind erosion. A detailed soil sampling across the site to a depth of 5.4 meters (18 feet) showed that the difference in vegetation types were related to the soil moisture and soil salinity found in the root zone, which were, in turn, related to discontinuities in the soil profile. Sandy loam is the predominant soil type within the plow layer, but it is underlain by clay lenses, caliche (cemented calcium carbonate), and sandy gravel. In the mesquite bosques a low-permeability clay lense was found at 3 meters and this held the soil moisture at approximately 13%, which was sufficient to allow the mesquite trees to thrive. In contrast, the bare ground contained approximately 3% moisture, whereas saltbush was located in areas with 7% moisture and high salinity levels. Obviously, any restoration plan for this site has to address the issue of retaining water within the upper part of the soil

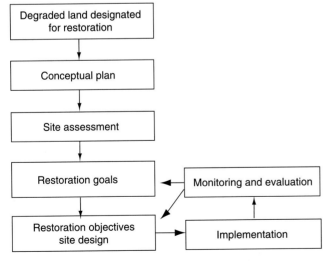

FIGURE 19.1 Flow chart of the restoration process. Built-in feedback loops are needed to allow for project modification.

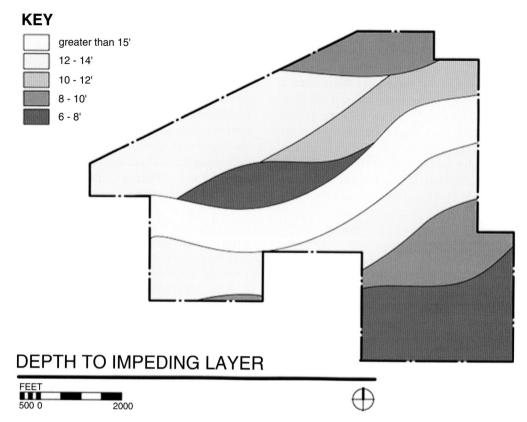

DEPTH TO IMPEDING LAYER

FIGURE 19.2 The depth to the impeding layer in the soil profile at a restoration site in Maricopa County, Ariz.

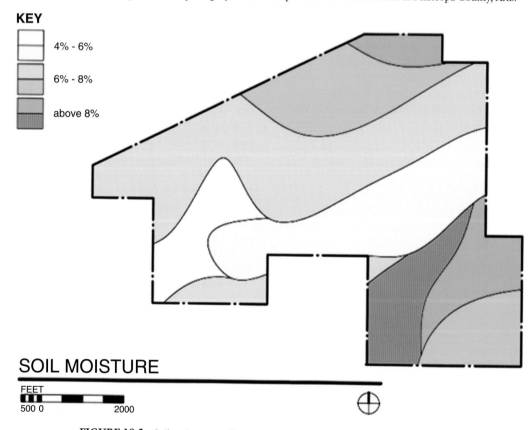

SOIL MOISTURE

FIGURE 19.3 Soil moisture gradients at a restoration site in Maricopa County, Ariz.

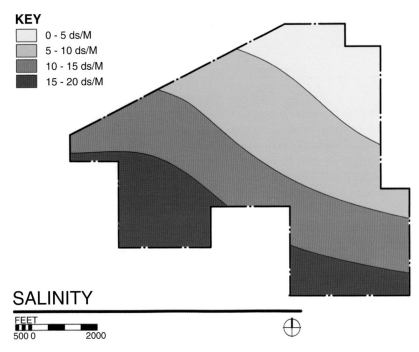

KEY

	0 - 5 ds/M
	5 - 10 ds/M
	10 - 15 ds/M
	15 - 20 ds/M

SALINITY

FEET
500 0   2000

FIGURE 19.4 Soil salinity gradient at a restoration site in Maricopa County, Ariz.

profile and, where possible, leaching salts below the root zone.

PLANT SURVEYS

The occurrence and relative abundance of certain plant species and their physiological and ecological tolerances provide information about environmental conditions that are of importance for understanding the nature of a site, and potential human health and ecological risks, plus the feasibility of different restoration alternatives. Typically, plant ecology investigations include four types of studies: (1) *plant species survey*; (2) estimates of the *percent cover* and age structure of dominant, perennial plant species; (3) evaluation of the composition, relative abundance, and distribution of *plant associations*; and (4) *vegetation mapping*.

The plant species survey is conducted by traversing a site, usually on foot, and noting each species present. Sometimes the survey is confined to perennial species only, and unknown species are collected for later identification at an herbarium. In formal surveys *voucher specimens* of each plant species are collected, pressed, and mounted on cardboard herbarium sheets to be deposited in a university or other recognized herbarium. Sufficient information is included on the label accompanying the specimen so that others can relocate the collection site, if necessary.

The percent-cover study attempts to quantify the percent of the site that is covered by bare soil or individual plant species. Generally, a *line intercept* method is used, employing a *baseline and transect* sampling scheme (Bonham, 1989). First, the plant community to be described is delineated on a map, and then 30-meter transect lines are chosen where actual plant counts will take place. In the field a 30-meter tape is stretched out, and the total distance intercepted by each plant species is recorded and used to calculate the percent cover of each species. For example, a transect might consist of 12% fourwing saltbush, 10% black greasewood, and 78% bare soil. The results from all transects were averaged to give an estimate of percent cover over the whole site. Each individual plant encountered along the transects can be further measured to determine height, width, leaf area, and age (if it forms annual rings), allowing a vegetation history of the site to be developed.

The plant association and vegetation mapping studies are used to delineate land management units with respect to ecological condition and potential for enhancement by revegetation. An association is a unit of classification that defines a particular plant community, and generally has a consistent floristic composition, a uniform appearance, and a distribution that reflects a certain mix of environmental factors that can be shown to be different from other associations. The association is a synthesis of local examples of vegetation called *stands*. There are several methods for delineating and mapping plant associations. One of the simplest to use is the *relevé* method (Barbour *et al.*, 1987), where stands are characterized, then grouped into associations using simple *ordination* and *gradient analysis* techniques.

SITE RESTORATION

Setting realistic goals for a restoration project is probably one of the most difficult parts of the process. The tendency is to aim at recreating an ideal habitat or ecosystem, one that mirrors adjacent undisturbed areas in biodiversity, ecological function, and services. The inability to achieve this level of restoration leaves many projects labeled as failures, when in fact, there are many incremental successes. We as human beings can design a restoration project and put it on a desired trajectory, but ecological processes are not static; rather they are in a constant state of change and readjustment in response to human and natural perturbations. Current trends in the field of restoration ecology suggest that the "dynamic nature of ecosystems be recognized, and accept that there is a range of potential short- and long-term outcomes of restoration projects" (Hobbs and Harris, 2001) and the focus should be on "desired characteristics of the (eco)system *in the future*, rather than in relation to what these were in the past (Pfadenhauer & Grootjans, 1999). Realistic goals, then, should specify ecological changes or outputs at a project site within the realm of its intrinsic dynamic nature. The better these goals are defined with respect to habitat creation, biodiversity, and socioeconomic needs, the greater the need to define inputs and intervening processes (Box, 1996), and the higher the success rate of the project. Restoration goals will always follow the key concepts of ecological systems: sustainability, resistance to erosion and invasion of alien species, productivity, nutrient retention, and a degree of biodiversity sufficient to support multilevel biotic interactions (microbes, plants, and animals). Often, specific ecosystem requirements are built into the restoration goals. For example, constructed wetlands are designed as replacement habitat for specific, sensitive bird species in areas where natural wetlands have been drained.

The methods chosen for the restoration of a particular site will be determined by the nature of the site, the level of existing degradation, and the desired outcome over time. The underlying causes of the degradation must be identified as either biotic, abiotic, or a combination of both. The restoration ecologist may ask, for example, whether the degradation of land was caused by simple overgrazing, or if the physiochemical soil and hydrological processes were changed to the point where biodiversity has been compromised. In highly polluted sites it may be necessary to remove contaminants from the soil and/or water before beginning restoration work. For example, mining activities may have contaminated the soil to the extent that it is toxic to humans, animals, and plants, in which case it must be removed and replaced by clean soil at the beginning of the restoration process. Thus for some sites, soil and groundwater remediation activities (Chapter 18) may be a component of an ecosystem restoration project. Yet, in other cases, minimal work such as managing grazing will be all that is necessary to make the site suitable for plant establishment and growth. The degree of intervention required in restoring a site ranges from "natural" restoration through passive to active restoration. The information gathered for the conceptual plan and from the site assessment should paint a fairly clear picture of the path a restoration project will follow, and will shape the achievable goals through carefully constructed project objectives, design, and implementation strategies.

RESTORATION OBJECTIVES

A project's implementation plan spells out the activities that are required to achieve the restoration goals, and thus is an integral part of the overall project design. Up to this point in the process, the main participants involved in developing restoration goals for a site will be scientists, politicians, federal and state employees, and possibly, local residents. However, once there has been a consensus on the restoration goals, and the planning stage moves to setting the objectives, a myriad of other players can become involved, depending on the complexity of the work. As we will see in the Monument Valley case study, *Phase 1* involved removing radioactive topsoil, then transporting, and capping it in a different and presumably safer location. The list of objectives to carry out this one goal was obviously very complex and explicit. Overall, objectives have to be measurable, and performance standards established that represent milestones of accomplishments for each portion of the project. It has already been noted that ecosystems are a dynamic entity and each small, natural, or anthropogenic change to a given site can cause unintended reactions. By documenting and measuring each objective in an unbiased way, it will not only be easier to pinpoint problems and make adjustments, but it will also be easier to explain minor setbacks to concerned stakeholders.

It is advantageous to integrate into a restoration project scientific studies that will quantitatively measure abiotic and biotic factors spatially and temporally. It is at this stage of the project that research experiments need to be carefully designed so they can become an integral part of the implementation process. A large body of literature addresses the experimental design and analysis of data for ecological and agronomic studies; however, the complexity of interactions, and the logistical challenges found in ecosystem restoration are often viewed as insurmountable, and data is simply not collected (Michner, 1997). New methods of handling and analyzing data allow more complex ecological interactions to be interpreted, which will benefit the field of restoration ecology immensely.

PROJECT IMPLEMENTATION

Just as an architect draws plans for a house and presents construction documents to a builder, so must a restoration ecologist draw a project plan, and compile a set of documents that identify all the actions and treatments needed to satisfy each project objective. These can include, but are not limited to, equipment, personnel, supplies, seeds, and plants. In some cases the reintroduction of animals may be a stated objective. The need for specific, sequential work orders is of paramount importance at this stage, since retrofitting a particular task will generally be more costly, and will delay the implementation of other parts of the project. General maintenance should also be scheduled into the overall plan.

Most of the installation in a restoration project will be completed in the first few years, but other objectives may not be met for some time. In constructing a wetland the reintroduction of aquatic life may have to wait until the site matures and optimal conditions regarding water quality and primary production will ensure the survival of the reintroduced species. As work proceeds, performance standards should be evaluated and recorded for each part of a restoration objective. This continuous monitoring will allow for adjustments if unforeseen problems arise. The flexibility to modify the plan as needed during the implementation phase is crucial to ecological restoration projects. Unlike a traditional engineered plan for, say, a new bridge where a predictable outcome is guaranteed, restoration projects are less predictable.

SITE MONITORING

"Success criteria need to relate clearly back to specific restoration goals" (Hobbs and Harris, 2001) and restoration objectives. Up until a few years ago, most projects were monitored in the field using typical agronomic and plant ecology techniques and in controlled greenhouse experiments that measured soil, microbial, and plant interactions. Although these methods are considered vital in describing the status of plant establishment, water quality, changes in soil chemistry, and microbial populations, the current trend is to try to integrate this

information into a broader ecological picture. With increased computing power and sophisticated data collection techniques, real-time data can more easily be obtained without additional personnel, thus giving restoration ecologists larger data sets with which to work.

SHORT- AND LONG-TERM MONITORING

Most restoration projects rely on short-term monitoring to assess the success of a project. Sponsors want to see the results of their investments, and the public expects immediate results. The reality of the situation is that ecological processes can take from decades to centuries to achieve a level of maturity; a time span that is not economically compatible with monitoring programs. Instead, shorter-term monitoring programs collect data, which are then extrapolated to predict generalized ecological patterns of change against a referenced ecosystem.

Restoration monitoring investigations must be sufficiently quantitative so that differences before and after restoration can be detected, and comparisons made between disturbed and referenced sites. In general, sampling methods must be sufficiently robust to detect differences of 10% to 15% with 95% certainty. Data should be collected on each specific project objective, which, for example, could include the efficacy of soil surface preparation, the addition of soil amendments, variable seeding rates, and survival of transplants under dry land or irrigated conditions. The analysis of this type of data from short-term monitoring is crucial in the evaluation process.

RESTORATION EVALUATION

At the beginning of this chapter we introduced ecological restoration as the process of assisting the recovery and management of ecological integrity, which includes: a critical range of variability in biodiversity, ecological processes and structures, regional and historical context, and sustainable cultural practices. With these parameters in mind, the evaluation of a restoration project should be sufficiently thorough to address the issues of ecological function, biodiversity, and sustainability over time. Commonly, projects are evaluated on a limited set of criteria, which are then compared to a reference ecosystem. This implicitly creates a success/failure scenario without taking into account ecosystem dynamics. Evaluations therefore should not only include quantitative data on specific project performance standards, but they should also incorporate how the project has fit into the greater regional, historic, and social landscape (Case Study 19. 1).

CASE STUDY 19.1 *Monument Valley, Arizona (Glenn et al., 2001)*

This case study involves a former uranium mill site on the Navajo Indian Reservation in Monument Valley, Ariz., that the U.S. Department of Energy had placed in its Uranium Mill Tailing Remedial Action (UMTRA) program. Each step in this restoration process follows a formal procedure that has been well documented. Starting in the 1950s, the Atomic Energy Commission encouraged the mining of uranium ore in the southwestern United States to provide fuel for the nuclear power industry and material for weapons. The milling process produced large masses of crude ore and tailings that covered many acres. These waste areas were surrounded by unlined evaporation and leaching ponds, from which "yellow cake," a crude form of uranium, was extracted. In the 1970s the price of yellow cake collapsed, and most of the mills went bankrupt and were subsequently abandoned. The owners made no effort to clean up the sites. In the 1980s, the Department of Energy was given the responsibility for restoring these sites. The primary problem associated with these sites is the piles of crushed ore and tailings, each pile covering several acres, which are mildly radioactive. In addition, toxic chemicals (heavy metals, nitrates, ammonia, and sulfates) have leached into the soil, and ultimately, the groundwater at many of these sites.

The first task in the restoration process was to determine the history of the Monument Valley site. A search was made of company records, former workers were interviewed, and archives of aerial photographs were assembled. An overview of how the mill operated was developed, and areas of concern for remediation and restoration were pinpointed. A map of the site was made, showing where the different processes in the milling operation took place. The second task was to determine the current extent and state of contamination. Intensive soil sampling for radioactivity, heavy metals, and other potentially toxic chemicals was undertaken. Bore holes were drilled into the water table to determine if the underlying aquifer was contaminated, and vegetation cover across the site was assessed. Maps were produced that detailed the extent of the contamination not only on the site, but also on adjacent land.

Once the site had been characterized, and the extent of contamination problems determined, a baseline risk assessment report was released that evaluated the potential for human and environmental damage if the site was not repaired. The Monument Valley site was given a high priority for remediation, and further studies were conducted to develop a set of restoration goals and objectives. Federal, state, and private stakeholders reviewed these goals in a series of public meetings. Those attending the meetings included representatives from the Department of Energy, the EPA, the Navajo Nation UMTRA, the Navajo Nation EPA, and the local community. Local residents were vocal in opposing plans that would negatively impact their traditional uses of the land. Stakeholder participation during this planning phase was critical to the ultimate acceptance of the plan by the community. The restoration plan for this UMTRA program is shown in Table 19.1.

Phase I of the restoration plan generated little controversy. It was quickly decided that the ore and tailings had to be removed from the site, and that the soil around the site had to be removed down to the level at which there was no more radioactivity. Subsequently, a fence was placed around the property to prevent grazing animals from entering. A graded road to the site was constructed across 20 miles of desert, and a fleet of trucks was commissioned to haul away the contaminated material. Local citizens were trained as truckers and equipment operators for the project. The contaminated material was taken to the nearby town of Mexican Hat, where it was spread over an impermeable bedrock surface and covered with three layers of material to prevent radon gas and radioactivity from escaping. The first layer was compacted clay (from a local site); the second layer was bedding sand; and the third layer was made up of large rocks. The rock layer was thick enough that plants could not easily establish themselves on the surface of the containment cell. The design of the containment cell was such that it is expected to prevent contaminants from leaking for at least 1000 years.

Phase II, still under development, involves first, repairing the damage to the land from Phase I, and second, dealing with a plume of contaminated water that is migrating underground away from the site. In removing surface contamination, over 100 acres of the site was denuded of native vegetation. It was necessary to replace this vegetation. The main chemical of concern in the contaminated groundwater plume is nitrate, originating as nitric acid that was used to leach uranium from the ore. Nitrate levels in the groundwater greatly exceed EPA standards for drinking water ($44 \, \text{mg L}^{-1}$), and this nitrate must somehow be removed. How to deal with the two problems was analyzed through a process called *value engineering*. All possible alternatives for restoring vegetation and cleaning up contaminated water were listed after preliminary analysis by the study team. The list was shortened to those that appeared to be both likely to succeed and were cost effective.

Options for restoring vegetation ranged from relatively low-cost measures, such as application of mulch and seed to the land in a liquid spray (*hydroseeding*), to higher cost measures, such as transplanting to the site native shrubs that were originally grown in a greenhouse, and providing irrigation for several years while they established a root system. In desert ecosystems such as Monument Valley, revegetation success generally increases in direct proportion to the amount of irrigation provided. In general, direct seeding cannot be relied on in areas receiving less than 250 mm of rainfall per year. (The Monument Valley site receives less than 200 mm per year.) Options for remediating the groundwater were even more expensive. Conventional treatment methods required that the water be pumped to the surface and passed through a water treatment plant, using either *deionization, evaporation,* or *distillation* to separate nitrates from the water. This process is known as "pump and treat" (see the "Pump and Treat" section in Chapter 18).

Further analysis, however, showed that the revegetation component and the plume remediation component could be integrated into a single solution.

By using the plume water as a source of irrigation water for native plants and forages crop that could be planted over the bare areas of the site, the nitrate in the plume water would serve as a fertilizer for the plants to be consumed by the grazing livestock. Using plants to solve environmental problems is called *phytoremediation* (see the "Phytoremediation" section in Chapter 18), and this became the preferred alternative at the Monument Valley UMTRA site because it provided a combined solution to two problems, and did not require construction and operation of an expensive water treatment plant. As of this writing, the phytoremediation option is undergoing review by stakeholders, and a demonstration phytoremediation plot has been established on site.

The planning process undertaken at the Monument Valley UMTRA site illustrates the numerous checks and balances built into a restoration plan. Many different disciplines are involved, and everyone with a possible stake in the outcome of restoration is brought into the process. In most cases a restoration strategy is not adopted until it achieves *consensus* among stakeholders as the best possible choice. Many years of study and planning may precede actual restoration. This is acceptable, as long as there is no imminent hazard, because land restoration is expensive and careful planning may help prevent costly mistakes.

APPROACHES TO ECOSYSTEM RESTORATION

The following sections provide some strategies toward the implementation of restoration goals and objectives; however, it must be remembered that each site will provide a unique set of characteristics and ecological challenges. As you read this section, refer to the case studies for Monument Valley and the Mission Copper Mine.

Natural Restoration

Natural restoration is essentially the process of allowing the ecosystem to heal itself without active management or

TABLE 19.1
Restoration Plan for the UMTRA Program

Phase	Description
1a)	Removal of contaminated material
b)	Containment of contaminated material
2a)	Restoration of damaged land
b)	Remediation of contaminated water

human interference. Essentially this is the same concept as "intrinsic bioremediation" (see Chapter 18). Depending on site-specific characteristics, natural restoration may not be a viable alternative. Natural ecosystems develop over long periods through the process of *ecological succession*. Think of a lava flow, such as those that still occur on the island of Hawaii on the slopes of Kilavea, burning through portions of the native rain forest. The lava cools quickly but lays barren for many years, too hostile an environment to support life. Eventually rain and wind erosion create tiny fissures in the lava where life can establish a foothold. Microorganisms, usually bacteria, are often the first forms of life to become established, followed by lichens. In fact, very few sites are microbiologically sterile. Microorganisms are a prerequisite for plant growth. Lichens and cyanobacteria are the next colonists to establish themselves on the flow. Continual breakdown of the parent lava by acids secreted by the lichens produces a thin layer of soil in which the first higher plants can root; first small ferns, followed by grasses and shrubs as the fissures widen due to the action of the plant roots. Cyanobacteria fix nitrogen, which supports the plant life. Each stage of succession conditions the lava substrate to favor the next stage, and finally the rain forest is restored. This process may take many hundreds or even thousands of years.

The process of succession is slower and less predictable in harsh environments than it is in tropical rain forests. Unfortunately, many of our damaged lands are in severe environments, such as arctic tundra or deserts. These environments pose a special challenge for restoration ecologists. Left alone, these lands often deteriorate further, through wind and water erosion, rather than gradually improving. These lands are most in need of human intervention to make them productive. The challenge for restoration ecologists is to finds ways to encourage the process of succession so it is predictable and takes place in a reasonable time span.

PASSIVE ECOLOGICAL RESTORATION

Passive ecological restoration projects primarily apply to land whose ecosystem is still functionally intact, but which has lost vegetative cover and biodiversity from such activities as overgrazing or habitat fragmentation. The implicit goals of these projects are to reduce or eliminate the causes of degradation, while encouraging the growth of indigenous plants to increase the productivity of the area in a sustainable way. In general, minimal soil preparation is needed, soil amendments and irrigation are not required, and seeds are simply broadcast in a designated area. Restoration under these conditions is usually coupled with land conservation objectives.

In many cases, however, improving soil surface conditions and increasing infiltration rates are necessary to "jump start" the restoration process. Table 19.2 outlines some common methods of soil preparation that improve surface conditions and increase infiltration rates. An added benefit of improving soil surface conditions is that the roughness on the surface provides "safe sites" for seeds and captures airborne and waterborne organic material. The focus here is to optimize existing conditions for successful germination and plant growth at minimal cost.

In semiarid and arid areas of the world where water is the limiting factor to successful restoration, the success of dry seeding methods generally declines as aridity increases. Some consider irrigation essential in areas that receive less than 250 mm of annual precipitation, but the need for irrigation, the amount, and application mode have been debated. However, numerous low-cost techniques can be used to capture and retain precipitation where it falls. The simplest and cheapest methods involve placing logs, rocks, or mulch on bare ground to capture moisture, nutrients, seeds, and soil from the surrounding area. Contour furrows, pits, and small depressions in surface soils play the same role in capturing essential elements for plant establishment. Figure 19.5 illustrates a furrow in an abandoned saline cotton field where plant establishment has occurred naturally. In the Sonoran, Mojave, and Chihuahuan deserts of the southwestern United States and Mexico, researchers have shown that "islands of fertility" exist under the canopies of shrubs, whereas the intershrub spaces show little biotic activity or nutrient retention (Schlesinger and Pilmanis, 1998). Beneath the canopies of shrubs and trees in these islands of fertility, nutrient levels were elevated, leaf litter had accumulated, and eolian soil had been captured and retained in mounds. Although this process can take decades, it could be surmised that the initial germination of the shrubs or trees occurred in small depressions in the soil surface where sufficient moisture occurred to ensure plant establishment. Over time, conditions beneath and adjacent to the plant canopies changed to allow recruitment of seeds from the local seed bank, thus increasing the base biodiversity of the area. These scattered plant communities are also found to have healthy populations of soil microbes, as well as attracting a variety of insects, reptiles, small mammals, and birds. In arid areas it is these islands of fertility that need to be recreated as part of any restoration project.

TABLE 19.2
Soil Preparation Methods

Soil Preparation	Advantages	Disadvantages	Cost
Ripping to 1 meter	Improves infiltration by breaking up compacted soil.	Soil horizons are often inverted. In some soils toxic compounds or salts could be deposited on the surface.	Moderate
Chiseling to 15 cm	Roughens surface to soil erosion and improves infiltration. Soil horizons undisturbed.	Soil surface is disturbed and provides good sites for seeds of invasive plants.	Low
Disking	Breaks the upper 2–3 cm of soil crust to improve infiltration.	Soil surface is disturbed and provides good seed sites for invasive plants.	Low
Contour furrows	Minimal surface disturbance. Acts as catchment for airborne nutrients, seeds, and organic material.	Surface compaction by machinery.	Low

FIGURE 19.5 A naturally vegetated farm furrow on an abandoned field in Maricopa County, Ariz.

ACTIVE ECOLOGICAL RESTORATION

Where the abiotic and biotic functions of an ecosystem have fallen apart, the cost of restoration will rise in proportion to the damage incurred. Active ecological restoration projects can be complex, costly endeavors, as illustrated by the aforementioned Monument Valley, Ariz., case study. However, many other sites fall under this category because they require more intervention than simply seeding designated areas. Anthropogenic damage to an area can alter the biogeochemical function of an ecosystem, and in many cases it is necessary to reintegrate the damaged land with the surrounding landscape; particularly with regard to the hydrological cycle. For example, farmers will level an area for new fields, irrigation canals or ditches will be constructed, and farm roads will likely circumscribe the property. In the process of adding new cultivated fields, the surrounding wildland has become fragmented, surface water has most likely been redirected, and topographical features flattened. Once this land is abandoned, it becomes a candidate for restoration. It will be necessary to reconnect this land to the surrounding wildland, restore the physiochemical functions to the soils, and seed or plant the area to maximize plant establishment and growth. These projects thus become active and generally high cost, because they involve other procedures in addition to simple seeding. Table 19.3 shows some of the methods employed for these higher cost restoration projects. The restoration goals and objectives developed for a project will specify the degree of work necessary to achieve the desired goals.

In general, the methods associated with active restoration fall into several categories: (1) landscape integration; (2) soil surface conditioning; (3) soil amendments; (4) water harvesting and irrigation; and (5) seeding and planting techniques. Landscape integration techniques will not be addressed in this chapter, and soil surface conditioning was covered in the "Passive Ecological Restoration" section. A comprehensive soil analysis of the site will determine the type of amendments necessary to

TABLE 19.3
Potential Procedures for Active Ecological Restoration

Parameter	Procedure
Landscape integration	Removal of irrigation canals, roads
	Reconnect natural drainage features
	Construction of berms, swales, and gabions to redirect runoff and control erosion
	Landshaping to naturalize site
Soil	Grading to enhance soil stability, reduce erosion, and enhance water harvesting
	Creation of microcatchments for water harvesting
	Ripping to diminish compaction
	Tilling to incorporate soil amendments
	Mulching to enhance water retention
	Fertilizer additions to encourage plant growth
	Amendments to adjust soil pH
Plant	Mulching to enhance germination of direct seeding
	Transplants to enhance establishment
	Hydroseeding
	Mycorrhizal inoculations
Other	Supplemental irrigation
	Wire cages to protect against wildlife

restore a favorable chemical balance to the soil. In arid and semiarid areas of the world, a build-up of salts, usually sodium, is the major barrier to plant establishment, and reclaiming this soil is generally a lengthy process. Normally this requires the addition of sufficient water to leach the salts below the root-zone. Since water is scarce in these areas, some of the best strategies involve harvesting rain and using salt-tolerant plants to create islands of fertility. Water harvesting techniques include constructing swales and gabions with contoured berms to catch and disperse water over a large area and redirecting rainfall and sheet-flow to small basins called microcatchments. Microcatchments have been used extensively around the world, with early reports dating back to the nineteenth century where olive trees were grown in Tunisia (Pacey and Cullis, 1986). A microcatchment is a small depression in the soil, often delineated with a small berm, into which water flows during a rainstorm. The size of the micro-catchment depends on the porosity of the soil and rainfall characteristics (amount, intensity, and distribution). However, for the purposes of ecological restoration, $100 \, m^2$ is adequate for planting several trees and shrubs.

At sites that can provide water, irrigation can dramatically increase germination rates and survival of transplants. In a study at Tuba City, Ariz. (rainfall approximately $200 \, mm \, yr^{-1}$), transplanting and deep-irrigation of saltbush shrubs over the first growing season resulted in a 90% survival rate and a sixfold increase in plant volume after 3 years. Irrigation methods can include standard commercial practices, such as center-pivot sprinklers, drip irrigation, microirrigation, and siphon tubes from canals. Irrigation is only used for seed germination and plant establishment, since one of the objectives of any restoration project is sustainability over time, and it is doubtful that continuous irrigation can be seen as meeting this criterion.

Plant selection is generally dictated by site conditions, native plants in the area, and availability of seeds or transplants. Popular opinion stresses the use of native plants; however, there are some circumstances where soil conditions are not conducive to their establishment and growth. In these situations plants will be chosen that are most likely to survive. For example, in highly salinized soils, native plants cannot tolerate the high salts in the soil, but a plant such as four-wing saltbush (*Atriplex canescens*) will not only thrive, but have the ability to sequester salts in specialized vacuoles in its leaves, thereby removing salt from the ground. Table 19.4 includes different methods of seeding sites, ranging from broadcasting to physically planting seedlings.

Many other specialized restoration methods have been developed for specific habitat types. Constructed wet-lands have become a popular means for polishing municipal sewage effluent and for providing habitat for waterfowl and other birds. Often, however, the constructed wetlands do not serve as well as natural wetlands in supporting diverse plant and animal communities. In the western United States, considerable attention has been paid to restoring riparian zones, by removing the invasive plant, salt cedar, and replacing it with native trees, such as cottonwood and willow. Often these projects do not succeed because the hydrological conditions of the river have been so altered that they no longer favor the native species (Stromberg, 2001).

ECOLOGICAL RESTORATION USING ORGANIC AMENDMENT

Highly disturbed sites often result in surface soils being devoid of organic matter. This can occur from a variety of human activities including: strip mining (where the surface top soil is removed), mine tailing storage (crushed and processed mineral ores deposited over existing topsoil), or from soil erosion. In all cases where organic matter is sparse or entirely absent, there are extremely low microbial populations, and it is common for these sites to have extreme pH, low permeabilities, and high soluble metal concentrations. These conditions are not suitable for sustainable plant growth, and they generally require some form of organic amendment to enhance the restoration process. Problems that occur due to low organic materials are shown in Table 19.5, and common sources of organic materials used to enhance ecosystem restoration are illustrated in Table 19.6.

The concept of organic amendments to enhance plant growth has been utilized for centuries as in the use of "night soil" (human feces and urine) to fertilize agricultural land. The use of raw waste material can spread disease (see Chapter 17), but in the United States, the solid material (*biosolids*) left after treatment of municipal sewage is further refined to eliminate potential pathogens before being applied to agricultural land. The long-term implications of such practices have been well documented (Sloan *et al.*, 1997; Artiola and Pepper, 1992). Biosolids have also been successfully applied to areas containing mining tailings or smelter waste that contain high or even phytotoxic levels of heavy metals (Li and Chaney, 1998). Biosolids that have undergone lime stabilization are particularly useful for such restoration because the increase in pH reduces the bioavailability of metals to plants. Biosolids and composts have also been used to restore diverse ecosystems, such as mountain slopes in the Washington Cascades or stabilize sand dunes in south-eastern Colorado.

TABLE 19.4
Comparison of Seeding Methods

General Method	Specific Method	Characteristics	Efficacy	Benefits	Drawbacks
Seeding	Broadcasting	Native seeds spread over an area using farm machinery or airplane. Seeds may or may not be covered with soil.	Low due to seed loss by predation, wind, and water.	Inexpensive.	Seeding rate two times that of other methods.
	Imprinting	Seeds pressed into furrowed soil depressions.	Plant establishment successful over time, particularly in arid and semiarid regions with no supplemental irrigation.	Relatively inexpensive. Not as damaging to soil as other methods.	Cannot control depth of seeding. Poor in very sandy soils.
	Drilling	Seeds are dropped into holes or furrows and covered with soil.	Successful with supplemental irrigation or adequate rainfall.	Minimal soil disturbance.	Not feasible with severe topographic limitations and rocky terrain.
	Hydroseeding	A mixture of water, seed, mulch, fertilizer and a tackifier sprayed onto soil surface.	Three to ten times more effective than free broadcasting of seed.	Erosion control for slopes. Not as damaging to existing vegetation as other methods.	Very expensive. Only suitable for areas close to existing rights of way.
	Pelleting	Seed is coated with powdered soil, clay, or glue and silica sand. Seed can be broadcast using farm machinery or airplane.	More effective in regions with high precipitation.	Less seed used than in broadcasting. Seed can be stored for a long time in dormancy.	Expensive and time-consuming process.
Planting	General evaluation	Trees and shrubs planted individually.	Effective for introducing late successional species of trees and shrubs.	Does not disturb existing vegetation.	More expensive than seeding. Inputs of labor and materials higher. Generally requires supplemental irrigation in dryland areas.

TABLE 19.5
Problems Related to Low Organic Matter Surface Materials

Parameter	Problem
Poor aggregation of primary particles	Compaction
	Low infiltration rates
	Low water holding capacity
	Limited aeration
Low nutrient status	Infertile soils
	Low microbial populations
Extreme pH	Affects chemical and biological properties
High metal concentrations	Toxicity

TABLE 19.6
Sources of Organic Materials for Ecosystem Restoration

Human and Animal Wastes	Industrial Wastes
Animal manures	Paper mill sludges
Biosolids	Sawdust
Composted wastes	Wood chips

In all cases of organic amendment added to poor soil, the critical parameter appears to be the magnitude of organic material applied. This is particularly important in desert ecosystems where high temperatures result in rapid decomposition and mineralization of organic compounds and materials. If insufficient organic matter is added to a disturbed site, the beneficial effect is not maintained for a sufficiently long time to allow stable revegetation to occur. The case study detailing the revegetation of mine tailings at the Mission Mine in Arizona illustrates the successful use of biosolids for site restoration (Case Study 19.2).

CASE STUDY 19.2 *Mission Copper Mine, Arizona*
Reclamation and Revegetation of Mine Tailings
Using Biosolid Amendment

In the United States, mining is a large industry that provides valuable raw material and creates economic benefit for local communities. However, the potential environmental damage incurred from this industry ranges from unsightly mine tailings to the leaching of toxic elements into nearby waterways and aquifers. In addition, wind-blown tailings can result in air pollution. Removal of the original vegetation, soil, and bedrock exposes the valuable mineral veins. The mineral ore is then processed to remove copper. Finally, the crushed rock is redeposited on land as a thick slurry. Typically, tailings piles are 30- to 40-m thick.

The physiochemical characteristics of mine tailings are totally unlike the displaced top soil that once supported vegetation in any given area. By removing and crushing bedrock from the mines and placing it on the surface, minerals will oxidize when exposed to the atmosphere. For example, pyrite (FeS_2) common around coal mines oxidizes to sulfuric acid (H_2SO_4) and iron oxide ($Fe[OH]_3$). Acid mine drainage (H_2SO_4), the leachate from tailings, can then contaminate surface and groundwaters, in addition to increasing the solubility of toxic metals. Mining tailings are not the ideal medium on which to grow plants. The crushed rock consists of large and small fragments with large void spaces in between. There is no organic material present, the cation exchange capacity (CEC) is very low, the water holding capacity of the material is poor to nonexistent, and there are few macronutrients (N.P.K.) available for the plants. Soil biota, in the form of bacteria and fungi, are present in low numbers, and finally the pH is usually low, which increases the likelihood that toxic metals are available to be taken up by the plants. The goals of reclaiming mine tailings therefore have to include the application of materials to amend the crushed rock substrate, and provide an adequate environment for plant growth. One potential solution is the use of biosolids.

In 1994 the Arizona Mined Land Reclamation Act was passed, which required reclamation of all mining disturbances on private land to a predetermined postmining use. In 1996 the Arizona Department of Environmental Quality (ADEQ) adopted new rules allowing for the use of biosolids during reclamation. The Arizona Mining Association (AMA) has estimated that there are 33,000 acres of active mine sites that can be reclaimed with the use of biosolids. In southern Arizona, biosolids have been used for almost two decades for land application on agricultural land and the commercial growth of cotton. However, due to large amounts of acreage being sold and retired from agricultural usage, there is currently a shortage of land for the application of biosolids. Therefore the concept that is emerging is that of utilizing one waste, namely biosolids, to reclaim another waste material, namely mine tailings. The issue, of course, is whether or not this can be done in an environmentally sound manner via a process that is economically viable.

Current Arizona regulations limit the amount of biosolids that can be used to reclaim sites. The restrictions are due to concerns over potential nitrogen and heavy metal leaching, both of which could impact underground aquifers and therefore compromise human health and welfare. However, greenhouse tests have shown that large amounts of biosolids are necessary to effectively reclaim mine tailings. Specifically, up to 275 dry tons per hectare may be necessary to promote active vegetation of mine tailings (Bengson, 2000). Thompson and Rogers (1999) conducted greenhouse tests utilizing 67 metric tons per hectare (dry weight basis) of biosolids on three different types of mine tailings, ranging from acidic to neutral. The biosolids are effective in promoting vegetative growth and increasing groundcover. In these studies there was little evidence of significant nitrate below 30 cm, nor was there any evidence of heavy metal increases due to biosolid application. Current ADEQ regulations limit the lifetime loading rate of biosolid applications to mine tailings to 400 dry tons per hectare. However, this amount can be applied as one application. Here we describe a case study illustrating the use of biosolids to restore and stabilize mine tailings.

OVERALL OBJECTIVE: To evaluate the efficacy of dried biosolids as a mine tailing amendment to enhance site stabilization and revegetation.

SPECIFIC OBJECTIVES
1. To evaluate the benefits of land application of dried biosolids to mine tailings, with respect to reclamation and stabilization.
2. To evaluate the hazards of metals and nitrate associated with the application of dried biosolids to mine tailings. Note that pathogens were not monitored because of the use of "exceptional quality" biosolids for the project. Exceptional quality biosolids normally contain very low concentrations of pathogens.

CASE STUDY 19.2 *(Continued)*

Experimental Plan

Study Site. A 2-hectare copper mine tailing plot located near Mission Mine, south of Tucson, was designatedfor this study. Biosolids were applied at a rate of about 220 dry tons per hectare across the site in December 1998, and then seeded with a variety of desert adaptive grass species including: oats, barley, Lehman lovegrass, bufflegrass, and bermuda grass. Supplemental irrigation was not used.

Results

Soil Microbial Response to Biosolids

Pure mine tailings contain virtually no organic matter and very low bacterial populations of approximately 10^3 CFU per gram of tailings, see "The Most Probable Number (MPN) Test" section in Chapter 17 for methods of determination of bacterial populations. A large population of heterotrophic bacteria is essential for plant growth and revegetation, and therefore monitoring soil microbial populations gives an insight into the probability of revegetation success. Biosolids routinely contain very high concentrations of organic matter, including the macroelements carbon and nitrogen, which are essential for promoting microbial growth and metabolism. Following biosolid amendment of the mine tailings, heterotrophic bacterial populations increased at the surface to approximately 10^7 CFU per gram (Table 19.7). Bacteria decreased with increasing depth from the surface, indicating the influence of the biosolid surface amendment on bacterial growth. Table 19.7 also shows that the microbial populations have at this point been stable for 33 months. Overall, the microbial data show the success of biosolid amendment in changing mine tailings into a true soil-like material.

Physical Stabilization: One of the main objectives in reclaiming mine tailings is *erosion control*. This is generally best accomplished through a revegetation program because the root structures of the plants help to hold soil particles in place. In this experiment the application of biosolids and the subsequent broadcast of grass seeds were the primary activities to promote site stabilization. In the desert Southwest, high summer temperatures and limited rainfall are normal; however, despite these extreme conditions, grasses have become established on these tailings. Table 19.8 shows the results of vegetation transect surveys conducted on this site 14 months, 21 months, and 33 months after initial seeding. Note the intrinsic variability in the transect data, and the need for multiple transects to be taken to

determine a realistic average. The vegetation cover increased from 18% at the 14-month survey to 78.2% after 33 months. At 14 months the predominant plant species were bermuda grass *(Cynodon dactylon)* and the invasive weed, Russian thistle or tumbleweed *(Salsola iberica)*, but by the 33rd month bufflegrass *(Pennisteum ciliare)* and Lehman lovegrass *(Erangrostis lehmanniana)* had replaced the Russian thistle. Figures 19.6 to 19.8 show the progressive increase of vegetation on this site over time. In this case the use of biosolids for enhanced revegetation and stabilization of mine tailings would be considered a success.

Evaluation of Potential Hazards—Soil Metal Concentrations: At Site 1, soil nitrate (Table 19.9) and total organic carbon (TOC) (Table 19.10) are very high at the surface, but decrease to the levels found in pure mine tailings at lower depths. The fact that nitrate and TOC concentrations are correlated is important because it creates substrate and terminal electron acceptor concentrations suitable for denitrification. Data presented in Table 19.9 show the nitrate concentrations from June 2000 to July 2001. Nitrate concentrations increased during the monsoon rainy season of 2000, most likely due to enhanced ammonification and subsequent nitrification. However, within the soil profile, nitrate concentrations decreased with depth. By the winter and spring of 2001, the nitrate concentrations at all soil depths had decreased. There was no evidence of the leaching of nitrate because concentrations at the 90 to 120 cm depth were always minimal. Therefore the most likely explanation for decreased nitrates within the soil profile is the process of denitrification. Soil nitrate concentrations became extremely high at both sites in the summer of 2001, again, most likely due to nitrogen mineralization and seasonal nitrogen cycling. Specifically during the warmer summer months, rainfall events appear to trigger microbial mineralization of nitrogen as nitrate. Double-digit values of nitrate at the 120 to 150 cm depth illustrate that there is the potential for some nitrate leaching during some portions of the year.

The application of biosolids to a project site brings some concern about the introduction of heavy metals to the environment. Data from this study show that metal concentrations are fairly consistent with soil depth (Table 19.11), indicating that the tailings are the major source of metals, not the biosolids. Further evidence of this is shown by the high molybdenum and copper valves typical of mine tailings. At this site there is little evidence of metals leaching through the soil profile. Additional data were collected on the

(Continued)

CASE STUDY 19.2 *(Continued)*

concentration of molybdenum, copper, and zinc in three plants on the site: Russian thistle, salt cedar, and bermuda grass. Table 19.12 shows the uptake of metals by these plants were extremely high and soil sampled beneath vegetation revealed a decrease in soil metal concentrations.

Conclusion

This study on the application of biosolids to mining tailing at the Mission Mine in Arizona shows that soil stabilization has been encouraged through revegetation techniques, and that the leaching of nitrate and heavy metals to important water resources has not been observed. This case study gives an indication of the extensive monitoring that is necessary to understand the restoration process, and the necessary duration of the monitoring process. With careful attention paid to subsurface geological and hydrological features at other sites, the application of biosolids can be a feasible restoration strategy.

TABLE 19.7

Plate Counts of Heterotrophic Bacteria at Mission Mine Project Site

| Sample Date | Depth of Sample (cm) | | | | |
	0–30	30–60	60–90	90–120	120–150
	CFU g^{-1}				
06/26/00	1.01×10^7	1.72×10^6	3.11×10^5	1.54×10^5	ND
09/11/00	3.16×10^6	4.44×10^5	8.12×10^5	3.54×10^4	ND
01/22/00	2.74×10^7	2.05×10^7	5.56×10^5	2.31×10^5	ND
03/26/00	3.76×10^7	2.25×10^6	1.99×10^4	7.83×10^4	5.48×10^4
06/11/01	7.74×10^6	6.86×10^5	6.72×10^4	1.22×10^5	6.36×10^4
10/29/01	1.30×10^7	8.36×10^5	2.40×10^5	1.41×10^5	1.19×10^5
09/10/01	1.45×10^6	5.83×10^5	4.10×10^4	1.42×10^5	1.40×10^5

ND, Not done.

TABLE 19.8

Vegetation Transects at the Mission Mine Project Site

	Basal Cover (%)	Crown Cover (%)	Total Cover (%)	Rock (%)	Litter (%)	Bare (%)
	02/23/00 (14 months)					
T-1	15	28	43	3	0	54
T-2	14	16	30	3	8	59
T-3	7	1	8	0	1	91
T-4	1	8	9	2	4	85
T-5	0	0	0	0	5	95
Average	**7.4**	**10.6**	**18**	**1.6**	**3.6**	**76.8**
	09/25/00 (21 months)					
T-1	13	39	52	5	6	37
T-2	11	20	31	6	6	57
T-3	6	45	51	2	1	46
T-4	13	48	61	3	2	34
T-5	1	52	53	0	4	43
Average	**8.8**	**40.8**	**49.6**	**3.2**	**3.8**	**43.4**
	09/24/01 (33 months)					
T-1	4	66	70	2	4	24
T-2	6	62	68	0	6	26
T-3	4	73	77	0	4	19
T-4	0	84	84	0	1	15
T-5	0	92	92	0	0	8
Average	**2.8**	**75.4**	**78.2**	**0.4**	**3**	**18.4**

FIGURE 19.6 Mine tailings before biosolid amendment.

FIGURE 19.7 Mine tailings after biosolid application.

FIGURE 19.8   Mine tailings 3 years after biosolid application.

TABLE 19.9

Nitrate Concentrations at the Mission Mine Project Site from June 2000 to July 2001

Sample Date	Depth of Sample (cm)				
	0–30	30–60	60–90	90–120	120–150
	mg kg^{-1}				
06/26/00	650	250	40	5	ND
07/10/00	1520	120	60	5	ND
07/26/00	1030	200	170	70	ND
02/05/01	480	250	330	150	60
03/26/01	190	40	40	15	5
06/11/01	2350	310	140	260	110
07/13/01	2350	590	220	205	50

ND, Not done.

TABLE 19.10

Total Organic Carbon (TOC)[a] at the Mission Mine from February 2001 to July 2001

Sample Date	Depth of Sample (cm)				
	0–30	30–60	60–90	90–120	120–150
	%				
02/05/01	1.4	0.2	0.3	0.4	0.3
03/26/01	1.0	0.1	0.2	0.2	0.1
06/11/01	1.6	0.2	0.1	2.0	0.1
07/13/01	1.6	0.4	0.1	0.2	0.2

[a]Mean values.

TABLE 19.11

Total Soil Metal Concentrations at the Mission Mine Project Site

Sample Depth (cm)	Total Metals[a] (mg kg^{-1})						
	Mo	Pb	As[b]	Cr	Zn	Cu	Ni
0–30[c]	68	23	<50	14	170	414	8.0
30–60	197	19	<50	19	111	1480	8.2
60–90	196	27	<50	15	168	2400	9.0
90–120	180	33	<50	15	130	1320	8.0
0–30	88	19	<34	12	154	247	<33.0

[a]Data are mean of samples collected on 06/26/00, 07/10/00, 07/26/00, and 06/11/01.
[b]Below detection limit.
[c]Biosolid amendment all within 0–30 foot depth.

TABLE 19.12

Total Metal Concentrations in Plant Tissue Samples at the Mission Mine Project Site

Plant Type	Total Metal (mg kg^{-1} dry weight basis)		
	Mo	**Zn**	**Cu**
Russian thistle	872	1230	35
Salt cedar	655	94	63
Bermuda grass	100	116	43

Samples taken 02/02/01.

QUESTIONS

1. Differentiate among: (a) rehabilitation; (b) revegetation; and (c) reclamation.

2. Based on your microbial expertise gained from Chapter 17, what soil microorganisms would sequentially be activated after land application of biosolids?

3. Identify the abiotic factors that influence the approach to ecosystem restoration on any particular project.

4. What factors primarily determine whether active or passive ecological restoration should be undertaken?

5. What is the influence of organic amendments on soil physical and chemical properties?

REFERENCES AND ADDITIONAL READING

Artiola, J.F., Pepper, I.L. (1992) Long-term influence of liquid sewage sludge on the organic carbon and nitrogen content of a furrow irrigated desert soil. *Biol. Fert. Soils* **14**, 30–36.

Barbour, M., Burk, J., Pitts, W. (1987) Terrestrial Plant Ecology, 2nd Edition. Benjamin/Cummins Publishing Co., Menlo Park, CA. 634 pp.

Bengson, S.A. (2000) Reclamation of copper tailings in Arizona utilizing biosolids. Mining, Forest and Land Restoration Symposium and Workshop, Golden Colorado, July 17–19 2000.

Bonham, C. (1989) Measurements for terrestrial vegetation. Wiley & Sons, Inc., New York. 338 pp.

Box, J. (1996) Setting Objectives and Defining Outputs for Ecological Restoration and Habitat Creation. *Restoration Ecology.* **4**, 427–432.

Glenn, E.P., Waugh, W.J., Moore, D., McKeon, C., Nelson, S.G. (2001) Revegetation of an abandoned uranium millsite on the Colorado Plateau, Arizona. *J. Environ. Qual.* **30**, 1154–1162.

Hobbs, R.J., Harris, J.A. (2001) Restoration Ecology: Repairing the Earth's Ecosystems in the New Millennium. *Restoration Ecology* **9**, 239–246.

Li, Y.M., Chaney, R.L. (1998) Phytostabilization of Zn smelter contaminated sites—The Palmerton Case pp. 197–210. *In* J. Vangronsveld and S.D. Cunningham (eds). Metal Contaminated Soils: *In situ* Inactivation and Phytorestoration. *Landes Bioscience*, Austin, TX.

Merriam-Webster, Inc. (1993) Collegiate Dictionary, 10th Edition. Springfield, MA.

Michener, W.K. (1997) Quantitatively Evaluating Restoration Experiments: Research Design, Statistical Analysis, and Data Management Considerations. *Restoration Ecology* **5**, 324–337.

Middleton, N., Thomas, D. (eds.). (1997) World Atlas of Desertification. Arnold Press, London. 182 pp.

Pacey, A., Cullis, A. (1986) Rainwater Harvesting: The Collection of Rainfall and Runoff in Rural Areas. Intermediate Technology Pub., London.

Pfadenhauer, J., Grootjans, A. (1999) Wetland restoration in Central Europe: aims and methods. *Appl. Veg. Sci.* **2**, 95–106.

Scharp, M. (2000) Sand dune stabilization and reclamation in South Eastern Colorado. Proc. Mining, Forest and Land Restoration Symposium, July 17–19, 2000.

Schlesinger, W. H., Pilmanis, A.M. (1998) Plant-soil interactions in deserts. *Biogeochem.* **42**, 169–187.

Sloan, J.J., Dowdy, R.H., Dolen, M.S., Linden, D.R. (1997) Long-term effects of biosolid applications on heavy metal bioavailability in agricultural soils. *J. Environ. Qual.* **26**, 966–974.

Stromberg, J. (2001) Restoration of riparian vegetation in the southwestern United States: importance of flow regime and fluvial dynamism. *J. Arid Environ.* **49**, 17–34.

Thompson, T. L., Rogers, M. (1999) Reclamation of acidic copper mine tailings using municipal biosolids. Research Report to The Arizona Department of Environmental Quality.

20

RISK ASSESSMENT AND ENVIRONMENTAL REGULATIONS

C.P. GERBA

Environmental monitoring data is often used in risk assessment to create environmental regulations. However, the task of interpreting data is critical in making decisions about potential health risks and corrective actions. Risk assessment is a process that has been formalized to estimate the risk of adverse health effects caused by exposure to harmful chemicals and microorganisms. The goal is to express risk caused by exposure to a contaminant in terms of probability of illness or mortality. In this format such information can be better utilized by decision makers to determine the magnitude of the problem and weigh the costs and benefits of prevention or corrective action. The purpose of this chapter is to provide a general background on the topic of risk analysis, and how it can be used in the problem-solving process.

RISK ASSESSMENT

Risk, which is common to all life, is inherent in everyday human existence. Risk assessment or analysis, however, means different things to different people. For example, Wall Street analysts assess financial risks, and insurance companies calculate actuarial risks, whereas regulatory agencies estimate the risks of fatalities from nuclear plant accidents, the incidence of cancer from industrial emissions, and habitat loss associated with increases in human populations. All these seemingly disparate activities have in common the concept of a measurable phenomenon called risk that can be expressed in terms of probability. Thus we can define *risk assessment* as the process of estimating both the probability that an event will occur and the probable magnitude of its adverse effects—economic, health safety–related, or ecological—

over a specified period. For example, we might determine the probability that an atomic reactor will fail, and the probable effect of its sudden release of radioactive contents on the immediate area, in terms of injuries and property loss over a period of days. In addition, we might estimate the probable incidence of cancer in the community where the radioactivity was released over a period of years. Or, in yet another type of risk assessment, we might calculate the health risks associated with the presence of pathogens in drinking water or pesticides in food.

The risk assessment process consists of four basic steps:

- *Hazard identification*—Defining the hazard and nature of the harm; for example, identifying a chemical contaminant, such as lead or carbon tetrachloride, and documenting its toxic effects on human beings.
- *Exposure assessment*—Determining the concentration of a contaminating agent in the environment and estimating its rate of intake in target organisms; for example, finding the concentration of aflatoxin (a fungal toxin) in peanut butter and determining the dose an "average" person would receive.
- *Dose–response assessment*—Quantitating the adverse effects arising from exposure to a hazardous agent based on the degree of exposure. This assessment is usually expressed mathematically as a plot showing a response (i.e., mortality) in living organisms to increasing doses of the agent.
- *Risk characterization*—Estimating the potential impact of a hazard based on the severity of its effects and the amount of exposure.

Once the risks are characterized, various regulatory options are evaluated in a process called *risk management*, which includes consideration of social, political, and economic issues, as well as the engineering problems inherent in a proposed solution. One important component of risk management is *risk communication*, which is the interactive process of information and opinion exchange among individuals, groups, and institutions. Risk communication includes the transfer of risk information from expert to nonexpert audiences. To be effective, risk communication must provide a forum for balanced discussions of the nature of the risk, lending a perspective that allows the benefits of reducing the risk to be weighed against the costs.

In the United States the passage of federal and state laws to protect public health and the environment has expanded the application of risk assessment. Major federal agencies that routinely use risk analysis include the Food and Drug Administration (FDA), the Environmental Protection Agency (EPA), and the Occupational Safety and Health Administration (OSHA). Together with state agencies, these regulatory agencies use risk assessment in a variety of situations:

- Setting standards for concentrations of toxic chemicals or pathogenic microorganisms in water or food.
- Conducting baseline analyses of contaminated sites or facilities to determine the need for remedial action and the extent of cleanup required.
- Performing cost/benefit analyses of contaminated-site cleanup or treatment options (including treatment processes to reduce exposure to pathogens).
- Developing cleanup goals for contaminants for which no federal or state authorities have promulgated numerical standards; evaluating acceptable variance from promulgated standards and guidelines (e.g., approving alternative concentration limits).
- Constructing "what if" scenarios to compare the potential impact of remedial or treatment alternatives and to set priorities for corrective action.
- Evaluating existing and new technologies for effective prevention, control, or mitigation of hazards and risks.
- Articulating community and public health concerns and developing consistent public health expectations among different localities.

Risk assessment provides an effective framework for determining the relative urgency of problems and the allocation of resources to reduce risks. Using the results of risk analyses, we can target prevention, remediation, or control efforts toward areas, sources, or situations in which the greatest risk reductions can be achieved with the resources available. However, risk assessment is not an absolute procedure carried out in a vacuum; rather, it is an evaluative, multifaceted, comparative process. Thus to evaluate risk, we must inevitably compare one risk to a host of others. In fact, the comparison of potential risks associated with several problems or issues has developed into a subset of risk assessment called *comparative risk assessment*. Some commonplace risks are shown in Table 20.1. Here we see, for example, that risks from chemical exposure are fairly small relative to those associated with driving a car or smoking cigarettes.

Comparing different risks allows us to comprehend the uncommon magnitudes involved, and to understand the level, or magnitude, of risk associated with a particular hazard. But comparison with other risks cannot, itself, establish the *acceptability* of a risk. Thus the fact that the chance of death from a previously unknown risk is about the same as that from a known risk does not necessarily imply that the two risks are equally acceptable. Generally, comparing risks along a single dimension is not helpful when the risks are widely perceived as qualitatively different. Rather, we must take account of certain qualitative factors that affect risk perception and evaluation when

TABLE 20.1

Examples of Some Commonplace Risks in the United States

Risk	Lifetime Risk of Mortality
Cancer from cigarette smoking (one pack per day)	1:4
Death in a motor vehicle accident	2:100 (1:50)
Homicide	1:100
Home accident deaths	1:100
Cancer from exposure to radon in homes	3:1000
Exposure to the pesticide aflatoxin in peanut butter	6:10,000
Diarrhea from rotavirus	1:10,000
Exposure to typical EPA maximum chemical contaminant levels	1:10,000–1:10,000,000

Based on data in Wilson and Crouch (1987) and Gerba and Rose (1992).

selecting risks to be compared. We must also understand the underlying premise that *voluntary risk is always more acceptable than involuntary risk*. For example, the same people who cheerfully drive their cars every day—thus incurring a 2:100 lifetime risk of death by automobile—may refuse to accept the 6:10,000 involuntary risk of eating peanut butter contaminated with aflatoxin.

Understandably, regulatory authorities are reluctant to be explicit about an "acceptable" risk. (How much aflatoxin would you consider acceptable in *your* peanut butter and jelly sandwich?) But it is generally agreed that a lifetime risk on the order of one in a million (or in the range of 10^{-6} to 10^{-4}) is trivial enough to be acceptable for the general public. Although the origins and precise meaning of a one-in-a-million acceptable risk remain obscure, its impact on product choices, operations, and costs is very real—running into hundreds of billions of dollars in hazardous waste site cleanup decisions alone, for example. The levels of acceptable risk can vary within this range. Levels of risk at the higher end of the range (10^{-4} rather than 10^{-6}) may be acceptable if just a few people are exposed, rather than the entire populace. For example, workers dealing with chemical manufacturing are allowed to tolerate higher levels of risk than the public at large. These higher levels are justified because workers tend to be a relatively homogeneous, healthy group, and because employment is voluntary.

PROCESS OF RISK ASSESSMENT

HAZARD IDENTIFICATION

The first step in risk assessment is to determine the nature of the hazard. For pollution-related problems, the hazard in question is usually a specific chemical, a physical agent (e.g., irradiation), or a microorganism identified with a specific illness or disease. Thus the hazard identification component of a pollution risk assessment consists of a review of all relevant biological and chemical information bearing on whether or not an agent poses a specific threat.

Clinical studies of disease can be used to identify very large risks (between 1:10 and 1:100). Most epidemiological studies can detect risks down to 1:1000, and very large epidemiological studies can examine risks in the 1:10,000 range. However, risks smaller than 1:10,000 cannot be studied with much certainty using epidemiological approaches. Since regulatory policy objectives generally strive to limit risks below 1:100,000 for life-threatening diseases, such as cancer, these lower risks are often estimated by extrapolating from the effects of high doses given to animals. However, this approach is not without controversy, see the "Dose–Response Assessment" section.

EXPOSURE ASSESSMENT

Exposure assessment is the process of measuring or estimating the intensity, frequency, and duration of human exposures to an environmental agent. Exposure to contaminants can occur via inhalation, ingestion of water or food, or absorption through the skin upon dermal contact. Contaminant sources, release mechanisms, transport, and transformation characteristics (see Chapter 16) are all important aspects of exposure assessment, as are the nature, location, and activity patterns of the exposed population.

An *exposure pathway* is the course that a hazardous agent takes from a source to a receptor (e.g., human being or animal) via environmental carriers or media—generally, air (volatile compounds, particulates) or water (soluble compounds) (Table 20.2). An exception is electromagnetic radiation, which needs no medium. The *exposure route*, or intake pathway, is the mechanism by which the transfer occurs—usually by inhalation, ingestion, and/or dermal contact. Direct contact can result in a local effect at the point of entry and/or in a systemic effect.

The quantitation of exposure, intake, or potential dose can involve equations with three sets of variables:

- concentrations of chemicals or microbes in the media
- exposure rates (magnitude, frequency, duration)
- quantified biological characteristics of receptors (e.g., body weight, absorption capacity for chemicals; level of immunity to microbial pathogens)

TABLE 20.2

EPA Standard Default Exposure Factors

Land Use	Exposure Pathway	Daily Intake	Exposure Frequency (days yr^{-1})	Exposure Duration (yr^{-1})
Residential	Ingestion of potable water	2 L day^{-1}	350	30
	Ingestion of soil and dust	200 mg (child)	350	6
		100 mg (adult)		24
	Inhalation of contaminants	20 m^3 (total)	350	30
		15 m^3 (indoor)		
Industrial and commercial	Ingestion of potable water	1 L	250	25
	Ingestion of soil and dust	50 mg	250	25
	Inhalation of contaminants	20 m^3 (workday)	250	25
Agricultural	Consumption of homegrown produce	42 g (fruit)	350	30
		80 g (vegetable)		
Recreational	Consumption of locally caught fish	54 g	350	30

Modified from Kolluru (1993).

Exposure concentrations are derived from measured, monitored, and/or modeled data. Ideally, exposure concentrations should be measured at the points of contact between the environmental media and current or potential receptors. It is usually possible to identify potential receptors and exposure points from field observations and other information. However, it is seldom possible to anticipate all potential exposure points and measure all environmental concentrations under all conditions. In practice, a combination of monitoring and modeling data, together with a great deal of professional judgment, is required to estimate exposure concentrations.

To assess exposure rates via different exposure pathways, we have to consider and weigh many factors. For example, in estimating exposure to a substance via drinking water, we first have to determine the average daily consumption of that water. But this isn't as easy as it sounds. Studies have shown that daily fluid intake varies greatly from individual to individual. Moreover, tapwater intake depends on how much fluid is consumed as tapwater and how much is ingested in the form of soft drinks and other non-tapwater sources. Tapwater intake also changes significantly with age, body weight, diet, and climate. Because these factors are so variable, the EPA has suggested a number of very conservative "default" exposure values that can be used when assessing contaminants in tapwater, vegetables, soil, and the like (see Table 20.2).

One important route of exposure is the food supply. Toxic substances are often bioaccumulated, or concentrated, in plant and animal tissues, thereby exposing human beings who ingest those tissues as food.

DOSE–RESPONSE ASSESSMENT

Contaminants are not equal in their capacity to cause adverse effects. To determine the capacity of agents to cause harm, we need quantitative toxicity data. Some toxicity data are derived from occupational, clinical, and epidemiological studies. Most toxicity data, however, come from animal experiments in which researchers expose laboratory animals, mostly mice and rats, to increasingly higher concentrations, or doses, and observe their corresponding effects. The result of these experiments is the *dose–response relationship*—a quantitative relationship that indicates the agent's degree of toxicity to exposed species. Dose is normalized as milligrams of substance or pathogen ingested, inhaled, or absorbed (in the case of chemicals) through the skin per kilogram of body weight per day (mg kg^{-1} day^{-1}). Responses or effects can vary widely—from no observable effect, to temporary and reversible effects (e.g., enzyme depression caused by some pesticides or diarrhea caused by viruses), to permanent organ injury (e.g., liver and kidney damage caused by chlorinated solvents, heavy metals, or viruses), to chronic functional impairment (e.g., bronchitis or emphysema arising from smoke damage), to death.

The goal of a dose–response assessment is to obtain a mathematical relationship between the amount (concentration) of a toxicant or microorganism to which a human being is exposed and the risk of an adverse outcome from that dose. The data resulting from experimental studies is presented as a dose–response curve, as shown in Figure 20.1. The abscissa describes the dose, whereas the ordinate measures the risk that some adverse health effect will occur. In the case of a pathogen, for instance, the ordinate

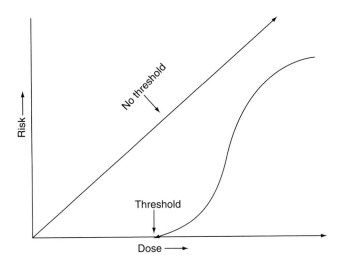

FIGURE 20.1 Relationship between a threshold and nonthreshold response.

may represent the risk of infection, and not necessarily illness.

However, dose–response curves derived from animal studies must be interpreted with care. The data for these curves are necessarily obtained by examining the effects of large doses on test animals. Because of the costs involved, researchers are limited in the numbers of test animals they can use—it is both impractical and cost prohibitive to use thousands (even millions) of animals to observe just a few individuals that show adverse effects at low doses (e.g., risks of 1:1000 or 1:10,000). Researchers must therefore extrapolate low-dose responses from their high-dose data. Dose–response curves are subject to controversy because their results change depending on the method chosen to extrapolate from—the high doses actually administered to laboratory test subjects to the low doses human beings are likely to receive in the course of everyday living.

This controversy revolves around the choice of several mathematical models that have been proposed for extrapolation to low doses. Unfortunately, no model can be proved or disproved from the data, so there is no way to know which model is the most accurate. The choice of models is therefore strictly a policy decision, which is usually based on understandably conservative assumptions. Thus for noncarcinogenic chemical responses, the assumption is that some *threshold* exists below which there is no toxic response; that is, no adverse effects will occur below some very low dose (say, one in a million) (see Figure 20.1). Carcinogens, however, are considered *nonthreshold*—that is, the conservative assumption is that exposure to any amount of carcinogen creates some likelihood of cancer. This means that the only "safe" amount

of carcinogen is zero, so the dose–response plot is required to go through the origin (0), as shown in Figure 20.2.

There are many mathematical models to choose from, including the *one-hit model*, the *multistage model*, the *multihit model*, and the *probit model*. The characteristics of these models for nonthreshold effects are listed in Table 20.3.

The effect of models on estimating risk for a given chemical is shown in Table 20.4. As we can see, the choice of models results in order-of-magnitude differences in estimating the risk at low levels of exposure.

The dose–response effects for noncarcinogens allow for the existence of thresholds; that is, a certain quantity of a substance or dose below which there is no observed adverse effect (NOAEL) by virtue of the body's natural repair and detoxifying capacity. The lowest dose administered that results in a response is given a special name: the lowest-observed-effect-level (LOEL). Examples of toxic substances that have thresholds are heavy metals. These thresholds are represented by the *reference dose*, or *RfD*, of a substance, which is the intake or dose of the substance per unit body weight per day (mg kg^{-1} day^{-1}) that is likely to pose no appreciable risk to human populations, including such sensitive groups as children. A dose–response plot for carcinogens therefore goes through this reference point.

In general, substances with relatively high slope factors and low references doses tend to be associated with higher toxicities. The RfD is obtained by dividing the NOAEL by an appropriate uncertainty factor, sometimes called a *safety factor*. A tenfold uncertainty factor is used to account for differences in sensitivity between the most sensitive individuals in an exposed human population.

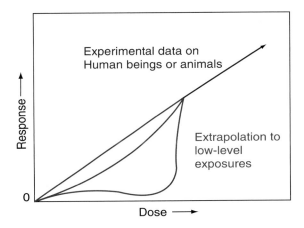

FIGURE 20.2 Extrapolation of dose–response curves. (Adapted from U.S. EPA, 1990.)

TABLE 20.3

Primary Models Used for Assessment of Nonthreshold Effects

Model[a]	Comments
One-hit	Assumes (1) a single stage for cancer; (2) malignant change induced by one molecular or radiation interaction *Very conservative*
Linear multistage	Assumes multiple stages for cancer *Fits curve to the experimental data*
Multihit	Assumes several interactions needed before cell becomes transformed *Least conservative model*
Probit	Assumes probit (log-normal) distribution for tolerances of exposed population *Appropriate for acute toxicity; questionable for cancer*

[a]All these models assume that exposure to the pollutant will always produce an effect, regardless of dose.

Modified from Cockerham and Shane (1994).

TABLE 20.4

Lifetime Risks of Cancer Derived from Different Extrapolation Models

Model Applied	Lifetime Risk (1.0 mg kg^{-1} day^{-1}) of Toxic Chemical[a]
One-hit	6.0×10^{-5} (1 in 17,000)
Multistage	6.0×10^{-6} (1 in 167,000)
Multihit	4.4×10^{-7} (1 in 2.3 million)
Probit	1.9×10^{-10} (1 in 5.3 billion)

[a]All risks are for a lifetime of daily exposure. The lifetime is used as the unit of risk measurement because the experimental data reflect the risk experienced by animals over their lifetimes. The values shown are upper confidence limits on risk.

From U.S. EPA, 1990.

These include pregnant women, young children, and the elderly, who are more sensitive than "average" people. Another factor of 10 is added when the NOAEL is based on animal data that are extrapolated to human beings. In addition, another factor of 10 is sometimes applied when questionable or limited human and animal data are available. The general formula for deriving an RfD is:

$$RfD = \frac{NOAEL}{VF_1 \times VF_2 \dots} \quad \text{(Eq. 20.1)}$$

where VF_i are the uncertainty factors. As the data become more uncertain, higher safety factors are applied. For example, if data are available from a high-quality epidemiological study, a simple uncertainty factor of 10 may be used by simply dividing the original value for RfD by 10 to arrive at a new value of RfD, which reflects the concern for safety. The RfDs of several noncarcinogenic chemicals are shown in Table 20.5. The relationship among RfD, NOAEL, and LOAEL is shown in Figure 20.3.

TABLE 20.5

Chemical RfDs for Chronic Noncarcinogenic Effects of Selected Chemicals

Chemical	RfD (mg kg^{-1} day^{-1})
Acetone	0.1
Cadmium	5×10^{-4}
Chloroform	0.01
Methylene chloride	0.06
Phenol	0.04
Polychlorinated biphenyl	0.0001
Toluene	0.3
Xylene	2.0

From EPA, IRIS integrated risk information.

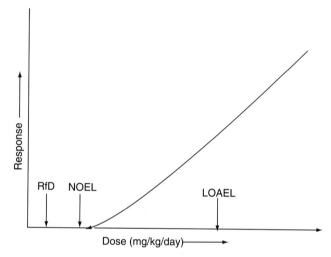

FIGURE 20.3   Relationships between RfD, NOAEL, and LOAEL for noncarcinogens.

The RfD can be used in quantitative risk assessments by using the following relationship:

$$\text{Risk} = \text{PF (CDI} - \text{RfD)} \quad \text{(Eq. 20.2)}$$

where CDI is the chronic daily intake, and the *potency factor (PF)* is the slope of the dose–response curve. Table 20.6 contains potency factors for some potential carcinogens:

$$\text{CDI(mg kg}^{-1}\text{ day}^{-1}) = \frac{\text{Average daily dose (mg day}^{-1})}{\text{Body weight(kg)}} \quad \text{(Eq. 20.3)}$$

But this type of risk calculation is rarely performed. In most cases the RfD is used as a simple indicator of potential risk in practice. That is, the chronic daily intake is simply compared with the RfD; then, if the CDI is below the RfD, it is assumed that the risk is negligible for almost all members of an exposed population.

TABLE 20.6
Toxicity Data for Selected Potential Carcinogens

Chemical	Potency Factor Oral Route mg kg day^{-1}
Arsenic	1.75
Benzene	2.9×10^{-2}
Carbon tetrachloride	0.13
Chloroform	6.1×10^{-3}
DDT	0.34
Dieldrin	30
Heptachlor	3.4
Methylene chloride	7.5×10^{-3}
Polychlorinated biphenyls (PCBs)	7.7
2,3,7,8-TCDD (dioxin)	1.56×10^5
Tetrachlorethylene	5.1×10^{-2}
Trichloroethylene (TCE)	1.1×10^{-2}
Vinyl chloride	2.3

From EPA, www.epa.gov/iris.

RISK CHARACTERIZATION

The final phase of risk assessment process is risk characterization. In this phase exposure and dose–response assessments are integrated to yield probabilities of effects occurring in human beings under specific exposure conditions. Quantitative risks are calculated for appropriate media and pathways. For example, the risks of lead in water are estimated over a lifetime assuming: (1) that the exposure is two liters of water ingested over a 70-year lifetime and (2) that different concentrations of lead occur in the drinking water. This information can then be used by risk managers to develop standards or guidelines for specific toxic chemicals or infectious microorganisms in different media, such as the drinking-water or food supply.

CANCER RISKS

If the dose-response curve is assumed to be linear at low doses for a carcinogen then:

$$\text{Incremental lifetime risk of cancer} = (CDI)\,(PF) \quad \text{(Eq. 20.4)}$$

The linearized multistage model assumptions (see Table 20.3) estimates the risk of getting cancer, which is not necessarily the same as the risk of dying of cancer, so it should be even more conservative as an upper-bound estimate of cancer deaths. Potency factors can be found in the EPA database on toxic substances called the Integrated Risk Information System (IRIS). Table 20.6 contains the potency factor for some of these chemicals.

NONCANCER RISKS

Noncancer risk is expressed in terms of a hazard quotient (HQ) for a single substance, or hazard index (HI) for multiple substances and/or exposure pathways.

$$\text{Hazard Quotient(HQ)} = \frac{\begin{array}{c}\text{Average daily dose}\\\text{during exposure period}\\(\text{mg kg}^{-1} \text{ day}^{-1})\end{array}}{\text{RfD (mg kg}^{-1} \text{ day}^{-1})} \quad \text{(Eq. 20.5)}$$

Unlike a carcinogen the toxicity is important only during the time of exposure, which may be one day, a few days, or years. The HQ has been defined so that if it is less than 1.0, there should be no significant risk or systemic toxicity. Ratios above 1.0 could represent a potential risk, but there is no way to establish that risk with any certainty.

When exposure involves more than one chemical, the sum of the individual hazard quotients for each chemical is used as a measure of the potential for harm. This sum is called the HI:

$$\text{Hazard Index} = \text{Sum of hazard quotients} \quad \text{(Eq. 20.6)}$$

MICROBIAL RISK ASSESSMENT

Outbreaks of waterborne disease caused by microorganisms usually occur when the water supply has been obviously and significantly contaminated (see Chapter 17). In such high-level cases, the exposure is manifest and cause and effect are relatively easy to determine. However, exposure to low-level microbial contamination is difficult to determine epidemiologically. We know, for example, that long-term exposure to microbes can have a significant impact on the health of individuals within a community, but we need a way to measure that impact.

For some time, methods have been available to detect the presence of low levels (one organism per 1000 liters) of pathogenic organisms in water, including enteric viruses, bacteria, and protozoan parasites. The trouble is that the risks posed to the community by these low levels of pathogens in a water supply over time are not like those posed by low levels of chemical toxins or carcinogens. For example, it takes just one amoeba in the wrong place at the wrong time to infect one individual, whereas that same individual would have to consume some quantity of a toxic chemical to be comparably harmed. Microbial risk assessment is therefore a process that allows us to estimate responses in terms of the *risk of infection* in a quantitative fashion. Although no formal framework for microbial risk assessment exists, it generally follows the steps used in other health-based risk assessments—hazard identification, exposure assessment, dose–response, and risk characterization. The differences are in the specific assumptions, models, and extrapolation methods used.

Hazard identification in the case of pathogens is complicated because several outcomes—from *asymptomatic* infection to death are possible, and these outcomes depend on the complex interaction between the pathogenic agent (the "infector") and the host (the "infectee"). This interaction, in turn, depends on the characteristics of the host, as well as the nature of the pathogen. Host factors, for example, include preexisting immunity, age, nutrition, ability to mount an immune response, and other nonspecific host factors. Agent factors include type and strain of the organism, as well as its capacity to elicit an immune response.

Among the various outcomes of infection is the possibility of *subclinical illness*. Subclinical (asymptomatic) infections are those in which the infection (growth of the microorganism within the human body) results in no obvious illness such as fever, headache, or diarrhea; that is, individuals can host a pathogen microorganism—and transmit it to others—without ever getting sick themselves. The ratio of clinical to subclinical infection varies from pathogen to pathogen, especially in viruses, as shown in Table 20.7. Poliovirus infections, for instance, seldom result in obvious clinical symptoms; in fact, the proportion of individuals developing clinical illness may be less than 1%. However, other enteroviruses, such as the

coxsackie viruses, may exhibit a greater proportion. In many cases, as in that of rotaviruses, the probability of developing clinical illness appears to be completely unrelated to the dose an individual receives via ingestion. Rather, the likelihood of developing clinical illness depends on the type and strain of the virus, as well as host age, nonspecific host factors, and possibly preexisting immunity. The incidence of clinical infection can also vary from year to year for the same virus, depending on the emergence of new strains.

Another possible outcome of infection is the development of clinical illness. Several host factors play a major role in this outcome. The age of the host is often a determining factor. In the case of hepatitis A, for example, clinical illness can vary from about 5% in children less than 5 years of age to 75% in adults. Similarly, children are more likely to develop rotaviral gastroenteritis than are adults. Immunity is also an important factor, albeit a variable one; that is, immunity may or may not provide long-term protection from reinfection, depending on the enteric pathogen. It does not, for example, provide long-term protection against the development of clinical illness in the case of the Norwalk virus, or *Giardia*. However, for most enteroviruses, and for the hepatitis A virus, immunity from reinfection is believed to be life long.

The ultimate outcome of infection—mortality—can be caused by nearly all enteric organisms. The factors that control the prospect of mortality are largely the same factors that control the development of clinical illness. Host age, for example, is significant. Thus mortality for hepatitis A and poliovirus is greater in adults than in children. In general, however, we can say that the very young, the elderly, and the immunocompromised are at the greatest risk of a fatal outcome of most illnesses. For example, the case fatality rate (%) for *Salmonella* in the general population is 0.1%, whereas it has been observed to be as high as 3.8% in nursing homes. In North America and Europe the reported cases:fatalities (i.e., the ratio of cases to fatalities reported as a percentage of persons who die) for enterovirus infections range from less than 0.1% to 0.94%, as shown in Table 20.8. The case:fatality rate for common enteric bacteria ranges from 0.1% to 0.2% in the general population.

Unlike chemical-contaminated water, microorganism-contaminated water does not have to be consumed to cause harm. Individuals who do not actually drink, or even touch, contaminated water also risk infection because pathogens—particularly viruses—may be spread by person-to-person contact or subsequent contact with contaminated inanimate objects (e.g., toys). This phenomenon is described as the *secondary attack rate*, which is reported as a percentage (Figure 20.4). For example, one person infected with poliovirus virus can transmit it to 90% of the persons with whom he or she

TABLE 20.7

Ratio of Clinical to Subclinical Infections with Enteric Viruses

Virus	Frequency of Clinical Illness[a] (%)
Poliovirus 1	0.1–1
Coxsackie	
A16	50
B2	11–50
B3	29–96
B4	30–70
B5	5–40
Echovirus	
overall	50
9	15–60
18	Rare–20
20	33
25	30
30	50
Hepatitis A (adults)	75
Rotavirus	
(adults)	56–60
(children)	28
Astrovirus (adults)	12.5

Modified from Gerba and Rose (1993).

[a]The percentage of those individuals infected who develop clinical illness.

TABLE 20.8

Case:Fatality Rates for Enteric Viruses and Bacteria

Organism	Case:Fatality Rate (%)
Virus	
Poliovirus 1	0.90
Coxsackie	
A2	0.50
A4	0.50
A9	0.26
A16	0.12
Coxsackie B	0.59–0.94
Echovirus	
6	0.29
9	0.27
Hepatitis A	0.60
Rotavirus	
(total)	0.01
(hospitalized)	0.12
Norwalk	0.0001
Adenovirus	0.01
Bacteria	
Shigella	0.2
Salmonella	0.1
Escherichia coli 0157:H7	0.2
Campylobacter jejuni	0.1

From Gerba and Rose (1992) and Gerba *et al.* (1995).

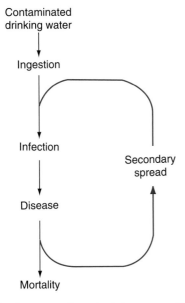

FIGURE 20.4 Outcomes of enteric viral exposure. (From Pollution Science @ 1996, Academic Press, San Diego.)

associates. This secondary spread of viruses has been well documented for waterborne outbreaks of several diseases, including that caused by Norwalk virus, whose secondary attack rate is about 30%.

The question of dose is another problem in exposure assessment. How do we define "dose" in this context? To answer this question, researchers have conducted a

number of studies to determine the infectious dose of enteric microorganisms in human volunteers. Such human experimentation is necessary because determination of the infectious dose in animals and extrapolation to humans is often impossible. In some cases, for example, humans are the primary or only known host. In other cases, such as that of *Shigella* and/or Norwalk virus, infection can be induced in laboratory-held primates, but it is not known if the infectious-dose data can be extrapolated to humans. Much of the existing data on infectious dose of viruses have been obtained with attenuated-vaccine viruses or with avirulent laboratory-grown strains, so that the likelihood of serious illness is minimized.

Next, we must choose a dose–response model whose abscissa is the dose, and whose ordinate is the risk of infection. The choice of model is critical so that risks are neither greatly over- or underestimated. A modified exponential (beta-Poisson distribution) or a log-probit (simple log-normal, or exponential, distribution) model may be used to describe this relationship for many enteric microorganisms (Haas, 1983). For the [β] model the probability of infection from a single exposure, P, can be described as follows:

$$P = 1 - (1 + N/\beta)^{-\alpha} \qquad \text{(Eq. 20.7)}$$

where N is the number of organisms ingested per exposure, and [α] and [β] represent parameters characterizing the host–virus interaction (Haas, 1983). Some values for [α] and [β] for several enteric waterborne pathogens are shown in Table 20.9; these values were determined from human studies. For some microorganisms an exponential model may better represent the probability of infection.

TABLE 20.9

Best-fit Dose–Response Parameters for Enteric Pathogen Ingestion Studies

Microorganism	Best Model	Model Parameters
Echovirus 12	Beta-poisson	$\alpha = 0.374$
		$\beta = 186.69$
Rotavirus	Beta-poisson	$\alpha = 0.26$
		$\beta = 0.42$
Poliovirus 1	Exponential	$r = 0.009102$
Poliovirus 1	Beta-poisson	$\alpha = 0.1097$
		$\beta = 1524$
Poliovirus 3	Beta-poisson	$\alpha = 0.409$
		$\beta = 0.788$
Cryptosporidium	Exponential	$r = 0.004191$
Giardia lamblia	Exponential	$r = 0.02$
Salmonella	Exponential	$r = 0.00752$
Escherichia coli	Beta-poisson	$\alpha = 0.1705$
		$\beta = 1.61 \times 10^6$

Modified from Regli *et al.* (1991).

$$P = 1 - \exp(-rN) \qquad \text{(Eq. 20.8)}$$

In this equation r is the fraction of the ingested microorganisms that survive to initiate infections (host-microorganism interaction probability).

When we use these models, we are estimating the probability of becoming infected after ingestion of various concentrations. The risk of acquiring a viral infection from consumption of contaminated drinking water containing various concentrations of enteric viruses is first determined by using Equation 20-7 or 20-8. Annual and lifetime risks can also be determined, again assuming a Poisson distribution of the virus in the water consumed (assuming daily exposure to a constant concentration of viral contamination), as follows:

$$P_A = 1 - (1 - P)^{365} \qquad \text{(Eq. 20.9)}$$

where P_A is the annual risk (365 days) of contracting one or more infections, and

$$P_L = 1 - (1 - P)^{25,550} \qquad \text{(Eq. 20.10)}$$

where P_L is the lifetime risk (assuming a lifetime of 70 years = 25,550 days) of contracting one or more infections.

Risks of clinical illness and mortality can be determined by incorporating terms for the percentage of clinical illness and mortality associated with each particular virus:

$$\text{Risk of clinical illness} = PI \qquad \text{(Eq. 20.11)}$$
$$\text{Risk of mortality} = PIM \qquad \text{(Eq. 20.12)}$$

where P is the probability of infection (see equation 20.7), I is the percentage of infections that result in clinical illness, and M is the percentage of clinical cases that result in mortality.

Application of this model allows us to estimate the risks of becoming infected, development of clinical illness, and mortality for different levels of exposure. As shown in Table 20.10, for example, the estimated risk of infection from 1 rotavirus in 100 liters of drinking water (assuming ingestion of 2 liters per day) is 1.2×10^{-3}, or almost 1 in 1000 for a single-day exposure. This risk would increase to 3.6×10^{-1}, or approximately one in three on an annual basis. Risks of the development of clinical illness and mortality also appear to be significant for exposure to low levels of rotavirus in drinking water.

The EPA has recently recommended that any drinking-water treatment process should be designed to ensure that human populations are not subjected to risk of infection greater than 1:10,000 for a yearly exposure. To achieve this goal, it would appear from the data shown in Table 20.10 that the virus concentration in drinking water would have to be less than 1 per 1000 liters. Thus if the average concentration of enteric viruses in untreated water is 1400/1000 liters, then treatment plants should be designed to remove at least 99.99% of the virus present in the raw water.

In summary, we can see that risk assessment is a major tool for decision making in the regulatory arena. This approach is used to explain chemical and microbial risks, as well as ecosystem impacts. The results of such assessments can be used to inform risk managers of the probability and extent of environmental impacts resulting from exposure to different levels of stress (contaminants). Moreover, this process, which allows for the quantitation and comparison of diverse risks, lets risk managers utilize the maximum amount of complex information in the decision-making process. This information can also be used to weigh the cost and benefits of control options and to develop standards or treatment options.

REGULATORY OVERVIEW

In the United States most environmental legislation is federal legislation that has been enacted within the last 30 years. In addition, there are many state environmental laws and regulations that have been patterned after—and work together with—the federal programs. Federal legislation is, however, the basis for the development of regulations designed to protect our air, water, and food supply, and to control pollutant discharge. Enactment of

TABLE 20.10

Risk of Infection, Disease, and Mortality for Rotavirus

Virus Concentration per 100 Liters	Risk	
	Daily	Annual
Infection		
100	9.6×10^{-2}	1.0
1	1.2×10^{-3}	3.6×10^{-1}
0.1	1.2×10^{-4}	4.4×10^{-2}
Disease		
100	5.3×10^{-2}	5.3×10^{-1}
1	6.6×10^{-4}	2.0×10^{-1}
0.1	6.6×10^{-5}	2.5×10^{-2}
Mortality		
100	5.3×10^{-6}	5.3×10^{-5}
1	6.6×10^{-8}	5.3×10^{-5}
0.1	6.6×10^{-9}	2.5×10^{-6}

Modified from Gerba and Rose (1993).

new legislation empowers the executive branch to develop environment-specific regulations and to implement their enforcement. Among other things, regulations may involve the development of standards for waste discharge, requirements for the cleanup of polluted sites, or guidelines for waste disposal. The EPA is the federal agency currently responsible for development of environmental regulations and enforcement.

Federal authority to establish standards for drinking water systems originated with the Interstate Quarantine Act of 1893. This provision resulted in the first water-related regulation in 1912. Thus in 1914 the first official drinking water standard—a bacteriological standard—was adopted. From 1914 to 1975, federal, state, and local health authorities, in conjunction with water works officials, used this standard to improve the nation's community water systems and to protect the public against waterborne disease.

A brief sketch of subsequent federal environmental legislation is presented here and summarized in Table 20.11.

Safe Drinking Water Act

In 1974 the passage of the *Safe Drinking Water Act (SDWA)* gave the federal government overall authority for the protection of drinking water (see Tables 9.1, 9.2, and 16.1). Previous to that time, the individual states had primary authority for development and enforcement of standards. Under this authority, specific standards were promulgated for contaminant concentrations and minimum water treatment (Table 20.12). *Maximum contaminant levels (MCLs)*, or *maximum contaminant level goals (MCLGs)*, were set for specific contaminants in drinking water. Whereas an MCL is an achievable,

TABLE 20.11
Scope of Federal Regulations Governing Environmental Pollution in the United States

Federal Regulation	Purpose/Scope
Policy	
National Environmental Policy Act (NEPA). Enacted 1970.	This act declares a national policy to promote efforts to prevent or eliminate damage to the environment. It requires federal agencies to assess environmental impacts of implementing major programs and actions early in the planning stage.
Pollution Prevention Act of 1990.	The basic objective is to prevent or reduce pollution at the source instead of an end-of-pipe control approach.
Water	
Clean Water Act. Enacted 1948.	Eliminates discharge of pollutants into navigable waters. It is the prime authority for water pollution control programs.
1977 Amendment	Covers regulation of sludge application by federal and state government.
Safe Drinking Water Act (SDWA). Enacted 1974.	Protects sources of drinking water and regulates using proper water-treatment techniques using drinking water standards based on maximum contaminant levels (MCLs).
Clean Air	
Clean Air Act. Enacted 1970.	This act, which amended the Air Quality Act of 1967, is intended to protect and enhance the quality of air sources. Sets a goal for compliance with ambient air quality standards.
Amendments of 1977.	To define issues to prevent industries from benefitting economically from noncompliance.
Amendments of 1990.	Basic objectives of CA 90 are to address acid precipitation and power plant emissions.
Hazardous Waste	
Comprehensive Environmental Response, Compensation, and Liability (CERCLA). Enacted 1980. Amended in 1980 to include Superfund.	The act, known as Superfund or CERCLA, provides an enforcement agency the authority to respond to releases of hazardous wastes. The act amends the Solid Waste Disposal Act.
Superfund Amendments and Reauthorization Act (SARA). Enacted 1986.	This act revises and extends CERCLA by the addition of new authorities, known as the Emergency Planning and Community Right-to-Know Act of 1986. Involves toxic chemical recall reporting.
Toxic Substances Control Act. Enacted in 1976.	This act sets up the toxic substances program administered by the EPA. The act also regulates labeling and disposal of PCBs.
Amendment of 1986.	Addresses issues of inspection and removal of asbestos.
Resource Conservation and Recovery Act (RCRA). Enacted in 1976. Amended in 1984.	As an amendment that completely revised the Solid Waste Disposal Act. As it exists now it is a culmination of legislation dating back to the passage of the Solid Waste Disposal Act of 1965. Defines hazardous wastes. It requires tracking of hazardous waste. Regulates facilities that burn wastes and oils in boilers and industrial furnaces. Requires inventory of hazardous waste sites.
Federal Insecticide, Fungicide, and Rodenticide Act (FIFRA). Enacted 1947.	Regulates the use and safety of pesticide products.
Amendments of 1972.	Intended to ensure that the environmental harm does not outweigh the benefits.

required level, an MCLG is a desired goal that may or may not be achievable. For example, the MCLG for enteric viruses in drinking water is zero because only one ingested virus may cause illness. We cannot always reach zero, but we are obliged to try. Within the provisions of the SDWA, there are a number of specific rules such as the *Surface Treatment Rule*, which requires all water utilities in the United States to provide filtration and disinfection to control waterborne disease from *Giardia* and enteric viruses (see Table 9.1). These rules are developed through a process that allows input from the regulated community, special interest groups in the general public, and the scientific community. This process is outlined in Figure 20.5. To aid utilities, the EPA has published guidance manuals on the types of treatment processes that have been developed. Other key sections of the SDWA provide for the establishment of state programs to enforce the regulations. Individual states are responsible for developing their own regulatory programs, which are then submitted to the EPA for approval. State programs must set drinking water standards equal to, or more stringent than, the federal standards. Such programs must also issue permits to facilities that treat drinking water supplies and develop wellhead protection areas for groundwater drinking supplies. Secondary drinking water standards have also been set for test, odor, and aesthetic properties (see Table 9.3).

In the revisions of the SDWA in 1996 the EPA was required to establish a list of contaminants to aid priority-setting. In establishing the list, the EPA has divided the contaminants, among which are priorities for additional research, those which need additional occurrence data, and those which are priorities for consideration for rule making. This contaminant candidate list (CCL) was published in 1998 (EPA, 1998). The EPA is required to select five or more contaminants and determine whether to regulate them. The list includes both chemicals and microorganisms.

COMPREHENSIVE ENVIRONMENTAL RESPONSE, COMPENSATION AND LIABILITY ACT

In the late 1970s, when the now-infamous Love Canal landfill in upstate New York was revealed as a major environmental catastrophe, the attendant publicity spurred Congress to pass the *Comprehensive Environmental Response, Compensation and Liability Act (CERCLA)* of 1980. This act makes owners and operators of hazardous waste disposal sites liable for cleanup costs and property damage. Transporters and producers must also bear some of the financial burden. This legislation also established a $1.6 billion cleanup fund (known as Superfund) to be raised over a 5-year period. Of that sum, 90% was to come from taxes levied on the production of oil and chemicals by U.S. industries and the rest from taxpayers. This level was later increased to $8.5 billion.

CERCLA establishes *strict liability* (liability without proof of fault) for the cleanup of facilities on "responsible parties." The courts have found under CERCLA, that these parties are "jointly and severally" liable; that is, each and every ascertainable party is liable for the full cost of removal or remediation, regardless of the level of "guilt" a party may have in creating a particular polluted site.

The Superfund is exclusively dedicated to clean up sites that pose substantial threats to human health and habitation (i.e., *imminent* hazards). Moreover, the Superfund provides only for the cleanup of contaminated areas and compensation for damage to property. It cannot be used to reimburse or compensate victims of illegal dumping of hazardous wastes for personal injury or death. Victims must take their complaints to the courts.

In an effort to address some of the liability concerns and fairness issues of CERCLA, as well as to help revitalize communities affected by contamination, the EPA launched its *Brownfields Economical Redevelopment Ini-*

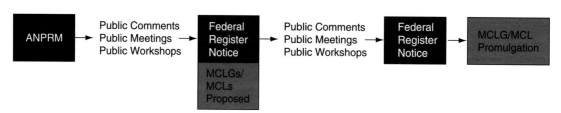

FIGURE 20.5 Regulatory development processes under the Safe Drinking Water Act. A series of steps is followed in the development of new regulations for contaminants in drinking water or processes for their control. These steps usually involve a series of public notices in the publication, *Federal Register*, and meetings to allow for comment from the public, environmental groups, and the regulated industry, before the final regulation is developed. *ANPRM*, Announcement of proposed rule-making. (From Pollution Science @ 1996, Academic Press, San Diego.)

TABLE 20.12
Primary Drinking Water Maximum Contaminant Levels (mg L^{-1}), Organized by Category

Inorganic Chemicals	Maximum Contaminant Levels (mg L^{-1})	Inorganic Chemicals	Maximum Contaminant Levels (mg L^{-1})
Arsenic	0.05	Mercury	0.002
Barium	2.	Nickel	0.1
Cadium	0.005	Nitrate (as N)	10.0
Chromiuim (total)	0.1	Nitrite (as N)	1.0
Copper	TT^a	Nitrate + nitrite	10.0
Fluoride	4.0	Selemium	0.05
Lead	TT^a	Thallium	0.002
Asbestos	7 million fibers/liter (longer than 10 µm)		

Radionuclides	Maximum Contaminant Levels (mg L^{-1})	Radionu-clides	Maximum Contaminant Levels (mg L^{-1})
Radium 226	20 pCi L	Beta particle and photon radioactivity	4 mrem yr
Radium 228	20 pCi L^{-1}	Radon	300 pCi L^{-1}
Gross alpha particle activity	15 pCi L^{-1}	Uranium	20 µg L^{-1}

^{a}Treatment technique (TT) requirement, rather than an MCL.

tiative in 1995. Brownfields are defined to be "abandoned, idled, or under-used industrial and commercial facilities where expansion or redevelopment is complicated by real or perceived environmental contamination." The Initiative is based on the belief that public fear of contamination and lender concerns for liability are causing companies to leave "brownfields" in the urban core and head for "greenfields" outside the cities.

Key components of the Brownfields initiative include removing eligible sites from the Superfund site tracking system, known as the Comprehensive Environmental Response, Compensation, and Liability Information System (CERCLIS). Initial actions removed 27,000 such sites out of the 40,000 that were being tracked. Removing sites from CERCLIS is intended to convey assurance that the EPA will no longer pursue Superfund action at these sites. A second, very important component of the Initiative is a clarification of liability issues that frequently cause potential buyers and lenders to avoid contaminated properties. For example, the EPA will not take action against owners when hazardous substances enter the property from aquifers contaminated elsewhere, as long

as the landowner did not in any way cause the problem, and as long as the property owner did not have a contractual arrangement with the polluter.

The brownfields issue is helping to focus attention on risk-based corrective actions in which the most feasible remediation technologies are coupled with best management practices to provide cost-effective cleanup without compromising the protection of public health, water quality, and the environment.

CLEAN AIR ACT

The Clean Air Act, originally passed in 1970, required the EPA to establish *National Ambient Air Quality Standards* (NAAQS) for several outdoor pollutants. NAAQS have been established by the EPA at two levels: primary and secondary. Primary standards are required to be set at levels that will protect public health, and include an "adequate margin of safety," regardless of whether the standards are economically or technologically achievable. Primary standards must protect even the most sensitive individuals, including the elderly and those suffering from respiratory disorders. Secondary standards are meant to be even more stringent than primary standards. Secondary standards are set to protect public welfare (e.g., buildings, crops, animals, fabrics). Because of the difficulty in achieving primary standards, secondary standards have had almost no role in air pollution control policy.

National Ambient Air Quality Standards now exist for six criteria pollutants: carbon monoxide (CO), lead (Pb), nitrogen dioxide (NO$_2$), ground-level ozone (O$_3$), sulfur dioxide (SO$_2$), and particulate matter (see Chapter 10). The Clean Air Act requires that the list of criteria pollutants be reviewed periodically and that standards be adjusted according to the latest scientific information. For a given region of the country to be in compliance with NAAQS, the concentration cannot be exceeded more than once in a calendar year.

The EPA is also required to establish emission standards for certain large stationary sources, such as power plants, incinerators, cement plants, petroleum refineries, and smelters.

This Act also sets air pollution control requirements for various geographic areas of the United States, which deals with the control of tailpipe emissions for motor vehicles. Requirements compel automobile manufacturers to improve design standards to limit carbon monoxide, hydrocarbon, and nitrogen oxide emissions. For cities or areas where the ozone and carbon monoxide concentrations are high, reformulated and oxygenated gasolines are required. This act also addresses power plant emissions of sulfur dioxide and nitrogen oxide, which can generate acid rain.

EXAMPLE 20.1. Application of Hazard Index and Incremental Carcinogenic Risk Associated with Chemical Exposure.

A drinking water supply is found to contain 0.1 mg L^{-1} of acetone and 0.1 mg L^{-1} of chloroform. A 70-kg adult drinks 2 L per day of this water for 5 years. What would be the hazard index and the carcinogenic risk from drinking this water?

First we need to determine the average daily doses (ADDs) for each of the chemicals and then their individual hazard quotients.

For Acetone

$$ADD = \frac{(0.1\,mg\,L^{-1})\,(2\,L\,day^{-1})}{70\,kg}$$
$$= 2.9 \times 10^{-3}\,mg\,kg^{-1}\,day^{-1}$$

From Table 20.5, the RfD for acetone is 0.1 mg kg^{-1} day^{-1}

$$Hazard\ quotient = \frac{2.9 \times 10^{-3}\,mg\,kg^{-1}\,day^{-1}}{0.1}$$
$$= 0.029$$

For Chloroform

$$ADD = \frac{(0.1\,mg\,L^{-1})\,(2\,L\,day^{-1})}{70\,kg}$$
$$= 2.9 \times 10^{-3}\,mg\,kg^{-1}\,day^{-1}$$

From Table 20.5 the RfD value for chloroform is 0.01 mg kg^{-1} day^{-1}

$$Hazard\ quotient = \frac{2.9 \times 10^{-3}\,mg\,kg^{-1}\,day^{-1}}{0.01}$$
$$= 0.029$$

Thus

$$Hazard\ Index = 0.029 + 0.29 = 0.319$$

Since the Hazard Index is less than 1.0 the water is safe. Notice that we did not need to take into consideration that the person drank the water for 5 years.

The incremental carcinogenic risk associated with chloroform is determined as follows:

$$Risk = (CDI)\,(Potency\ Factor)$$

$$CDI = \frac{(0.1\,mg\,L^{-1})\,(2\,L\,day^{-1})\,(365\,days\,yr^{-1})\,(5\,yrs)}{(70\,kg)\,(365\,days\,yr^{-1})\,(70\,yrs)}$$
$$= 4.19 \times 10^{-5}\,mg\,kg^{-1}\,day^{-1}$$

From Table 20.6, the potency factor for chloroform is 6.1 × 10^{-3}

$$Risk = (CDI)\,(Potency\ factor)$$

$$Risk = (4.19 \times 10^{-5}\,mg\,kg^{-1}\,day^{-1})$$
$$(6.1 \times 10^{-3}\,mg\,kg^{-1}\,day^{-1}) = 2.55 \times 10^{-7}$$

From a cancer risk standpoint the risk over this period of exposure is less than the 10^{-6} goal.

EXAMPLE 20.2. Application of a Virus Risk Model to Characterize Risks from Consuming Shellfish.

It is well known that infectious hepatitis and viral gastroenteritis are caused by consumption of raw or, in some cases, cooked clams and oysters. The concentration of echovirus 12 was found to be 8 plaque-forming units per 100 g in oysters collected from coastal New England waters. What are the risks of becoming infected and ill from echovirus 12 if the oysters are consumed? A person usually consumes 60 g of oyster meat in a single serving or

$$\frac{8\,PFU}{100\,g} = \frac{N}{60\,g} \quad N = 4.8\,PFU\ consumed$$

From Table 20.9, $\alpha = 0.374$, $\beta = 186.64$. The probability of infection from equation 20–7 is then

$$P = 1 - \left(1 + \frac{4.8}{186.69}\right) = 0.374$$

If the risk of clinical illness is 50% from equation 20–11 then

$$Risk\ of\ clinical\ illness = (9.4 \times 10^{-3})$$
$$(0.50) = 4.7 \times 10^{-3}$$

If the case–fatality rate is 0.001%, then from equation 20–12

$$Risk\ of\ mortality = (9.4 \times 10^{-3})\,(0.50)\,(0.001)$$
$$= 4.7 \times 10^{-6}$$

CASE STUDY 20.1 *How Do We Set Standards for Pathogens in Drinking Water?*

In 1974 the U. S. Congress passed the Safe Drinking Water Act, which gave the U. S. Environmental Protection Agency the authority to establish standards for contaminants in drinking water. Through a risk analysis approach, standards have been set for many chemical contaminants in drinking water. Setting standards for microbial contaminants proved more difficult because (1) methods for the detection of many pathogens are not available; (2) days to weeks are sometime required to obtain results; and (3) costly and time-consuming methods are required. To overcome these difficulties, coliform bacteria had been used historically to assess the microbial quality of drinking water. However, by the 1980s it had become quite clear that coliform bacteria did not indicate the presence of pathogenic waterborne *Giardia* or enteric viruses. Numerous outbreaks had occurred in which coliform standards were met, because of the greater resistance of viruses and *Giardia* to disinfection. A new approach was needed to ensure the microbial safety of drinking water.

To achieve this goal a treatment approached was used. As part of the Surface Treatment Rule (STR), all water utilities that use surface waters as the source would be required to provide filtration to remove *Giardia* and enough disinfection to kill viruses. The problem facing the EPA was how much removal should be required. To deal with this issue, the EPA for the first time used a microbial risk assessment approach. The STR established that the goal of treatment was to ensure that microbial illness from *Giardia lamblia* infection should not be any greater than 1 per 10,000 exposed persons annually (10^{-4} per year). This value is close to the annual risk of infection from waterborne disease outbreaks in the United States (4×10^{-3}). Based on the estimated concentration of *Giardia* and enteric viruses in surface waters in the United States from the data available at the time, it was required that all drinking water treatment plants be capable of removing 99.9% of the *Giardia* and 99.99% of the viruses. In this manner it was hoped that the risk of infection of 10^{-4} per year would be achieved. The STR went into effect in 1991.

To better assess whether the degree of treatment required is adequate, the EPA developed the Information Collection Rule, which requires major drinking water utilities that use surface waters to analyze these surface water for the presence of *Giardia*, *Cryptosporidium*, and enteric viruses for almost 2 years. From this information the EPA hopes to assess whether the treatment currently required is enough to ensure the 10^{-4} yearly risk. Potentially, utilities that have heavily contaminated source water may require greater levels of treatment in the future.

If a person consumes oysters 10 times a year with 4.8 PFU per serving, then the risk of infection in 1 year $= 1 - (1 - (0.374)^{10} = 9.9 \times 10^{-1}$ (from Equation 20.9).

QUESTIONS

1. List the four steps in a formal risk assessment.

2. Why do we use safety factors in risk assessment?

3. What is the most conservative dose–response curve? What does it mean?

4. What is the difference between risk assessment and risk management?

5. What are some of the differences between the risks posed by chemicals and those posed by microorganisms?

6. Suppose a 50-kg individual drinks 2 L day^{-1} of drinking water containing 1.0 mg L^{-1} of chloroform and 0.1 mg L^{-1} phenol. What is the hazard index? Is there cause for concern?

7. Estimate the cancer risk for a 70-kg individual consuming 1.5 liters of water containing trichloroethylene (TCE) per day for 70 years.

8. Calculate the risk of infection from rotavirus during swimming in polluted water. Assume 30 ml of water is ingested during swimming, and the concentration of rotavirus was 1 per 100 liters. What would the risk be in a year if a person went swimming 5 times and 10 times in the same water with the same concentration of virus?

REFERENCES AND ADDITIONAL READING

Cockerham, L.G., and Shane, B.S. (1994) *Basic Environmental Toxicology.* CRC Press, Boca Raton, FL.

EPA. (1998) *Environmental Protection Agency. Announcement of the Drinking Water Contaminant Candidate List.* Federal Register, March 2, 1998. Vol. 63, No. 40, pp. 10273–10287.

Gerba, C.P., Rose, J.B. (1993) Estimating viral disease risk from drinking water. In *Comparative Environmental Risks* (C.R. Cothern, Ed.), pp. 117–135. Lewis Publishers, Boca Raton, FL.

Gerba, C.P., Rose, J.B., Haas, C.N. (1995) Waterborne disease—Who is at risk? In *Water Quality Technology Proceedings,* pp. 231–254. American Water Works Association, Denver, CO.

Haas, C.N., Rose, J.B., Gerba, C.P. (1999) *Quantitative Microbial Risk Assessment.* John Wiley and Sons, New York.

Kolluru, R.V. (1993) *Environmental Strategies Handbook.* McGraw-Hill, New York.

Landis, W.G., Yu, M.H. (1995) *Introduction to Environmental Toxicology.* Lewis Publishers, Boca Raton, FL.

National Research Council (1983) *Risk Assessment in the Federal Government: Managing the Process.* National Academy Press. Washington, DC.

National Research Council. (1989) Improving Risk Communication. National Academy Press. Washington, DC.

Regli, S., Rose, J.B., Haas, C.N., Gerba, C.P. (1991) Modeling the risk from *Giardia* and viruses in drinking water. *J. Am. Water Works Assoc.* **83,** 76–84.

Rodricks, J.V. (1992) *Calculated Risks.* Cambridge University Press, New York.

Wilson, R., Crouch, E.A.C. (1987) Risk assessment and comparisons: an introduction. *Science* **236,** 267–270.

INDEX